LARGE MARINE ECOSYSTEMS

Stress, Mitigation, and Sustainability

LARGE MARINE ECOSYSTEMS

Stress, Mitigation, and Sustainability

Kenneth Sherman
Lewis M. Alexander
Barry D. Gold
editors

AAAS
PRESS

A publishing division of the
American Association for the
Advancement of Science

Library of Congress Cataloging-in-Publication Data

Large marine ecosystems V: stress, mitigation and sustainability of large marine
ecosystems / Kenneth Sherman, Lewis Alexander, Barry Gold, editors.
   p. cm. —
Includes bibliographical references and index.
ISBN 0-87168-506-X
1. Marine ecology—Congresses.   2. Marine resources—Management—Congresses. I. Sherman, Kenneth, 1932– . II. Alexander, Lewis M., 1921– .
III. Gold, Barry D., 1952– .   IV. Series.
   Q181.A1A68 no. 92-39S
[QH541.5.S3]
500 s—dc20
[574.5'2636]   92-38527
                                                   CIP

The findings, conclusions, and opinion stated or implied in this publication are
those of the authors. They do not necessarily reflect the views of the Board of
Directors, Council, or membership of the American Association for the
Advancement of Science.

AAAS Publication: 92-39S
International Standard Book Number: 0-87168-506-X
Printed in the United States of America on acid-free paper

Contributors List

Lewis M. Alexander, *University of Rhode Island, Kingston, RI*

Jurgen Alheit, *POLARMAR, Bremerhaven, Germany*

Andrew Bakun, *Pacific Fisheries Environment Group, NOAA/NMFS, Monterey, CA*

Carlos Bas, *Facultad de Ciencias del Mar, Las Palmas, Spain*

Martin H. Belsky, *Albany Law School, Albany, NY*

Patricio Bernal, *Instituto de Fomento Pesquero, Santiago, Chile*

Denis Binet, *ORSTOM/IFREMER, Nantes, Cedex, France*

Johan Blindheim, *Institute of Marine Research, Bergen, Norway*

Giovanni Bombace, *Istituto Di Ricerche Sulla Pesca Marittima (IRPEM), Ancona, Italy*

L. A. Borets, *VNIRO, All Union Research Institute of Fisheries and Oceanography, Moscow, Russia*

Daniel L. Bottom, *Research and Development Section, Oregon Dept. of Fish and Wildlife, Corvallis, OR*

Robin F. Brown, *Marine Science Division, Oregon Dept. of Fish and Wildlife, Newport, OR*

John F. Caddy, *FAO/UN Fisheries Department, Rome, Italy*

Villy Christensen, *ICLARM, Manila, Philippines*

Biliana Cicin-Sain, *University of Delaware, Center for the Study of Marine Policy, Newark, Delaware*

Niels Daan, *Netherlands Institute for Fishery Investigations, Ijmuidan, Netherlands*

Arthur Lyon Dahl, *UNEP Regional Office for Europe, Geneva, Switzerland*

S. N. Dwivedi, *Government of India, Indian Council of Agricultural Research, New Delhi, India*

Graciela Garcia-Moliner, *Physical Oceanography Group, University of Rhode Island, Narragansett, RI*

Barry D. Gold, *House Committee on Science, Space and Technology Subcommittee on Science, Washington, DC*

Bruce P. Hayden, *Dept. of Environmental Sciences, University of Virginia, Charlottesville, Virginia*

Gotthilf Hempel, *Center for Tropical Marine Ecology, Bremer, Germany*

Martin W. Holdgate, *IUCN The World Conservation Union, Gland, Switzerland*

D. V. Holliday, *TRACOR Applied Sciences, Inc., San Diego, California*

Kim K. Jones, *Research and Development Section, Oregon Dept. of Fish and Wildlife, Corvallis, OR*

Graeme Kelleher, *Great Barrier Reef Marine Park Authority, Canberra, Australia*

John Knauss, *U.S. Dept. of Commerce/NOAA, Washington, DC*

V. V. Kuznetsov, *VNIRO, All Union Research Institute of Fisheries and Oceanography, Moscow, Russia*

Robert W. Knecht, *University of Delaware, College of Marine Studies, Newark, Delaware*

Simon A. Levin, *Dept. of Ecology and Evolutional Biology, Princeton University, Princeton, NJ*

Emile Marchal, *ORSTOM, Institut Océanographique Paris, France*

Alasdair D. McIntyre, *University of Aberdeen, Dept. of Zoology, Aberdeen, Scotland*

Joseph R. Morgan, *East-West Center, University of Hawaii, Honolulu, Hawaii*

Daniel Pauly, *ICLARM, Manila, Philippines*

Dennis A. Powers, *Hopkins Marine Station, Stanford University, Pacific Grove, California*

Victor Prescott, *University of Melbourne, Parkville, Victoria, Australia*

H.S.H. Prince Rainier III of Monaco, *The Principality of Monaco*

G. Carleton Ray, *Dept. of Environmental Sciences, University of Virginia, Charlottesville, Virginia*

Michael R. Reeve, *National Science Foundation, Division of Oceanography, Washington, DC*

Jeffrey D. Rodgers, *Research and Development Section, Oregon Dept. of Fish and Wildlife, Corvallis, OR*

R. Tucker Scully, *Office of Oceans and Polar Affairs, U. S. State Department, Washington, DC*

Kenneth Sherman, *Ecosystems Dynamics Branch, Narragansett Laboratory, Northeast Fisheries Science Center, NMFS, NOAA; University of Rhode Island, Narragansett, RI*

V. P. Shuntov, *VNIRO, All Union Research Institute of Fisheries and Oceanography, Moscow, Russia*

Hein Rune Skjoldal, *Institute for Marine Research, Bergen, Norway*

Qisheng Tang, *Huanghai Sea Fisheries Research Institute, Qingdao, People's Republic of China*

James A. Yoder, *Physical Oceanography Group, University of Rhode Island, Narragansett, RI*

Table of Contents

Part Three: Sustainability and Management of Large Marine Ecosystems

Part Four: Technology Applications to the Monitoring Process in Large Marine Ecosystems

Message From the Editors

The multidisciplinary group of participants assembled at Monaco was there to consider the utility of the large marine ecosystem (LME) concept as a means for furthering national actions to ensure the conservation and sustainable use of marine resources. Dr. Martin Holdgate, Director General, International Union for Conservation of Nature and Natural Resources (IUCN)-The World Conservation Union; Dr. John Knauss, U.S. Under Secretary of Commerce for Oceans and Atmosphere; and the conference host, His Serene Highness, the Prince of Monaco, underscored the importance of this theme in their addresses to the partici-

pants. In recognition of the importance of the ideas of policymakers in encouraging the application of sound scientific principles to the growing environmental problems facing world leaders, we are pleased to include these messages of concern and encouragement as a Foreword and Preface to this volume.

Kenneth Sherman
Lewis Alexander
Barry Gold

Narragansett, RI
September 1992

Preface

The meeting on which this volume is based took place in Monaco, a country with a long and distinguished ocean tradition. It was most fitting that His Serene Highness, Prince Rainier, graciously agreed to host this conference, because the Grimaldi family was instrumental in supporting the great age of ocean discovery during the nineteenth century and the early part of the twentieth century. The tradition continues with support for the Ocean Science Center of Monaco, the International Hydrographic Bureau, the International Laboratory of Marine Radioactivity, and the world famous Musée Oceanographique. In Monaco, where old traditions are facing new challenges, this conference appropriately served as a starting point for launching a new international effort to mitigate the effects of increasing stresses on ocean resources.

Because the authors in this volume conclude that the concept of large marine ecosystems (LMEs) makes ecological sense, I believe it is high time we get on with the business of monitoring large marine ecosystems with the goal of trying to understand how they work. We know LMEs are subject to large year-to-year and decade-to-decade variability. We know that much of this variability is independent of man-induced effects. One of the goals of LME programs must be to distinguish between the variability that is "natural" and that which is anthropogenic.

These are exciting times in which to be an environmental scientist. Ready or not, we are in the midst of a major, worldwide, ecological experiment. This planet is stressed by a population of 5 billion people, and I expect that stress is going to increase during the next century as the population heads toward 10, perhaps 15, billion people. Almost all of us share the goal of living better than our grandparents did. As a consequence, we are seeing loss of species, change in climate, and change in our environment. One can make a good case that if present trends continue, we will be passing through a geological boundary—and we are responsible.

Like many of you, I have followed the debate of the past few years as to whether the Cretaceous-Tertiary boundary, the K-T boundary that "did in" the dinosaurs, was caused by a giant meteorite, an outpouring of volcanoes, or some combination of both. To the best of my knowledge, no one has blamed that boundary on the dinosaurs.

We are probably the first species on this earth capable of causing a geological boundary. We are certainly the first species capable of recognizing what we are doing.

Although we can see the effects of man almost everywhere in our environment, differentiating between and quantifying those changes in our environment caused by man and those resulting from natural variability is often difficult. Consider climate change, a subject that has captured the attention of the public and politicians, as well as scientists. The atmosphere of this earth has slowly warmed and cooled over time scales of decades and centuries. We are reasonably certain that a major key to variations on this time scale is a result of ocean-atmosphere interactions. We can show this is true in a statistical sense in our large computer-driven global climate models. If we let our models run for the equivalent of hundreds of years, they show the correct range and time scale of temperature fluctuations. No two years are alike. The temperature changes from year to year and decade to decade, but our models are not yet good enough to allow us to input this year's values and predict next year's climate.

Thus, there is debate about global warming. There is little disagreement that a doubling of greenhouse gases such as carbon dioxide and methane will affect our climate. All models suggest some warming of the atmosphere, but the positive and negative feedbacks associated with that perturbation are very complicated, and our models are very crude. One cannot determine whether the slight warming trend of the last century is natural or man-driven. The present debate is not about whether a doubling of greenhouse gases will cause some warming of the atmosphere, but about how much and how soon.

As limited as is our knowledge of atmospheric variability and the reasons for it, our knowledge of ocean variability is even more rudimentary. Some of our best evidence is indirect. Most experts conclude that any variability in the atmospheric climate of more than a year must be accompanied by an analogous change in the ocean climate. To be more precise, the changes in the atmospheric climate are probably driven in large part by changes in the ocean. In contrast to the reasonably good records of atmospheric temperature that exist for the past several hundred years, our ocean records are very limited. We can simply infer changes in the ocean by noting changes in the atmosphere. We do not know the amplitude and shape of the spectrum in the ocean, let alone the processes that determine it.

What about biological variability in the ocean? We continue to gather an increasing number of examples of biological fluctuations resulting from overfishing and from localized pollution, but well-documented examples of natural variability are more rare. I expect the best examples are the high-frequency, year-to-year changes in the recruitment of commercial fish species. There can be orders of magnitude of variation in yearly recruitment levels, which, in the absence of other pressures, lead to a relatively stable population of mature fish.

There is some evidence of low-frequency (greater than 10 years) fluctuations. Perhaps the best-documented is the periodic dominance of first the sardine and then the anchovy populations off the west coast of California. There has been speculation that not all the recent spectacular crashes of commercial fisheries, such as the Alaskan king crab and the Peruvian anchovy, have been solely caused by overfishing. We suspect that the declines in these fisheries may have been helped by natural environmental variability. We simply do not know enough about the processes involved, nor about the complexity of the interactions, to make meaningful predictions of trends. This we must do, if we are to manage our living marine resources wisely.

How does all of this apply to LMEs? This planet is being stressed by a growing population and a growing standard of living. It is clear that there are fluctuations in the atmosphere and in the ocean, as well as in terrestrial and marine ecosystems, that are independent of the influence of man. We have an increasing number of man-induced changes in our physical environment: the ozone hole in the stratosphere, smog and acid rain in the atmosphere, pollution in our estuaries, and eutrophication in some of our enclosed seas. It is often difficult, however, to separate the natural variation from that which is man-induced in many of our physical systems, particularly in the oceans. In large part, I believe this is because our level of understanding is inadequate. With rare exceptions, the monitoring of physical, chemical, and biological parameters in LMEs is weak. Without exception, our understanding of how the systems work in any LME is weak. Perhaps it is time to do something about it.

I expect that before the end of the twentieth century we will have the beginnings of an international ocean-observing system, but that system will be designed to monitor large-scale ocean-atmosphere interactions. Its purpose will be climate prediction, and I expect the emphasis, at least in the beginning, will be centered on such areas as the tropical Pacific Ocean and Antarctica, where these types of interactions are expected to be strongest. As contemplated, it is not a system keyed to LMEs. I expect an LME system would be more concerned with pollution and other man-induced changes; I expect it would be more ecologically oriented.

If one set out to design a coastal ocean-monitoring system, are LMEs the correct geographical unit? If they are, would it be useful to organize a set of regional programs, each designed for a specific LME? Such programs would have as a goal the monitoring of the system and understanding how the system works and how man is perturbing the system. Those responsible for the program of each LME could meet locally on a regular basis. Perhaps every few years they could come together internationally to compare notes and report on the health of the ocean. This program could be organized through one, or some combination, of the United Nations' agencies, perhaps with advice from the International Council of Scientific Unions (ICSU). I believe the experience of ICSU's International Geosphere Biosphere Program (IGBP) may be useful. Their objective is "to describe and understand the interactive physical, chemical, and biological processes that regulate the total Earth system, the unique environment it provides for life, the changes that are occurring in that system, and the manner by which these changes are influenced by human action" (Abelson, 1986). Substituting "large marine ecosystem" for "total Earth system" may provide a concise objective for a program of the type a number of the authors in this volume are interested in developing. The development of such a program would be appropriate for consideration in Brazil in 1992 at the United Nations Conference on Environment and Development.

The long-term goal of such a program, or its ultimate goal, will be to acquire sufficient understanding of how the system works so as to predict its future state; that will not be easy. The short-term goal, which I believe is achievable, is no less important: It is to monitor the health of the ocean.

John Knauss

Reference

Abelson, P. H. 1986. The International Geosphere-Biosphere Program (editorial). Science 234:657.

The International Union for Conservation of Nature and Natural Resources (IUCN)-The World Conservation Union plays a significant role in promoting conservation of the large marine ecosystems with which this volume is concerned. I use the word "conservation," of course, in the usual sense—the maintenance of the natural systems of the oceans, as an important component of planetary biological diversity and ecological function, and as an important resource for sustainable use.

A subsidiary concern here is the transferability of concepts of conservation and sustainable use, largely developed on land, into the oceans. It is possible to argue that concepts have, from time to time, been loosely transferred and, as a result, have led to suboptimal formulation of policy goals and management practices.

For example, the word "development" can be defined reasonably precisely for terrestrial resources as the modification of the structure and functioning of environmental systems so that an increasing proportion of their productivity is provided in a form directly useful to humanity. Development may involve changing forests growing on fertile soil to farmlands capable of sustainable use, and producing food, fiber, or other useful products at an enhanced level. Development can also mean the controlled conversion of productive living resources into nonproductive-built environments; it is often used uncritically to refer to the expansion of the technosphere at the expense of the biosphere, regardless of long-term sustainability.

One can contend that the oceans are not the scene of development in the sense that the term can be used on land. The transformation of marine ecosystems to yield an increasing quantity of a product, directly useful to humanity, has largely been a feature of coastal zones. Mariculture for algae, molluscs, crustaceans, and fish has been a growth industry in the last two decades, but accounts for only about 5% or less of global fishery landings. Much development affecting the marine environment has been in the form of the encroachment of the built-environment on the natural, with engineering alterations of coastlines and the discharges of industrial wastes—in neither case, conducive to the balance of sustainable productivity of marine ecosystems.

Over much of the ocean, sustainability has not been the determinant motivation in human-use patterns. As we are all aware, early fisheries treated their quarry as an "open-access resource" with a strong element of competition between the exploiters, and these two factors were directly responsible for the overconsumption, and consequent depletion, of such target organisms as whales, fur seals, and a number of stocks of food fishes. In 1982, it was estimated that world fishery landings were 15 to 20% less than they might have been had fishing effort and offtake been adjusted to the optimum productivity of the systems. Since then, there has been, if anything, an increase in the nonselective and destructive impact of fisheries, despite the regulatory effort. Industrial fishery and driftnetting are both threatening the balance of ecological systems, and tending, through their nonselectivity, to a great degree of waste through the blind catches of nontarget organisms that they are capturing and discarding.

The problems created by the "open-access approach" have been exacerbated by the fact that most offshore fisheries are managed as separate entities from the nearshore environments that provide the breeding sites and nursery areas of these "offshore" species. It is becoming more widely accepted that we can no longer ignore such interrelationships. Our management strategies must take account of the dependence of offshore fisheries on nearshore environments, as well as the dependence of different components of the marine food web on each other.

In a similar way, over much of history the discharge of pollutants to the sea was regarded as the ideal disposal method because of the vastness of the oceans; the apparently large dilution that they provided; and the belief that by putting a waste into the sea, it was being removed from the immediate environment of humanity. Virtually no country in the world had, until recently, a considered policy to regulate its discharges to the oceans at a level well within the capacity of the marine system to disperse those wastes without threat to ecological balance, and without the accumulation of undesirable residues in living organisms. Very similarly, engineering changes on the coastline have normally been planned without much thought of their impacts on the ecology and productivity of the systems affected—even though it has been known for decades among marine biologists that shallow and productive inshore waters, most at risk from engineering development and pollution, were important nursery areas for species that later moved offshore and became the targets of commercial fisheries.

Of course, there have been attempts at a regulatory effort. A number of fisheries commissions, attempting to give guidance on optimal yields, were established in the early part of this century. Since they were based on voluntary consent, however, they generally had no means of restricting catches to appropriate levels, and many states withdrew from the agreements. In the 1950s, the concept of regional fisheries commissions found favor once again, so that now they exist for most ocean basins and seas. The main value of these efforts has been a better monitoring of the state of stocks, and at least some recognition of the need for regulation of the fishery effort. In the northeast Atlantic, both the information base and the controls have been relatively good, whereas in some other areas, like the North Pacific, only some species have been effectively managed.

The most recent regional fisheries convention, the Convention on the Conservation of Antarctic Marine Living Resources (CCAMLR), is unique in that for the first time, it incorporated in its provisions a requirement that the taking of krill and other target species should be regulated. This requirement was established not only to remove the threat to the species concerned, but to sustain the overall balance of the ecosystem and to promote the recovery of depleted species.

Even so, there are questions as to whether CCAMLR will effectively regulate international effort on the high seas in Antarctic waters any better than the other fisheries conventions have proved their capacity to adjust fishing to the optimal level, and thereby, provide maximal yields without ecological imbalance. It is commonly stated that the salvation of marine fisheries has, in fact, been the extension of national jurisdictions to 200 miles, thereby giving each single sovereign state the opportunity to regulate the use made of resources within its limits. There is nothing surprising about this. On land, too, the establishment of a unique economic interest in a resource, together with the powers to manage the use of that resource, has often been proved the soundest base for conservation and sustainability. It is a sad fact of modern life that international instruments are generally less effective than national ones.

Marine pollution has also been a focus for international action. The London, Paris, and Oslo Conventions of the early 1970s were established, first and foremost, because of international concern about the dumping of chemical wastes in the epicontinental seas. Regional conventions, spearheaded by UNEP's Regional Seas Office (now Ocean and Coastal Areas Programme Activities Centre), have been a distinctive feature of the 1970s and 1980s. They provide for international action to agree on management objectives for areas of coastal seas in many parts of the world that are important for their fisheries and at risk from pollution. Inevitably, however, these conventions can only operate if the party sovereign states impose strict controls on their land-based activities, since the overwhelming bulk of damaging pollution entering the sea comes from the shore, and progress has been slow. It was only in 1987, for example, that the "precautionary principle" was accepted by ministers of all the riparian states as a guiding precept in the regulation of pollution of the North Sea. Also, it was only then that ministers undertook to use the best available technology to curb the inputs of persistent and dangerous pollutants to those vulnerable waters.

IUCN is concerned with the sustainable use of natural resources and the maintenance of global biodiversity and global ecological functions as essential preconditions for a habitable planet. We are naturally unhappy about the way in which ocean resources have been used. As the draft successor volume to the World Conservation Strategy says, despite the expenditure of billions of dollars and thousands of lifetimes worldwide in order to understand and regulate human impact on the sea and its resources, the efforts have not even approached what is needed. We need to form much clearer policies for the future, and these will have to encompass both national and international agreements and actions, since while a greater part of the ocean lies beyond national jurisdiction, all impacts upon the oceans either arise within the territory of sovereign states and their contiguous Exclusive Economic Zones (EEZs), or are carried out as a result of national exploitative activities. The international agreement and instrument is only a frame to guide the national.

I suggest that there are two particular needs as we approach the future:

The first broad requirement is to codify the objectives that we should pursue in order to ensure conservation and sustainable use of ocean and marine resources. There are five specific objectives that we could consider in this context:

(1) The integrity, diversity, and functioning of marine ecosystems should be maintained.

(2) All harvests of marine organisms should be held below sustainable yield, and the harvest of an individual species should not endanger stocks of other species or the overall balance and functioning of the ecosystem. To this end, catch methods should be as selective as practicable, and the capture devices used

should not be capable of continuing to fish, should they break adrift from human control.

(3) Inputs of sediment and chemicals, borne especially on rivers and land-based discharges, must not endanger the balance and functioning of marine ecosystems.

(4) National activities in managing 200-mile EEZs must not have adverse impacts on the ecology of adjacent 200-mile EEZs belonging to other states or on the balance of oceanic systems outside the limits of national jurisdiction.

(5) International cooperation should be promoted in order to ensure that ecosystems and organisms that occur within areas of ocean shared between states are managed in a compatible way.

The second broad requirement is to agree on national and international actions that transform these broad objectives into practical conservation and sustainable-use methods. These should include sound methods for the study of marine and oceanic ecosystems; for monitoring especially the status of stocks of exploited species or incidentally affected species; and for the collection, international exchange, and joint evaluation of the resulting data.

Substantial steps have already been taken in this direction, through such organizations as the Intergovernmental Oceanographic Commission (IOC) and the International Council for the Exploration of the Sea (ICES), but these should be extended by the following measures:

(1) The development of agreed methods for the management of marine resources. It is axiomatic that an ecosystem approach should be required; that inshore and offshore fisheries management should be regulated as part of one planned procedure; that the overall use patterns for large areas of ocean should be negotiated and agreed among all states concerned; and that areas that are especially vulnerable to misuse or need specially coordinated international activities should be identified and managed on such a basis. Beyond this, the roles of the various national and regional institutions now existing, and especially the commissions of the various conventions, should be reviewed so that there is the optimal international coordination of action.

(2) Action to prevent the pollution of the sea from land-based sources must have high priority, as numerous ministerial and international meetings have stated in recent decades. This will only be a practical proposition, however, if there is international exchange of information on the precise technologies that achieve industrial goals while minimizing the production of persistent and hazardous waste. Very similarly, knowledge of the best practices to avoid excessive release to the sea of nutrients from intensive agriculture and livestock husbandry need to be exchanged. It is useless for marine conservationists to simply require that inputs of toxic materials to the ocean be greatly reduced or prevented, unless it is fully understood that the best available technology should be freely shared between nations, in the overall interests of the conservation of the biosphere.

(3) Marine conservation areas should have, as part of their objectives, the protection of habitats that are important recruitment and breeding areas. They should be considered and established as a part of the overall strategy for the safeguarding of marine biological diversity and ecosystem processes.

The concept of the protected area has perhaps been transferred from the terrestrial to the marine environment rather uncritically. There are real differences between terrestrial and marine ecology that need to be taken into account before protected areas are established and managed. Among these ecological differences are the following:

(1) Very little marine primary productivity passes to the decomposers. Most marine plankton is ultimately consumed by herbivorous zooplankton or other consumers, and the benthic detritus feeders are substantially nourished by fecal material, rather than dead plant material.

(2) Turnover rates are also high in the zooplankton and, indeed, the whole dynamic of marine systems differs from that of the terrestrial.

(3) Recruitment strategies for marine organisms are often quite different from those of their terrestrial counterparts. Many marine organisms, including benthic and littoral sedentary organisms, have planktonic larvae. Many fish species have eggs that are widely dispersed on ocean currents, but even species of more restricted dispersion commonly have inshore nursery grounds and offshore areas exploited by adults.

(4) Marine waters mix and mingle, so that nutrients move over large areas. For example, the nutrients that well to the surface in Antarctic waters are borne from areas as far afield as the North Atlantic. The result of these movements is that it is impractical to define and bound marine protected areas in ecological terms with the same precision that terrestrial protected areas can be defined.

The establishment of boundaries for marine protected areas is likely to be of very limited ecological significance. Certainly, these boundaries will be transgressed by water movements, with their pollutant load, and by planktonic and wide-ranging pelagic organisms.

While defined areas may have ecological distinction, this will commonly result from habitat features such as depth range, temperature regimes, and the impact of terrestrial systems.

The purpose in defining marine protected areas is primarily to regulate human activities, but even these cannot be bounded exactly. Certain activities that may affect the integrity of an area directly, like dynamite fishing, coral mining, the dumping of wastes, the impact of fishery, the impact of tourism, or the disturbance of the habitat by modification of adjacent coastlines, can be regulated within the area. Other impacts will, however, originate well outside any marine protected area, for example, through the long-distance transfer of pollution by water or air, the input of substances in drainage from adjacent land, or the impact of human activities on adjacent marine areas. In establishing marine protected areas, it is essential to judge which key actions need to be regulated, and if these occur outside the protected area itself, then such controls need to be a part of a wider overall policy.

In conclusion, I would like to comment on the usefulness of this volume. IUCN needs the judgments it brings to bear in evaluating which policies we should be promoting for the sustainable use of the oceans. We need them also to determine where the priorities for action lie among the numerous proposals in "Caring for the World: A Strategy for Sustainability," the document designed to supplement the World Conservation Strategy of 1980.

The contributions to this volume are scientifically authoritative and environmentally precise. The authors have not hesitated to criticize existing policy and to propose substitutes, where it is clear that we need to do things differently, and very probably, with a greater sense of urgency than has characterized the human attitude toward the oceans over past decades.

Martin W. Holdgate

Introduction

Opening Remarks by His Serene Highness, Prince Rainier III, of Monaco

As president of the International Council for the Scientific Exploration of the Mediterranean Sea, we are particularly happy that the Oceanographic Museum in the Principality of Monaco is hosting this international conference, the objective of which is to introduce the large marine ecosystem (LME) concept and its application to the management of living resources within the context of regional planning.

The immensity and the urgency of the problems facing our marine coastal states are highlighted by the joint efforts of the World Conservation Union, UNESCO's Intergovernmental Oceanographic Commission, and the National Oceanic and Atmospheric Administration of the U.S. Department of Commerce.

The presence at this meeting of Dr. Martin Holdgate, Director General of the World Conservation Union, of Dr. Gunnar Kullenberg, Secretary General of UNESCO's Intergovernmental Oceanographic Commission, and of Dr. John Knauss, Under Secretary for Oceans and Atmosphere of the U.S. Department of Commerce, bears witness to the commitment of these institutions to a sustained effort to put forward foundations for a new international order which we feel must be built upon the conservation and restoration of the marine world's living resources.

The scientific and economic significance of this conference has drawn the support of several bodies, namely: the United Nations Environment Program (UNEP), the World Bank, the American Association for the Advancement of Science (AAAS), the Council for Ocean Law, the National Science Foundation (NSF), the International Food and Agriculture Organization (FAO), the International Coastal and Oceans Organization, the Marine Mammal Commission, and the Scientific Committee on the Oceanic Research (SCOR), to which we have pleasure in extending our gratitude and our congratulations to each and to all for their discernment.

For its part, the International Council for the Scientific Exploration of the Mediterranean Sea is most happy to have been responsible for the organization of this fifth conference which expresses converging concerns. We have no doubt that your discussions will encourage dialogue at the talks which are to take place in a few days' time at this institution's XXXII Plenary Assembly-Congress whose deliberations we will follow personally.

Let us not forget that the present conference is the product of an intense period of preparation led by Danny Elder, Coordinator of the World Conservation Union's Marine Program; Thomas L. Laughlin, Head of the International Liaison Office of NOAA; and Kenneth Sherman, Director of the Narragansett Laboratory. We wish also to acknowledge the efforts of the Oceanographic Museum and the work of Professor François Doumenge and his colleagues, to whom we extend particular thanks.

We are delighted to see eminent specialists from the world's most prestigious oceanographic centers gathered here for this international conference. Our only regret is the absence of Professor Takahisa Nemoto, Director of the Oceanographic Institute of the Imperial University of Tokyo, who so generously welcomed us aboard the research vessel *Hakuko Maru* when berthed in the principality last January, and who died suddenly a few weeks ago.

As sovereign of a state for which the maritime environment is vital, we are most particularly aware of the necessity to preserve the ocean's resources. By signing the "Ramoge" agreement with France and Italy, the Principality of Monaco has committed itself to a regional cooperation policy as early as year 1976.

Therefore, the Principality of Monaco has taken a leading role in keeping with the subsequent emergence of a management policy implicitly founded on the LME concept.

More than ever before, the spreading of industrial pollution, the disposal of urban waste at sea, the long-range contamination of aerosols, as well as the global problem of overexploitation of fish resources which is threatening extinction of numerous species, demonstrate that all states are concerned and require a response from mankind as a whole.

Major disasters can be provoked by climatic, hydrological, or geological accident as well as by the overexploitation of fish resources or agricultural, urban, or industrial pollution. In the middle term, human activities are likely to disrupt the ocean-atmosphere relationships and affect the protective ozone layer as well as the partial pressure of carbon dioxide in air and water. Appropriate responses will only be found through an international cooperation and a multidisciplinary approach, namely in the disciplines of physics and biology, as well as economics and law. Hence, we have the Principality of Monaco's support to the Council of Europe's Open Partial Agreement on the prevention of, protection against, and organization of relief in major natural and technological disasters.

In this respect, the Scientific Center of Monaco, founded in 1960, has just undergone reorganization to comply with the demands of increased efficiency. The European Oceanographic Observatory, created within the framework of this reorganization, will contribute to the forecasting, prevention, and handling of major disasters affecting the marine life. Its research laboratories will develop appropriate methods and means to detect the risks and to speed up the restoration of disrupted environments.

Furthermore, the last conference dedicated to LMEs highlighted energy transfers and the relationships between different elements of the marine life. Also emphasized was the urgent need to set up a model management system to avoid disastrous crises which jeopardize numerous communities over vast areas. The increased incidence of eutrophication which can affect entire seas (the Adriatic in 1989), and high death tolls among seals in the northwest Atlantic, great sharks, reptiles, and marine mammals caught in nets drifting in the southern South Pacific, the northern Pacific, the Bay of Biscay, and, unfortunately, the Mediterranean, underline the quite obvious global dimension of the dangers to which the ocean is exposed.

It is our fervent hope that scientists will work together to arouse a universal awareness of the risks incurred if states do not act quickly to set up multi- and supranational systems to provide for the implementation of vital safeguards and conservation measures.

The Convention for the Conservation of Antarctic Marine Living Resources is, we feel, an example which should be followed, as are the concerted efforts in the U.S. at federal and state levels, including federal institutions such as NOAA, to establish a framework of appropriate measures for the protection and the conservation of resources and the environment through a dynamic biogeographical approach in the context of LMEs.

We are happy that the organizers of this international conference on the LME concept and its application to regional marine resource management have chosen the Principality of Monaco to hold this first meeting outside the U.S. in association with the International Council for the Scientific Exploration of the Mediterranean Sea, founded by my noble ancestor, Prince Albert I. Henceforth, the international law of the sea must seek its foundations in the results of scientific research.

In this spirit, we feel it is vital that highly qualified fora such as yours make the voice of wisdom and reason heard calling for widespread international dialogue to lay down standards protecting species and their habitats. The devastation which has long gone unpunished, wrought by a series of mining and fishing processes, has already affected wide areas which were still unsullied 30 or 40 years ago. We will namely mention the African, American, and Southeast Asian intertropical zones as well as the subarctic seas of the Atlantic and the Pacific.

The southern part of the planet, in its turn, is being affected, with damage to the Kerguelen shelf and plateau of Patagonia and Cape and especially the influences of driftnet fishing in the southern Pacific.

The time has come to react by proposing an alternative to corporate and national egoism. Our good wishes are, therefore, with you throughout your deliberations in the hope that they will rapidly lead to the international dialogue which alone can provide a solution for an anxious and expectant mankind.

Acknowledgments

This volume is the fifth in a series on large marine ecosystems (LMEs). The first four LME conferences were held in the United States during annual meetings of the American Association for the Advancement of Science (1984, 1987, 1988, and 1990). This volume is based on the first international conference on LMEs, held in Monaco in 1990. In the volumes, emphasis is placed on LMEs as a unifying concept for aiding in the long-term sustainability of marine resources by focusing research, monitoring, and management efforts within the boundaries of distinct ecologically based units of ocean space. Specialists generally publish separately on the use and management of ocean resources. It is not often that results of multidisciplinary scientific and management studies are brought together in a single volume as they are here.

The LME approach brings multidisciplinary marine studies to bear on regional-scale concepts of resource sustainability by examining the causes of variability in the productivity of those regions around the margins of the world's oceans from which 95% of the annual yields of usable fisheries biomass is harvested. Emphasis is placed on identification of the primary, secondary, and tertiary driving forces controlling the large-scale variability of biomass yields within and among LMEs.

This volume represents the collective efforts of oceanographers, biologists, ecologists, modelers, managers, lawyers, and geographers. Their contributions are aimed at bringing multidisciplinary specialties together in an effort to address the advantages of a more holistic approach to the research, monitoring, and management of marine resources.

The editors are indebted to the contributors for their willingness to take time out of busy schedules to develop the expert syntheses and reviews needed to move marine resources science and management forward. We are pleased to acknowledge the interest and financial support of the National Oceanic and Atmospheric Administration (NOAA) and the National Marine Fisheries Service (NMFS). We gratefully acknowledge support for the conference from the United Nations Environmental Program (UNEP), World Bank, American Association for the Advancement of Science, Council for Ocean Law, National Science Foundation, Food and Agriculture Organization of the United Nations, International Coastal and Oceans Organization, Marine Mammal Commission, and Scientific Committee on Oceanic Research (SCOR).

The LME conference and this volume would not have been possible without the willing and capable cooperation of many people who gave unselfishly of their time and effort. We are indebted to Donna Busch and Karen Heise Gentile, NMFS, Narragansett, for their extraordinary dedication, care, and expertise in technically editing the entire volume that now reads well in English; given the diversity in the international contributions, their's was a challenge well met. We thank Patricia Morgan, Director, and Louise Rosenblatt Goines and other members of the editorial staff at AAAS Press for their care in the final production of the volume. The early planning of logistics for the conference was very capably handled by Laura Hedrick, NMFS, Narragansett. We also would like to acknowledge the fine efforts of Jennie Dunnington and Janet Kelly, also of NMFS, Narragansett, whose contributions in the production stages of the volume were invaluable. We extend special thanks to Tom Laughlin of NOAA Headquarters and Danny Elder of the International Union for Conservation of Nature and Natural Resources (IUCN)-The World Conservation Union, Gland, Switzerland, for their extraordinary effort in managing the conference venue and its organization. We acknowledge the tireless efforts on behalf of the conference extended by Dr. François Doumenge, who brought to the planning stages his wide experience in marine studies and generously provided the use of the facilities of the Musée Oceanographique, Monaco. A special debt of thanks is extended to His Excellence, Prince Rainier, who, as host of the conference, provided not only a meeting place rich in marine history and magnificent in location, overlooking the Mediterranean Sea, but also for his inspiring opening address to the participants.

The Editors

September 1992

Part One:
Sustainability of
Large Marine Ecosystems

Large Marine Ecosystems as Global Units for Marine Resources Management—An Ecological Perspective

Kenneth Sherman

A new paradigm in ocean use was initiated in 1982 when the United Nations (U.N.) Law of the Sea Convention established Exclusive Economic Zones (EEZs) up to 200 nautical miles from the baselines of territorial seas, granting coastal states sovereign rights to explore, manage, and conserve the natural resources of the zones. Within and extending seaward beyond the boundaries of the zones are large marine ecosystems (LMEs) that are being subjected to increased stress from growing exploitation of fish and other renewable resources, coastal zone damage, river basin runoff, dumping of urban wastes, and fallout from aerosol contaminants. LMEs are relatively large regions, on the order of 200,000 km^2 or larger, characterized by distinct bathymetry, hydrography, productivity, and trophically dependent populations (Sherman and Alexander, 1986). The theory, measurement, and modeling relevant to monitoring the changing states of LMEs are imbedded in reports on multistable ecosystems (Holling, 1973; 1986; Beddington, 1986) and pattern formation and spatial diffusion in ecosystems (Levin, 1978; 1990).

The ecological concept that critical processes controlling the structure and functioning of biological communities can best be addressed on a regional basis (Ricklefs, 1987; Levin, 1990; Graham et al., 1991) is consistent with the LME approach to monitoring living marine resources, the "health" of their habitats, and their management. Since the 1960s, the number of reports of changes in the structure and community dynamics of marine ecosystems attributed to habitat loss, increasing pollution levels, and the effects of excessive fishing mortality has grown (Sherman et al., 1990). Overfishing has caused multimillion

metric ton biomass flips among the dominant pelagic components of the fish community off the northeastern U.S. (Sissenwine, 1986). The biomass flip, wherein a dominant species rapidly drops to a low level to be succeeded by another species, can generate cascading effects among other components of the ecosystem, including marine birds (Powers and Brown, 1987), mammals, and zooplankton (Overholtz and Nicolas, 1979; Payne et al., 1990). Other human-induced perturbations to marine ecosystem populations are the incidental catches of marine mammals during fishing operations and the growing impacts of pollution. Efforts to reduce stress and mortality on marine mammal by-catch are being pursued (Bonner, 1982; Loughlin and Nelson, 1986; Waring et al., 1990). Pollution problems at the continental margins of marine ecosystems that impact natural productivity cycles, including eutrophication, the presence of toxins in poorly treated sewage discharge, and loss of wetland nursery areas to coastal development, need to be addressed (GESAMP, 1990). The growing awareness that biomass yields are being influenced by multiple, but different, driving forces in marine ecosystems around the globe has accelerated efforts to broaden research strategies to encompass food chain dynamics and the effects of environmental perturbations and pollution on living marine resources from an ecosystem perspective. Mitigating actions to reduce stress on living resources in the oceans are required to ensure the long-term sustainability of biomass yields.

The biomass constituting 95% of the annual yield of marine fisheries is caught within the geographic limits of 49 LMEs. The boundaries of the LMEs are depicted in Figure 1, along

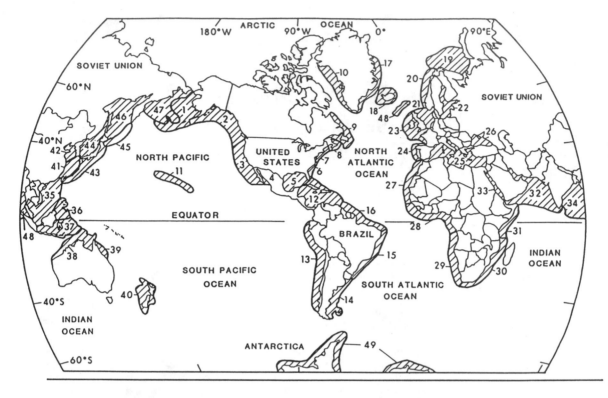

WORLD MAP OF LARGE MARINE ECOSYSTEMS

1. Eastern Bering Sea
2. Gulf of Alaska
3. California Current
4. Gulf of California
5. Gulf of Mexico
6. Southeast U.S. Continental Shelf
7. Northeast U.S. Continental Shelf
8. Scotian Shelf
9. Newfoundland Shelf
10. West Greenland Shelf
11. Insular Pacific--Hawaiian
12. Caribbean Sea
13. Humboldt Current
14. Patagonian Shelf
15. Brazil Current
16. Northeast Brazil Shelf
17. East Greenland Shelf
18. Iceland Shelf
19. Barents Sea
20. Norwegian Shelf
21. North Sea
22. Baltic Sea
23. Celtic-Biscay Shelf
24. Iberian Coastal

25. Mediterranean Sea
26. Black Sea
27. Canary Current
28. Guinea Current
29. Benguela Current
30. Agulhas Current
31. Somali Coastal Current
32. Arabian Sea
33. Red Sea
34. Bay of Bengal
35. South China Sea
36. Sulu-Celebes Seas
37. Indonesian Seas
38. Northern Australian Shelf
39. Great Barrier Reef
40. New Zealand Shelf
41. East China Sea
42. Yellow Sea
43. Kuroshio Current
44. Sea of Japan
45. Oyashio Current
46. Sea of Okhotsk
47. West Bering Sea
48. Faroe Plateau
49. Antarctic

Figure 1. Boundries of 49 large marine ecosystems.

with a list of the LMEs. Criteria used to geographically delimit LMEs are consideration of distinct bathymetry, hydrography, productivity, and trophically dependent populations. Several LMEs are semi-enclosed seas, such as the Black Sea, the Mediterranean Sea, and the Caribbean Sea. The LMEs can be divided into domains, or subsystems, such as the Adriatic Sea, a subsystem of the Mediterranean Sea LME. In other LMEs, geographic limits are defined by the scope of continental margins. Among these are the U.S. Northeast Continental Shelf, the East Greenland Sea, and the Northwestern Australian Shelf. The seaward limit of the LMEs extends beyond the physical outer limits of the shelves to include all or a portion of the continental slopes as well. Care was taken to limit the seaward boundaries to the areas affected by ocean currents, rather than relying simply on the 200-mile EEZ or fisheries zone limits. Among the ocean current LMEs are the Humboldt, Canary, and Kuroshio Currents. The biomass consisting of 95% of the annual yields of marine fisheries is listed by LME in Table 1.

For nearly 75 years, following the turn of the twentieth century, fishery scientists have been preoccupied with single-species stock assessments, while during this same period biological oceanographers did not achieve any great success in predicting fish yield based on food chain studies. As a result, through the mid-1970s, the predictions of the levels of biomass yields for different regions of the world ocean were open to disagreement (Ryther, 1969; Alverson *et al.*, 1970; Lasker, 1988). A milestone in fishery science was achieved in 1975 with the convening of a symposium by the International Council for the Exploration of the Sea that focused on fluctuations in the fish stocks of the North Sea and their causes. The symposium, which considered the North Sea as an ecosystem—following the lead of Steele (1974), Cushing (1975), Andersen and Ursin (1977), and others—was prompted by a rather dramatic shift in the dominance of the finfish species of the North Sea from a balanced pelagic and demersal finfish community prior to 1960 to a demersally dominated community from the mid-1960s through the mid-1970s. Although no consensus on cause and effect was reached by the participants, it was suggested by the convener (Hempel, 1978) that the previous studies of 7-1/2 decades may have been too narrowly focused and that future studies should take into consideration fish stocks, their competitors, predators and prey, and interactions of the fish stocks with their environments, the fisheries, and pollution from an ecosystems perspective.

The utility of a more holistic ecosystems approach to resource management is being recognized as an important recent development among marine scientists, geographers, economists, government representatives, and lawyers (Byrne, 1986; Christy, 1986; Alexander, 1989; Belsky, 1989; Crawford *et al.*, 1989; Morgan, 1989; Prescott, 1989). Effective management from an ecosystems perspective will be contingent on the identification of the primary, secondary, and tertiary driving forces causing large-scale changes in biomass yields. Management of species responding to strong environmental signals will be enhanced by improving the understanding of the physical factors forcing biological changes, whereas in other LMEs where the prime driving force is predation, either by natural predators or by human predation expressed as excessive fishing mortalities, options can be explored for implementing adaptive management strategies. Mitigating actions are required to ensure that the "pollution" of the coastal zone of LMEs is reduced and does not become a principal driving force in any LME. Concerns remain regarding the socioeconomic and political difficulties in management across national boundaries, as in the case of the Sea of Japan ecosystem where the fishery resources are shared by five countries (Morgan, 1988), the North Sea ecosystem, or the 38 nations sharing the resources of the Caribbean Sea ecosystem.

Although development of research, monitoring, and management strategies for LMEs is an evolving process (Morgan, 1988; Alexander, 1989), sufficient progress has been made to allow useful comparisons of the principal forces driving variability in fisheries biomass yields among selected LMEs around the globe. The large-scale spatial dimensions of LMEs precludes a strictly controlled experimental approach to their study. However, as regional ecosystems, they are amenable to the comparative method of science as described by Mayr (1982) and practiced by Bakun (1986; 1990) and Bakun and Parrish (1980; 1990). Since 1984, 29 case studies investigating the major causes of large-scale perturbations in biomass yields of LMEs have been completed (Table 2). Changes in the ocean climate of the northern North Atlantic during the late 1960s and early 1970s have been considered by several marine scientists as the dominant cause of change in food chain structure and fisheries biomass yields of three LMEs—the Norwegian Shelf ecosystem (Ellertsen *et al.*, 1990), the Barents Sea ecosystem (Skjoldal and Rey, 1989; Borisov, 1991), and the West Greenland Sea ecosystem (Hovgaard and Buch, 1990). Excessive fishing mortality was a secondary factor in the decline of the cod and capelin stocks of these LMEs.

Several LMEs are presently being subjected to ecosystem management. Among these

Table 1. Contributions by country and LME representing 95% of the annual global catch in 1987.

Country	Percentage of world marine nominal catch*	LMEs producing annual biomass yield	Cumulative percentages
Japan	14.43	Oyashio current, Kuroshio Current, Sea of Okhotsk, Sea of Japan, Yellow Sea, East China Sea, W. Bering Sea, E. Bering Sea, Scotia Sea	
USSR	12.63	Sea of Okhotsk, Barents Sea, Norwegian Shelf, W. Bering Sea, E. Bering Sea, Scotia Sea	
USA	7.03	Northeast US Shelf, Southeast US Shelf, Gulf of Mexico, California Current, Gulf of Alaska, E. Bering Sea	
China	6.72	W. Bering Sea, Yellow Sea, E. China Sea, S. China Sea	
Chile	5.98	Humboldt Current	
Peru	5.65	Humboldt Current	50
Korea Republic	3.50	Yellow Sea, Sea of Japan, E. China Sea, Kuroshio Current	
Thailand	2.48	South China Sea, Indonesian Seas	
Indonesia	2.45	Indonesian Seas	
Norway	2.40	Norwegian Shelf, Barents Sea	
India	2.09	Bay of Bengal, Arabian Sea	
Denmark	2.07	Baltic Sea, North Sea	
Iceland	2.02	Icelandic Shelf	
Korea D. P. Rep.	1.99	Sea of Japan, Yellow Sea	
Philippines	1.78	S. China Sea, Sulu-Celebes Sea	
Canada	1.75	Scotian Shelf, Northeast US Shelf, Newfoundland Shelf	75
Spain	1.69	Iberian Coastal Current and Canary Current	
Mexico	1.55	Gulf of California, Gulf of Mexico, California Current	
South Africa	1.12	Benguela Current, Agulhas Current	
France	1.00	North Sea, Biscay-Celtic Shelf, Mediterranean Sea	80
Ecuador	0.84	Humboldt Current	
UK-Scotland	0.82	North Sea	
Poland	0.80	Baltic Sea	
Viet Nam	0.77	South China Sea	
Malaysia	0.74	Gulf of Thailand, Andaman Sea, Indonesian Seas, South China Sea	
Brazil	0.72	Patagonian Shelf, Brazil Current	
Turkey	0.72	Black Sea, Mediterranean Sea	
Argentina	0.69	Patagonian Shelf	
Namibia	0.64	Benguela Current	
Italy	0.62	Mediterranean Sea	
Morocco	0.61	Canary Current	

Table 1. *(continued)*

Country	Percentage of world marine nominal catch*	LMEs producing annual biomass yield	Cumulative percentages
New Zealand	0.54	New Zealand Shelf Ecosystem	
Netherlands	0.53	North Sea	
Portugal	0.49	Iberian Shelf, Canary Current	
Faeroe Islands	0.44	Faeroe Plateau	90
Pakistan	0.42	Bay of Bengal	
Ghana	0.40	Gulf of Guinea	
Senegal	0.35	Gulf of Guinea, Canary Current	
Venezuela	0.34	Caribbean Sea	
Ireland	0.31	Biscay-Celtic Shelf	
UK Eng., Wales	0.30	North Sea	
Bangladesh	0.29	Bay Bengal	
Hong Kong	0.28	South China Sea	
Sweden	0.26	Baltic Sea	
Australia	0.25	N. Australian Shelf, Great Barrier Reef	
Cuba	0.25	Caribbean Sea	
Romania	0.25	Black Sea	
German Dem. Rep.	0.22	Baltic Sea, Scotia Sea	
Panama	0.21	California Current, Caribbean Sea	
Sri Lanka	0.19	Bay of Bengal	
Nigeria	0.18	Gulf of Guinea	
Uruguay	0.17	Patagonian Shelf	
Finland	0.16	Baltic Sea	95

*Percentages based on fish catch statistics from FAO Yearbook, vol. 64, 1987.

are the Yellow Sea ecosystem, where the principal effort is underway by the People's Republic of China (Tang, 1989); the multispecies fisheries of the Benguela Current ecosystem, under the management of the government of South Africa (Crawford *et al.*, 1989); the Great Barrier Reef (Bradbury and Mundy, 1989; Kelleher, this volume) and Northwest Australian Continental Shelf ecosystems, under the management of the state and federal governments of Australia (Sainsbury, 1988); and the Antarctic marine ecosystem, under the management of the Commission for the Conservation of Antarctic Marine Living Resources (CCAMLR) and its 21-nation membership (Scully *et al.*, 1986; Sherman and Ryan, 1988). Scientists and managers of the states of Oregon and California have developed a plan for research, monitor-

ing, and management of marine resources of the Northern California Current ecosystem (Bottom *et al.*, 1989).

At present, no single international organization is authorized to monitor the changing ecological states of LMEs and to reconcile the needs of individual nations where mitigation actions are necessary to reverse the deleterious impacts of stress on productivity and biomass yields (Myers, 1990). The need for a regional approach, rather than a global one, for implementation of marine research, monitoring, and stress mitigation has been recognized (Taylor and Groom, 1989; Malone, 1991).

It is within the nearshore coastal domains of the LMEs that the human-induced stress on ecosystems requires mitigating actions to ensure the continued productivity and economic

Table 2. List of 29 LMEs and subsystems for which syntheses relating to primary, secondary, or tertiary driving forces controlling variability in biomass yields have been completed for inclusion in LME volumes.

Large Marine Ecosystem	Volume No.*	Authors
U.S. Northeast Continental Shelf	1	M. Sissenwine
	4	P. Falkowski
U.S. Southeast Continental Shelf	4	J. Yoder
Gulf of Mexico	2	W. Richards and M. McGowan
	4	B. Brown et al.
California Current	1	A. MacCall
	4	M. Mullin
	5	D. Bottom
Eastern Bering Shelf	1	L. Incze and J. Schumacher
West Greenland Shelf	3	H. Hovgård and E. Buch
Norwegian Sea	3	B. Ellertsen et al.
Barents Sea	2	H. Skjoldal and F. Rey
	4	V. Borisov
North Sea	1	N. Daan
Baltic Sea	1	G. Kullenberg
Iberian Coastal	2	T. Wyatt and G. Perez-Gandaras
Mediterranean-Adriatic Sea	5	G. Bombace
Canary Current	5	C. Bas
Gulf of Guinea	5	D. Binet and E. Marchal
Benguela Current	2	R. Crawford et al.
Patagonian Shelf	5	A. Bakun
Caribbean Sea	3	W. Richards and J. Bohnsack
South China Sea-Gulf of Thailand	2	T. Piyakarnchana
Yellow Sea	2	Q. Tang
Sea of Okhotsk	5	V. Kusnetsov et al.
Humboldt Current	5	J. Alheit and P. Bernal
Indonesia Seas-Banda Sea	3	J. Zijlstra and M. Baars
Bay of Bengal	5	S. Dwivedi
Antarctic Marine	1 & 5	R. Scully et al.
Weddell Sea	3	G. Hempel
Kuroshio Current	2	M. Terazaki
Oyashio Current	2	T. Minoda
Great Barrier Reef	2	R. Bradbury and C. Mundy
	5	G. Kelleher
South China Sea	5	D. Pauly and V. Christensen

*Vol. 1, Variability and Management of Large Marine Ecosystems. Edited by K. Sherman and L. M. Alexander. AAAS Selected Symposium 99. Westview Press, Inc., Boulder, CO, 1986.
Vol. 2, Biomass Yields and Geography of Large Marine Ecosystems. Edited by K. Sherman and L. M. Alexander. AAAS Selected Symposium 111. Westview Press, Inc., Boulder, CO, 1989.
Vol. 3, Large Marine Ecosystems: Patterns, Processes, and Yields. Edited by K. Sherman, L. M. Alexander, and B. D. Gold. AAAS Symposium. AAAS, Washington, DC, 1990.
Vol. 4, Food Chains, Yields, Models, and Management of Large Marine Ecosystems. Edited by K. Sherman, L. M. Alexander, and B. D. Gold. AAAS Symposium. Westview Press, Inc., Boulder, CO, 1991.
Vol. 5, Stress, Mitigation, and Sustainability of Large Marine Ecosystems. Edited by K. Sherman, L. M. Alexander, and B. D. Gold. AAAS Press, Washington, DC, 1992 (this volume).

vitality of marine resources. Although the management of LMEs is an evolving scientific and geopolitical process, sufficient progress has been made to allow for useful comparisons among the primary, secondary, and tertiary driving forces influencing large-scale changes in the biomass yields and long-term sustainability of LMEs. Results from a series of summary reports are presented in this volume. They depict (i) the principal driving forces shaping the structure and function of LMEs, (ii) new technological applications for monitoring changes within LMEs, and (iii) the effects of these driving forces on the sustainability and management of living marine resources of several LMEs.

Studies on the stress, mitigation, and sustainability of LMEs presented in this volume have been initiated in part from reports and discussions at the fifth in a series of LME symposia and conferences. The first symposium was held during the 1984 American Association for the Advancement of Science (AAAS) meeting in New York; the second at the 1987 AAAS meeting in Chicago; and the third at the 1988 meeting of the AAAS in Boston. The fourth symposium, which focused on "Food Chains, Yields, Models, and Management of Large Marine Ecosystems," was convened in New Orleans at the 1990 AAAS meeting (Sherman *et al.*, 1991). The fifth symposium, on which the present volume is based, stems from an international conference on LMEs held in Monaco in 1990. The Monaco conference on LMEs was convened by several national and international organizations, including the National Oceanic and Atmospheric Administration (NOAA), IUCN-The World Conservation Union, Intergovernmental Oceanographic Commission (IOC), United Nations Educational, Scientific, and Cultural Organization (UNESCO), and the International Council for the Scientific Exploration of the Mediterranean (ICSEM). The conference was sponsored by the U.N. Environmental Program (UNEP); American Association for the Advancement of Science (AAAS); Council for Ocean Law; National Science Foundation; Food and Agriculture Organization of the U.N.; World Bank; International Coastal and Oceans Organization; Marine Mammal Commission; and Scientific Committee on Oceanic Research (SCOR). The hosting organization was the Musée Oceanographique de Monaco and the Principality of Monaco.

The overall objective of the conference was to introduce the Large Marine Ecosystem Research and Management Approach to a broadly representative group of scientists, managers, and administrators concerned with the sustainable use of marine resources. This was accomplished with the cooperation of the 140 participants representing scientists, managers, and administrators from 37 countries. With regard to the application of the LME approach to principal driving forces affecting resource yields, the conference was fortunate to have had expert and comprehensive presentations on several important LMEs.

In addition, comparative analyses on the effects of basin-wide physical and biological parameters influencing biomass yields were presented on the Barents Sea, Norwegian Sea, and West Greenland Sea ecosystems by Drs. J. Blindheim and H.R. Skjoldal of the Institute of Marine Research, Bergen, Norway (chapter 17), and on the Benguela Current, California Current, and Patagonian Shelf ecosystems by Dr. A. Bakun, NOAA, Monterey, California (chapter 18).

The three opening-session speakers, H.S.H. Prince Rainier of Monaco; Dr. Martin Holdgate, Director General of the World Conservation Union; and Dr. John Knauss, U.S. Under Secretary of Commerce for Oceans and Atmosphere, provided the conference with guidance and encouragement for moving forward with the development and implementation of strategies for the monitoring and management of marine ecosystems. The deliberations during this 6-day conference brought the participants closer to achieving this goal.

The evolution of the LME concept and its applicability to scientific and management issues involving fisheries, coastal zones, pollution, and global warming was reported by Dr. G. Hempel of the Institute for Tropical Biology, Bremen, Germany, in a coauthored presentation (chapter 3). Dr. M. Reeve of the National Science Foundation suggested ways for incorporating process-oriented ocean studies within the framework of LMEs, particularly with regard to the Global Ecosystem Dynamics (GLOBEC) program of NSF (chapter 5).

The steps to be taken to integrate LME scientific studies into the management of ocean resources and ocean space were addressed by Professors R. Knecht and B. Cicin-Sain of the University of Delaware (chapter 21). A well-designed plan for the research and management of the Northern California Current system LME was put forth by Dr. D. Bottom of the Oregon Fish and Game Department (chapter 24). With regard to management of LMEs, Dr. N. Daan of the Institute for Fishery Investigations, Ijmuiden, The Netherlands (chapter 23), reported on the need for an integrated management approach for LMEs in general, and the North Sea ecosystem in particular. Results of his recent model of the effects of closing areas of the North Sea to fishing to enhance fish stock production showed that closed areas had little effect. Dr. G. Kelleher emphasized the utility of a holistic approach to the management of entire LMEs in his excellent presentation on the sustainable development of the Great Barrier Reef as an LME (chapter 25).

The feasibility was demonstrated for supporting actions to increase the sustainability of usable biomass production of LMEs. In the Yellow Sea, fishermen now harvest catches of 10,000 metric tonnes of juvenile shrimp a year. The Chinese government first introduced juvenile shrimp for grow-out purposes in 1984, based on the premise that the reduction of the natural predator field of demersal fish stocks provided an opportunity to introduce a higher economic-yield shrimp resource, as reported by Dr. Q. Tang of the Huanghai Sea Fisheries Research Institution, Qingdao, People's Republic of China (chapter 10). A second example is the experimental introduction of artificial reefs or substrates for incorporating superfluous primary production to enhance biomass yields of benthic molluscs, fish, and crustaceans in the Adriatic ecosystem, presented by Dr. G. Bombace of the Institute of Marine Fisheries Research, Ancona, Italy (chapter 13). With regard to pollution studies, the importance of controlling the sources of pollution in the river systems and wetlands of the coastal margins of LMEs was emphasized by Dr. A. McIntyre, Aberdeen, Scotland, who noted the lack of significant contamination of the offshore, more open waters of LMEs (chapter 4).

The influence of ocean physics on the pelagic stock yields and overfishing on the demersal stock yields were considered as important driving forces in the Sea of Okhotsk ecosystem by Dr. V. Kuznetsov of the All-Union Research Institute of Marine Fisheries and Oceanography, Moscow, and his colleagues V. Shuntov and L. Borets of PINRO Vladivostock, former U.S.S.R. (chapter 9). The importance of ocean physics as the principal force driving variability in biomass yields of the Humboldt Current ecosystem was stressed by Drs. J. Alheit of Polarmar, Bremerhaven, Germany, and P. Bernal of the Institute for Fisheries Studies in Santiago, Chile (chapter 8). Similarly, changes in velocity and direction of the Equatorial Undercurrent is the principal force driving variability in the yields of the pelagic fisheries of the Gulf of Guinea ecosystem, according to Drs. D. Binet of France, and E. Marchal (ORSTOM), France (chapter 12); ocean physics also is a prime cause of variability in biomass yield of the Canary Current ecosystem as reported by Dr. C. Bas of the Spanish Institute of Oceanography, Las Palmas, Spain (chapter 11). In both LMEs, the effects of fishing-induced, large-scale variability in biomass yields is considered secondary.

In contrast, loss of habitat in the coastal zone around the Bay of Bengal ecosystem has had a detrimental impact on the nursery grounds of shrimp, whereas the annual yields from offshore fisheries of the ecosystem have increased from 1.9 million metric tons (mmt) in 1981 to 2.4 mmt in 1987, according to Dr. S. N. Dwivedi of the Department of Ocean Development, New Delhi, India (chapter 7). Given the growing stresses imposed on the ecosystem by the dense population along its coastal margin, Dr. Dwivedi encouraged the conference participants to consider convening a symposium for marine scientists and students from India and other countries bordering the Bay of Bengal.

From the case studies, new and useful managerial insights were gained into the principal driving forces of LMEs. The desirability for improved monitoring of physics-driven LMEs, such as the Humboldt Current and Patagonian Shelf ecosystems, was indicated with respect to fisheries yields and with regard to early warning of the impacts of global warming (Dr. A. Bakun, chapter 18).

From the theoretical presentations on LMEs, it was apparent that the problems of scale and variability need to be addressed to improve understanding of the processes occurring within an LME. Dr. S. Levin of Cornell stressed the importance of recognizing that more effort is required to understand the relationships among scales and levels of biological and physical organization within LMEs (chapter 6).

The LME as a geographic unit was discussed as a model for ocean management. In this regard, Dr. M. Belsky, Albany Law School, argued that the U.N. Law of the Sea can be interpreted as mandating an ecosystem approach to resource and environmental management (chapter 20). R. T. Scully of the U.S. State Department provided the conference with an example of the ecosystem approach to the management of living marine resources in the Antarctic, supported by a 21-country commission and an active scientific committee (chapter 22). Dr. A. Dahl of UNEP indicated to the conference that the LME concept is compatible with the U.N. regional seas approach to ocean research and management (chapter 2).

A comparison of LMEs around the Pacific Rim, including two different LMEs—the Sea of Japan and the Humboldt Current, was given by Dr. J. Morgan of the East-West Center, University of Hawaii (chapter 27), followed by a discussion of the role of political factors in the management of the Canary Current, Gulf of Guinea, and Benguela Current ecosystems by Dr. V. Prescott of the University of Melbourne, Australia (chapter 26). The importance of human intervention on the sustained biomass yields of the Mediterranean and Black

Sea ecosystems, based on changes in fisheries landings, was presented by Dr. J. Caddy of Food and Agriculture Organization of the United Nations (FAO), Rome, Italy (chapter 14). Important areas for further consideration in the development and application of the LME concept, including the clarification of the number and extent of LMEs on a global basis, were noted by Dr. L. Alexander of the University of Rhode Island (chapter 19).

Progress in LME studies will be made more rapidly with the application of new technology now available in hydroacoustics to deal with the difficult problem of sampling the zooplankton component of LMEs, as described by Dr. D. V. Holliday of the TRACOR Corporation, San Diego, California (chapter 28).

The combination of new molecular biological techniques for species and stock identification, along with available laboratory robotics systems, should allow for more efficient biological sampling designs and processing of samples at sea and in the laboratory, as outlined by Dr. D. Powers of Stanford University (chapter 29). On a larger scale, it was clear from the presentation by Dr. J. Yoder of the University of Rhode Island that ocean features and processes can be studied from a combination of satellites and surveys using color scanning and other photo-optical sensors for measuring productivity, standing stock, and flux at the LME scale for the oceans of the world (chapter 30).

Based on the deliberations held during the Monaco conference, the concept of linking LME research, monitoring, and management projects between a developed country and a developing country has taken the form of action in a planned symposium on "The Status and Future of Large Marine Ecosystems of the Indian Ocean" being organized jointly by Kenya and Belgium to be convened in Mombasa, Kenya, in 1993. Other regional LME symposia and monitoring projects are being considered for funding by the World Bank Global Environmental Facility for the Gulf of Guinea and South China Sea ecosystems as follow-on activities to the U.N. Conference on Environment and Development held in Rio de Janiero in June 1992.

References

Alexander, L. M. 1989. Large marine ecosystems as global management units. In: Biomass yields and geography of large marine ecosystems. pp. 339–344. Ed. by K. Sherman and L. M. Alexander. AAAS Selected Symposium 111. Westview Press, Inc., Boulder, CO. 493 pp.

Alverson, D.L., Longhurst, A.R., and Gulland, J.A. 1970. How much food from the sea? Science 168:503–505.

Andersen, K. P. and Ursin, E. 1977. A multispecies extension to the Beaverton and Holt Theory of Fishing, with accounts of phosphorus circulation and primary production. Medd. Dan. Fisk. Havunders. N.S. 7:319–435.

Bakun, A. 1986. Definition of environmental variability affecting biological processes in large marine ecosystems. In: Variability and management of large marine ecosystems. pp. 89–108. Ed. by K. Sherman and L. M. Alexander. AAAS Selected Symposium 99. Westview Press, Inc., Boulder, CO. 319 pp.

Bakun, A. 1990. Global climate change and intensification of coastal ocean upwelling. Science 247:198–201.

Bakun, A., and Parrish, R. H. 1980. Environmental inputs to fishery population models for eastern boundary current regions. In: Workshop on the effects of environmental variation on the survival of larval pelagic fishes. pp. 67–104. Ed. by G. D. Sharp. IOC Workshop Report 28. UNESCO, Paris.

Bakun, A., and Parrish, R. H. 1990. Comparative studies of coastal pelagic fish reproductive habitats: The Brazilian sardine (*Sardinella aurita*). J. Cons. Int. Explor. Mer 46:269–283.

Beddington, J. R. 1986. Shifts in resource populations in large marine ecosystems. In: Variability and management of large marine ecosystems. pp. 9–18. Ed. by K. Sherman and L. M. Alexander. AAAS Selected Symposium 99. Westview Press, Inc., Boulder, CO. 319 pp.

Belsky, M. H. 1989. The ecosystem model mandate for a comprehensive United States ocean policy and law of the sea. San Diego L. Rev. 26(3):417–495.

Bonner, W. N. 1982. Seals and man, a study of interactions. Univ. Washington Press, Seattle. 170 pp.

Borisov, V. 1991. The state of the main commercial species of fish in the changeable Barents Sea ecosystem. In: Food chains, yields, models, and management of large marine ecosystems. pp. 193–203. Ed. by K. Sherman, L. M. Alexander, and B. D. Gold. AAAS Symposium Proceedings. Westview Press, Inc. Boulder, CO. 320 pp.

Bottom, D. L., Jones, K. K., Rodgers, J. D., and Brown, R. F. 1989. Management of living resources: A research plan for the Washington and Oregon continental margin. Oregon Dept. Fish Wildl. NCRI–T–89–004. 80 pp.

Bradbury, R. H., and Mundy, C. N. 1989. Large-scale shifts in biomass of the Great Barrier Reef ecosystem. In: Biomass yields and geography of large marine ecosystems. pp. 143–167. Ed. by K. Sherman and L. M. Alexander. AAAS Selected Symposium 111. Westview Press, Inc., Boulder, CO. 493 pp.

Byrne, J. 1986. Large marine ecosystems and the future of ocean studies. In: Variability and management of large marine ecosystems. pp. 299–308. Ed. by K. Sherman and L. M. Alexander. AAAS Selected Symposium 99. Westview Press, Inc., Boulder, CO. 319 pp.

Christy, F. T., Jr. 1986. Can large marine ecosystems

be managed for optimum yields? In: Variability and management of large marine ecosystems. pp. 263 – 267. Ed. by K. Sherman and L. M. Alexander. AAAS Selected Symposium 99. Westview Press, Inc., Boulder, CO. 319 pp.

Crawford, R. J. M., Shannon, L. V., and Shelton, P. A. 1989. Characteristics and management of the Benguela as a large marine ecosystem. In: Biomass yields and geography of large marine ecosystems. pp. 169 – 219. Ed. by K. Sherman and L. M. Alexander. AAAS Selected Symposium 111. Westview Press, Inc., Boulder, CO. 493 pp.

Cushing, D. H. 1975. Marine ecology and fisheries. Cambridge Univ. Press, London. 278 pp.

Ellertsen, B., Fossum, P., Solemdal, P., Sundby, S., and Tilseth, S. 1990. Environmental influence on recruitment and biomass yields in the Norwegian Sea ecosystem. In: Large marine ecosystems: Patterns, processes, and yields. pp. 19 – 35. Ed. by K. Sherman, L. M. Alexander, and B. D. Gold. American Association for the Advancement of Science, Washington, DC. 242 pp.

GESAMP (Group of Experts on the Scientific Aspects of Marine Pollution). 1990. The state of the marine environment. UNEP Regional Seas Reports and Studies No. 115. Nairobi.

Graham, R. L., Hunsaker, C. T., O'Neill, R. V., and Jackson, B. L. 1991. Ecological risk assessment at the regional scale. Ecol. Applications 1:196 – 206.

Hempel, G. (Editor). 1978. Symposium on North Sea fish stocks—Recent changes and their causes. Rapp. P.-v. Reun. Cons. int. Explor. Mer 172. 449 pp.

Holling, C. S. 1973. Resilience and stability of ecological systems. Institute of Resource Ecology, Univ. British Columbia, Vancouver, Canada. 23 pp.

Holling, C. S. 1986. The resilience of terrestrial ecosystems: Local surprise and global change. In: Sustainable development of the biosphere. pp. 292 – 317. Ed. by W. C. Clark and R. E. Munn. Cambridge Univ. Press, London. 491 pp.

Hovgaard, H., and Buch, E. 1990. Fluctuation in the cod biomass of the West Greenland Sea ecosystem in relation to climate. In: Large marine ecosystems: Patterns, processes, and yields. pp. 36 – 43. Ed. by K. Sherman, L. M. Alexander, and B. D. Gold. American Association for the Advancement of Science, Washington, DC. 242 pp.

Kelleher, G. 1993. Sustainable development of the Great Barrier Reef as a large marine ecosystem. This volume, chapter 24.

Lasker, R. 1988. Food chains and fisheries: An assessment after 20 years. In: Toward a theory on biological-physical interactions in the world ocean. pp. 173 – 182. Ed. by B. J. Rothschild. NATO ASI Series. Series C: Mathematical and Physical Sciences, Vol. 239. Kluwer Academic Publishers, Dordrecht, The Netherlands.

Levin, S. A. 1978. Pattern formation in ecological communities. In: Spatial pattern in plankton communities. pp. 433 – 470. Ed. by J. A. Steele. Plenum Press, New York.

Levin, S. A. 1990. Physical and biological scales, and modelling of predator-prey interactions in large marine ecosystems. In: Large marine ecosystems: Patterns, processes, and yields. pp. 179 – 187. Ed. by K. Sherman, L. M. Alexander, and B. D. Gold. American Association for the Advancement of Science, Washington, DC. 242 pp.

Loughlin, T. R., and Nelson, R., Jr. 1986. Incidental mortality of northern sea lions in the Shelikof Strait, Alaska. Mar. Mammal Sci. 1:14 – 33.

Malone, T. C. 1991. River flow, phytoplankton production and oxygen depletion in Chesapeake Bay. In: Modern and ancient continental shelf anoxia. pp. 83 – 93. Ed. by R. V. Tyson and T. H. Pearson. Geological Society Spec. Publ. No. 58.

Mayr, E. 1982. The growth of biological thought. Harvard Univ. Press, Cambridge, MA. 974 pp.

Morgan, J. R. 1988. Large marine ecosystems: An emerging concept of regional management. Environment 29(10):4 – 9, 26 – 34.

Morgan, J. R. 1989. Large marine ecosystems in the Pacific Ocean. In: Biomass yields and geography of large marine ecosystems. pp. 377 – 394. Ed. by K. Sherman and L. M. Alexander. AAAS Selected Symposium 111. Westview Press, Inc., Boulder, CO. 493 pp.

Myers, N. 1990. Working towards one world (book review). Nature 344:499 – 500.

Overholtz, W. J., and Nicolas, J. R. 1979. Apparent feeding by the fin whale, *Balaenoptera physalus*, and humpback whale, *Megoptera novaeangliae*, on the American sand lance, *Ammodytes americanus*, in the northwest Atlantic. Fish. Bull. U.S. 77:285 – 287.

Payne, P. M., Wiley, D. N., Young, S. B., Pittman, S., Clapham, P. J., and Jossi, J. W. 1990. Recent fluctuations in the abundance of baleen whales in the southern Gulf of Maine in relation to changes in selected prey. Fish. Bull. U.S. 88:687 – 696.

Pimm, S. L. 1984. The complexity and stability of ecosystems. Nature 307:321 – 326.

Powers, K. D., and Brown, R. G. B. 1987. Seabirds. In: Georges Bank. pp. 359 – 371. Ed. by R. H. Backus. MIT Press, Cambridge, MA. 593 pp.

Prescott, J. R. V. 1989. The political division of large marine ecosystems in the Atlantic Ocean and some associated seas. In: Biomass yields and geography of large marine ecosystems. pp. 395 – 442. Ed. by K. Sherman and L. M. Alexander. AAAS Selected Symposium 111. Westview Press, Inc., Boulder, CO. 493 pp.

Ricklefs, R. E. 1987. Community diversity: Relative roles of local and regional processes. Science 235:167 – 171.

Ryther, J. H. 1969. Relationship of photosynthesis to fish production in the sea. Science 166:72 – 76.

Sainsbury, K. J. 1988. The ecological basis of multispecies fisheries, and management of a demersal fishery in tropical Australia. In: Fish population dynamics, 2nd ed. pp. 349 – 382. Ed. by J. A. Gulland. John Wiley and Sons, New York. 422 pp.

Scully, R. T., Brown, W. Y., and Manheim, B. S. 1986.

The convention for the conservation of Antarctic marine living resources: A model for large marine ecosystem management. In: Variability and management of large marine ecosystems. pp. 281 – 286. Ed. by K. Sherman and L. M. Alexander. AAAS Selected Symposium 99. Westview Press, Inc., Boulder, CO. 319 pp.

Sherman, K., and Alexander, L. M. (Editors). 1986. Variability and management of large marine ecosystems. AAAS Selected Symposium 99. Westview Press, Inc., Boulder, CO. 319 pp.

Sherman, K., and Alexander, L. M. (Editors). 1989. Biomass yields and geography of large marine ecosystems. AAAS Selected Symposium 111. Westview Press, Inc., Boulder, CO. 493 pp.

Sherman, K., Alexander, L. M., and Gold, B. D. (Editors). 1990. Large marine ecosystems: Patterns, processes, and yields. AAAS Symposium. AAAS, Washington, DC. 242 pp.

Sherman, K., Alexander, L. M., and Gold, B. D. (Editors). 1991. Food chains, yields, models, and management of large marine ecosystems. AAAS Symposium. Westview Press, Inc. Boulder, CO. 320 pp.

Sherman, K., and Ryan, A. F. 1988. Antarctic marine living resources. Oceanus 31(2):59 – 63.

Sissenwine, M. P. 1986. Perturbation of a predator-controlled continental shelf ecosystem. In: Variability and management of large marine ecosystems. pp. 55 – 85. Ed. by K. Sherman and L. M. Alexander. AAAS Selected Symposium 99. Westview Press, Inc., Boulder, CO. 319 pp.

Skjoldal, H. R., and Rey, F. 1989. Pelagic production and variability of the Barents Sea ecosystem. In: Biomass yields and geography of large marine ecosystems. pp. 241 – 286. Ed. by K. Sherman and L. M. Alexander. AAAS Selected Symposium 111. Westview Press, Inc., Boulder, CO. 493 pp.

Steele, J. H. 1974. The structure of marine ecosystems. Harvard Univ. Press, Cambridge, MA. 128 pp.

Tang, Q. 1989. Changes in the biomass of the Yellow Sea ecosystems. In: Biomass yields and geography of large marine ecosystems. pp. 7 – 35. Ed. by K. Sherman and L. M. Alexander. AAAS Selected Symposium 111. Westview Press, Inc., Boulder, CO. 493 pp.

Taylor, P., and Groom, A. J. R. (Editors). 1989. Global issues in the United Nations' framework. Macmillan, London. 371 pp.

Waring, G. T., Payne, P. M., Parry, B. L., and Nicolas, J. R. 1990. Incidental take of marine mammals in foreign fishery activities off the northeast United States, 1977 – 88. Fish. Bull. U.S. 88:347 – 360.

Appendix to Chapter 1

List of Symposia and Workshops on LMEs since 1984.

Symposium on Variability and Management of Large Marine Ecosystems, Annual Meeting of the American Association for the Advancement of Science (AAAS), New York, NY, May 28–29, 1984. (Volume published in 1986: Variability and Management of Large Marine Ecosystems; Westview Press, Inc., Boulder, CO; 319 pp; Edited by K. Sherman and L. M. Alexander.)

Symposium on Biomass and Geography of Large Marine Ecosystems, Annual Meeting of the AAAS, Chicago, IL, February 16–17, 1987. (Volume published in 1989: Biomass and Geography of Large Marine Ecosystems; Westview Press, Inc., Boulder, CO; 493 pp; Edited by K. Sherman and L. M. Alexander.)

Symposium on Perturbation and Yield of Large Marine Ecosystems, in the Seminar Program, Frontiers of Marine Ecosystem Research, Annual Meeting of the AAAS, Boston, MA, February 12–14, 1988. (Volume published in 1990: Large Marine Ecosystems: Patterns, Processes, and Yields; AAAS, Washington, DC; 242 pp; Edited by K. Sherman, L. M. Alexander, and B. D. Gold.)

Symposium on Food Chains, Yields, Models, and Management of Large Marine Ecosystems, Annual Meeting of the AAAS, New Orleans, LA, February 16, 1990. (Volume published in 1991: Food Chains, Yields, Models, and Management of Large Marine Ecosystems; Westview Press, Inc., Boulder, CO; 320 pp; Edited by K. Sherman, L. M. Alexander, and B. D. Gold.)

International Conference on The Large Marine Ecosystem (LME) Concept and Its Application to Regional Marine Resource Management, Monaco, October 1–6, 1990. (this volume: Stress, Mitigation, and Sustainability of Large Marine Ecosystems; AAAS Press, Washington, DC; Edited by K. Sherman, L. M. Alexander, and B. D. Gold.) (Report published in 1992: Conference Summary and Recommendations—The Large Marine Ecosystem (LME) Concept and Its Application to Regional Marine Resource Management; 51 pp; U.S. Department of Commerce, NOAA, Technical Memorandum NMFS-F/NEC–91; Edited by K. Sherman, and T. Laughlin.)

Symposium on Science and Management of Large Marine Ecosystems, Annual Meeting of the AAAS, Washington, DC, February 18, 1991.

Meeting of the Ad Hoc Committee on Large Marine Ecosystems, UNESCO Headquarters, Paris, France, March 22–23, 1991. (Report published in 1992: Ad Hoc Committee on Large Marine Ecosystems Meeting Report; 31 pp; U.S. Department of Commerce, NOAA, Technical Memorandum NMFS-F/NEC–92; Edited by K. Sherman and T. Laughlin.)

Large Marine Ecosystems Monitoring Workshop, Cornell University, Ithaca, NY, July 13–14, 1991. (Report published in 1992: Large Marine Ecosystems Monitoring Workshop Report; 78 pp; U.S. Department of Commerce, NOAA, Technical Memorandum NMFS-F/NEC–93; Edited by K. Sherman and T. Laughlin.)

Symposium on The Northeast Shelf Ecosystem: Stress, Mitigation, and Sustainability, convened by National Oceanic and Atmospheric Administration (NOAA), National Marine Fisheries Services (NMFS), Northeast Fisheries Science Center

(NEFSC), at the University of Rhode Island, Narragansett, RI, August 12–15, 1991. (Report published in 1992: The Northeast Shelf Ecosystem: Stress, Mitigation, and Sustainability; 47 pp; U.S. Department of Commerce, NOAA, Technical Memorandum NMFS-F/NEC–94; Edited by K. Sherman and T. Laughlin.)

Emerging Theoretical Basis for Monitoring Changing States (Health) of Large Marine Ecosystems. (Summary Reports of two workshops: Narragansett, RI, April 1992, Cornell University, July 1992 [in press]; U.S. Deparment of Commerce, NOAA, Technical Memorandum NMFS-F/NEC–__; Edited by K. Sherman.)

Chapter 2

The Large Marine Ecosystem Approach to Regional Seas Action Plans and Conventions: A Geographic Perspective

Arthur Lyon Dahl

Introduction

Today's environmental concerns reflect not only an awareness that our local environment is deteriorating, but a realization of the increasing scale of human impacts, even to the planetary level. The problems of the ozone hole and greenhouse gasses are clearly global, yet most other environmental problems are regional or are best addressed at a regional level.

The challenge is how to respond to these problems that go beyond the state and the nation and require intergovernmental cooperation for their solutions. There is a need to build legal and institutional frameworks for action that correspond to the scale of the problems.

The large marine ecosystem (LME) concept is defining a new scale of system that often presents challenges for management. The Regional Seas Programs of the United Nations Environment Program (UNEP) and comparable programs of other organizations have many years of experience in building intergovernmental cooperation at scales that may be appropriate to the management requirements of LMEs. They represent a variety of existing mechanisms for governments to work together to assess and resolve common problems in shared seas and coastal areas. The LME concept may provide the basis for a new element of cooperation within these programs.

Regional Seas Programs

There are presently 10 UNEP Regional Seas Programs, with two others in preparation and two similar regional organizations outside UNEP's framework. Many of these regions cor- respond with, or may include, LMEs. Most have a legal framework for cooperation, including a convention and appropriate protocols. There is an action plan agreed-upon at an intergovernmental meeting and implemented through an appropriate work program. A trust fund or other financing mechanism supports agreed upon regional activities. The secretariat may be UNEP or a regional organization, and the larger programs may have subsidiary structures for project implementation. While UNEP provides overall coordination, many other parts of the United Nations system cooperate in the regional programs, including United Nations Educational, Scientific, and Cultural Organization (UNESCO), Intergovernmental Oceanographic Commission (IOC), Food and Agriculture Organization of the United Nations (FAO), Intergovernmental Maritime Organization (IMO), International Atomic Energy Agency (IAEA), World Health Organization (WHO), and other organizations such as IUCN-The World Conservation Union.

The Mediterranean Action Plan, started in 1974, is the oldest and one of the best developed. Under the Barcelona Convention and its protocols, UNEP provides the secretariat with a regional coordinating unit in Athens which oversees the pollution assessment, the control programs, and data management. Several regional activity centers have been established, including the Priority Actions Program in Yugoslavia, which oversees coastal zone management and environmental impact assessment projects, among others; a Regional Marine Pollution Emergency Response Center in Malta; a center for long-range planning (the Blue Plan) in France; and a Regional Activity Center for Specially Protected Areas in Tunisia.

The Caribbean Environment Program, covering the wider Caribbean under the Cartagena Convention, has its regional coordinating unit operated by UNEP in Jamaica. This program is responsible for a wide range of activities, from marine pollution monitoring to specially protected areas and wildlife.

Along the Pacific coast of South America from Chile to Panama, there is the South-East Pacific Action Plan under the Lima Convention. Its secretariat is the Permanent Commission for the South Pacific (CPPS), located in Santiago de Chile. Its major focus has been pollution assessment and monitoring, but it is now expanding to include such concerns as coastal zone management, environmental impact assessment, and coastal erosion.

The South Pacific Regional Environment Program (SPREP) developed from regional activities started in 1974. An action plan was adopted in 1982 under the joint auspices of the South Pacific Commission (SPC), the South Pacific Forum, UNEP, and the U.N. Economic and Social Commission for Asia and the Pacific (ESCAP). The South Pacific Commission has been the secretariat, but the participating governments have recently decided that SPREP should evolve into an autonomous intergovernmental organization. SPREP has become the principal channel for the coordination and implementation of environmental activities over its vast region. Although LMEs may be less obvious at the scale of the tropical Pacific (apart from the Australian coast), there are probable linkages between coral reef areas through larval transport, which share some characteristics of LMEs.

In the East Asian Seas, the Association of Southeast Asian Nations (ASEAN) countries have adopted an action plan and a trust fund, with UNEP as the secretariat, as well as a number of regional projects. The action plan for the South Asian Seas is still awaiting adoption, but some regional activities have already begun. In the Persian Gulf region, the Kuwait Regional Convention and Action Plan are implemented by the Regional Organization for the Protection of the Marine Environment (ROPME) in Kuwait, but activities were temporarily interrupted by the invasion of Kuwait in 1991. Another action plan and convention have been adopted for the Red Sea and the Gulf of Aden with a secretariat, the Red Sea and Gulf of Aden Environment Program (PERSGA), in Saudi Arabia.

In Eastern Africa, the coast from Somalia to Mozambique and the offshore islands are covered by the Nairobi Convention signed in 1985 and by an action plan and trust fund with UNEP as the secretariat. A major regional marine pollution assessment and control program is just starting, with a coastal zone management project due to begin next year. The West and Central African Action Plan, under the Abidjan Convention, has emphasized marine pollution assessment and coastal erosion control. UNEP provides the secretariat, and a regional coordinating unit is to be established soon.

New regions in which regional seas action plans are to be developed include the Black Sea and the Northwest Pacific, including China, Japan, the two Koreas, and Russia.

Two other regions have conventions independent of UNEP: the Baltic Sea, under the Helsinki Commission; and the North Sea and eastern North Atlantic, under the Oslo and Paris Commissions. The latter is considering a possible expansion across the North Atlantic.

Regional Seas Program Elements

The activities in each regional sea are decided by the governments concerned and depend on their priorities. Thus, there is considerable diversity among the programs. The following elements are frequently found in these regional programs:

- Marine pollution assessment and control, including assistance in establishing and maintaining monitoring laboratories, applying standard methods, assuring quality control, studying pollution effects, and managing the resulting data, leading to recommendations for pollution control measures.
- Contingency planning for oil spills and other accidents involving hazardous materials.
- Assessment and control measures for coastal erosion.
- Establishment of marine and coastal protected areas and measures to protect threatened species.
- Educational activities and training to build the human capacity in the region.
- Task teams to review the effects of expected climatic changes and sea-level rise on the regions.
- Implementation of environmental impact assessments for coastal and marine projects.
- The integration of all these activities into comprehensive coastal zone management.

Most regional programs started with a focus on chemical parameters, because this was the easiest way to address marine pollution problems. We are now attempting to strengthen the scientific bases necessary to address biological parameters concerning habitats and ecosystems, because it is important to know what is happening to those basic resources that are under increasing human pressure, especially in coastal areas. This process will require determining what to measure and at what geographic and temporal scales, in order to obtain significant and cost-effective results without being swamped by natural variability. We hope that the experience being gained from the LME concept will be of help here. There needs to be an increasing dialogue between the scientific community doing basic research on these systems and the managers who must design mechanisms to manage and develop them in sustainable ways.

The LME Approach and Regional Seas Programs

The Regional Seas Programs can provide mechanisms for intergovernmental action on many concerns of LMEs that fall within their scope, including pollution controls, protection of critical species such as marine mammals, identification and protection of critical habitats, and where appropriate, the possibility of adopting new protocols for LMEs. The fisheries component of LMEs is probably best addressed through the various regional fisheries councils and conventions.

If governments are to learn to manage LMEs effectively, they will need the following kinds of information:

- The presence and nature of a coherent system at the scale of an LME, and the boundaries of that system.
- The natural variability of the ecosystem against which any human effects must be judged.
- The human effects that have been observed on the system, including causal factors, consequences for the system, and the economic significance for existing or potential human uses of the system or its resources.
- The management requirements for sustainable use of the system, and the cost of proposed management measures.

Given the scale and complexity of LMEs,

it will not be easy to meet the above requirements. We must work together to build both the scientific understanding of these large-scale phenomena in the oceans and the response strategies necessary to manage them effectively.

Bibliography

Permanent Commission for the South Pacific (CPPS)/United Nations Environment Program (UNEP). 1983. Action plan for the protection of the marine environment and coastal areas of the South-East Pacific. UNEP Regional Seas Reports and Studies. No. 20. UNEP, Geneva. 16 pp.

South Pacific Commission (SPC)/South Pacific Bureau for Economic Cooperation (SPEC)/U.N. Economic and Social Commission for Asia and the Pacific (ESCAP)/UNEP. 1983. Action plan for managing the natural resources and environment of the South Pacific region. UNEP Regional Seas Reports and Studies. No. 29. UNEP, Geneva. 16 pp.

UNEP. 1983. Action plan for the Caribbean environment program. UNEP Regional Seas Reports and Studies. No. 26. UNEP, Geneva. 26 pp.

UNEP. 1983. Action plan for the protection and development of the marine and coastal areas of the East Asian region. UNEP Regional Seas Reports and Studies. No. 24. UNEP, Geneva. 16 pp.

UNEP. 1983. Action plan for the protection and development of the marine environment and coastal areas of the West and Central African region. UNEP Regional Seas Reports and Studies. No. 27. UNEP, Geneva. 13 pp.

UNEP. 1983. Action plan for the protection of the marine environment and coastal areas of Bahrain, Iran, Iraq, Kuwait, Oman, Qatar, Saudi Arabia, and the United Arab Emirates. UNEP Regional Seas Reports and Studies. No. 35. UNEP, Geneva. 15 pp.

UNEP. 1985. Action plan for the protection, management, and development of the marine and coastal environment of the Eastern African region. UNEP Regional Seas Reports and Studies. No. 61. UNEP, Nairobi. 9 pp.

UNEP. 1985. Mediterranean action plan. UNEP, Geneva. 28 pp.

UNEP. 1986. Action plan for the conservation of the marine environment and coastal areas of the Red Sea and Gulf of Aden. UNEP Regional Seas Reports and Studies. No. 61. UNEP, Nairobi. 15 pp.

Chapter 3

Scientific and Organizational Aspects of Large Marine Ecosystems Research

Gotthilf Hempel and Kenneth Sherman

During the last decade, ocean scientists and managers have gained important new insights into the processes that govern the ecology of coastal and ocean species and their interactions. This has led to the conceptual framework of the large marine ecosystem "LME" approach, which provides the basis for developing new strategies for marine resource management and research. These new strategies have broad application to the ways in which countries manage their Exclusive Economic Zones (EEZs) and to the regionally focused conventions and action plans aimed at environmentally sound management of marine resources. Typical fishery-related questions to be answered by the LME approach are concerned with the physical (current systems, stratification), biological (food supply, predation), and man-made (fisheries, etc.) impacts affecting fisheries.

Recently, marine science has gained much recognition by the public and by politicians. At the same time, much of the public interest has shifted from marine resources toward concern about the health of the ocean, particularly of the coastal regions. The LME approach focuses on the integration of basic science and applied research in exploitation as well as protection of large marine areas. In a small number of countries, particularly in North America, Europe, and eastern Asia, marine scientists in the 1990s are in a most enviable position both compared with their predecessors and with their colleagues in other parts of the world. Since the 1960s, the number of vessels and institutions engaged in marine studies has grown impressively. The explosion of information is evident in the growing number of volumes and journals in marine science. However, the track record in applying the information to

the pragmatic needs of society is not such a clear success. This is where the LME concept comes into the picture—through its attempt to link oceanography, marine ecology, fishery research, and resource management in regions of multiple use of the ocean.

Over the past decade, there has been much debate on the assimilative capacity of the oceans for pollution and the dumping of wastes. In some countries, science has supported the banning of ocean dumping, yet in other countries, the practice is allowed. We observe that fish stocks continue to decline in the face of high exploitation rates. The coastal zone margins of the oceans are, in some parts of the globe, overbuilt, polluted, and unsuitable as spawning and nursery habitats for marine species.

Historical Background

The LME concept has three major roots in the history of marine science: (i) the multidisciplinary research in the North Sea under the auspices of the International Council for the Exploration of the Sea (ICES), (ii) the Food and Agriculture Organization of the United Nations' (FAO) assessment of the fish resources of the ocean, and (iii) the American Association for the Advancement of Science's (AAAS) approach linking ecosystem analysis, complex modeling, and management.

The ICES root

At the beginning of the twentieth century, the North Sea was the cradle of international cooperation in fishery biology and hydrography, mainly through standardized surveys carried out four times a year. In those days, scientists had already realized "that isolated investigations in limited areas of the sea were of little value and that cooperation in the international field was highly desirable" (Went, 1972).

The emergence of the ecosystem approach in its present form can be traced in the series of symposia held by ICES, as published in the council's Rapports et Proces-Verbaux:

- 1975: North Sea Fish Stocks—Recent Changes and Their Causes (Hempel, 1978)
- 1976: Marine Ecosystems and Fisheries Oceanography (Parsons *et al.*, 1978)
- 1980: The Canary Current: Studies of An Upwelling Ecosystem (Hempel, 1982)
- 1982: The Biological Productivity of North Atlantic Shelf Areas (Zijlstra, 1984)
- 1984: Contaminant Fluxes Through the Coastal Zone (Kullenberg, 1986)
- 1987: Oceanography and Biology of Arctic Seas (Hempel, 1989)

The North Sea Symposium in 1975 was prompted by the large-scale changes in the abundance of fish stocks during the 1960s and early 1970. Various attempts were made to explain the causes of those changes. In the summary of the symposium, it was stated: "The recent changes in the North Sea fisheries, fish stocks, and their environment, as revealed at the symposium emphasized the need for increased international collaborative, multidisciplinary work, and the continued dialogue between fishery biologists and 'pure' biological and physical oceanographers who jointly have to define what information is indispensable for the management of multispecies fisheries in the North Sea, under changing environmental conditions. The symposium posed a bundle of questions concerning the understanding of the functioning of the North Sea as an ecosystem, which is heavily exploited by man and is influenced by the exchange of water masses with the open Atlantic and by the surrounding heavily industrialized land masses. Oil and gas drilling activities have become an additional factor, the effects of which on the North Sea ecosystem are only just beginning to be studied" (Hempel, 1978). Meanwhile, overfertilization of the hinterland has been recognized as a further driving force of ecological change in the North Sea.

Since 1975, a new approach to the problem of identifying major driving forces of the North Sea ecosystem has been adopted by a number of ICES countries. The approach emphasizes ecosystem modeling, detailed studies of the early life history, and the feeding relationships of various fish stocks. The study of the changes in North Sea fish stocks faces a problem common to many ecosystem studies. These stocks require monitoring over large areas and decades in order to describe correlations. On the other hand, the events are small-scale in space and time, which determine year-to-year differences in recruitment and, hence, the buildup and decline of stocks.

The FAO root

The second root of the LME approach is linked to the development of the overall attitude of society to natural resources, which for a long time were considered unlimited on a global scale. In the late 1950s, economists started to take stock of the world resources and their consumption by a rapidly increasing human population. The report of the Club of Rome and the first oil crisis changed the way of thinking of many politicians all over the world. Meanwhile, we know that many of the predictions were wrong because of underestimates of the mineral resources and because of new technologies that are much less demanding on raw materials.

Even more difficult was the prediction of the future supply of food for mankind. The world fisheries catch increased at an annual rate of several percent in those years as a result of exploration of new fishing grounds all over the world and increased fishing for industrial purposes. Extrapolations of those trends led to very optimistic figures. At the same time, European, Japanese, and U.S. scientists were confronted with the decline of overfished stocks in their home waters. These scientists were pessimistic regarding further increases in world fisheries. In 1966, FAO, through its Advisory Committee on Marine Resources Research (ACMRR), decided to come up with a set of new estimates individually made for each of the large statistical areas of the World Ocean. The final product was the famous book on *The Fish Resources of the World Ocean*, edited by J. Gulland (1971). The FAO approached the question of potential exploitable resources of each statistical area from different angles, starting with descriptions of the bottom topography, width of the shelf, steepness of the slope, sediments, current systems, and salinity and temperature regimes. Biomass and productivity of phytoplankton and zooplankton, as well as estimates of benthos, were meant to provide information on the biological richness of the area, particularly on the shelf. Those estimates, however, were very hard to translate into fish production and potential catch. Nevertheless, they served as first indications for areas with little or no fishery statistics. The second level of information came from surveys of fish eggs and larvae, from acoustic fish surveys, and from exploratory fishing. Laboratories and experts from Europe, the U.S., and Japan provided the expertise for those surveys in many parts of the world. The reliability of the methods employed was a matter of concern to various international working groups of FAO and the Scientific Committee for Oceanographic Research (SCOR). The third level of the FAO assessment consisted of fish stock assessments based on fishery statistics applying models of fish population dynamics. The concepts of growth overfishing and recruitment overfishing of individual stocks were later replaced by the concept of human impact on an ecosystem level affecting entire multispecies fish populations in their trophic and competitive relationships. Other impacts, such as habitat destruction and pollution, also came into focus when public concern turned from worries about resources to worries about the environment.

The authors of the FAO assessments were aware of the weakness of using statistical areas rather than current systems and natural production zones. The FAO areas coincided with natural regions in only a few instances—for example, the Mediterranean, where two regional councils for marine science and for fisheries had been in existence for a long time.

A more natural division of the world's ocean and its coastal regions was needed in order to describe and compare typical ecosystems in terms of their food web structure, productivity at various trophic levels, and their fishery potentials under different fishing strategies, such as fishing for human consumption only or for trash fish for meal and oil. Longhurst's book on selected marine ecosystems was a milestone in this respect (Longhurst, 1981).

The AAAS root and present-day LME approach studies

The third root of the LME concept goes back to modeling the changes of fishing effort, fishing methods, and target species in the North Sea by Anderson and Ursin (1977), and the northeast U.S. shelf by Clark and Brown (1977), Grosslein et al. (1979), Sissenwine et al. (1984), and Fogarty et al. (1987). They explained the changes in fish populations and individual stocks in a holistic way and produced predictive models that allowed advanced management of the fisheries under the new regimes of EEZs. The concept of LMEs as large management units was developed as a result of those approaches on both sides of the Atlantic. The

further development of the LME concept can be followed in three AAAS symposia published in the AAAS Selected Symposium series (Sherman and Alexander, 1986, 1989; Sherman et al., 1990).

The symposia reviewed the large-scale ecosystem perturbations affecting biomass yields that have been documented around the Pacific for the East Bering Sea, Kuroshio Current, Oyashio Current, Yellow Sea, Gulf of Thailand, Northwest Australia Coast, and Gulf of Alaska ecosystems. In the Atlantic, large-scale population shifts have been reported for the Northeast U.S. Continental Shelf, East Greenland Sea, Barents Sea, Norwegian Sea, North Sea, Baltic Sea, and the Iberian Coastal and Benguela Current ecosystems.

The present conference goes one step further by bringing together marine scientists and ocean managers to synthesize available information on the scientific, economic, legal, and geographic bases for the management of entire marine ecosystems. The goal of the LME concept in terms of applied science is a holistic, ecologically sound management strategy (Sherman and Alexander, 1989). With the recent development of LME research, there is a tendency to fractionate the large ecosystems into smaller ecological and/or management units. On the other hand, the recognition of the global character of environmental changes, as well as of economical and technical development, calls for approaches on a large scale to maintain cognizance of common trends.

Although the LME approach, as outlined in the AAAS symposia, was primarily the response of U.S. fishery science to the growing demand for ecologically sound fishery management, it has great potential also for marine ecology in general. Comparative LME studies may elucidate the evolution of ecosystems, in the same way that comparative anatomy and physiology are keys to the understanding of the evolution of species. Comparative ecosystem studies might give insight into the basic functioning of these ecosystems and permit guesses about the future development of certain systems under various scenarios of direct human impacts and regional and global environmental changes.

The future of comparative LME studies is promising, as more systems become analyzed. Studies of polar ecosystems may serve as examples. The Commission for the Conservation of Antarctic Marine Living Resources (CCAMLR) is the first international organization that applies the ecosystem concept to the living resources of a large marine area. The Scientific Committee of CCAMLR has been overseeing scientific efforts designed to monitor the ecosystem as a prerequisite to support the management of Antarctic living resources. Apart from the studies in life history and stocks of the small fish populations of the Antarctic, the research has focused on the importance of krill as a key species—as prey for marine fish, mammals, and birds, while also being utilized as a human food source. The Biological Investigations of Marine Antarctic Systems and Stocks (BIOMASS) Program was in fact the largest international effort in LME research so far, as it lasted for 14 years and involved a dozen countries.

Also in the Arctic, international cooperation, supported by powerful icebreaking research vessels, permitted systematic studies of plankton, sea-ice biota, and benthos, linked with nutrient chemistry in the ice-covered area. Scientists are now studying the Greenland Sea ecosystem to understand the formation, circulation, and exchange of its water masses. Remote sensing by aircraft and satellites provides detailed broad-scale images of the surface waters and sea ice on a continuous basis. The experimental and observational approaches of ocean physics in the Greenland Sea study are linked with biological fieldwork and with ecosystem models that include components for integrating water column, ice cover, and seabed topography with the entire food web. Another initiative of the Arctic Ocean Sciences Board (AOSB) is the International Polynya study. Three large summer polynyas have been chosen—off northeast Greenland, off Baffin Island (North Water), and off Lawrence Island in the Bering Sea—for a comparative study of the ecosystems in those seasonally ice-free areas surrounded by large masses of sea ice. The newly founded International Arctic Science Committee (IASC) will provide even broader international participation and opening of the entire Arctic Ocean for cooperative ecosystem studies. This is particularly important in view of the fact that global changes in climate will have one of the most immediate impacts on Arctic waters and their ice cover.

Although considerable effort is now under way by the marine science community in studying regional ecosystems, a more organized system for encouraging comparative studies of LMEs has proven useful. The LME concept has provided a means for increasing the awareness of existing international organizations by closing the information gap through the activities of an ad hoc committee on LME research and management. The committee has served as a communications bridge to relevant existing organizations (e.g., Intergovernmental Oceanographic Commission [IOC], International Union for the Conservation of Nature and Natural Resources [IUCN], SCOR, and FAO) and is presently examining the causes and effects of changing states of LMEs in relation to biomass yields and global change and how they relate to management options. Pertinent information is synthesized by national experts at LME symposia. The results are peer-reviewed and published. A list of recently published reports and papers presently in press is listed in the appendix to Chapter 1. The published reports address scientists, administrators, teachers, managers, and students interested in marine resources reviews on the states of LMEs. With the increasing use of most EEZs for fisheries, aquaculture, mineral exploitation, land reclamation, transport, habitation, tourism, and as sanctioned or unsanctioned dumping sites, there is a growing need for LME research all over the world. Global change adds yet another dimension to large-scale comparisons and long-term monitoring. Those research demands, although economically and scientifically well founded, cannot be met by the majority of coastal states that are short of training, personnel, and adequate facilities to carry out full-scale LME research. In fact, up to now the LME approach has been adopted by a number of laboratories in a few countries only. Scientists and administrators in the majority of countries still consider it as too demanding in terms of personnel and equipment. The disequilibrium in the distribution of oceanographic research capability in the world is still alarming, in spite of all the efforts by the countries themselves and by international and national organizations assisting countries and laboratories in their struggle for adequate research capacities and working conditions.

Scientists might be discouraged by the LME approach and consider it too far out of their reach. They are afraid to lose sight of the fact that conventional fishery research is by no means passé and is needed to meet the immediate requirements of the fishing industry. However requisite fishery statistics, fish stock assessments, and life-history studies are to any LME study, they are particularly valuable for ecological science and for management if conducted in a standardized way that allows the integration of data from various sources.

Governmental agencies might also be discouraged from supporting broad LME research because of its low immediate economic payoff or even short-term sacrifices in fishing yield or other activities. Bilateral and international funding agencies might be reluctant to finance research projects that are not visibly profitable in the very near future.

There are certain steps to be taken to bring many countries and their marine science communities closer to general LME goals:

- National programs in fishery research and basic marine biology and oceanography should be considered as essential parts of LME research and should be adjusted and standardized to meet its requirements.

- Better communication, coordination, and cooperation among neighboring institutions and countries could lead toward a coherent regional LME program of national activities.
- A strengthening of the scientific and technical staff through training programs is essential. Therefore, assistance from outside the region should be primarily in terms of training. In some cases, advice is needed for the design of LME research. Equipment and its maintenance might also have to be obtained from outside sources.
- Any further development in LME research in many parts of the world depends on partnership between scientific institutions within and outside the region. Partnership has to be based on the mutual recognition of interests. Otherwise, it will become increasingly difficult to recruit experts from outside the region. The main bottleneck lies in the shortage of first-class personnel willing to work under suboptimal conditions and train local people on-site.
- Pilot LME projects in less-developed regions may attract foreign scientists to bring their skills and equipment with them and provide training on the job. Those projects might become a tool or incentive to revitalize bilateral and international cooperation in a targeted and sophisticated manner, thus attracting additional international and bilateral funding. Regional pilot projects should be on the agenda of the governing bodies of the United Nations and European community.

J. Alheit, who has been responsible for the IOC/Ocean Science and Living Resources (OSLR) program, has proposed the following steps with regard to pilot programs for very few suitable LMEs:

(1) Convene a global workshop with representatives (scientists, managers) of the selected LMEs, with the following objectives:
- develop a common concept of LME studies
- agree on a shopping list of data requirements
- demonstrate the use of modeling
- find partners (international agencies, and giving national institutions, research institutes) in the developed countries who would assist in regional studies
- identify training needs

- develop a plan for collating the data available (approximately 1 year) to have a basis for regional workshops

2) Convene regional workshops (1 year after first workshop), with the following objectives:
- identify key processes to be studied (based on data made available over the previous 12 months)
- develop a plan for integrated ecosystem studies on a regional and national level
- organize these studies

3) Establish a training package for national use within the European Economic Community (EEC) for ecosystem management. Emphasis should be on fisheries. Provide the opportunity for training in LME concepts through short- and long-term university-level courses (master's degree level).

4) Follow-up activities should include workshops to analyze the new databases. Ongoing FAO fisheries programs can be strengthened by refocusing them around the natural boundaries of regional LMEs. The United Nations Environmental Program (UNEP) Regional Seas programs can be enhanced by taking a more holistic ecosystems approach to resource sustainability issues as part of an overall effort to improve the health of the oceans. The LME research and monitoring strategies are compatible with the proposed Global Ocean Observing System (GOOS) of IOC and will strengthen GOOS (UNESCO, 1990) by adding an ecosystem module to the existing physical and meteorological modules. The LME approach is compatible with the National Science Foundation (NSF) Global Ocean Ecosystem Dynamics Studies (GLOBEC) program and can provide useful data inputs to the Joint Global Ocean Flux Study (JGOFS). The concept has been discussed at ICES, International Council for Scientific Exploration of the Mediterranean (ICSEM), FAO, IOC, IUCN, and UNEP with generally favorable responses for further development within the United Nations framework of ongoing programs. The LME approach is compatible with the FAO interest in studying "catchment basins" and quantifying their impacts on enclosed and semi-enclosed seas. Observations presently underway under the OSLR program now operating within the California Current, Humboldt Current, and Iberian Coastal ecosystems provide an important framework for expanded LME studies of these systems in relation to not only fisheries issues, but also problems of pollution and coastal zone management.

Why the LME Approach?

One of the positive findings of LME studies is the

recognition by a growing number of experts in marine science that the large marine ecosystem is a useful area of ocean space for linking point-source problems on a local scale (e.g., pollution, shellfish closures, coastal zone damage) to large-scale issues on a global scale (e.g., global warming, ozone depletion, overexploitation). It is at the regional scale that physics, chemistry, and biology interact to shape the character of ocean productivity and biomass yields. This has placed the focus on LMEs as a useful scale for regional ecosystem research, monitoring, and management. The LME concept provides a practical means for overcoming the present "sectorization" of ocean studies and management by focusing attention on entire marine ecosystems and programs pertinent to their long-term development and sustainability within the already stressed environments and resources of areas around the margins of ocean basins.

Mitigation actions are required to reduce the stresses on long-term sustainability. More risk-averse management practices based on sound ecosystem principles will provide the basis for reducing losses from inefficient management practices, while promoting long-term sustainability and greater economic stability among the industries dependent on marine resources.

References

Anderson, K. P., and Ursin, E. 1977. A multispecies extension to the Beverton and Holt theory of fishing, with accounts of phosphorus circulation and primary production. Medd. Dan. Fisk. Havunders. N.S. 7:319 – 435.

Clark, S. H., and Brown, B. E. 1977. Changes of biomass of finfishes and squids from the Gulf of Maine to Cape Hatteras, 1963 – 74, as determined from research vessel survey data. Fish. Bull. U.S. 75:1 – 21.

Fogarty, M., Sissenwine, M., Grosslein, M. D. 1987. Fish population dynamics. In: Georges Bank. pp. 494 – 509. Ed. by R. H. Bakus. MIT Press, Cambridge, MA. 593 pp.

Grosslein, M., Brown, B. E., and Hennemuth, R. C. 1979. Research, assessment and management of a marine ecosystem in the northwest Atlantic—a case study. In: Statistical ecology, vol. 2: Environmental biomonitoring, assessment, prediction, and management—certain case studies and related quantitative issues. pp. 289 –

357. Ed. by J. Cairns, Jr., G. P. Patil, and W. E. Waters. International Cooperative Publishing House, Fairland, MD.

Gulland, J. A. 1971. The fish resources of the world ocean. Fishing News (Books) Ltd., West Byfleet, UK. 255 pp.

Hempel, G. (Editor). 1978. North Sea fish stocks—Recent changes and their causes. Rapp. P.-v. Reun. Cons. int. Explor. Mer 172. 449 pp.

Hempel, G. (Editor). 1982. The Canary Current: Studies of an upwelling ecosystem. Rapp. P.-v. Reun. Cons. int. Explor. Mer 180. 455 pp.

Hempel, G. (Editor). 1989. Oceanography and biology of arctic seas. Rapp. P.-v. Reun. Cons. int. Explor. Mer 188. 254 pp.

Kullenberg, G. (Editor). 1986. Contaminant fluxes through the coastal zone. Rapp. P.-v. Reun. Cons. int. Explor. Mer 186. 485 pp.

Longhurst, A. R. (Editor). 1981. Analysis of marine ecosystems. Academic Press, Inc., Ltd., London, UK. 741 pp.

Parsons, T. R., Jansson, B.-O., Longhurst, A. R., and Saetersdal, G. (Editors). 1978. Marine ecosystems and fisheries oceanography. Rapp. P.-v. Reun. Cons. int. Explor. Mer 173. 240 pp.

Sherman, K., and Alexander, L. M. (Editors). 1986. Variability and management of large marine ecosystems. AAAS Selected Symposium 99. Westview Press, Inc., Boulder, CO. 319 pp.

Sherman, K., and Alexander, L. M. (Editors). 1989. Biomass yields and geography of large marine ecosystems. AAAS Selected Symposium 111. Westview Press, Inc., Boulder, CO. 493 pp.

Sherman, K., Alexander, L. M., and Gold, B. D. (Editors). 1990. Large marine ecosystems: Patterns, processes, and yields. American Association for the Advancement of Science, Washington, DC. 242 pp.

Sissenwine, M. P., Cohen, E. B., and Grosslein, M. D. 1984. Structure of the Georges Bank ecosystem. Rapp. P.-v. Reun. Cons. Perm. Int. Explor. Mer 183:243 – 254.

United Nations Educational, Scientific, and Cultural Organization (UNESCO). 1990. UNEP-IOC-WMO Report of meeting of experts on long-term global monitoring system of coastal and near-shore phenomena related to climate change. Paris, 10 – 14 December 1990. 23 pp.

Went, A. J. 1972. Seventy years agrowing. A history of the International Council for the Exploration of the Sea, 1902 – 1972. Rapp. P.-v. Reun. Cons. int. Explor. Mer 165. 252 pp.

Zijlstra, J. J. (Editor). 1984. The biological productivity of North Atlantic shelf areas. Rapp. P.-v. Reun. Cons. int. Explor. Mer 183. 284 pp.

Chapter 4

Application of
Large Marine Ecosystems Management to
Global Marine Pollution

Alasdair D. McIntyre

Introduction

A good idea, when simply formulated, sometimes seems so obviously right and sensible that we either wonder why it had not been thought of earlier, or we tell ourselves that of course we had always known it, had implicitly accepted it, and had been using it for a long time without bothering to spell it out in detail. This is probably the reaction of many people to large marine ecosystems (LMEs). The concept, once expressed early in the 1980s, quickly came into general use, and many of these systems have now been identified and defined. In the studies of LMEs published so far, considerable emphasis has been placed on the changes and variability associated with them, as well as on the production of renewable resources within specific LMEs, with particular reference to the effects of overexploitation and ways to manage this problem. However, like all good generalizations, the concept is flexible and open to development. Overexploitation of living resources, of course, does generate significant stress on the system, but there are many other sources of stress on LMEs. This chapter takes a broader look at these areas and considers how the LME concept might be relevant to a more general management approach.

If we accept the general definition of LMEs as extensive areas, usually greater than 200,000 km^2, with unique hydrographic regimes, submarine topography, and trophically linked populations (Sherman and Alexander, 1986), we could locate them almost anywhere in the oceans. However, the reference to submarine topography perhaps indicates a shelf area, and indeed most of the 20 recently recognized LMEs are associated with the coast (Sherman, 1989).

Sources of Stress In the Coastal Region

Before looking specifically at LMEs, one might ask what are the stresses relevant to the coastal zone in general. In this context, the most recent global review is contained in a United Nations' report (GESAMP, 1990); its conclusions can be summarized briefly under six headings.

Coastal development

Coastal development includes a great diversity of activities, ranging from the construction of ports and harbors, industrial installations, sea defenses, and metropolitan areas to the development of a variety of tourist and recreational facilities such as hotel complexes and marinas. This will clearly have significant impact on shallow-water ecosystems, and the problem is directly related to the movement of large numbers of people to the coast, either as permanent residents or as visitors. This problem is highlighted by the fact that three-quarters of the world's cities with more than 4 million people are located along coasts.

Hinterland activities

While these developments at the edge of the sea can be damaging, some very significant effects at the coast can result from activities far inland, sometimes hundreds of miles from the sea. Two general types of activity are particularly relevant. The manipulation of hydrological cycles, including activities such as building dams and barriers for power generation, flood control, or irrigation, which reduce the flow of water to the sea, alter the salinity regimes and ecology of estuaries and change the erosion characteristics of the coast—sometimes with disastrous results. Various land-use practices can have adverse effects of the opposite kind. Perhaps the best example is the felling of forests, which can result in soil erosion inland and an increased transport of sediment to the coast.

Urban and wastewater discharges

Because of increased coastal population and related human activity, the sea is subjected to a much greater input of wastewater, which is often composed of untreated sewage and industrial effluents. The sewage presents a public health risk from the human pathogens it contains, threatening both recreational users of the water and consumers of potentially contaminated seafood. Industrial effluents, containing such diverse constituents as metals, oil, and synthetic organic compounds, can result in an accumulation of toxic residues in seafood near the source of the input and may lead to deterioration of environmental quality over a wider area, with an associated reduction in biological diversity.

Nutrient inputs

Nutrient input problems are caused mainly by nitrogen and phosphorus, and although these are present in both the sewage and industrial effluents referred to above, there is significant additional input from agriculture and from intensive stock rearing. Some coastal areas show markedly enhanced nutrient levels with associated effects, including unusual algal blooms, deoxygenation of the water, and fish kills.

Exploitation of nonliving resources

At present, the major exploited nonliving resources are oil and gas, which are being exploited in some places up to the edge of the continental shelf. The environmental effects of this activity are well documented.

Also important are aggregates such as sand and gravel, various metal-rich minerals, phosphorites and polymetallic nodules, encrustations, and sulphides. Included also in this category is the use of the sea as a renewable energy source in developing thermal, tidal, or wave energy. Although this is not a major activity at present, it will become increasingly attractive as fossil fuels are depleted.

Exploitation of living resources

Overfishing, which has already been mentioned, is an example of exploitation of living resources. Besides the direct threat it poses to the resource, there are problems of alteration of the natural food web by the removal of populations that constitute a significant part of the overall structure of ecosystems and the direct damage that fishing gear causes to the seabed and its fauna. A more recent issue is associated with the global expansion of mariculture and the increasing threat this activity presents.

Impacts on the Coastal Region

Most of the problems in the sea arising from man's activities can be discussed under one of the above six headings. Given this background, we may now turn to large marine ecosystems. I will consider how our knowledge of global marine pollution and the associated activities of man are relevant to the way we manage LMEs and thus might alter our approach to defining and assessing them.

It is not unreasonable to suggest that because much of the LME discussion in the past has focused on physical oceanography and on fish stocks, the perspective has been from the oceans. However, when we consider pollution, a new perspective is required. In the open ocean, and even on the outer shelf, the major contaminating inputs are from the atmosphere, and therefore diffuse at low concentrations, or from shipping, and thus are confined to the major transport lanes. Most marine pollution originates directly from the continents, so our perspective should be from the land. A number of the topics discussed above therefore assume particular importance and should be examined more closely in relation to their effects.

Hinterland activities must rank high in this consideration. For example, most of the main rivers in western and central Africa have been dammed. The Indus River in southern Asia is almost entirely diverted for irrigation. The Nile, Danube, Rhine, and Colorado all provide further examples of rivers with waters variously utilized and flows restricted (Halim, 1990). It is suggested that by the end of the twentieth century about 66% of the world's total stream flow will be controlled by dams. Water discharged from major impoundments will differ from natural river waters in its composition, sediment, organic loading, and seasonal flow pattern.

The coastal consequences are immense, as a few examples will show. The active delta of the Indus has shrunk from 2,500 km^2 to a mere triangle of 250 km^2, and hypersalinity has damaged fisheries. The almost complete withdrawal of Nile waters has altered the sediment budget, with the result that summer deposition no longer compensates for winter erosion, producing extreme shoreline changes. In addition, both the fishery that coincided with the September-November flood season and the spawning of benthic invertebrates, which is related to the Nile bloom, have been affected. In the United States, a most dramatic illustration is found in the San Francisco Bay and delta. Upstream impoundments, diversions, and pumped exports have greatly reduced river flows in this area and have resulted in major effects, including loss of marshlands, reduced sediment discharges, increased salinity of bay waters, and consequent adverse effects on commercial fish species such as salmon, shad, striped bass, and sportfishing in general.

The other category of hinterland activities includes deforestation that, among other things, causes increased runoff and sediment transport, which results in an increase in turbidity, suspended material, and, eventually, sedimentation in coastal regions. This process causes adverse effects on species that require clear water, such as corals. There is good evidence, for example, that the death of coral reefs off the Philippines in recent years is a result of sedimentation caused by deforestation far inland.

With respect to activities on the coast itself, the various development operations I have mentioned have led to extensive draining of wetlands and the concreting of large sections of the coast, so that coastal habitats have been lost all over the world; this process is continuing. It is particularly evident in regions such as the Mediterranean, where both industrial and tourist development is important. In Spain, 42% of the east coast is affected by tourist construction; in Yugoslavia, urban and tourist growth affects 20% of the coast; around the centers of higher population at Marseilles, Naples, Athens, Istanbul, and Alexandria, 90% of the coast is under concrete.

The population expansion associated with these developments results in a marked increase in sewage and industrial waste discharges. In the past, these were seen as local issues, but with the spread of urbanization, they are now of regional concern in many parts of the world. In addition to the obvious dangers of contaminated seafood, there is the worry that more subtle and long-term changes may be taking place in water and sediment quality, which will affect spawning areas and nursery grounds and result in damage to sensitive egg, larval, and juvenile stages of fish and shellfish.

The nutrient inputs also are being reassessed and upgraded from a local to a regional, if not a global, issue. Areas with enhanced nutrient loading can now be identified in many parts of the world. In the Baltic Sea, this has caused detectable changes in the benthic fauna of many bays, inlets, and estuaries. In the shallow southern part of the North Sea, which is influenced by large industrialized rivers, the concentrations of nitrate and phosphate have increased by factors of 4 and 1.5, respectively (Lancelot et al. 1987); the species composition of the phytoplankton has changed, and fish and benthos have been killed by deoxygenation of the water.

There are records of eutrophication from locations in the Mediterranean where river outflows and urban discharges occur. Some areas, such as the Gulf of Lyons and the Bay of Ismir, show extreme effects, but the most extensive changes are in the northern Adriatic. That area historically has been subject to periodic excessive algal blooms; however, in recent years they have increased and extended with unfortunate impacts on fisheries and tourism. A similar story could be told for many other parts of the world, and although the source of the inputs varies from place to place, a profile can be produced of the type of site likely to be at risk. A sea area that is at least semi-enclosed, has no good access to flushing by clean water from the open sea, and has large coastal populations devoted to industry, agriculture, and tourism is highly susceptible to eutrophication.

The two remaining activities I have listed are concerned with the exploitation of resources, nonliving and living. The exploitation of minerals on the continental shelf has little direct effect over extensive areas in normal circumstances. Oil and gas platforms have a marked impact on the seabed within a radius of 2 – 3 km^2, and where several platforms are located near each other, the area of effect is extended. However, in the North Sea, for example, where 165 platforms were operating in 1988, the impacted seabed covered less than 1% of the area. Sand and gravel dredging may also cause serious problems (for example, if herring spawning grounds are damaged), but in general the effects are local. However, it is again from the land that a significant impact should be noted in the context of mining. In many parts of the world, the exploitation of inland minerals, from China clay to metals, leads to discharges into rivers that can contaminate major stretches of coast.

The impact of living resource exploitation is even more varied. Overfishing has already been referred to, and abusive fishing (e.g., the use of poisons or of explosives on coral reefs) should be mentioned, but the threats from the spread of mariculture deserve special

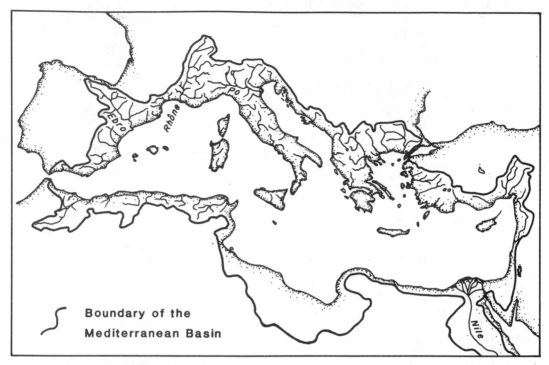

Figure 1. The Mediterranean Sea and its catchment area.

attention. Culture of a wide range of fish species in cages and other floating enclosures in coastal waters result in inputs of various kinds, including feces, waste food, and chemicals used in production. However, there are also other factors such as the reduction of visual amenity and the restriction of access to other users of the sea. Whereas these are mainly local effects, there is a more significant regional impact associated with the construction of coastal ponds for the culture of fish and shellfish (particularly shrimp), namely, the measurable annual loss of habitat in several parts of the world.

Shrimp farming is expanding, particularly in the southeast Pacific. In Panama, the estimated depletion of mangroves is 1% per year, and in Columbia and Ecuador, the situation is similar. Also, in some countries such as Ecuador, the supply of cultivated shrimp larvae is enhanced by harvesting wild specimens, using very fine mesh nets that produce an enormous bycatch of other juveniles and may have a general impact on coastal productivity. In India, thousands of acres of mangroves have been reclaimed, while in Bangladesh about 8% of the mangrove area has been taken over for shrimp culture. In Indonesia, nearly 30% of the mangrove forests were converted to agricultural land or shrimp ponds in the decade up to 1980, and estimates suggest that an additional 25%

will be lost by the end of the twentieth century. These wetlands act as a buffer between the hinterland and the offshore ecosystems of seagrasses and coral reefs. Their removal allows inland sediments in runoff waters to penetrate coastal areas, diminishing light and smothering nearshore ecosystems.

Just as the culture, harvesting, and processing of marine living resources can damage the seas, there are also effects from comparable activities relating to terrestrial living resources. The adverse effects from runoff of agrochemicals have already been described, but effluents from the processing of timber can also be a major source of pollution in many parts of the world. This has been particularly well documented in Canada and Sweden, where fjords and coastal waters are significantly affected by effluents and other waste production from pulp mills.

Implications for LMEs

Having briefly reviewed these effects, let's consider how they influence our views on the definition and management of LMEs. One major point that emerges is that a significant part of coastal pollution, in the broadest sense, can be traced back to activities in the hinterland, which means that the environmental quality of LMEs having large coastal sections—and this includes most LMEs—will be influenced by operations in the watersheds of rivers flowing into the area. This suggests two important implications for the management of such LMEs.

First, LMEs should be delineated not just in terms of their marine characteristics, but also in terms of their associated river systems. The continuity of the aquatic realm, from freshwater mountain streams to the edge of the continental shelf, should be recognized. It is interesting to consider the possibility that river outflow, concentrating the rainfall from a wide area as it does, may play a more important indirect role in channeling atmospheric input to the marine environment than is played by direct atmospheric deposit to the ocean. Figure 1 shows the appearance of the Mediterranean Sea if the watershed is included; Figure 2 gives a comparable picture for the North Sea. It is clear that the geographical area of concern is greatly increased when the watershed is included.

Second, defining an LME to include the associated watershed will require considerable expansion of current approaches to regulating and controlling pollution. It will require that all major activities inland, such as the construction of dams, power stations, and other facilities, be examined at the initial planning stage for potential effects at the coast. Consideration of such remote effects should be a mandatory component of associated environmental impact studies. However, it is not sufficient to introduce this procedure for single construction projects or for specific individual activities. To be effective, planning should be done on the basis of managing whole river systems. Such planning is currently envisaged for the River Po in Italy and will, hopefully, contribute to improved conditions in the upper Adriatic Sea. For the Po, the management concerns are not extremely complex because only one country is concerned. However, for rivers like the Rhine, the problems of integrated management are much greater, because several highly industrialized countries are involved.

Apart from these activities, which are clearly related to the hinterland, there are others on the list which, although not fully in the sea, have immediate and major effects on marine ecosystems. The loss of mangroves, seagrass beds, and other wetlands as a result of mariculture development and the concreting of coasts through industrial, urban, and tourist development are important examples. Another is the reduction in environmental quality in both the water and sediment compartments caused by a diversity of effluents associated with development. As has already been emphasized, these threats from the land are much more important in the context of marine pollution than any contamination from the

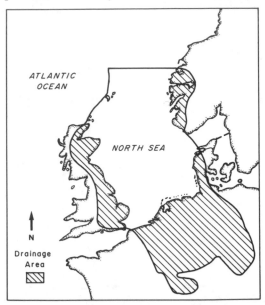

Figure 2. The North Sea and its catchment area.

open ocean. Indeed, impingement of ocean waters into coastal areas usually represents a flushing and cleansing that usually is highly beneficial to the coastal region.

In conclusion, how can these considerations contribute to a more integrated management of LMEs in the context of pollution? I suggest that we extend their delineation to include the watersheds of inflowing rivers. We should identify the characteristics of the coastal strip in each LME, noting the presence of potential pollution sources, and ask a number of questions. Are there significant cities and large areas of industrialization? Is there a single predominant industry such as oil or pulp mills, or is industry diversified so as to produce a wide range of effluents? Is the area suitable for mariculture development? Is the nature of the coastline such that effluents are contained and not flushed out to the open sea? Is there an important tourist industry and is it highly seasonal? Are there national agreements on pollution regulations and control? Are they implemented and are they effective? This approach, added to existing single-species analyses of fishery resources, will provide a more comprehensive picture and contribute to more effective management of LMEs.

References

Group of Experts on the Scientific Aspects of Marine Pollution (GESAMP). 1990. The state of the marine environment. UNEP Regional Seas Reports and Studies No. 115. UNEP, Nairobi.

Halim, Y. 1990. Manipulations of hydrological cycles. Annex to the GESAMP report on the state of the marine environment. UNEP, Nairobi.

Lancelot, C., Billen, G., Sournia, A., Weisse, T., Colijn, F., Veldhuis, M. J. W., Davies, A., and Wassman, P. 1987. *Phaeocystis* blooms and nutrient enrichment in the continental coastal zones of the North Sea. Ambio 16:38 – 46.

Sherman, K. 1989. Large marine ecosystems: A case study. In: New developments in marine science and technology. pp. 97 – 114. Ed. by L. M. Alexander, S. Allen, and L. C. Carter. Law of the Sea Institute, U. of Hawaii, Manoa, Honolulu, HI. 530 pp.

Sherman, K. and Alexander, L. M. (Editors). 1986. Variability and management of large marine ecosystems. AAAS Selected Symposium 99. Westview Press, Inc., Boulder, CO. 319 pp.

Chapter 5

Application of International Global Change Research Programs, Including GLOBEC, to Long-Term Large Marine Ecosystems Management

Michael R. Reeve

Introduction

The central scientific question for biological oceanographers and living resource managers over the coming years concerns climate and its interaction with the physics and chemistry of the oceans to affect and control variability of ocean ecosystems and their component populations. The importance of this question now is in the context of potential unidirectional global climate change. Will a global warming trend over the next few decades and its consequent modification of ocean circulation result in non-reversible ecological changes? Will overall ocean biomass and secondary production increase or decrease? Will major resource species be replaced by others unsuitable for harvesting? Will the structure of marine ecosystems be radically altered, or will some ecosystems, such as coral reefs, be eliminated?

The second important question concerns feedback mechanisms. What effects would such changes in ocean ecology have, in turn, on biogeochemical cycles to influence climate through, for instance, the modification of the CO_2 cycle in the oceans?

Ecologists working in the marine environment have always been aware of the great variability of ecosystems and their component populations, but have largely concentrated attention on the annual cycle of seasonal change, the causes of which are readily understandable. Biologists have been slow to come to grips with most spatial and temporal scales of variability because of the enormous difficulties in obtaining the necessary biological data. Adequate technology is lacking, and the cost of research is potentially high. Beyond this, they lack the kinds of models and theoretical approaches to

focus the limited resources and constrain the potentially unlimited amount of data collection, laboratory experimentation, and process-oriented field studies.

Our sister disciplines of atmospheric and ocean climate and circulation, as well as ocean biogeochemistry, have been planning and, in some cases, embarking on field observational programs on the global physical and chemical interconnections that drive ecosystem and population change. The purpose of this chapter is to underline the fact that much of the information biological oceanographers and resource managers will need will be generated over the next 10 or more years. Equally importantly, the computing resources needed to integrate biological parameters into models of circulation and biogeochemistry will become readily available through technological advances.

Global Change and Global Change Programs

It is well accepted among scientists that a rapid change is already under way in world climate, including the trend for progressively elevated temperatures in the ocean. Long time-series data over the past several decades have demonstrated that temperature variability on the order of only 0.5-1.0°C has been correlated with large interannual to decadal fluctuations of animal populations. The well-known relationship between El Niño Southern Oscillation (ENSO) events and the biological productivity of the Peruvian upwelling system is an example. The current unprecedentedly fast unidirectional change predicted may, therefore, have enormous consequences for ocean food chains and the regional resource stocks they support. This change will affect both biomass levels and species composition.

Other disciplines have not been slow to recognize the enormity of the research challenges facing them. One of the first planning efforts, born out of the very strong El Niño of 1982 – 1983, was the Tropical Ocean—Global Atmosphere (TOGA) program (NRC, 1990). This program, an international effort to identify the role of the coupled tropical ocean and atmospheric dynamics in the interannual variability of global climate and weather, is already halfway through its 10-year implementation phase. Scientists expect that by the year 2000, it will be possible to predict the occurrence and strength of El Niño phenomena accurately enough to provide forecasts of the ensuing regional ocean physical dynamics. Much less research has been expended on the biological consequences of these physical phenomena, although there have been indications that events such as these affect LMEs—for instance, the distribution and biomass of populations ranging from tropical coral reefs to as far north as Alaska in the Pacific.

At least a decade ago, physical oceanographers realized that the only way that they would understand the physical circulation of a specific region, at least in terms of how it might evolve in the future, is to understand the circulation of the world ocean as a whole. Thus, the World Ocean Circulation Experiment (WOCE) was born. It is only now, after a decade of planning, strategy debate, and, most practically, the search for international financial support, that its field program is becoming operational (U.S. WOCE, 1990).

The goal of WOCE is to obtain a "snapshot" of the world ocean circulation now, which would provide a baseline for comparing future changes. In doing so, massive amounts of oceanographic and satellite-based data will be used to greatly improve the regional precision of general circulation models. Physical circulation is a primary driving force of biomass changes of LMEs. A new hypothesis, which characterizes ocean circulation in terms of the "ocean conveyor belt," suggests that in recent geological times there have been rapid, major switches in ocean circulation patterns on a timescale of tens to hundreds of years (Broecker and Denton, 1989). Man-induced global change might be expected to accelerate such phenomena, with violent short-term consequences both for physics and biology.

The prospect of the atmosphere and ocean being called upon to receive greatly increased loads of CO_2 and other gases prompted the initiation of the Global Ocean Flux Studies (GOFS) program, now a large-scale international effort called "JGOFS" for "Joint GOFS" (SCOR, 1990). The mechanisms and rates of movement of carbon dioxide from the atmosphere into the ocean, and its long-term deposition in the ocean sediments or return to the atmosphere, are largely unknown, yet vital to computations of potential global warming. These processes are intimately bound with the activities of the ocean biota. Because the JGOFS program is primarily concerned with mass balances and fluxes of biogenic gases, it has chosen to focus on the variability and control of primary productivity—the input term—for atmospheric CO_2 and the vertical transport and transformation of subsequently produced biological particulate detrital material. The program can be expected, like the WOCE and TOGA programs, to lead the way to the next generation of predictive models, particularly for regional primary production, a vital component of any ecosystem model.

Taken as a whole, the government resources that are being set aside for these programs are an indication of the immense amount of potentially relevant research in ocean ecosystems dynamics that will be forthcoming. In the United States alone, the outlays amounted to over $50 million in fiscal year 1991 and may well double within 5 years.

International governmental and nongovernmental organizations, including the Intergovernmental Oceanographic Commission (IOC), the World Meteorological Organization (WMO), the International Geosphere/Biosphere Program (IGBP) of the International Council of Scientific Unions (ICSU), and its scientific committees such as that for ocean sciences (Scientific Committee on Oceanographic Research [SCOR]), have helped to forge successful planning and implementation activities for these programs. To date, there does not exist any comparable international organization for research on the impact of global change on ocean ecosystems. Only within the past year (1991) have IOC and SCOR begun to consider such a possibility. Within the United States, the federal government Committee on Earth and Environmental Sciences (CEES) has developed the complex, interagency Global Change Research Program, which began in 1989 (CEES, 1990). This program does include a focus on ocean ecosystem dynamics.

Global ecosystems

There have been two main approaches taken to understanding ocean ecosystems. In one, the oceanographic approach, emphasis has been on interrelating the physics and biology of the ocean. This approach tends to focus on the effects of physics on recruitment processes in the plankton. On the other hand, the comparative approach of the LME concept tends to focus on the end product—the fish or other living resources and their population management by fishing and other forms of harvesting.

The latter approach is the subject of this volume and is extensively treated. I will briefly review the basic oceanographic approach (which focuses on the interactions in nature that produce the fish stocks in the first place) and describe its history in the United States over the past decade as an example.

In the early 1980s, three workshops were organized that included both oceanographers and fisheries scientists. These workshops influenced subsequent global change program planning at the U.S. National Science Foundation (NSF). Ecosystems and recruitment processes were in its original suite of long-range global change plans in 1985. The National Academy of Sciences (NAS) became involved in the mid-1980s, and the National Oceanic and Atmospheric Administration (NOAA) and NSF entered into an agreement to cooperate on the funding of ocean ecosystem research. This collaboration has led, beginning with the 1991 fiscal year, to the establishment of funded programs by both agencies as part of their global change research programs, with the intention of jointly achieving an annual level of effort approaching $20 million within 5 years.

In the meantime, a large workshop with international representatives was convened in 1988 at Wintergreen, Virginia, which resulted in the formation of a steering committee to plan a program of research on the effects of change on global ocean ecosystems dynamics (GLOBEC), on behalf of U.S. academic scientists. Their activities are well advanced and an initial science plan has been published (JOI, 1991a).

Recently, SCOR and IOC agreed to examine the status of ecosystems dynamics in relation to global change, with a view to making recommendations on a course of action for the international community.

What is Needed?

On the basis of the Wintergreen workshop report (JOI, 1989), as well as several other more recent informal international discussions, it is clear that there are several pressing areas that need to be studied; these are summarized below.

Technology

There has been a consensus over the past few years that our present technology for collect-ing and analyzing the zooplankton, which has changed little since the nineteenth century, is woefully inadequate. Routine methods such as using plankton nets and microscopic counting are not compatible with modern real-time technology of physics, chemistry, or even those for phytoplankton biomass assessment. A breakthrough in this area would be comparable to the introduction of the Conductivity Temperature Depth (CTD) sensor or in situ fluorometer. Ideally, there also should be some large aerial sensing mechanism comparable to ocean satellite infrared and color sensors. This latter objective is currently remote but might be achievable in the 1990s (see below).

In the 1980s, the Office of Naval Research (ONR) Oceanic Biology Program and NSF Ocean Sciences Research Section collaborated in funding the development of acoustic systems that seek to push size discrimination downward from the traditional adult fish target to the smallest copepod stage, that is, 100 µm and below. This was successfully achieved with the deployment of the Multifrequency Acoustic Profiling System (MAPS) sampler of Holliday (see chapter 28, this volume). The potential power of such an instrument is that it can be routinely deployed as part of a CTD or other profiling instrument to obtain real-time data about the biomass and size distribution of zooplankton. At present, it might take weeks or months to reconstruct such information from net samples, and the opportunity to investigate phenomena of interest has long gone.

A GLOBEC working group report (JOI, 1991b) envisages the need for a profiling system similar to the MAPS instrument for detailed studies, a high-speed towed instrument for wide-scale surveys, and systems that can be used on drifting buoys and long-term moorings. Acoustic sensing techniques will always need ground-truthing by actually withdrawing animals from their medium by nets or pumps for direct visual inspection. Eventually, however, we must develop operational image analysis and artificial intelligence techniques to speed and automate the process on shipboard and ultimately in situ.

Other technologies that are being pursued (see Powers, chapter 29, this volume) include the application of biotechnology and molecular biology techniques to estimate physiological condition (e.g., nutritional or reproductive) and for rapid identification of species and their young. This determination is currently not feasible for some groups of organisms, even using laborious morphological examinations.

Finally, there is a need to make better use of current and expected satellite technology to relate ocean color and circulation to regional secondary production models, because mere knowledge of primary production distribution, while a general indicator of ocean productivity (including that of fish), is a poor indicator of local and short-term secondary production events (Lasker, 1988). Multispectral sensors, planned to be aboard the earth-observing satellites (EOS) by the end of the twentieth century, offer the potential for sensing large concentrations of ocean surface biomass characterized by reflective wavelengths not currently sensed. It is possible that some optical characteristics of dense, surface planktonic animal populations might be remotely sensed.

LMEs and time series

The purpose of advancing technology is to enhance the sophistication of field measurements and observations to be on a par with the other oceanographic disciplines. This is important both for long time-series observations and short-term process studies.

Inherent in the LME concept is the recognition of the importance of long-term data series collected for management purposes. These data series are necessary not only to measure trends and the changing health of the system, but also to develop predictive capabilities. Also vital to the LME concept is the recognition that no LME is an independent entity, but is, to a greater or lesser extent, dependent on physical and biological events in all the other LMEs. The case is often made that most of the world's fisheries yield comes from coastal LMEs, and fisheries management is necessarily a localized activity on political, if not strictly scientific, grounds. It must also be readily apparent, however, that events at times and places distant from any single LME (e.g., an ENSO event) may well control a fishery by modifying productivity and sequential timing of food chain events thousands of kilometers away.

In order to develop the LME concept and its practical application fully, plans should be made to first subdivide the whole ocean realm into its component LMEs, and not merely those readily recognizable at the ocean margins. The next step is to come to an agreement on a basic set of long-term observations to be made and maintained that would serve to measure change.

This process has already started in the terrestrial context. For the past 10 years in the United States, NSF and other agencies have supported a gradually increasing number of "representative" terrestrial research sites from the Arctic to the subtropical. This program will be expanded to the Antarctic soon. While supporting individual research, which varies among the sites depending on their special nature and the interests of local researchers, these Long-Term Ecological Research Site teams are networked together, meet frequently to coordinate and determine policy objectives, and are linked by computer to a common database.

In the marine environment, the JGOFS program has set up two mid-ocean long time-series stations, one off Hawaii and one off Bermuda, as pilot efforts to explore the feasibility and problems inherent in such an enterprise. Although there are a few existing well-known marine long time-series (besides fishery catch statistics), the data are not readily accessible or even completely analyzed. They lack any common rationale or international backing for joint sponsorship.

Modeling

More biologists should be trained to understand the predictive numerical modeling done by physicists in order to incorporate the biological terms that will eventually provide population dynamics outputs. Massive efforts are being planned in physical (TOGA, WOCE) and biogeochemical (JGOFS) modeling.

For instance, physical modelers, working with supercomputers such as the Cray XMP at the U.S. National Center for Atmospheric Research (NCAR), recently ran a North Atlantic basin circulation model for a multiyear period, which required about 1,000 hours of supercomputer time. The JGOFS program already has a modeling activity that is beginning to incorporate biogeochemical parameters into physical models to make predictions about bulk biological parameters in the ocean such as primary production, biomass, and carbon distribution and flow. These modeling programs will gradually become more powerful as computing resources permit greater resolution and field programs provide more data for improved theoretical approaches. In turn, this will provide essential groundwork for the eventual incorporation of biological terms leading to population predictive outputs.

Theory

One of the problems we face is a basic lack of understanding of how the marine food chain operates, particularly at the secondary level. This is directly attributable to the archaic sampling technology referred to above, which has severely limited our detailed view of complex interactions. There are three avenues of research that require attention.

The first set of problems involves questions of scale. Patchiness in time and space appears to exist down to the smallest scales of present-day resolution (cm-mm) and probably smaller. Should the focus be at the individual organism level and its relationship to its environment, in order to scale upward to understand LMEs? This approach is advocated by Rothschild and Osborn (1988) and would be supported intuitively by experimentalists who point out that the "units of currency" are the individual acts of prey and predation or growth and fecundity. Or should scientists devise models that cascade down in space and time from global or basin-scales only far enough for the particular level of interpretation desired, that is, LME or population of interest? Physicists have primarily taken the latter approach, biologists the former. In integrating physics with biology, it is not obvious which of these two approaches, or perhaps some intermediate, will be most fruitful.

Another important question concerns the biological average versus the extreme. Many marine animals, including most fish, produce hundreds to millions of offspring per parent. Recruitment of them to adult populations could always be considered a failure in human terms. Success and huge adult biomass might be the result of 1% survival instead of 0.01% or less. The survivors are probably not representative of the average genetic composition of the spawning, but an extreme subset, dependent on the ambient environmental conditions. Sampling and experimental approaches have traditionally been concerned with reporting average events, biomass, or physiological rates. Individual offspring, even from the same parents, can vary widely in their genetic characteristics. Populations of newly hatched animals of the ctenophore *Mnemiopsis* (a larval fish predator), for instance, can vary by several orders of magnitude in biomass after 10 days of growth, even at the same feeding level. "Average" animals might be the least likely to survive in variable food and/or predator environments, and our average data, therefore, might not be very relevant in model building.

Finally, our inadequate sampling technology has caused us to chronically underestimate the biomass and trophic significance of plankton predators, both those that feed directly on young fish and those that remove the copepod food on which the young fish depend. Many of these predators are the larger gelatinous organisms that have never been sampled successfully and are very difficult to work with experimentally, but whose patch formation potential is very great. They often have daily biomass-doubling potential, very high fecundity, and

very short life cycles. Some, such as the offshore ctenophores, are impossible to sample as adults because they dissolve in formalin. All ctenophore young are destroyed without trace in preservatives and, to my knowledge, have never been reported in plankton samples. Such poorly sampled ephemeral populations, most likely to exert their maximum influence during the peak of recruitment and production events, are also likely to be significant in the microbial food loop and in geochemical cycles.

Process-oriented studies and LMEs

One of the strong connections I foresee between a well-established network of LMEs, in which long-term observations are being made, and research programs such as GLOBEC would be the ability to make intensive studies of specific processes in the context of LMEs, where much background data and continuing long-term observations have been and are being made. Such a study might involve several ships, using advanced technology to test specific theoretical approaches and models over the course of ephemeral, but important, periods—for example, the spawning and subsequent fate of a fish larval population of interest.

As an example of a recent major process study, in 1989 the JGOFS program studied the course of the spring bloom in the North Atlantic as it spread northward and the subsequent fate of the particulate carbon formed. The experiments and observations extended over 6 months and involved the scientists and ships of seven nations.

A more pertinent example, in the current context, is the proposed Cod and Climate Change program, currently in the planning stage and involving U.S. GLOBEC, NOAA, and scientists of other North Atlantic International Council for the Exploration of the Sea (ICES) countries (such as Canada, Norway, and the United Kingdom). The program's purpose is to examine the potential effects of climate change and physics on the North Atlantic food chains and ultimately on the cod stocks. The North Atlantic Ocean basin is, in effect, an LME with its ocean margin subcomponents such as Georges Bank. There already exists for the North Atlantic perhaps the largest amount of biological oceanographic and fisheries data of any large-scale region. One can easily envisage the development of a large-scale cooperative investigation in which one or two nations might take responsibility for a particular subelement (e.g., the United States and Canada for Georges Bank and adjacent regions) within integrated large-scale ocean experiments on the basis of agreed-upon scientific priorities.

Ideally, JGOFS and ecosystems dynamics experiments should be coordinated within the same time frame, to maximize the utility and interpretation of both kinds of experiments. It is hoped that this will occur in the future as ecosystems planning matures.

Conclusion

Governments and other organizations with a vital interest in the management and optimization of resources of LMEs face the requirement of predicting the future course of events and developing responses to them. They face the twin problems of natural and human-induced variability and the added impacts of fishing stress. Over the longer term of global change, the latter cannot be predicted without the ability to predict the former.

It is clear that within 50 years the ocean will be routinely instrumented for continuous monitoring, just as the atmosphere is today for weather forecasting. It is imperative that biologists take part in this effort and make the case for its importance. We currently do not have the operational instruments for this task, or even a clear consensus on what ought to be measured.

As ecosystems change, they will impact the rest of the biogeochemical and climate system. Quantification of the nature of such changes will be sought from ecologists by other scientists interested in these feedback phenomena. The ocean CO_2 system, and hence the atmospheric CO_2 system and global warming, for instance, will be affected by changes in secondary production.

We will be continuously barraged with questions of a "what if?" nature and expected to provide guidance. One such example has already appeared (ASLO, 1991): it is being strongly advocated that by stimulating ocean primary production over massive areas (by the addition of shiploads of the apparently limiting nutrient iron) ocean uptake of CO_2 from the atmosphere will counterbalance anthropogenic increases and alleviate global warming. Irrespective of whether the hypothesis is viable, what kind of primary production would be stimulated? How would "normal" phytoplankton species fare? How would this change (or perhaps disrupt) ocean food chains and ultimately resource populations? Would adverse effects be readily reversible?

It is time to organize a concerted program, involving LME managers and the governments and academic scientists of interested nations, to address the massive influx of ocean data and models that will be forthcoming from programs such as WOCE, TOGA, and JGOFS. It is essential that we develop plans for using this information to enhance our understanding of how ocean ecosystems work and how they will change.

References

ASLO (American Society of Limnology and Oceanography). 1991. What controls phytoplankton production in nutrient-rich areas of the open sea? Report of the ASLO Symposium. February 22 – 24, 1991, San Marcos, CA. 17 pp.

Broecker, W. S., and Denton, G. H. 1989. The role of ocean-atmosphere reorganizations in glacial cycles. Geochim. Cosmochim. Acta 53:2465 – 2501.

CEES (Committee on Earth and Environmental Sciences). 1990. Our changing planet: The FY 1991 research plan. The U.S. Global Change Research Program. Committee on Earth and Environmental Sciences, Reston, VA.

JOI (Joint Oceanographic Institutions, Inc.). 1989. GLOBEC: Global Ocean Ecosystems Dynamics. Joint Oceanographic Institutions, Inc., Washington, DC.

JOI. 1991a. GLOBEC: The initial science plan. Joint Oceanographic Institutions, Inc., Washington, DC.

JOI. 1991b. GLOBEC: Workshop on acoustical technology and the integration of acoustical and optical sampling methods. Joint Oceanographic Institutions, Inc., Washington, DC.

Lasker, R. 1988. Food chains and fisheries: An assessment after 20 years. In: Toward a theory on biological-physical interactions in the world ocean. pp. 173 – 182. Ed. by B. J. Rothschild. NATO ASI Series. Series C: Mathematical and Physical Sciences, Vol. 239. Kluwer Academic Publishers, Dordrecht, The Netherlands.

NRC (National Research Council). 1990. TOGA: A review of progress and future opportunities. National Academy Press, Washington, DC.

Rothschild, B. J., and Osborn, T. R. 1988. Small scale turbulence and plankton contact rates. J. Plankton Res. 10:465 – 474.

SCOR (Scientific Committee on Oceanographic Research). 1990. The Joint Global Ocean Flux Study Science Plan. Scientific Committee on Ocean Research, Halifax, Canada.

U.S. WOCE (World Ocean Circulation Experiment). 1990. World ocean circulation experiment—U.S. WOCE implementation report No. 1. U.S. WOCE Office, College Station, TX.

Chapter 6

Approaches to Forecasting Biomass Yields in Large Marine Ecosystems

Simon A. Levin

Introduction

The oldest and venerable tradition in mathematical ecology is that of describing the dynamics of marine fisheries. Nearly a century ago, Vito Volterra, one of the greatest of mathematicians, was inspired by his son-in-law, Umberto d'Ancona, the distinguished biologist, to turn his attention to the consideration of the regular oscillations observed in the fisheries statistics in the Adriatic. Volterra, who was then in the process of constructing his theory of integral equations, found that theory to be remarkably appropriate to the age-structured dynamics of the Adriatic fisheries; and in the early twentieth century, he developed both subjects, one basic and one applied, at a rapid pace and with mutual stimulation. To this day, Volterra's theory of integral equations remains at the core of the subject, and his dynamical systems approach to a theory of interacting unstructured biological populations remains (with Alfred Lotka's parallel development) the foundation of the theoretical literature. Despite the fact that the Lotka-Volterra equations are too simplistic to be applied literally to any real populations, the elegant demonstration that the implicit delay embodied in predator-prey dynamics is, by itself, sufficient to drive sustained oscillations has been among the most influential theoretical results in ecology, and indeed, in all biology.

The approach pioneered by Volterra and Lotka has spawned uncountable imitators and has helped shape much of our thinking about population interactions. Their most important contribution, however, was not in the detailed equations considered, but in the demonstration of the power of the quantitative and analytical approach. Later efforts, such as those of Beverton, Holt, and Ricker in exploring stock recruit relationships (Ricker, 1954; Beverton and Holt, 1957), and those of Clark (1985) in developing the theory of mathematical bioeconomics, were made possible in large part by the successes of Volterra's pioneering efforts.

For all of these approaches, various technical problems made application difficult. Even the simplest discrete dynamical systems could give rise to complicated and sometimes chaotic dynamics (Ricker, 1954; May, 1972). Parameter estimation, recognized as a difficulty in large-scale ecosystem models because of limited data sets, could also be problematical for even simple models. Age structure, spatial structure, and other manifestations of population heterogeneity introduce more biological realism and more parameters to estimate. In consideration of such complications, the problem of relevant detail (Levin, 1991) becomes paramount, and one must turn to efforts at aggregation and simplification (Ludwig and Walters, 1985; Iwasa et al., 1987, 1989) that reduce consideration to the smallest possible set of parameters necessary to capture the essential features of the system.

Aggregation and simplification methods, which explore the consequences of viewing the system at different scales of resolution, are thus intertwined with investigations of patchiness in space and time and the recognition that the variability of any system is particular to the spatial, temporal, and other scales chosen for investigation. In oceanography, that recognition is implicit in the Stommel diagram (Haury et al., 1978) and in Steele's discussion (Steele, 1978) of the perceptual bias inherent in any choice of scales. Furthermore, all organisms, through their life history and dispersal patterns, experience environmental variability with a perceptual bias conditioned by their own behaviors. Thus, it is essential to characterize the patterns and scales of variability in natural systems, to understand how those patterns are filtered by perceptual biases, and to elucidate how they serve to shape the ecological and evolutionary dynamics of natural populations. Patterns of variability and patchiness are fundamental to the survival of species and to interspecific interactions such as predation; thus, they are also fundamental to the dynamics of large marine ecosystems (LMEs).

It follows that efforts to model LMEs must deal explicitly with problems of scale and pattern. The patchiness in resource distributions will affect consumption rates and can bias biomass estimates fundamentally. Although techniques exist for dealing with some aspects of patchiness without explicit consideration of spatial distributions, the aggregated models that are used for this purpose are on shaky theoretical ground, unless they can be related to more explicit spatial models that can track the changes in distributions in relation to extrinsic and intrinsic forcing. Ship transect data (Weber *et al.*, 1986; Macaulay, 1988) can provide snapshots of the distribution of physical motions, phytoplankton, and zooplankton abundance, and such data can, in theory, be correlated with distributions of consumers at higher trophic levels. Such correlations provide a beginning, but only a beginning, for the development of mechanistic models that underlie reliable prediction and management.

The appropriate models to describe biomass dynamics will vary with the questions being asked, and multiple models, relevant to different scales, will often be needed. Hence, a key objective of modeling must be to relate processes across disparate scales of space, time, and organizational complexity. Through quantitative methods, we can characterize the patterns of heterogeneity and variability; through modeling and experimentation, we can relate those patterns to the processes responsible for structuring and maintaining them.

This approach has direct application to the study of marine ecosystems (Levin *et al.*, 1989). For example, transect data of phytoplankton and zooplankton abundance allow the description of patterns of spatial variability and the characterization of the patch structure of populations. Typically, one finds patchiness (nonuniform distribution) across the size spectrum, from meters to tens of kilometers (Morin *et al.*, 1988). Self-similarity (Mandelbrot, 1977) may be detectable across broad ranges and may provide rules for scaling; but this self-similarity may break down as the key factors regulate pattern change. Although systems are less open

to external inputs on broad scales than on fine scales, the scale of influence of physical processes is often broader than that of biological processes; typically, this will be reflected in the processes that determine the structure of patches. For example, for the distribution of Antarctic krill, large-scale patchiness appears to be determined by fluid motions, and small-scale processes by biological motions (Levin *et al.*, 1989). I will return to this point later.

The analysis of pattern involves a separation of biological and physical influences and a unification of what generally have been distinct literatures. Coherent physical structures, such as Langmuir cells or warm core rings, can impose patterns of patchiness that maintain integrity in the face of mixing processes. More generally, the trade-offs between physical mixing processes, such as turbulent diffusion, and biological growth processes stimulated by localized nutrient bursts can produce patches of characteristic sizes, determined by the ratio of diffusion to growth (Kierstead and Slobodkin, 1953; Okubo, 1980; Platt, 1985). Similar characteristic scales can be estimated for other processes (e.g., diffusive instabilities [Levin and Segel, 1976]) and compared with observed patterns as a step toward identifying the causes of patchiness.

The method outlined above has problems. The use of theory as a means to assemble a catalog of possible mechanisms underlying pattern is an excellent beginning, and the use of correlations as crude filters to reduce the lists of candidate mechanisms is a sensible complementary step. Yet, correlation does not imply causation, and causation does not ensure correlation. For example, change in a key variable (e.g., climate) can trigger qualitative shifts in system dynamics (Levin, 1978; Holling 1986, 1992); the resultant patterns likely will retain no evidence of the events that potentiated change. In such cases, there is causation, but no correlation. On the other hand, as is well known, spurious correlations can also mislead (Lehman, 1986); correlations may exist among variables that have no mechanistic relationship to one another. Thus, one must go beyond correlations to examine mechanistic underpinnings. Another problem is that the estimates available for key parameters are likely to be so rough that the process of using models to screen candidate mechanisms can be only a crude first step.

All of this argues for a mechanistic approach and for the development of models that relate the patterns observed on one level to processes that operate at other levels; relating phenomena on different scales is the essence of modeling and theory. The combination and integration of observation (on diverse scales) and modeling provides a powerful tool for understanding observed patterns and for developing the power to predict and manage.

A case in point involves the Southern Ocean LME. Available data (Weber et al., 1986; Macaulay, 1988) show patterns of patchiness in the distribution of krill across a range of scales. A priori, there is no reason to expect the determinants of patchiness to be the same on every scale; and indeed, cross-correlation of krill data with other factors (e.g., temperature and fluorescence) suggests that there are different determinants and different explanations on different scales (Weber et al., 1986; Levin et al., 1989). On broad scales, the characteristics of krill patterns do not seem to differ significantly from those of temperature and fluorescence, and a reasonable null hypothesis seems to be that the patterns are physically determined reflections of fluid motions. On fine scales, that hypothesis is unsupportable: krill clearly are more aggregated on those scales than can be explained by physical forcing. The explanation lies in the fact that krill are active swimmers and are known to aggregate into swarms and schools (Hamner et al., 1983; Hamner, 1984) on scales that are consistent with the observed fine-scale patchiness.

The obvious strategy for modeling LMEs is to combine these approaches by interfacing physical and biological models operating on very different spatial and temporal scales; yet such efforts are rare in the ecological literature. For the above example, Hoffmann (1988) and her collaborators have translated the Eulerian descriptions embodied in the Navier-Stokes equations into Lagrangian models that allow the description of passive forcing—for example, of the movements of eggs and larvae. Grunbaum (1992), building on the approaches of Okubo (1986) and Sakai (1973), extends Hoffman's passive forcing. His individual-based models consider both the forces imposed by the fluid dynamics and the additional "forces" that reflect the tendencies of individuals to orient their motions in relation to other individuals. In Grunbaum's model, an individual changes direction at random points in time, in response to physical forcing and information about local densities and population gradients. Individual movements are conditioned by an organism's evolutionarily mediated tendency to seek predetermined target densities, conditioned by estimates of abundance within some detection distance. Of course, this approach involves explicit assumptions about individual behaviors, and other assumptions must also be examined.

Approaches of this sort derive from the heritage of Volterra and Lotka, but deviate significantly in their explicit consideration of space, their attention to scale, and the indi-

vidual-based approach to model construction. Volterra and Lotka worked at a time when only very aggregated information about population fluctuations and distributions was available, when computers had not yet been imagined, and when quantitative ecology in general existed only to the extent that Volterra and Lotka had breathed life into it. We are at a very different stage in modeling and observation in the 1990s, able to build upon the firm foundations developed by those intellectual giants, and capable of using the sophisticated methodologies and technologies available to take the subject to new levels. The problem of scaling, and in particular of understanding how pattern in LMEs is related to the behaviors of individuals, is the key intellectual problem before us, and one that we now have the tools to address.

Acknowledgments

I gratefully acknowledge support from the U.S. Department of Commerce, National Oceanic and Atmospheric Administration, through grants NA88EA-H-00005 and NA90EA-H-AF002, and a University Research Initiative Program grant to the Woods Hole Oceanographic Institution from the Department of Defense, Office of Naval Research.

References

Beverton, R. J. H., and Holt, S. J. 1957. On the dynamics of exploited fish populations. Ministry of Agriculture, Fisheries and Food. London Fish. Invest. Ser. 2(19):1 – 533.

Clark, C. W. 1985. Bioeconomic modelling and fisheries management. Wiley Interscience, New York. 386 pp.

Grunbaum, D. 1992. Local processes and global patterns: Biomathematical models of Bryozoan feeding currents and density dependent aggregations in Antarctic krill. Ph.D. diss., Cornell University, Ithaca, New York.

Hamner, W. M. 1984. Aspects of schooling in Euphausia superba. Jour. of Crustacean Biol. 4 (Spec. No. 1):67 – 74.

Hamner, W. M., Hamner, P. P. Strand, S. W., and Gilmer, R. W. 1983. Behavior of Antarctic krill, Euphausia superba: Chemoreception, feeding, schooling, and molting. Science 220:433 – 435.

Haury, L. R., McGowan, J. A., and Wiebe, P. H. 1978. Patterns and processes in the time-space scales of plankton distributions. In: Spatial pattern in plankton communities. pp. 277 – 327. Ed. by J. H. Steele. Plenum, New York.

Hoffmann, E. E. 1988. Plankton dynamics on the outer southeastern U.S. continental shelf. Part III: A coupled physical-biological model. Jour. of Mar. Res. 46:919 – 946.

Holling, C. S. 1986. The resilience of terrestrial ecosystems: Local surprise and global change. In: Sustainable development of the biosphere. pp. 292 – 317. Ed. by W. C. Clark and R. E. Munn. Cambridge University Press, Cambridge, England. 491 pp.

Holling, C. S. In press. Cross-scale morphology, geometry and dynamics of ecosystems. Ecol. Monographs. 62(4).

Iwasa, Y., Andreasen, V., and Levin, S. 1987. Aggregation in model ecosystems. I. Perfect aggregation. Ecol. Modelling 37:287 – 302.

Iwasa, Y., Levin, S. A., and Andreasen, V. 1989. Aggregation in model ecosystems. II. Approximate aggregation. IMA Jour. of Mathematics Applied in Medicine and Biology 6:1 – 23.

Kierstead, H., and Slobodkin, L. B. 1953. The size of water masses containing plankton bloom. Jour. of Mar. Res. 12:141 – 147.

Lehman, J. T. 1986. The goal of understanding in limnology. Limnology and Oceanography 31:1160 – 1166.

Levin, S. A. 1978. Pattern formation in ecological communities. In: Spatial pattern in plankton communities. pp. 433 – 466. Ed. by J. H. Steele. Plenum, New York.

Levin, S. A. 1991. The problem of relevant detail. In: Differential equations—Models in biology, epidemiology and ecology. pp. 9 – 15. Ed. by S. Busenberg and M. Martelli. Lecture Notes in Biomathematics 92. Springer-Verlag, Berlin.

Levin, S. A., Morin, A., and Powell, T. H. 1989. Patterns and processes in the distribution and dynamics of Antarctic krill. pp. 281 – 299 In: Scientific committee for the Conservation of Antarctic Marine Living Resources (CCAMLR) Selected Scientific Papers, Part 1, SC-CAMLR-SSP/5, Hobart, Tasmania, Australia.

Levin, S. A., and Segel, L. A. 1976. An hypothesis for the origin of planktonic patchiness. Nature 259:659.

Ludwig, D., and Walters, C. J. 1985. Are age-structured models appropriate for catch-effort data? Can. Jour. of Fish. and Aq. Sci. 42:1066 – 1072.

Macaulay, M. C. 1988. Statistical problems in krill stock hydroacoustic assessments. In: Scientific committee for the Conservation of Antarctic Marine Living Resources (CCAMLR) Selected Scientific Papers, Part 1, SC-CAMLR-SSP/5, Hobart, Tasmania, Australia.

Mandelbrot, B. B. 1977. Fractals: Form, chance, and dimension. Freeman, San Francisco. 365 pp.

May, R. M. 1972. Will a large complex system be stable? Nature 238:413 – 414.

Morin, A., Okubo, A., and Kawasaki, K. 1988. Acoustic data analysis and models of krill spatial development. pp. 311 – 329 In: Scientific committee for the Conservation of Antarctic Marine Living Resources (CCAMLR) Selected Scientific Papers, Part 1, SC-CAMLR-SSP/5, Hobart, Tasmania, Australia.

Okubo, A. 1980. Diffusion and ecological problems: Mathematical models. Biomathematics 10. Springer-Verlag, Berlin, Germany. 254 pp.

Okubo, A. 1986. Dynamical aspects of animal grouping: Swarms, schools, flocks, and herds. Advanced Biophysics 22:1 – 94.

Platt, T. 1985. Structure of the marine ecosystem: Its allometric basis. In: Ecosystem theory for biological oceanography. pp. 55 – 75. Ed. by R. E. Ulanowicz and T. Platt. Can. Bull. of Fish. and Aq. Sci. 213.

Ricker, W. E. 1954. Stock and recruitment. Jour. of Fish. Res. Bd. of Can. 11:559 – 623.

Sakai, S. 1973. A model for group structure and its behavior. Biophysics (Japan) 13:82 – 90.

Steele, J. H. 1978. Some comments on plankton patches. In: Spatial pattern in plankton communities. pp. 1 – 20. Ed. by J. H. Steele. Plenum, New York.

Weber, L. H., El-Sayed, S. Z., and Hampton, I. 1986. The variance spectra of phytoplankton, krill and water temperature in the Antarctic Ocean south of Africa. Deep-Sea Res. 33:1327 – 1343.

Part Two:
Regional Case Studies—
Stress and Mitigation of
Large Marine Ecosystems

Long-Term Variability in the Food Chains, Biomass Yield, and Oceanography of the Bay of Bengal Ecosystem

S. N. Dwivedi

Introduction

The Bay of Bengal is a distinct tropical ecosystem situated in the monsoon belt. The southern part of the bay merges into the Indian Ocean and the northern part on the east and west is surrounded by the Asian continent. The major rivers drain 200 km^3 of water and 12.0 x 10^9 tons of silt during the monsoon season, which influence and govern the ecosystem's dynamics. The bay has positive water balance and high precipitation; this causes silting of harbors, formation of sand bars near river mouths, and creation of the Bengal Fan. Small subsystems, wetlands, marshes, mangroves, backwaters, and coastal lakes play an important role in the overall productivity. The last few decades have witnessed anthropogenic changes in coastal waters and also in the Bay of Bengal. Around the globe, studies of different LMEs indicate that these changes are caused by the shifting of populations, excessive stress on fish stocks from natural causes and fishing by man; and environmental conditions causing changes in currents, dynamics, productivity, and coastal pollution (Sherman, 1990). The morphology and ecology of lakes connected to the Bay of Bengal, such as Lakes Chilka (Orissa) and Pulicut (Tamil Nadu), have changed. The exchange of seawater has become restricted, the area and volume of the lakes have been reduced, and the biodiversity of fauna and flora has been adversely affected. The brackish water character is changing because of a predominantly freshwater influence. There is a danger that in a few decades, Pulicut Lake might become a wetland. On the other hand, because of a change in hydrology, some areas are becoming more saline. More than 25% of

the mangroves in Sunderbans have died. Multispecies communities are changing to single-species dominance. The pollution of coastal areas, which serve as nursery grounds for commercially valuable species of prawns, might ultimately affect the stocks of prawns in the bay. Along the Kerala coast, the catch of prawns has decreased because of pollution, and conservation measures are now necessary.

In the Hooghly estuary of West Bengal, some freshwater fishes spawn during the monsoon; endemic estuarine fauna and migratory marine species also come to spawn. In the last three decades (1960 – 1990) there have been catastrophic changes. *Hilsa ilisha*, an anadromous fish that used to constitute about 70% to 80% of total fish landings in the Hooghly, is disappearing. Changes in salinity and hydrology cause large shoals of *H. ilisha* to congregate near the head of the bay, but they do not enter the estuary. Marine catfish have become dominant in coastal areas. Considerable research has been conducted in the Bay of Bengal by coastal institutions, but data are lacking. In the absence of comprehensive oceanographic data on the bay, information from the east coast of India and the Andaman Sea have been used.

Freshwater input through the estuaries influences the coastal surface waters of the bay. During the southwest monsoon especially, the coastal circulation is influenced by the entry of freshwater- and wind-induced currents. This pattern changes during the northeast monsoon when the current direction changes to southwest. In this process, nutrients from terrestrial runoff increase the overall nutrient levels; however, this is not reflected by an increase in primary and secondary production. These changes

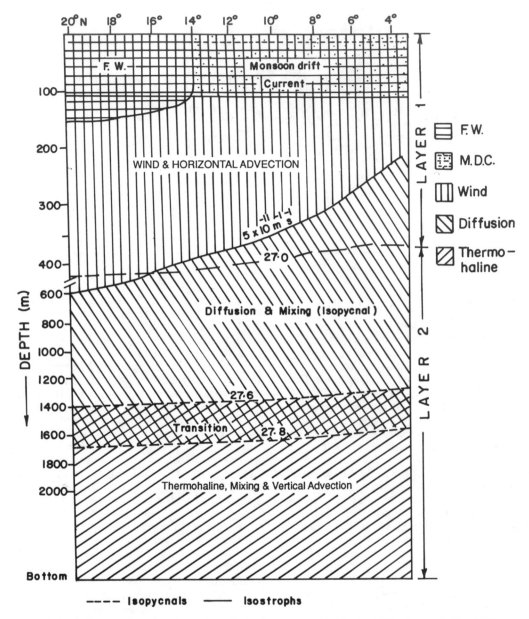

Figure 1. Schematic representation of the depth limits (in meters) of influences of freshwater influx (F.W.), monsoon drift current (M.D.C.), and wind forcing (Wind), together with the dominant processes operating in layers 1 and 2 of the Bay of Bengal.

are seen down to 100 m depth. At 200 m and 500 m deep, high-salinity, low-temperature waters that have low oxygen are present throughout the year (Fig. 1). Suryanarayana (1988) has described the influence of freshwater influx in the Bay of Bengal from the Indian coast.

The waters of the Bay of Bengal show considerable spatial and temporal variability—

the result of seasonal freshwater discharges from India's east coast rivers (Fig. 2). The volume of water discharged is very high, with the majority from the Ganga, Mahanadi, Godavari, and Krishna Rivers (Fig. 3). Eighty percent of the annual discharge enters the coastal environment from July to September during the southwest monsoon. This pattern of seasonal freshwater influx has remained practically un-

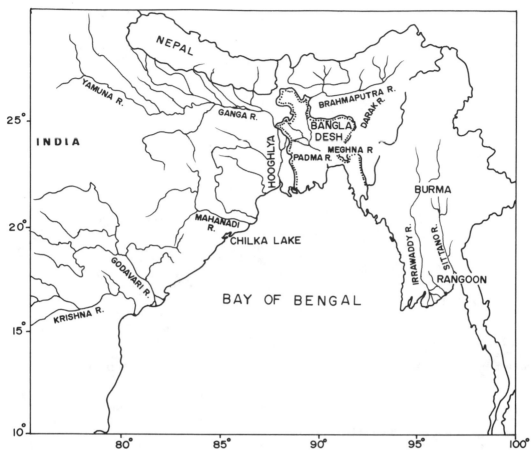

Figure 2. Freshwater drainage of the Northern Bay of Bengal.

changed over the last few decades (Fig. 3). The overall pattern shows a consistent average annual discharge. For the past 60 years, the average annual freshwater discharges from the Ganga, Godavari, and Krishna Rivers have been 15,000 m³/sec, 3,500 m³/sec, and 2,000 m³/sec, respectively. Similar patterns occur off Bangladesh and Burma. The impact of freshwater drainage on the salinity of the coastal and open-ocean waters is controlled by pools of relatively low-saline water, 40 m to 50 m deep.

Under the impact of continuous river discharges, waters with low salinity move toward the south-southwest. These relatively less-saline waters have been found to move offshore around 10° N, where their influence is significant (Suryanarayana, 1988). Within this region, the water density parallels the salinity distribution. The density of the waters at the surface has a larger range in the north (near the head of the bay) during the southwest monsoon and shows a decrease that levels off south of 16° N

latitude. During the northeast monsoon, the general trend is similar, despite minor differences. Data indicate that this pattern appears to have been consistent over the 5 decades from 1901–1960.

Based on the cross-shore component of the wind stress at eight coastal stations, the indices computed (Fig. 4) show a greater prevalence of upwelling in the central area than in the northern or southern areas (Suryanarayana, 1988). Although the forcing required for these vertical motions is favorable in the southern zones, constraints—such as continued freshwater input to the northern zone and the influx of open ocean waters on the narrow shelf of the southern zone—suppress these vertical upward movements. Thus, the major influence of freshwater influx appears to be in surface waters along the coast, primarily in the upper part of the Bay of Bengal. In the coastal areas off Orissa, West Bengal, Bangladesh, Burma, and parts of Thailand, low-salinity, warmer

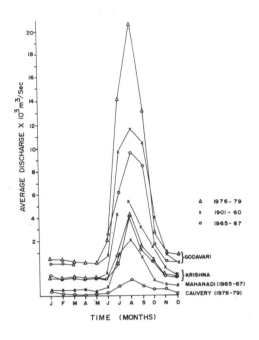

Figure 3. Average monthly freshwater discharge (10³m³/sec) through the Godavari, Krishna, Mahanadi, and Cauvery Rivers (after Suryanarayana, 1988).

waters that are rich in oxygen remain at the surface. At depths greater than 100 m, the influence of monsoon waters is negligible. Deeper than 200 m, the high-salinity, low-temperature water does not change significantly between southwest and northeast monsoons. The combination of surface layers with higher temperature, high oxygen, and low salinity in nearshore areas, and deep waters that are more saline, cooler, and poor in oxygen, have resulted in a unique ecosystem. Three groups of fish can tolerate this oxygen regime: one lives in low-oxygen, deeper waters; a second moves between low- and higher-oxygen waters from deep water to surface layers; and the third lives near the surface in high-oxygen waters.

Subsystems of the Bay

The presence of different water masses in coastal areas has produced small subsystems along the coast that differ in their environmental characteristics and community composition. These coastal areas are also subject to greater anthropogenically induced change, mainly caused by waste disposal and pollution. The subsystems are described below.

Subsystem A, along the Maldives, is char-

acterized by high-salinity oceanic waters with no estuarine influence and is rich in corals and tuna. Subsystem B, south of Sri Lanka and extending along Tamil Nadu, Andhra Pradesh, up to the Orissa border, has a narrow continental shelf with little estuarine influence and higher surface salinity. Between these two regions, there is a productive subsystem, Subsystem C, in Palk Bay, that has a sandy bottom and is rich in the production of clupeids and other pelagic fish. Many decades ago, this area was noted for the production of pearl oysters.

Subsystem D, along Orissa, West Bengal, and Bangladesh, is dominated by an estuarine influence caused by the addition of freshwater and silt. Here, there is a major fishery for a migratory fish, *Hilsa*. The subsystem is characterized by a relatively wide continental shelf, monsoon winds, high precipitation and run-off, surface water temperatures of 20° – 30° C, relatively low salinity of coastal waters, and the presence of low-oxygen, low-temperature, high-salinity waters below 100 m. *Hilsa*, clupeids, catfish, polynemids, and prawn constitute important fisheries in the region. In the same subsystem near Bangladesh, the estuarine influence on the coastal waters increases. The salinity is low (17 – 18 parts per thousand [ppt] in the monsoon season and 30-31 ppt in the dry season) and turbidity is high because of silt. The bottom is alluvial silt and mud out to 40 m depth, where it becomes sandy. There are about 450,000 estuarine acres, rich in mangroves, which serve as nursery and feeding grounds for juvenile prawn. The prawn are collected on a large scale for aquaculture in "Bheries." Collection of juvenile prawn in this manner is reported to adversely affect the prawn catches in the nearshore region. Although *Hilsa* constitutes an important artisanal fishery, precise estimates are lacking. In Bangladesh, the catch increased to approximately 150,000 tons or more, 6% of which comes from the sea. The catch on the Indian side is about 27,000 tons, of which only 22% comes from the sea. *Hilsa* used to be the major fishery in the 1970s, but recent catch levels have dropped drastically. In Burma, the *Hilsa* catch is around 4 to 5 thousand tons.

This general picture indicates a decline in the estuarine fishery caused by pollution and reduced water flow, but *Hilsa* has high fecundity and matures at the end of its first year. Therefore, the fishery has been saved from overexploitation so far. However, adverse changes in the environment and especially increased estuarine pollution are potential threats, and immediate management action is necessary.

The Myanmar (Burma)-Rakhoine coast is narrow, rough, and not actively fished. The

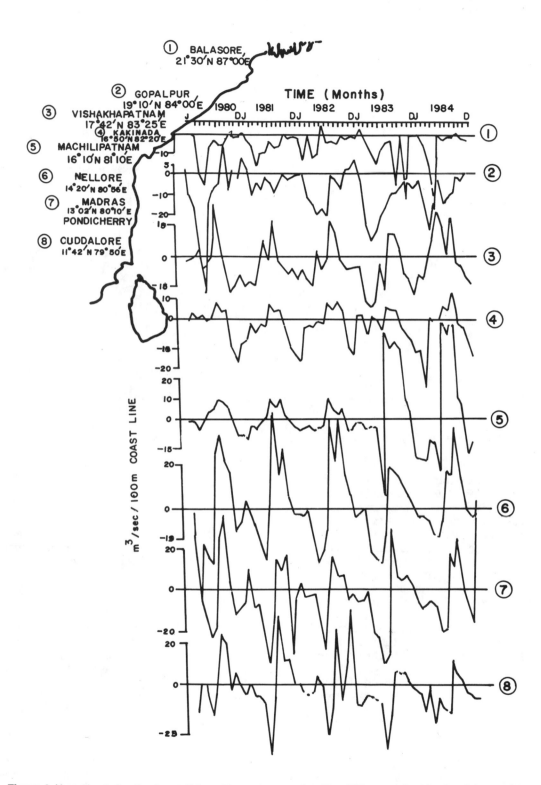

Figure 4. Upwelling index (onshore-offshore Ekman transport in m³/sec/100m coastline) for the eight coastal stations from 1980-1984 (after Suryanarayana, 1988).

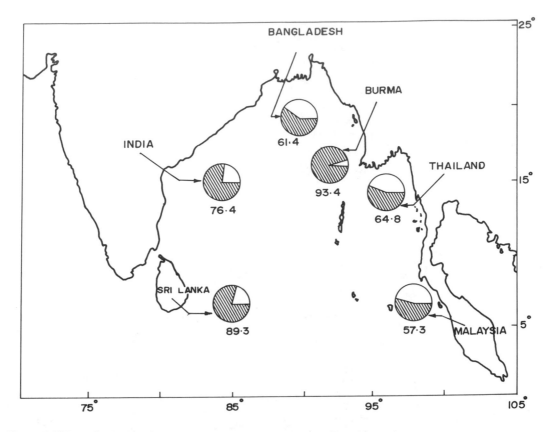

Figure 5. Fish production (%) from nearshore coastal subsystems—Bay of Bengal.

Mergui coast is rocky and hard bottomed. Only the areas up to 120 m deep and beyond 250 m are able to be trawled (Sivasubramanian, 1985). Subsystem E, off Thailand, has a rough bottom topography characterized by rocks and seamounts. The coastal areas support mangroves inshore and hard corals where the depth increases abruptly. In the Malaysia strait, the bottom is rough, with limited trawling areas. The influence of monsoons is not significant here. Oxygen profiles indicate stagnant bottom waters. This area has important pelagic mackerel fisheries that have localized stocks in the strait that spend their entire life cycle in a small region.

The ecosystem comprised of subsystems A to E, described above, has an extensive coastline and a large shelf.

Major Fisheries

The entire coastline along the Bay of Bengal is densely populated by fishermen from developing countries. They still use primitive, nonmechanized gear, and practice artisanal fishing. In all these countries, the artisanal fisheries account for the majority of the catch (89% in Sri Lanka, 76% in India, 61% in Bangladesh, and 57% in Malaysian regions; Fig. 5). This fishing effort is high. Conventional species are nearly overexploited; however, nontraditional species and offshore areas are underexploited.

In all the countries, the coastal prawn fishery is a major export earner. The fishery is conducted using outrigger boats on fishing grounds along the shelf and slope in less than 100 m (Fig. 6). Although reliable data on prawn stocks and their population dynamics are not available, they are being intensively fished, and appear to have almost reached the limit of optimal sustainable yield (OSY). The industry is concerned about overexploitation. Data from India indicate that the growth in total prawn landings has been marginal, thus the real gains for the industry result from an increase in prices. The modal size caught and catch-per-unit-effort (CPUE) indicate that the fishery has reached the OSY level and needs careful management. Prawn constitutes 12% to 20% of the trawl catch, and a bycatch of about 50% to 60% is thrown back into the sea, which is an impor-

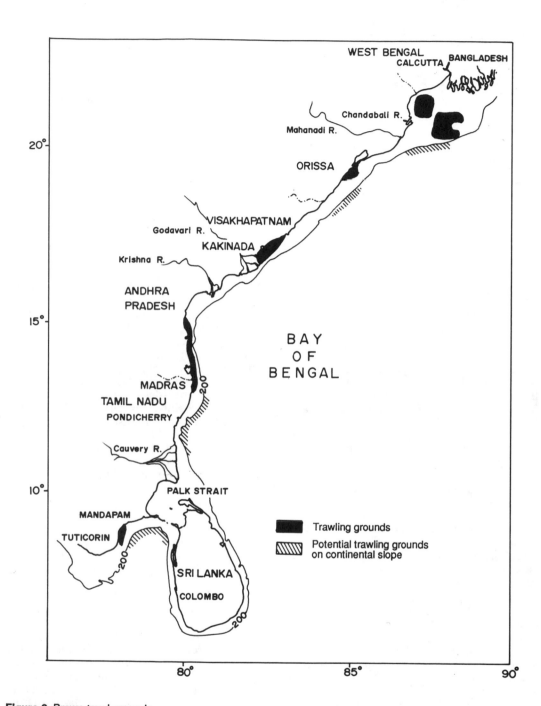

Figure 6. Prawn trawl grounds.

tant limitation. A suitable system for using low-market-value bycatch, a good source of inexpensive protein, has not yet been developed.

The other important fishery is tuna, including yellowfin (*Thunnus albacares*), bigeye (*T. obesus*), longtail (*T. tonggol*), skipjack (*K. pelamis*), little tuna (*E. affinis*), and frigate tuna (*Auxis* sp.). In the countries near the equatorial region—the Maldives, Sri Lanka, and the northwest coast of Sumatra—catches primarily consist of skipjack, little tuna, and yellowfin. Locations of commercial catches indicate that the distribution of tuna is closely associated with hydrographic characteristics of the ecosystem.

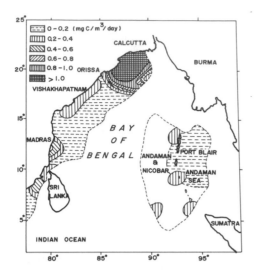

Figure 7. Primary productivity at the surface. (Source: Oceanographic Atlas of EEZ [Exclusive Economic Zone] of India, National Institute of Oceanography [NIO], 1982)

The fishery for oceanic species extends into the Bay of Bengal up to 15° N, the limit to which the bay is oceanic in character. In the areas further north, *E. affinis* and *Auxis* sp. dominate the catches.

Benthic, phytoplanktonic, and zooplanktonic production are also higher in coastal areas because they receive the nutrient-rich water from land sources (Fig. 7). The phytoplankton and zooplankton at the surface also show strong seasonal trends.

The rivers and estuaries also drain pollutants, agricultural pesticides, and industrial waste. Heavy deposits of pesticides, exceeding permissible limits, are found along the coast. Near coastal cities and ports, the deposition levels are higher, posing a threat to the coastal ecosystems.

There is some evidence that storm surges and cyclones are becoming more frequent and intense in nature. They pose a threat in the northeast and northwest parts of the bay. Storm surges and cyclones are accompanied by wind-induced water movements that result in upwelling and long-term changes in productivity. If there is a sea-level rise, the area most likely to be affected is coastal Bangladesh (Fig. 8).

Hydrography and Productivity

The surface water temperature along the coast varies from 24.8° – 30.8° C. Surface temperatures in Andaman and Nicobar waters range from 27.0° to 28.5° C. The average temperature

in a one-degree grid at 100 m ranged from 12° to 18° C. The salinity of the Bay of Bengal is lower and highly variable. Recently, with greater access to research vessels, new information about water masses in the bay has become available.

A review of the literature indicates that studies on the oceanic properties of the bay were initiated in the nineteenth century by Colonel Sewell onboard H.M.S. *Investigator*. Subsequently, the International Indian Ocean Expedition, during 1960 – 1965, collected hydrographic data to determine circulation patterns.

The Bay of Bengal receives large quantities of freshwater and also is subject to periodic, steady wind-forcing as discussed earlier. As the waters of the bay open to the south into the Indian Ocean, the equatorial current systems also influence the hydrography of the bay. To date, studies quantifying the depth influences of freshwater input, wind-forcing, and equatorial currents on the dynamics of these waters have not been attempted.

The research vessels, *Gaveshini* (1976) and *Sagar Kanya* (1983), conducted systematic surveys in the Bay of Bengal to collect seasonal hydrographic data to determine the thermohaline characteristics of the ecosystem. Analysis of data obtained through one survey during the 1984 southwest monsoon season onboard the research vessel *Sagar Kanya* enabled documentation of the following hydrographic features (Murty *et al.*, 1990).

1) Freshwater input influences waters of northern, eastern, and western bay regions. The vertical extent of the influence reaches to depths of 90 m at the head of the bay and west of the Andaman-Nicobar Islands, 30 m in the central bay, and 50 m in the western Bay of Bengal. As a result of intense influx of freshwater along the 88° E meridian, the dilution effects could be seen to 200 m in the north bay (Sastry *et al.*, 1985).

2) The wind-forcing over the bay is strong during the southwest monsoon and induces divergence (upwelling) in the central bay. The depthwise influence increases from 200 m in the south bay to 600 m in the north bay, where the upper 200 m is influenced by the freshwater input.

3) Below the wind-influenced layer, the water column can be separated into two layers with a well-defined transition. The diffusion and lateral mixing act together within the water column immediately below the depth of wind influence and 1400 m. A deep thermo-

haline layer exists below 1700 m where vertical advection dominates.

4) The influence of the equatorial monsoon current extends up to 14° N latitude, in the upper 100 m. This current enters the bay from the southwest, allowing high-salinity waters of the Arabian Sea to penetrate into the bay.

5) The surface circulation is characterized by a large cyclonic gyre influenced largely by the monsoon current south of 15° N. This gyre is surrounded by anticyclonic cells influenced by the freshwater input.

The average phosphate in coastal surface waters was found to be 1 g/l near Calcutta, with a range of 0.03 – 2.37 g/l. In Andaman and Nicobar waters, the range of phosphate concentrations at 200 m was 1.5 – 2.6 g/l. The average maximum nitrate concentration in surface waters of the east coast is a little higher than that of the west coast and Andaman and Nicobar Islands, with the maximum found along Orissa and the west Bengal coast. The bay is characterized by the presence of coastal eddies and the absence of large-scale upwelling, and consequently, it is less productive than the Arabian Sea. Coastal waters are under the estuarine influence of low-salinity and high-nutrient input that supports high primary production. The average water column primary production is 435 gC/m^2/yr on the east coast, 286 gC/m^2/yr for the Andaman Sea, and 348 gC/m^2/yr for the west coast of India, respectively. The Andaman Sea is less productive, which may be because of high-salinity, low-phosphate, and low-oxygen values. The coastal waters are more productive because of estuarine influence, but the absence of major upwelling processes does not permit a high level of production to be obtained. The central parts of the bay are less productive because of the absence of large-scale mixing or upwelling. Secondary production data indicate that average zooplankton production in the coastal areas of the bay is also poor.

A generalized picture of fish catch indicates that a depth range of 100 – 200 m is the most productive. The catch-per-hour (kg) is 24.3 up to 100 m; 3.6 from 100 to 200 m; and 29.3 from 200 to 300 m. The data also show a slight increase from south to north. The catch at 12° N is 25 kg/hr; at 15° N, 151 kg/hr; at 18° N, 223 kg/hr; and at 19° N, 310 kg/hr. Fish catch also indicates that the nearshore subsystem is productive because of general oceanographic conditions. The depth range of 100 – 200 m shows highest production. However, pelagic and water column productivity in waters from 0 to 50 m is higher than from 50 to 100 m. At 16° N, it was 185 and 115 kg/hr, and at 17° N it

was 86 and 30 kg/hr, respectively. The fish catch of the bay was 1.9 million tons in 1981 and increased to 2.4 million tons in 1987.

There have been ecological changes in the estuaries and coastal areas caused primarily by silting and pollution; however, these have not affected total production trends in the ecosystem. The differences became more apparent when we compared the Indian coastline with other areas. Fish catch from India's Exclusive Economic Zone (EEZ) in the Bay of Bengal has been consistently less than that from the Arabian Sea, a difference on the order of 50%. Catch rates at 12°, 16°, and 19° N, in depths of 0 – 50, 50 – 100, 100 – 200, and 200 – 300 m indicate that the 50 – 100 m zone is the most productive.

The benthic standing crop is an important component in the food chain and is used to assess productivity. The water column down to 30 m has high biomass production, with biomass decreasing with water depth. The mean benthic biomass is 7.32 gm/m^2 in the Andaman Sea and 5.32 gm/m^2 in the Bay of Bengal. In the shelf and nearshore regions, macrofauna dominates, whereas in the slope up to 100 m depth, microfauna is dominant. Biomass decreases from north to south. The high mean value of 45 gm/m^2 is found in the northernmost part of the bay and is attributed to enrichment from riverine flows. Lower biomass values in the bay are associated with the oceanographic regime. How-

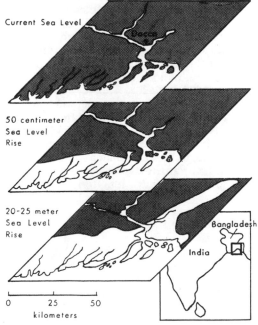

Figure 8. Effect of various heights of sea-level rise on low-lying Bangladesh. (Courtesy of United Nations Environment Fund)

ever, the area south of the Andaman Sea, between 10° and 11° N latitude, is an exception, with a higher biomass that may be caused by local factors. From 0 to 60 m, the macrobenthos account for 70% to 80% of the biomass, whereas beyond 200 m, the microbenthos constitute more than 70% of the standing crop. The areas off the estuary mouths also show high standing crops. The mean values off the Hooghly, Mahanadi, Godavari, and Krishna Rivers are 2.48, 16.04, 13.64, and 0.11 gm/m^2, respectively, and indicate higher production off the Mahanadi and Godavari, which may be attributed to inputs of organic matter from the backwaters and mangrove swamps.

The general pattern, which indicates that the coastal waters are more productive than the open ocean, has not changed. However, the health of the estuaries and coastal waters has deteriorated during the last 5 decades, as a result of pollution near cities and estuaries. More than 25% of the mangroves in the Sunderbans area have died, and species diversity has been reduced. Chilka Lake has become more shallow because of siltation, and, consequently, has become a "threatened subsystem." In the southern part of the Bay of Bengal, the fish catch is dominated by tunnies, seer fish, and carangids. During the last 3 decades, there have been annual variations, but the long-term changes in species composition and total fish catch are not noticeable. However, in the northern part, the catches of catfish have increased, which indicate a change in the species composition of the food chain. Meteorological data also show that there has been an increase in the number and intensity of storm surges and cyclones. These cyclones cause upwelling, which generally results in increased catch of shrimp production. The total fish catch from 1960 to 1989, as well as oceanographic parameters, nutrients, oxygen, temperature, and primary productivity, did not indicate any significant change in the bay proper; however, there have been significant changes in the nearshore region out to 50 m. The area between 50 and 100 m has remained substantially productive and did not show any long-term changes.

Conclusion

In the Bay of Bengal, hydrographic and ecological studies have been initiated, but there is a need for more intensive, integrated investigations of this ecosystem.

The Bay of Bengal is predominantly affected by monsoons, which are accompanied by strong coastal winds, and result in a large input of freshwater and silt. In the nearshore area, the deepwater layer, which is more saline and cooler, is present at more than 100 m and supports an upper low-salinity, warm-water layer. Consequently, productivity is not high.

The coastal waters are less saline, are subject to greater pollution, and are affected by storm surges and cyclones. In some nearshore areas, the ecosystem is under great threat from pollution, whereas offshore areas are less productive and underexploited, particularly at depths below 100 m. The coastal areas have large variations and constitute four small subsystems, each requiring different management approaches.

Oceanographic research in the Bay of Bengal has yielded only fragmented information. To develop a better understanding of the long-term changes in this interesting and complex ecosystem, which is likely to be more severely affected as a result of global warming and sea-level rise, it is essential that plans be initiated for a more comprehensive study of the ecosystem.

References

Murty, V. S. N., Sarma, Y. V. B., Suryanarayana, A., Babu, M. T., and Sastry, J. S. 1990. Report of the physical oceanographic characteristics of the Bay of Bengal during southwest monsoons. Technical Report. NIO/TR-8/90, National Institute of Oceanography, Goa, India.

Sastry, J. S., Rao, D. P., Murty, V. S. N., Sarma, Y. V. B., Suryanarayana, A., and Babu, M. T. 1985. Watermass structure in the Bay of Bengal. Mahasagar 18(2):153-162.

Sherman, K. 1990. Productivity, perturbations, and options for biomass yields in large marine ecosystems. In: Large marine ecosystems: Patterns, processes, and yields. pp. 206-219. Ed. by K. Sherman, L. M. Alexander, and B. D. Gold. AAAS Symposium. AAAS, Washington, DC. 242 pp.

Sivasubramanian, K. 1985. Marine fishery resources of the Bay of Bengal. BOBP/WP36 (RAS/91/051). Marine Fishery Resources Management, Sri Lanka. 66 pp.

Suryanarayana, A. 1988. Effect of wind and freshwater discharge on hydrography and circulation of the western Bay of Bengal. Ph.D. diss., Andhra University, Waltair, India. 91 pp.

Chapter 8

Effects of Physical and Biological Changes on the Biomass Yield of the Humboldt Current Ecosystem

Jurgen Alheit and Patricio Bernal

Introduction

The Humboldt Current ecosystem off western South America is one of the major upwelling systems of the world. It is dominated by four pelagic schooling fish species: anchovy, Eng*raulis ringens*; sardine, *Sardinops sagax*; horse mackerel, *Trachurus murphyi*; and mackerel, *Scomber japonicus*. Over the last 3 decades, fisheries in the Humboldt Current have experienced drastic changes in yields and species dominance. The most spectacular case, which is well documented in many fisheries textbooks, is that of the Peruvian anchovy fishery, once the largest fishery in the world with 12.5 million tons landed in 1970 (Tsukayama, 1983). According to Castillo and Mendo (1987), the real catch amounted to 15 million tons in 1970, taking into consideration that underreporting of the real catches was frequent.

The pelagic fish stocks of the Humboldt Current are exposed to three principal factors governing their strong fluctuations: (i) drastic short-term perturbations of their environment, typical for this ecosystem (El Niño-Southern Oscillation [ENSO]); (ii) fishing pressure; and (iii) subtle long-term changes in the environment.

The nature and consequences of ENSO events have been described by physical oceanographers and biologists in numerous articles (for key references, see Arntz and Tarazona, 1989). The Peruvian anchovy fishery has been analyzed recently from an ecosystem perspective in two books (Pauly and Tsukayama, 1987; Pauly *et al.*, 1989).

Descriptions of the fluctuations of the pelagic fisheries in the Humboldt Current have been somewhat limited because they often concentrated on a single national fishery. Furthermore, the impact of subtle long-term changes in the environment on fish stocks and fisheries has rarely been taken into account. Our objectives are (i) to view fluctuations of pelagic fisheries and their environment simultaneously for the entire Humboldt Current, including fisheries of Ecuador, Peru, and Chile, and (ii) to point out possible impacts of environmental changes other than El Niño.

Physical Characteristics

Currents

In common with the other large eastern boundary currents, the Humboldt Current is characterized by the principally equatorward flow of cold, low-salinity waters. However, this general description simplifies a rather complex set of flows and counterflows that occur semipermanently between the coast and 1,000 km offshore.

The oceanic sector is dominated by a sluggish, wide flow toward the equator with characteristic velocities of about 0.04 ms^{-1} coinciding with the boundary of the anticyclonic gyre of the Southeast Pacific and which has been recognized as the Chile-Peru Oceanic Current (Robles *et al.*, 1980). Toward the coast, a counterflow to the south, the Peruvian Oceanic Countercurrent (Wyrtki, 1967), has been identified at 79° W off Peru and between 76° and 77° W, 500 km off the Chilean coast. This counterflow has characteristic velocities of about 0.06 ms^{-1}. The faster Humboldt Current is located somewhat closer to the coast, between 300 and 400 km offshore, with a core at about 200 m depth. The Humboldt Current often reaches velocities greater than 0.18 ms[-1]. Between the core axis of the Humboldt Current and the coast, three other branches can occasionally be distinguished. They are relatively unstable and show clear seasonal fluctuations. The more permanent branch is a counterflow to the south with a velocity core at about 100 m depth off Peru and at 200 m depth off Chile. This counterflow is poor in oxygen (< 1 ml/l) and rich in nutrients. It corresponds to the Gunther Current or Peru-Chile Undercurrent described by several authors (Brockmann *et al.*, 1980; Robles *et al.*, 1980). This obvious heterogeneity of flows demonstrates that seasonality and mesoscale processes are dominant in this eastern boundary current system. Satellite imagery (Espinoza *et al.*, 1983; Fonseca and Farias, 1987) has shown a rich spectrum of plumes, eddies, filaments, and other transient structures that occur regularly in this region and point to intense horizontal mixing. Given the methods applied to study circulation in this region (mainly through geostrophy), some of the discrete flows identified in the past could in fact correspond to aspects of these complex mesoscale phenomena. Silva and Fonseca (1983) propose a more parsimonious interpretation of the circulation patterns off northern Chile, which are characterized by three main flows that fluctuate between seasons.

Upwelling and other localized processes

Coastal upwelling has been identified as the most important and relevant process in the eastern boundary currents, presumably being almost exclusively responsible for the extremely high levels of organic production. In Peru, coastal upwelling reaches a maximum during winter and a minimum during summer. In northern Chile (18° – 30° S) it peaks during spring, and off central Chile (30° – 38° S) it peaks during late spring and summer. This temporal progression of coastal upwelling from the north to the south results from the displacement of the subtropical center of high atmospheric pressure that intensifies and moves southward as summer progresses. Therefore, off Peru, upwelling occurs all year-round, whereas, in central Chile, it is restricted to spring and summer.

The upwelling paradigm has overshadowed the importance of other phenomena that control distribution and abundance of organisms and explain the location of spawning and feeding centers of fish and, thus, of fishing grounds. The eastern boundary of the Pacific has a comparatively narrow shelf off Peru and an extremely narrow shelf off Chile. However, wherever a significant shelf is present, it is associated with important mesoscale structures resulting from the interaction of circulation with bottom topography and the coastline. Frontal structures develop at the coastal upwelling centers as the pycnocline rises to the surface close to the coast and "breaks" the surface in a zone where hyperbaroclinic conditions are established (Mooers et *al.*, 1978). When upwelled waters are not modified fast enough by mixing, they will necessarily sink at the frontal boundary and so retain nutrients, plankton, and detritus within the upwelled plume. These phenomena have been observed along the entire coast of the Humboldt Current. They play an important role for the maintenance of the spawning grounds of the coastal neritic sardine and anchovy and for the feeding grounds of the more oceanic horse mackerel.

Eastern boundary currents have received the close attention of the international oceanographic community after the recognition of the global character of ENSO. ENSO causes a clear oceanographic signal along both eastern boundaries of the Pacific. The ENSO events of 1957 – 1958, 1972, and 1982 – 1983 showed a remarkable bihemispheric manifestation (Bernal, 1981; Wooster and Fluharty, 1985). When analyzing the long-term variability of the Humboldt Current system, the ENSO signal will, without doubt, appear as one of the dominant features. However, despite the interannual periodicity of its appearance, its duration is rather short (about a year). For the ecosystem, its evolution, and dynamics, ENSOs have to be viewed, therefore, as a typical perturbation—in other words, it is closer to a delta function than to a step function.

Biomass and Catch Data

The acquisition of reliable data on the total biomass of the four dominant pelagic species for the entire period of their fisheries in the Humboldt Current, which started in the early 1950s, proved to be a problem. Catch data are available; however, they are often based on serious underreports of the true catches (Menz and French, 1982; Castillo and Mendo, 1987). The official catch data, as published by the Food and Agriculture Organization of the United Nations (FAO) and the fisheries departments of the three countries involved, are used here. Although aware of the problem of underreporting, we believe that the official statistics constitute the best database available at present, and we assume that they reflect major trends in the fluctuations of fish stocks after the fishery reached a certain level. We are also aware of other factors having a strong influence on trends in catch data, such as improved fishing methods, changes in mesh size, and changes in vulnerability of fish caused by oceanographic factors.

Total catches of pelagics

The four dominant pelagic species are considered in this analysis. Other clupeiform pelagic species that might be of local importance on the northern and southern edge of the Humboldt Current are omitted from these studies.

Pelagic fisheries in the Humboldt Current began in the early 1950s, reached a peak in the early 1970s, and showed strong fluctuations during the remainder of the 1970s and 1980s (Fig. 1). The total pelagic yield was almost exclusively influenced by the rise of the anchovy fishery in the 1950s – 1960s, its collapse in 1972, and the subsequent rise of the fishery for sardine, horse mackerel, and mackerel. The usual explanation for these fluctuations is that a combination of extensive overfishing and the strong ENSO in 1972 – 1973 caused the collapse of the Peruvian anchovy, thus allowing the other three species to expand in an environment where the stocks of the dominant species were substantially reduced. In the following, catch data of anchovy and sardine stocks along the Ecuadorian, Peruvian, and Chilean coast are analyzed stock-by-stock in order to identify and compare major population trends.

Anchovy

Three anchovy stocks have been identified in the Humboldt Current ecosystem: (i) the northern and central stock in Peru, (ii) the stock in southern Peru and northern Chile, and (iii) the stock of Talcahuano in Chile (Serra, 1983) (Fig. 2). Anchovy was the dominant species in these three regions until about 1972. The stock in northern and central Peru, which supported the famous Peruvian anchovy fishery in the 1960s and 1970s, is by far the largest and most important stock. Analyses of fish scale abundance from sediment cores show that anchovy was always the dominant species in this area (DeVries and Pearcy, 1982). From 1960 until 1972, its mean biomass was 15 million tons on average (Tsukayama, 1983; Zuta et al.). Biomass peaked in 1967 at about 22 – 23 million tons. The stock decreased dramatically to 9 million tons at the beginning of 1972 and further to 4.1 million tons at the end of 1972 (Tsukayama, 1983; Zuta *et al*). Within a year, the stock lost about 5 million tons because of excessive fishing. The comparatively strong ENSO of 1972 concentrated the anchovy near the coast, where there were still some cold upwelling plumes; this made the anchovy extremely vulnerable to fishing. In 1972, 4.4 million tons of anchovy were caught within 3 months (Tsukayama, 1983). Similar trends in anchovy biomass were shown by a monthly virtual population analysis-(VPA) based time series that included predatory effects on anchovy,

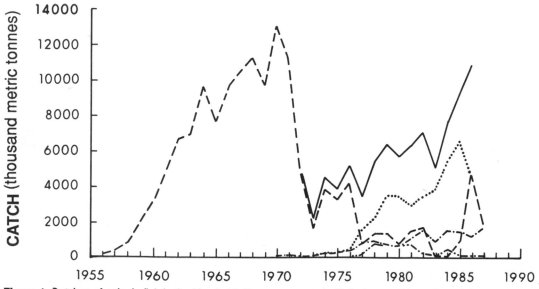

Figure 1. Catches of pelagic fish in the Humboldt Current ecosystem. Pelagics, total (——); anchovy (- - -); sardine (...); horse mackerel (-.-.-); mackerel (-..-..-).

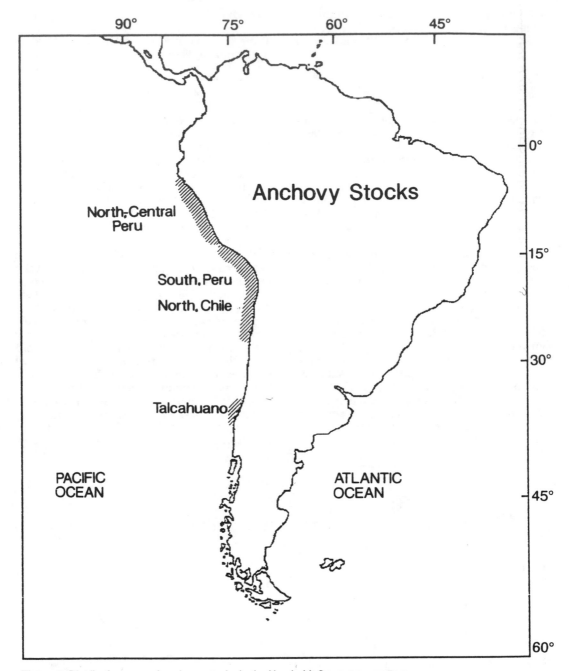

Figure 2. Distribution area of anchovy stocks in the Humboldt Current ecosystem.

developed by Pauly and Palomares (1989). According to their estimates, the anchovy biomass peaked in 1970 with 23 million tons.

For the northern and central stock, anchovy catches peaked in 1970 with 10.9 million tons (Fig. 3). The catches fell dramatically during the ENSO in 1972, remained between 1 and 6 million tons until 1982, and decreased to an extremely low level during the ENSO of 1982 – 1983, and particularly thereafter in 1984. However, in 1985 the stock recovered and catches rose to 3 million tons in 1986.

Several interesting aspects of this sequence in catch and biomass data should be noted:

(1) According to Zuta *et al.* (1983) and Tsukayama (1983), the biomass of

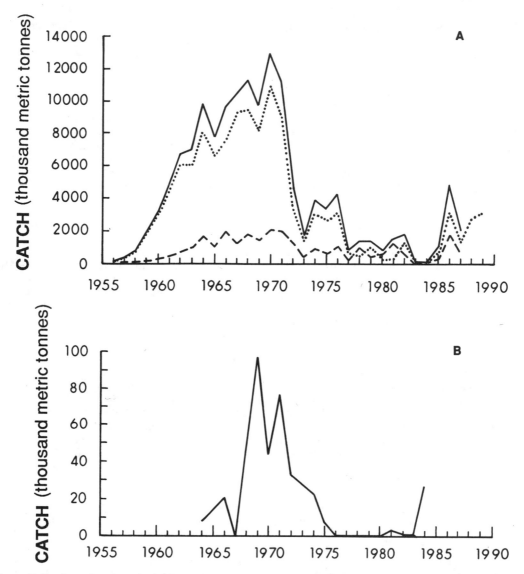

Figure 3. Catches of anchovy in the Humboldt Current ecosystem. A: Anchovy, total (———); northern/central stock off Peru (....); southern stock off Peru and northern stock off Chile (- - -). B: Stock off Talcahuano in Chile.

Peruvian anchovy was 22 – 23 million tons in 1967. It dropped rather rapidly to about 12 million tons in 1969. According to Pauly and Palomares (1989), the biomass peaked in 1970 with 23 million tons and dropped in 1971 to 9 – 14 million tons.

(2) The recruitment collapse of 1971 occurred *before* the onset of the 1972 – 1973 ENSO (Mendelsohn and Mendo, 1987; Mendelsohn, 1989).

(3) For the August-September 1981 period, the spawning biomass of the northern and central stock was esti-mated at 1.2 million tons (95% confidence limits: ± 58%) (Santander *et al.*, 1984) applying the egg production method (Alheit, 1985; Lasker, 1985).

(4) In 1982, the recruitment from the 1981 year-class was very poor (Zuta *et al.*, 1983) and 1.2 million tons of anchovy from the northern and central stock were caught (Fig. 3). It is, therefore, concluded that at the onset of the 1982 – 1983 ENSO, the biomass of this stock must have been extremely low, very likely far below 1 million tons, as a result of excessive fishing pressure

and poor recruitment. Tsukayama and Santander (1987) report that in 1983 Peruvian anchovy had the lowest biomass ever recorded.

(5) This extremely low stock in the northern and central region virtually exploded to several million tons in 1985 and 1986 (probably well above 6 million tons, according to Csirke, 1988), when 670,000 and 3 million tons, respectively, were caught (Fig. 3). It is not known where this excellent recruitment originated. Very few anchovies were caught during the 1982 – 1983 ENSO event. Arntz and Tarazona (1989) showed that some anchovies withdrew to deeper water (> 100 m) to survive. However, Barber *et al.* (1985) reported that phytoplankton concentrations at these depths were too low to support large number of anchovy for any length of time. An alternative explanation might be that the recruitment of the northern and central stock off Peru came from the stock off southern Peru and northern Chile, where the effects of the 1982 – 1983 ENSO were less dramatic.

Whatever caused this surprising recovery of the northern and central stock, it is obvious that the Peruvian anchovy can increase its population size rapidly and drastically when the environmental conditions become favorable. Although the anchovy population is reduced by the effect of ENSO or of "warm" years, ENSO does not seem to have such a long lasting impact on the anchovy, nor are its effects necessarily as drastic and deleterious in the long term, as previously assumed.

Comparing the annual catches of the three anchovy stocks in the Humboldt Current, the almost simultaneous development from 1956 to 1980 is striking (Fig. 3). Catches rose steadily each year from 1956 on and peaked initially in 1964. They dropped simultaneously in 1965 (ENSO). All three stocks supported high-catch levels from 1966 to 1971, and catches of the two large stocks peaked in 1970. Catches decreased again simultaneously in 1972 (ENSO) and entered a period of strong fluctuations until the 1980s. However, catches of the stock off southern Peru and northern Chile remained relatively high and did not show the dramatic, almost steady decline of the stock off northern and central Peru. Catches of all three stocks declined to extreme lows in response to the ENSO event in 1982 – 1983, but rose again considerably in 1986 (Talcahuano data not included). The anchovy landings in northern Chile even reached a historic record in 1986 (1.25 million tons), when a fully developed fishery switched its tar-

get species from sardine to anchovy and benefited from two sudden, strong recruitment pulses in 1984 and 1987.

Sardine

Four sardine stocks have been identified in the Humboldt Current ecosystem (Serra and Tsukayama, 1988): (i) the stock off Ecuador and northern and central Peru, (ii) the stock off southern Peru and northern Chile, (iii) the stock of Coquimbo in Chile, and (iv) the stock off Talcahuano in Chile (Fig. 4).

Sardine spawning (Zuta et al., 1983) and catches were insignificant (Serra, 1983) during the 1950s and 1960s. From 1964 to 1971, the only distinct spawning areas were in northern Peru and in northern Chile (Bernal et al., 1983). After 1971, the sardine expanded to the northern and southern extremities of both refuge areas. Sardine spawning off Peru from 1966 to 1968 was poor and limited to the region between 6° and 10° S (Zuta et al., 1983). After 1969, an increase in spawning was observed. After the ENSO of 1972 – 1973, sardine spawning increased strongly, and the spawning area expanded considerably, although the new spawning areas did not overlap with those previously occupied by the anchovy. From 1976 to 1980, spawning increased further (Zuta et al., 1983). From 1964 to 1973, no eggs or larvae were observed in Chile south of 25° S. However, a new spawning area off Talcahuano was established subsequently (Bernal et al., 1983; Serra 1983) (Fig. 5).

Sardine spawning increased and the geographic distribution of spawn expanded during warm years, such as in 1969 (Santander and Flores, 1983). This phenomenon appears to be related to the more frequent coastal advance of the subtropical surface waters (Santander and Flores, 1983; Tsukayama, 1983).

An increase in the relative abundance of sardine and horse mackerel was observed in 1970 (Subsecretaria de Pesca, 1983). Catches of sardines in the Humboldt Current ecosystem increased after the ENSO of 1972 – 1973 when anchovy biomass and landings dropped dramatically (Fig. 6). Off Coquimbo in 1973, sardine catches increased significantly, and in 1974 the sardine became the dominant species. Anchovies had never been caught in the region off Coquimbo (Mendez, 1987). Significant increases in the catches of the other three sardine stocks were observed several years after the 1972 – 1973 ENSO: in 1976 in the stocks off Ecuador and northern and central Peru, and in the stock off southern Peru and northern Chile, and in 1978 in the stock off Talcahuano (Fig. 6). Catches of the two large stocks rose steadily until 1985. Interestingly, during the ENSO in

1983, the landings from the stocks off southern Peru and northern Chile and off Coquimbo peaked, whereas the stock off northern and central Peru/Ecuador was at a low. This resulted from extensive southward migrations of the sardines apparently trying to escape the warm waters coming from the north and from sardines concentrating close to the coast, thus becoming more vulnerable to fishing (Mendez, 1987; Tsukayama and Santander, 1987). According to Mendez (1987), in 1983, it was no longer possible to separate the sardine stocks from northern and southern Peru. In 1983, during the strongest ENSO event of the twentieth century thus far, Chile landed the highest sardine catches ever recorded, whereas total

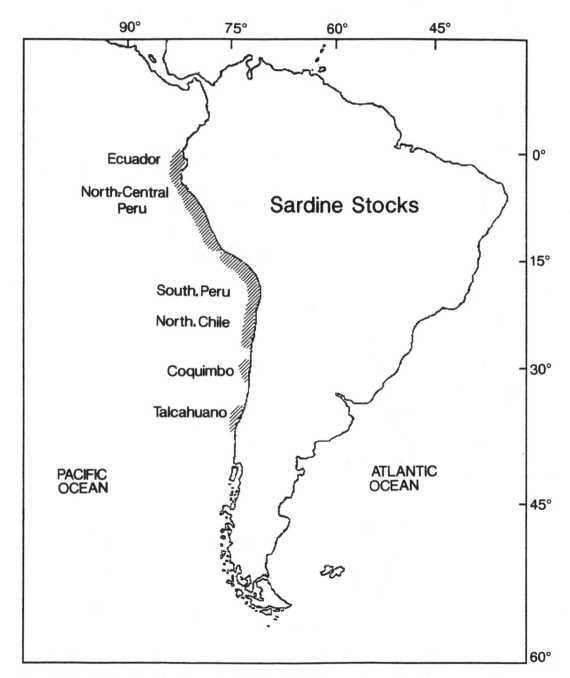

Figure 4. Distribution area of sardine stocks in the Humboldt Current ecosystem.

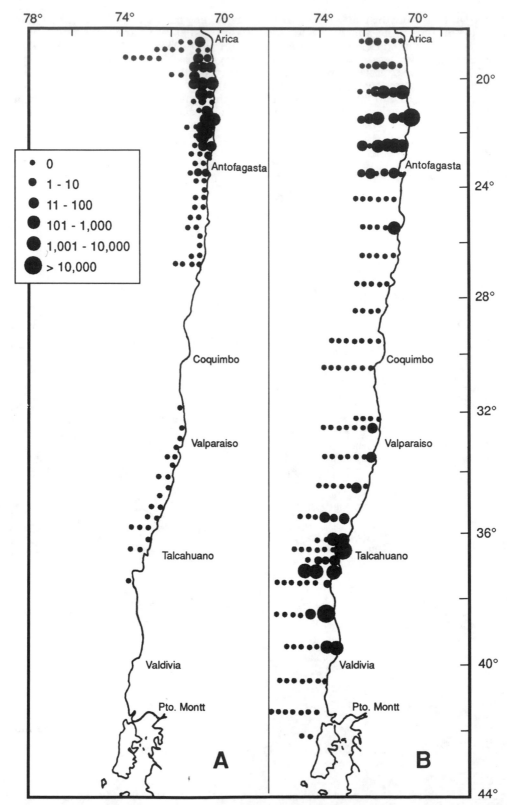

Figure 5. Geographic distribution and abundance of sardine eggs off Chile. Left side (A): August 1963 – 1972. Right side (B): August 1981.

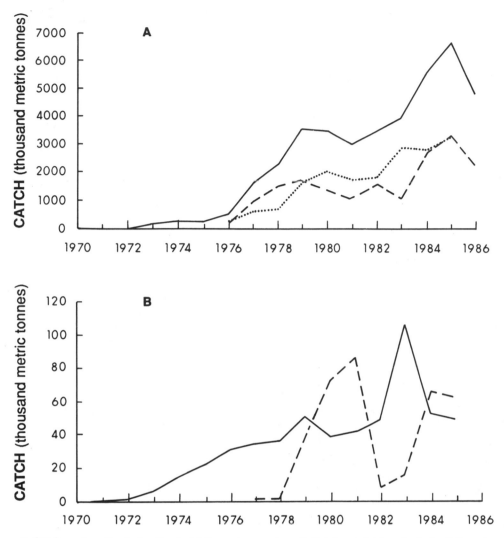

Figure 6. Catches of sardine in the Humboldt Current ecosystem. A: Sardine, total (——); stock off Ecuador and northern and central Peru (....); stock off southern Peru and northern Chile (- - - -). B: Stock off Coquimbo (——); stock off Talcahuano (- - - -).

catches of pelagic fish in Peru dropped considerably. Two years afterward, in 1985, when the sardines had returned northward again, Ecuador had a record year of sardine catches (Serra and Tsukayama, 1988). There is evidence that the sardines caught off northern Chile had migrated considerable distances southward from Peru. Their fat content was reduced by more than 40%, resulting in a low-quality, lower-priced fish meal (Romo, 1985).

Impact of Physical Factors

ENSO is commonly considered as the overriding physical factor influencing yield and spe-

cies composition of the pelagic catches in the Humboldt Current. The year 1972 saw a dramatic drop in anchovy biomass and catches in response to the ENSO event. The subsequent ENSO, in 1976, caused another drastic decrease in anchovy catches. Simultaneously, catches of sardine and horse mackerel surpassed the anchovy landings for the first time (Fig. 1), and after 1977, the sardine became the most important species in the fishery. The strong ENSO event of 1982–1983 reduced the anchovy to a historic low, with catches around 100,000 tons in 1983. Anchovy catches have subsequently risen to medium levels. Whenever there was a warm year, including ENSO years, the anchovy catches dropped and sardine catches increased.

Clearly, El Niño has extremely adverse effects on the anchovy, reducing the survival of all life history stages and increasing its vulnerability to the fishery, whereas it has a positive effect on sardines. Although ENSO events are relatively frequent (DeVries and Pearcy, 1982), the anchovy has always recovered. Therefore, it seems unlikely that ENSO events in the 1970s are the sole cause of the decrease of the anchovy populations and the shift in species dominance in the pelagic fish community of the Humboldt Current. The combined effect of the 1972 – 1973 ENSO, along with heavy fishing pressure, has often been suggested as an explanation for the decline of the anchovy in the 1970s (Csirke, 1988; Muck, 1989a). However, the fast recovery of the anchovy after the strong ENSO in 1982 – 1983 proves otherwise. Most likely, first-order, direct effects of ENSOs on the anchovy and sardine are not sufficient to explain the observed changes, because ENSOs affect all species in the ecosystem. Therefore, the outcome in terms of species composition at one or more trophic levels must be influenced by a complex set of interspecific, second-order interactions and direct effects (Pianka, 1987). These more subtle ecological effects are still poorly understood.

It has been demonstrated here that there are a number of changes in the pelagic fish community that began prior to the ENSO of 1972 – 1973: (i) the recruitment collapse of anchovy in 1971, before the onset of ENSO (Mendelsohn and Mendo, 1987; Mendelsohn, 1989); (ii) an increase in sardine spawning since 1969 (Zuta et al., 1983); and (iii) an increase in the relative abundance of sardine and horse mackerel in 1970 (Subsecretaria de Pesca, 1983).

In recruitment studies of the northern and central anchovy stock off Peru, Mendelsohn (1989) defined "recruitment" as the population estimates of the smallest class (length range = 3.75 – 4.25 cm) in the fishery (Pauly and Palomares, 1989). Mendelsohn's estimates increase from 1963 to about 1969 or 1970, after which they drop dramatically. Recruitment was fairly constant until 1969, peaked conspicuously in 1969, and decreased after 1971 to a new, low equilibrium (Mendelsohn, 1989). The 1960s may have been a period of exceptionally and consistently high recruitment (Pauly and Palomares, 1989).

Mendelsohn (1989) shows a number of time series of environmental variables which he considers in "all likelihood" as surrogates for the actual physical processes that affect the anchovy. Patterns in two of these data sets match the biological trends presented here: (i) a higher mean level of oceanic offshore transport in the 1970s – 1980s than in the 1960s,

which, in addition, appears to have been a decade of decreased variance in transport; and (ii) a slight upward trend in sea-surface temperature (SST) off Peru from 1971 onward.

In Chile, the increase of the sardine stock after the ENSO of 1972 – 1973 is matched by the upward trend of SST off Arica (Fig. 7). This upward trend is in turn reflected by the general warming of the coastal zone off northern Chile, which lasted for several years.

According to Muck (1989b), SST variability reflects the dynamics of other physical properties, such as temperature, upwelling, turbulence, and depth of mixed layer. He studied the standard deviation of SST from simultaneous monthly measurements of nine shore stations along the Peruvian coast. He reported obvious changes in the anomalies of the SST standard deviations after 1968, when the trend switched from mainly negative to positive. The anomalies of the standard deviations of SST determined for the month of October (spawning peak of Peruvian anchovy) were negative or moderately positive until 1968 and from then on mainly positive until 1983. The eight most positive years were in the period from 1969 to 1978 (Fig. 7).

Other groups of organisms have also undergone fundamental changes in their population sizes; the changes appear to have started in the late 1960s. Bernal et al. (1983) demonstrated a decrease in zooplankton biomass off northern Chile beginning in 1969 (Fig. 7). A time series of zooplankton biomass off Peru from 1964 to 1986 shows the same trend (Carrasco and Lozano, 1989). The high peaks of zooplank-

⟶

Figure 7. Time series of several parameters:

A - 12-month running mean of standard deviations associated with mean monthly sea-surface temperature values off Peru (5°-15° S), 1953-1984 (after Loeb and Rojas, 1988). Dotted line: long-term mean values.

B - Anomalies of mean annual standard deviation of sea-surface temperature from nine coastal stations off Peru (4°-17° S), 1953-1987 (after Muck, 1989a).

C - Sea-surface atmospheric pressure anomalies off Arica, Chile (18° S), 1960-1982 (after Loeb and Rojas, 1988).

D - Surface layer turbulent mixing index values off Peru (5°-15° S), 1953-1984 (after Bakun, 1987).

E - Zooplanton volume off Peru, 1964-1987 (after Carrasco and Lozano, 1989). Areas: E1 (3°30'-5°59' S); E2 (6° - 13°58' S); E3 (14° - 18°30' S).

F - Zooplankton biomass (standardized units) off Chile, 1964-1973 (after Bernal et al., 1983).

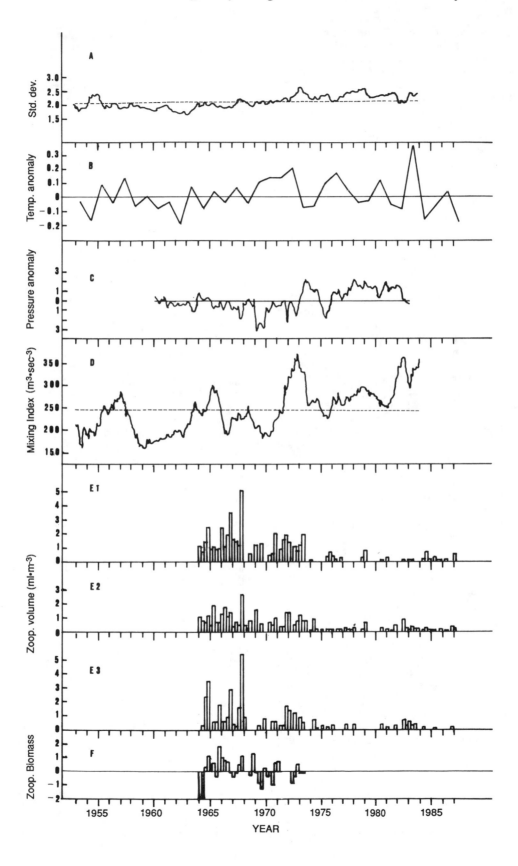

ton volume before 1971 were never again reached in subsequent years (Fig. 7).

The strongest evidence for biological changes occurring in the Humboldt Current prior to 1971 is presented by Loeb and Rojas (1988). They studied the interannual variation of ichthyoplankton composition and abundance relations off northern Chile from 1964 to 1983 and observed a marked shift in the relative abundance of larval fish of the nonfished mesopelagic species from 1969 to 1970. The abundance of mesopelagic species is not influenced directly by fishing activities and might, therefore, in contrast to anchovy and sardine, be a good indicator of environmental changes. Loeb and Rojas (1988) suggested that a subtle, but large-scale, low-frequency, environmental change occurred in the 1969–1970 period. This suggestion is supported by several long-term physical databases from Chile and Peru. From 1960 to 1972, the sea-surface atmospheric pressure anomalies that occurred off northern Chile were predominantly negative, particularly in 1969. After 1972, positive anomalies were predominant (Fig. 7). Off Peru, the wind-driven turbulent mixing index of surface waters generally increased during and after the ENSO of 1972–1973 (Fig. 7). Loeb and Rojas (1988) assumed that increased nearshore influence of the subtropical surface waters might be responsible for the increased larval survival and recruitment of sardines, as previously suggested by Santander and Flores (1983). Loeb and Rojas (1988) concluded that an atmospherically driven oceanic circulation change, beginning in the late 1960s and possibly involving onshore presence of subtropical or oceanic waters (or both), might be responsible for the shifts observed in zooplankton and larval and adult fish occurring in the late 1960s. They suggested that the fluctuations in anchovy and sardine populations are primarily governed by low-frequency hydrographic events augmented secondarily by fisheries.

Important changes in the Humboldt Current ecosystem, such as the collapse of the anchovy and the rise of the sardine after the ENSO of 1972–1973, have been associated by many authors with the occurrence of El Niño. We are suggesting that there are other, more subtle, trends in the physical oceanography of the system that might play a similar or greater role in the changes described for the Humboldt Current. These trends are of lower frequency than the occurrence of the ENSO, and therefore modify local conditions for longer periods of time. Such trends have often been overlooked because they are difficult to observe and extract from the noisy record of environmental variables. The frequency domain in time and space of these low-frequency changes is almost inaccessible in regions where geophysical time series are short, sporadic, or of low quality. In addition, we also suggest that the changes occur through direct and indirect effects among the several species inhabiting the system. Indirect effects include exploitative competition, predation, apparent competition, indirect mutualism, and other processes, including second- and third-order interactions (Pianka, 1987).

In spite of the remarkable interest of and research effort applied by the international oceanographic community to the physical processes governing the Humboldt Current ecosystem, there are still important gaps in our knowledge. For example, all data available indicate that during ENSO years global atmospheric forcing of currents dominates. During non-ENSO years, local forcing could play a more important role. The mechanisms of local forcing in the Humboldt Current are not well understood and definitely require more attention in the future (E. Hagen, Institute for Baltic Research, Warnemunde, Germany, pers. com., 1990). In the studies of global forcing, particular attention was usually given to the impact of equatorial currents on the Humboldt Current and the Peruvian Undercurrent. However, the impact of changes in the southern currents (e.g., Circumpolar Current and West Wind Drift) on the Humboldt Current has been greatly neglected so far (E. Hagen, pers. com., 1990). Basin-wide atmospheric forcing and its interannual variability must play a role in controlling these large-scale, low-frequency phenomena. Lluch-Belda *et al.* (1989) reached similar conclusions when comparing worldwide fluctuations of sardine and anchovy stocks. They suggested that long-term fluctuations of these species seem to be dominated by environmental variations that can give rise to large and prolonged changes in biomass and yield, thus causing persistent "regimes" of high and low abundance of the species. The apparent coherence in trends of various sardine stocks on an oceanic scale leads the authors to conclude that the mechanisms responsible for long-term changes in stock sizes, which are not yet clear, are operating also on oceanic scales.

The Role of Predation

The impact of predation on all life history stages of the anchovy has received considerable attention recently.

Egg mortality rates of Peruvian anchovies and sardines can be extremely high. Smith et *al.* (1989) showed that the instantaneous rates of daily egg mortality (Z) in August–September 1981 were, depending on the region, between 0.90 and 2.14 for anchovies and between

1.28 and 3.88 for sardines off Peru. At times, anchovies feed heavily on their own eggs (de Ciechomski, 1967; Hunter and Kimbrell, 1980; MacCall, 1981; Alheit, 1987; Valdes *et al.*, 1987). A considerable part of the high egg mortality in anchovy results from cannibalism. Twenty-two percent of the total egg mortality at peak spawning time of Peruvian anchovy in August – September 1981 was caused by cannibalism (Alheit, 1987); 28% in Californian anchovy (Hunter and Kimbrell, 1980; MacCall, 1981); and up to 70% in South African anchovy, in an area of particularly high adult density (Valdes *et al.*, 1987).

Whereas in August – September 1981, 76.4% of the Peruvian anchovy had anchovy eggs in their stomachs (Alheit, 1987), similar studies in February 1985, the period of the secondary spawning peak, revealed that only few anchovies had fed on their eggs (Alheit, unpubl.). Initially, this is surprising, because the anchovy biomass in 1981 was 1.2 million tons (Santander *et al.*, 1984) and around 3 million tons in February 1985 (Pauly and Palomares, 1989) and, as one might expect, lower cannibalism rates at lower biomass values operate as a compensatory mechanism. However, anchovy eggs were hardly encountered further offshore than 30 miles from the coast in 1981 (Santander *et al.*, 1984), but were found up to 90 miles offshore in 1985. Population density, rather than biomass, appears to be the decisive factor influencing egg cannibalism rates. In addition, anchovies tend to concentrate near the coast (Jordan, 1971) during peak spawning in August – September.

These data are interesting in view of the findings of Mendelsohn and Mendo (1987), who analyzed seasonal recruitment success of the Peruvian anchovy from 1953 to 1981. They found that the annual recruitment peak stems from the secondary spawning peak in January – February, whereas the peak spawning in August – September leads only to few recruits.

Pauly (1987) suggested that egg mortality from cannibalism is a strong density-dependent mechanism controlling recruitment of the Peruvian anchovy. Smith *et al.* (1989) discussed thoroughly whether or not interannual differences in the rate of egg cannibalism are sufficient to explain differences in recruitment, and they concluded that this would be unlikely.

In addition to anchovy, other filtering fishes, such as sardines, juveniles of other species, and invertebrates prey heavily on anchovy eggs. Predation by euphausiids on anchovy eggs off California accounts for 28% of the total egg mortality (Theilacker, 1988), and the mean number of anchovy eggs in sardine stomachs (per unit of fish weight) was more than double that in anchovy stomachs in Peruvian

waters (Alheit, 1987). Clearly, the impact of predation on eggs and larvae of anchovies and sardines needs further study.

Until recently, it was assumed that the guano birds were the major predators of anchovy in the Humboldt Current (Jordan, 1967; Schaefer, 1970; Murphy, 1972; Furness, 1982). However, Muck (1989a) recently showed that predation by horse mackerel, mackerel, and hake far exceeds the consumption of anchovy by birds. He assumed that predation of horse mackerel on the anchovy of the central and northern stock throughout the 1950s and in the mid-1970s surpassed the anchovy catches of the fishery (Muck and Sanchez, 1987). Because horse mackerel and mackerel are usually distributed farther offshore than anchovy, predation is particularly heavy in years of higher SST, which brings the horse mackerel closer to the shore and leads to increasing spatial overlap of both species. According to Muck (1989a), coastal SST appears to be an appropriate parameter for quantification of the distribution overlap between both species, and he suggested that SST-mediated distribution patterns of anchovy and horse mackerel largely determine the dynamics of the anchovy stocks. Consequently, he proposed a flexible exploitation scheme for Peru according to which anchovy should be fished particularly when its biomass is high and SST is low, and horse mackerel should be exploited heavily when SST is high.

Conclusion

After consideration of the impact of several physical and biological factors on the dynamics and yield of pelagic fish in the Humboldt Current, the question arises regarding management advice could be given. The situation is complicated by the fact that most of the stocks are shared by two countries. The alongshore migrations of the pelagics, which are governed by hydrographic conditions, often lead to unusually high concentrations of fish in one country, which might cause excessive (over)fishing by this country, much to the dismay of the other country. In addition, certain environmental situations, such as ENSO, that have adverse effects on the fishery in one country may increase yields in another country, as demonstrated by the decrease of anchovy catches in Peru and the simultaneous increase of sardine landings in Chile in 1982 – 1983. Furthermore, during the period when the yields of pelagic fish steadily declined in Peru in the 1970s, and early 1980s, total catches of pelagic fish in the northern fishery of Chile, as well as in the southern fishery off Talcahuano, more than doubled. Interestingly, up to 1989, total pelagic yield in the Humboldt Current had almost re-attained the high values of the early 1970s, with the difference that the fishery had changed from a monospecific to a multispecies fishery.

Because of the overriding importance of physical factors for biomass yields in the Humboldt Current, it seems to be of utmost necessity to establish an environmental monitoring system in which all three countries—Ecuador, Peru, and Chile—actively participate in a cooperative manner. This system might be established under the leadership of the Comisión Permanente del Pacifico Sur (CPPS) with co-sponsorship of United Nations' agencies such as the Intergovernmental Oceanographic Commission (IOC), FAO, or United Nations Environment Program (UNEP). A strong environmental database might allow future predictions of changes in the dynamics of the fishes in the Humboldt Current ecosystem and could contribute considerably to a rational fisheries management system for the region.

References

Alheit, J. 1985. Egg production method for spawning biomass estimates of anchovies and sardines. ICES C.M. 1985/H:41. 10 pp.

Alheit, J. 1987. Egg cannibalism versus egg predation: Their significance in anchovies. S. Afr. J. Mar. Sci. 5:467 – 470.

Arntz, W., and Tarazona, J. 1989. Effects of El Niño 1982 – 83 on benthos, fish and fisheries off the South American Pacific coast. In: Global ecological consequences of the 1982/83 El Niño-southern oscillation. pp. 323 – 360. Ed. by P. W. Glynn. Elsevier Oceanography Series.

Barber, R. T., Chavez, F. P., and Kogelschatz, J. E. 1985. Efectos biologicos de El Niño. Com. Perm. Pacif. Sur Bol. ERFEN. 14:3 – 29.

Bernal, P. 1981. A review of the low-frequency response of the pelagic ecosystem in the California current. Calif. Coop. Oceanic Fish. Invest. Rep. 22:49 – 62.

Bernal, P. A., Robles, F. L., and Rojas, O. 1983. Variabilidad fisica y biologica en la region meridional del sistema de corrientes Chile-Peru. FAO Fish. Rep. 291:683 – 711.

Brockmann, C., Fahrbach, E., Huyer, H., and Smith, R. L. 1980. The poleward undercurrent along the Peru coast: 5 – 15. Deep-Sea Res. 27A:847 – 856.

Carrasco, S., and Lozano, O. 1989. Seasonal and long-term variations of zooplankton volumes in the Peruvian Sea, 1964 – 1987. In: The Peruvian upwelling ecosystem: Dynamics and interactions. pp. 82 – 85. Ed. by D. Pauly, P. Muck, J. Mendo, and I. Tsukayama. ICLARM Conference Proceedings 18.

Castillo, S., and Mendo, J. 1987. Estimation of unregistered Peruvian anchoveta (Engraulis ringens) in official catch statistics, 1951 to 1982.

In: The Peruvian anchoveta and its upwelling ecosystem: Three decades of change. pp. 109 – 116. Ed. by D. Pauly and I. Tsukayama. ICLARM Studies and Reviews 15.

de Ciechomski, J. D. 1967. Investigations of food and feeding habits of larvae and juveniles of the Argentine anchovy Engraulis anchovita. Calif. Coop. Oceanic Fish. Invest. Rep. 11:72 – 81.

Csirke, J. 1988. Small shoaling pelagic fish stocks. In: Fish population dynamics. pp. 271 – 302. Ed. by J. A. Gulland. Wiley Interscience, Chichester. U.K. 422 pp.

DeVries, T. J., and Pearcy, W. G. 1982. Fish debris in sediments of the upwelling zone off central Peru: A late Quaternary record. Deep-Sea Res. 29:87 – 109.

Espinoza, F. R., Neshyba, S., and Mexicano, Z. 1983. Surface water motion off Chile revealed in satellite images of surface chlorophyll and temperature. In: Proceedings of the international conference on marine resources of the Pacific. pp. 41 – 57. Ed. by P. M. Arana. Universidad Catolica de Valparaiso.

Fonseca, T. R., and Farias, M. 1987. Estudio del proceso de surgencia en la costa chilena utilizando percepcion remota. Invest. Pesq. (Chile) 34:33 – 46.

Furness, R. W. 1982. Competition between fisheries and sea bird communities. Adv. Mar. Biol. 20:225 – 307.

Hunter, J. R., and Kimbrell, C. A. 1980. Egg cannibalism in the northern anchovy, Engraulis mordax. Fish. Bull. U.S. 78:811 – 816.

Jordan, R. 1967. The predation of guano birds on the Peruvian anchoveta (Engraulis ringens Jenyns). Calif. Coop. Oceanic Fish. Invest. Rep. 11:105 – 109.

Jordan, R. 1971. Distribution of anchoveta (Engraulis ringens) in relation to the environment. Invest. Pesq. 35:113 – 126.

Lasker, R. 1985. An egg production method for estimating spawning biomass of pelagic fish: Application to the northern anchovy, Engraulis mordax. U.S. Dept. of Comm. Nat'l. Oceanic and Atmospheric Admin. Washington, D.C. NOAA Technical Rep. NMFS 36. 99 pp.

Lluch-Belda, D., Crawford, R. J. M., Kawasaki, T., MacCall, A. D., Parrish, R. H., Schwartzlose, R. A., and Smith, P. E. 1989. World-wide fluctuations of sardine and anchovy stocks: The regime problem. S. Afr. J. Mar. Sci. :195 – 205.

Loeb, V. J., and Rojas, O. 1988. Interannual variation of ichthyoplankton composition and abundance relations off northern Chile, 1964 – 83. Fish. Bull. U.S. 86:1 – 24.

MacCall, A. D. 1981. The consequences of cannibalism in the stock-recruitment relationship of planktivorous pelagic fishes such as Engraulis. IOC, Paris IOC Workshop Rep. 28:201 – 220.

Mendelsohn, R. 1989. Reanalysis of recruitment estimates of the Peruvian anchoveta in relationship to other population parameters and the surrounding environment. In: The Peruvian upwelling ecosystem: Dynamics and interactions. pp. 364 – 385. Ed. by D. Pauly, P. Muck, J. Mendo, and I. Tsukayama. ICLARM Conference Proceedings 18.

Mendelsohn, R., and Mendo, J. 1987. Exploratory analysis of anchoveta recruitment off Peru and related environmental series. In: The Peruvian anchoveta and its upwelling ecosystem: Three decades of change. pp. 294 – 306. Ed. by D. Pauly and I. Tsukayama. ICLARM Studies and Reviews 15.

Mendez, R. 1987. Cambios bioticos y efectos sobre los recursos pesqueros y pesquerias en Chile. Rev. Com. Perm. Pacifico Sur 16:7 – 96.

Menz, A., and French, S. 1982. The fishery for small pelagic fishes off the coast of Ecuador: Its development and investigation. In: Segundo Seminario Taller. Bases biologicas para el uso y manejo de recursos naturales renovables: Recursos biologicos marinos. Ed. by J. C. Castilla. Monogr. Biol. 2:2 – 17.

Mooers, C. N. K., Flagg, C. N., Boicourt, W. C. 1978. Prograde and retrograde fronts. In: Oceanic fronts in coastal processes. Ed. by M. J. Bowman and W. E. Esaias. Springer-Verlag, Berlin.

Muck, P. 1989a. Major trends in the pelagic ecosystem off Peru and their implications for management. In: The Peruvian upwelling ecosystem: Dynamics and interactions. pp. 386 – 403. Ed. by D. Pauly, P. Muck, J. Mendo, and I. Tsukayama. ICLARM Conference Proceedings 18.

Muck, P. 1989b. Relationships between anchoveta spawning strategies and the spatial variability of sea surface temperature off Peru. In: The Peruvian upwelling ecosystem: Dynamics and interactions. pp. 168 – 173. Ed. by D. Pauly, P. Muck, J. Mendo, and I. Tsukayama. ICLARM Conference Proceedings 18.

Muck, P., and Sanchez, G. 1987. The importance of mackerel and horse mackerel predation for the Peruvian anchoveta stock (a population and feeding model). In: The Peruvian anchoveta and its upwelling ecosystem: Three decades of change. pp. 276 – 293. Ed. by D. Pauly and I. Tsukayama. ICLARM Studies and Reviews 15.

Murphy, G. I. 1972. Fisheries in upwelling regions with special reference to Peruvian waters. Geoforum 11:63 – 71.

Pauly, D. 1987. Managing the Peruvian upwelling ecosystem: A synthesis. In: The Peruvian anchoveta and its upwelling ecosystem: Three decades of change. pp. 325 – 342. Ed. by D. Pauly and I. Tsukayama. ICLARM Studies and Reviews 15.

Pauly, D., Muck, P., Mendo, J., and Tsukayama, I. (Editors). 1989. The Peruvian upwelling ecosystem: Dynamics and interactions. ICLARM Conference Proceedings 18. Instituto del Mar del Peru (IMARPE), Callao, Peru; Deutsche Gesellschaft fur Technische Zusammenarbeit (GTZ), GmbH, Eschborn, Federal Republic of Germany; and International Center for Living Aquatic Resources Management (ICLARM), Manila, Philippines. 438 pp.

Pauly, D., and Palomares, J. L. 1989. New estimates of monthly biomass, recruitment, and related statistics of anchoveta (*Engraulis ringens*) off Peru (4° – 14° S), 1953 – 1985. In: The Peruvian upwelling ecosystem: Dynamics and interactions. pp. 189 – 206. Ed. by D. Pauly, P. Muck, J. Mendo, and I. Tsukayama. ICLARM Conference Proceedings 18.

Pauly, D., and Tsukayama, I. (Editors). 1987. The Peruvian anchoveta and its upwelling ecosystem: Three decades of change. ICLARM Studies and Reviews 15. Instituto del Mar del Peru (IMARPE), Callao, Peru; Deutsche Gesellschaft fur Technische Zusammenarbeit (GTZ), GmbH, Eschborn, Federal Republic of Germany; and International Center for Living Aquatic Resources Management (ICLARM), Manila, Philippines. 351 pp.

Pianka, E. R. 1987. The subtlety, complexity and importance of population interactions when more than two species are involved. Revista Chilena de Historia Natural. 60:351 – 361.

Robles, F. L., Alarcon, E., and Ulloa, A. 1980. Water masses in the northern Chilean zone and their variations in the cold period (1967) and warm periods (1969, 1971 – 73). In: Proceedings of the workshop on the phenomenon known as "El Niño." pp 83 – 174. UNESCO.

Romo, D. 1985. Composicion quimica de la harina de pescado chilena durante el fenomeno el Niño 1982 – 83. Invest. Pesq. (Chile) 32:141 – 151.

Santander, H., Alheit, J., and Smith, P. E. 1984. Estimacion de la biomasa de la poblacion desovante de anchoveta peruana, *Engraulis ringens*, en 1981 por aplicacion del "Metodo de Produccion de Huevos." Bol. Inst. Mar Peru, Callao 8:208 – 250.

Santander, H., and Flores, R. 1983. Los desoves y distribucion larval de cuatro especies pelagicas y sus relaciones con las variaciones del ambiente marino frente al Peru. FAO Fish. Rep. 291:835 – 867.

Schaefer, M. B. 1970. Men, birds, and anchovies in the Peru current-dynamic interactions. Trans. Am. Fish Soc. 99:461 – 467.

Serra, J. R. 1983. Changes in the abundance of pelagic resources along the Chilean coast. FAO Fish. Rep. 291:255 – 284.

Serra, R., and Tsukayama, I. 1988. Sinopsis de datos biologicos y pesqueros de la sardina *Sardinops sagax* (Jenyns, 1842) en el Pacifico suroriental. FAO Sinopsis sobre la Pesca 13. FAO, Rome 60 pp.

Silva, N., and Fonseca, T. H. 1983. Geostrophic component of the ocean flow off northern Chile. In: Proceedings of the international conference on marine resources of the Pacific. pp. 59 – 70. Ed. by P. M. Arana. Universidad Catolica de Valparaiso. Valparaiso, Chile.

Smith, P. E., Santander, H., and Alheit, J. 1989. Comparison of the mortality rates of Pacific sardine, *Sardinops sagax*, and Peruvian anchovy, *Engraulis ringens*, eggs off Peru. Fish. Bull. U.S. 87:497 – 508.

Subsecretaria de Pesca. 1983. Recursos pelagicos y pesquerias en Chile. Rev. Com. Perm. Pacifico Sur 13:5 – 23.

Theilacker, G. H. 1988. Euphausiid predation on larval anchovy at two contrasting sites off California determined with an ELISPOT immunoassay. In: Immunochemical approaches to coastal, estuarine, and oceanographic questions. pp. 304 – 311. Ed. by C. M. Yentsch, F. C. Mague, and P. K. Horan. Springer-Verlag 25, Berlin.

Tsukayama, I. 1983. Recursos pelagicos y sus pesquerias en Peru. Rev. Com. Perm. Pacifico Sur 13:25 – 63.

Tsukayama, I., and Santander, H. 1987. Cambios bioticos y efectos sobre los recursos pesqueros y las pesquerias en Peru. Rev. Com. Perm. Pacifico Sur 16:97 – 166.

Valdes, E. S., Shelton, P. A., Armstrong, J. J., and Field, J. G. 1987. Cannibalism in South African anchovy: Egg mortality and egg consumption rates. S. Afr. J. Mar. Sci. 5:613 – 622.

Wooster, W. W., and Fluharty, D. L. 1985. El Niño North: Niño effects in the eastern subarctic Pacific Ocean. Washington Sea Grant Program Publication. Seattle, Washington. 312 pp.

Wyrtki, K. 1967. Circulation and water masses in the eastern equatorial Pacific Ocean. Int. J. Oceanol. Limnol. 1:117 – 147.

Zuta, S., Tsukayama, I., and Villaneuva, R. 1983. El ambiente marino y las fluctuaciones de las principales poblaciones pelagicas de la costa peruana. FAO Fish. Rep. 291:179 – 253.

Chapter 9

Food Chains, Physical Dynamics, Perturbations, and Biomass Yields of the Sea of Okhotsk

V. V. Kuznetsov, V. P. Shuntov, L. A. Borets

Introduction

There have been few studies of the ecology of the Far East seas (Shuntov *et al.*, 1990), and among these seas, the Sea of Okhotsk is probably the least investigated. Recently, Shuntov and coauthors have carried out extensive studies of the Sea of Okhotsk ecosystem (Shuntov, 1985; Shuntov *et al.*, in press; Shuntov and Dulepova, in press). New data have been obtained and results of previous investigations have been systematized and summarized in this paper. Consequently, qualitative estimates have been made to characterize trophic relationships and other aspects of the ecosystem. In some cases, lack of original information has forced the authors to use rough estimates based on selection of analogues and logical conclusions. This chapter represents the first attempt to describe the Sea of Okhotsk ecosystem as a whole.

Main Features of the Hydrological Regime of the Sea of Okhotsk

The Sea of Okhotsk is a continental marginal water body that occupies an area of about 1.6 million km², with an average depth of 891 m, maximum depth of 3,916 m, and water volume of 1,318 thousand km³ (Leonov, 1960; Dobrovolsky and Zalogin, 1982). The entire sea is located in a temperate zone; however, because of its large area, there are marked differences in climate, hydrographic regime, and biology of the northern and southern parts of the sea. Bottom relief is characterized by a complex shape, but on the whole, the northern area is shallow, whereas the southern area is deep (Udintsev, 1957). The Sea of Okhotsk is the coldest of the Far East seas.

In winter, the temperature of the surface layers is in the range of -1.5° to -1.8° C through almost the entire area, and most of the sea is covered with ice. In shelf areas, cold temperatures extend through the water column to the bottom, and in the deep water areas they extend down to 100 – 200 m. In summer, surface layers warm to 11° – 13° C, except off Hokkaido Island where surface temperatures can reach 19° C, but warming extends down to 30 – 75 m only. Below this warm layer, down to 150 m, there is a persistent cold layer where even in summer the temperature can be -1.7° C. Below 150 m, temperature increases with depth up to 2° – 2.5° C at 750 – 1,500 m and then decreases gradually to 1.8° C.

The Sea of Okhotsk has a complicated current system characterized by three large cyclonic gyres in the northwestern, northeastern, and southern areas of the sea (Fig. 1) (Chernyavsky, 1981). Waters of the Sea of Okhotsk exit into the North Pacific Ocean through the straits of the southern part of the Kuril Island chain. This loss is offset by the influx of Pacific waters through the straits of the northern part of the Kuril Islands. Within the range of macrocirculation system, a great number of meso- and the microcirculation formations can be distinguished.

Formation of Productive Zones and Interannual Variability

The Sea of Okhotsk is one of the highly productive areas of the world ocean. Waters off Kamchatka and northern and western areas of the sea are especially abundant in plankton

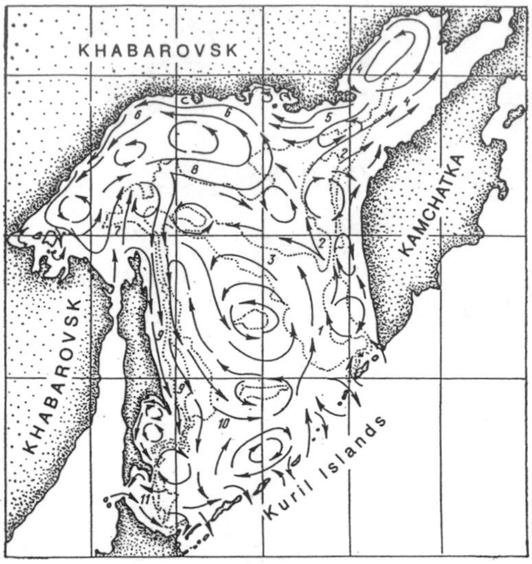

Figure 1. Generalized surface circulation in the Sea of Okhotsk ecosystem (Markina and Chernyavsky, 1984). Currents: (1) West Kamchatka; (2) Northern branch, West Kamchatka; (3) Middle; (4) Penzhinsk; (5) Yamsk; (6) Northern Okhotsk; (7) Amur; (8) Northern Okhotsk Countercurrent; (9) East Sakhalin; (10) Northeastern; (11) Soya.

(Markina and Chernyavsky, 1984). In contrast, the abundance of plankton in the central deep part of the sea is relatively poor. Distribution of benthos and plankton in the sea is distinctly patchy (Figs. 2, 3, and 4), and this is attributed to the complexities of circulation patterns. Productive zones form in upwelling areas which transport great quantities of nutrients into the photic zone. In addition, high plankton concentrations in the areas of downwelling are observed, primarily caused by mechanical accumulation of plankton.

The climate and hydrography of the Sea of Okhotsk are highly variable from year to year (Fig. 5), which causes changes in conditions for reproduction of marine organisms, including commercial fishery resources. The hydrological processes in the Sea of Okhotsk are related to characteristic features of atmospheric processes over the Northwest Pacific (Davydov, 1984). Variation in atmospheric cyclonic circulation strengthens or weakens heat advection via currents flowing into the Sea of Okhotsk through the northern straits of the Kuril Islands.

Trophic Relationships

The saprophytic chain in the Sea of Okhotsk

includes bacteria and microzooplankton. However, no information is available on bacteria and protozoans, though it is known that production of microheterotrophs can exceed primary production (Sorokin, 1977). No reliable data on biomass and production of phytoplankton are available either. To estimate the volume of phytoplankton production in the Sea of Okhotsk, Shuntov et al. (in press) used data on the Bering Sea which has been studied more extensively. Estimated primary production in wet weight for the entire Sea of Okhotsk is 15×10^9 tonnes per year. Similar rough estimates for bacteria and protozoans (microzooplankton), based on selection of analogues and extrapolation, were 5.2×10^9 and 2.1×10^9 tonnes per year, respectively.

Matter and energy formed by producers are distributed between pelagic and benthic components of the food web. The second trophic level in the pelagic zone is formed by phytophagous and euryphagous organisms that consume algae. Nonpredatory zooplankton can be separated into three size groups. Of plankters less than 1.5 mm, *Pseudocalanus minutus* and *Oithona similis* form the bulk of the biomass. Most of the biomass in the group of plankters 1.5 – 3.5 mm consists mainly of immature stages of *Metridia okhotensis, M. pacifica, Pseudocalanus minutus*, and sometimes of copepodites of *Calanus cristatus* and *C. plumchrus*, for example. The third group

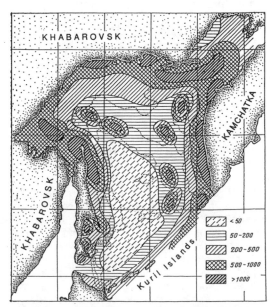

Figure 2. Average phytoplankton biomass distribution (mg/m³) for 0 – 100 m in the Sea of Okhotsk ecosystem. (Markina and Chernyavsky, 1984)

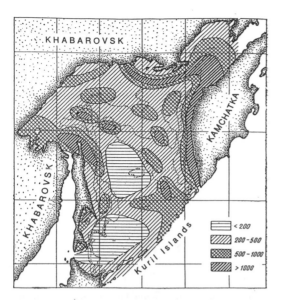

Figure 3. Average zooplankton biomass distribution (mg/m³) for 0 – 100 m in the Sea of Okhotsk ecosystem. (Markina and Chernyavsky, 1984)

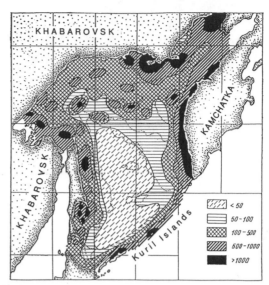

Figure 4. Average benthos biomass distribution (g/m²) in the Sea of Okhotsk. (Markina and Chernyavsky, 1984)

(over 3.5 mm) is composed primarily of the copepods *C. glacialis*, *C. cristatus*, *M. okhotensis*, *M. pacifica*, and the euphausiids, *Thysanoessa raschii*, *T. longipes*, *T. inermis*, and *Euphausia pacifica*. Predatory plankters, mainly from the largest size fraction (chaetognaths, hyperiids, and jellyfish), belong to the third trophic level of the food web.

Trophic levels in the zoobenthos are similar. Sestonovores and detritivores occupy the second level (nonpredatory benthos) and include foraminifera, sponges, priapulids, bivalve molluscs, barnacles, ophiurans, echinoderms, holothurians, and tunicates. Of the predatory benthos (third trophic level), actinidians, polychaetes, nemerteans, starfish, and decapods are the most abundant. The bulk of demersal species of fish occupy the third trophic level, but a few of them—for example, halibuts and large-sized sculpins—are higher carnivores and belong to the fourth trophic level.

The total annual production of zooplankton in the Sea of Okhotsk is estimated at 3×10^9 tonnes with benthic production estimated at 3.4×10^9 tonnes.

Biomass of the major commercial species, walleye pollock (*Theragra chalcogramma*), has been estimated to be 10 – 15 million tonnes over the last 5 years. Estimated biomass of demersal shelf fish has been relatively stable recently at about 2 million tonnes. Biomass of small mesopelagic fish, predominantly argen-

tine, *Leuroglossus schmidti*, is estimated to be 15 – 30 million tonnes (Iljinskij, 1988). The total fish biomass and production in the Sea of Okhotsk are at least 30 and 15 million tonnes, respectively. Squid biomass and production are correspondingly about 1 and 3 million tonnes. Biomass and production of commercially harvested demersal invertebrates are estimated to be 1.5 and 0.5 million tonnes, respectively. The corresponding figures for mammals are 0.17 and 0.03 million tonnes; for birds, the figures are 0.006 and 0.002 million tonnes.

The productivity dynamics in the Sea of Okhotsk are characterized by a relatively small role of herbivorous zooplankton (referring to the predatory chain), a substantial role of carnivorous plankton (which is not essential to nekton feeding), and the large portion of production by herbivorous plankton and demersal organisms that is converted to detritus (Fig. 6). The role of fish in the consumption of herbivorous plankton production is negligible. A substantially larger portion of herbivorous zooplankton is preyed upon by carnivores.

Commercial Catch

The potential fish yield in the Sea of Okhotsk is about ¼ – ⅕ of the total production of commercial fish (Shuntov and Dulepova, 1990),

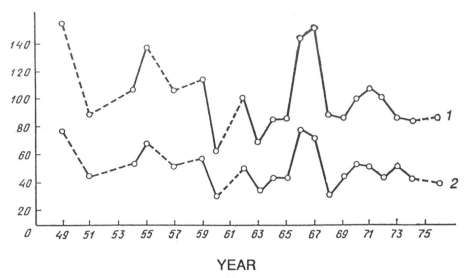

YEAR

Figure 5. Interannual variability of cold water mass (thousand km²) in the northern Sea of Okhotsk ecosystem in 1946 – 1976: (1) June; (2) October. (Chernyavsky, 1973)

which will total about 3.7 – 4.6 million tonnes. This estimate is considerably in excess of the present catch, which averages about 2.5 million tonnes.

Major commercial species in the Sea of Okhotsk are represented by cod (mainly walleye pollock), flounder, herring, and Pacific salmon. Recent years are characterized by abundant runs of West Pacific sardine into the sea in the summer and a high abundance of Pacific saury.

Walleye pollock (Theragra chalcogramma)

The walleye pollock is the most abundant and widely distributed fish species in the Sea of Okhotsk. Relatively intense exploitation of the walleye pollock stock began in the mid-1960s, and since the 1970s the catch has exceeded the total catch of other fishes (Fig. 7). The fishery for walleye pollock is conducted mainly in the northern are of the sea and especially off West Kamchatka (Fig. 8). The ecology of walleye pollock has not been sufficiently investigated. Fadeyev (1980) indicated that the abundance of this species in the Sea of Okhotsk has not changed and that the increase in catch was attributable exclusively to industrial and economic reasons. However, there is some convincing evidence that walleye pollock abundance of West Kamchatka more than doubled in the 1960s (Kachina, 1979; Kachina and Sergeeva, 1981). Fluctuations in walleye pollock abundance have been attributed to climatic

and oceanological factors, in particular, to fluctuations in warm- and cold-year periods. However, there has been no definitive study of this relationship. Existing descriptions of walleye pollock population structure are also rather inconsistent. Pushnikov (1982) indicated the existence of five different stocks in the sea, though others contend there are only one to three stocks.

Flounder

Commercial utilization of flounder (*Pleuronectidae*) stocks in the Sea of Okhotsk was most intensive in the second half of the 1950s – 1960s, when the Soviet catch alone constituted 80 – 140,000 tonnes. Subsequently, catches decreased and flounder stocks became depleted. Species composition and biomass of flounder in the Sea of Okhotsk shelf area are given in Table 1. From the table, it is evident that flounder is abundant in the northern shelf area; however, these estimates are rough and resources are rather dispersed. More than half of the biomass is represented by *Limanda sakhalinensis*, which, because of its small size, is scarcely exploited. Flounder stocks in the West Kamchatka shelf area were heavily exploited from 1956 to 1961, when the catch reached 75,000 – 130,000 tonnes (including the Japanese catch). Since 1976, Japan has ceased fishing for flounder and the Soviet fishery has been limited. Maximum sustainable yield for this shelf area when the stock is in good condition is estimated at 50,000 – 60,000 tonnes, out of which about 37,000 tonnes are commercial species (Shuntov, 1985). As a result of long-term overexploitation, the stocks of flounder, especially yellowfin sole, at the southeastern shelf of Sakhalin Island are depleted.

A substantial stock of Greenland halibut exists in the Sea of Okhotsk. It is harvested along the continental slope where it constitutes, on average, one-third of the catch per haul. Recent estimates (Borets, 1986; Shuntov, 1985) suggest that the total biomass of this species in the sea is about 750,000 tonnes. However, the fishery for this species is limited for several reasons. The potential yield is estimated at 25,000 tonnes, whereas actual catches are one-fourth this amount. Several other species, such as grenadier, eelpout, skate, and perch, are taken as bycatch in this fishery.

Flounder in the Sea of Okhotsk and other Far East seas have been shown to be confined to the local shallow areas with the formation of isolated stocks. Shuntov (1985), based on an analysis of catch composition of flounders and other demersal fish at various fishing intensities, has concluded that bottom communities are rather resistant to the impact of fisheries.

Herring

There are a number of herring (*Clupea pallasi pallasi*) stocks in the Sea of Okhotsk, among which Sakhalin-Hokkaido, Okhotsk (the north-eastern part of the sea), and Gizhiginsk-Kamchatsk stocks are the most important commercially. Until the 1920s, herring catches were relatively low throughout the entire area. Catches increased considerably in the early 1930s off Sakhalin (200,000 – 400,000 tonnes), but by the mid-1930s they exhibited a downward trend, catches being only 20,000 – 30,000 tonnes, and then declined further (Fig. 9). In the 1930s, the Soviet fishery for spawning herring of two northern stocks (Okhotsk and Gizhiginsk-Kamchatsk) was initiated (Fig. 10). Since the mid-1970s, herring catches sharply declined as a result of the depleted status of all the stocks. An increase in abundance of the Okhotsk herring stocks was observed in the 1980s. A relationship has been observed between type of year (cold or warm) and year-class strength (Tyurnin, 1980). In the cold years, prevailing winds (from the land) clear coastal areas of ice in spring, whereas southerly winds cause ice to drift to

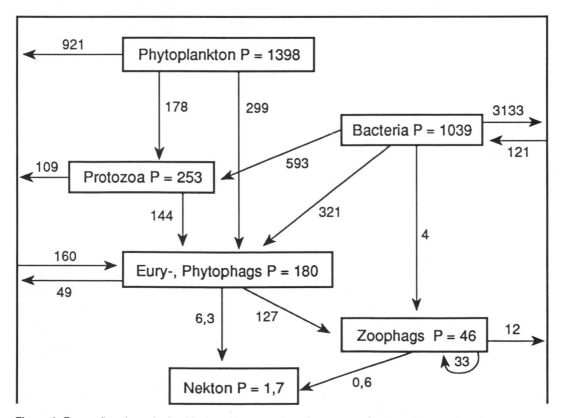

Figure 6. Energy flow through simplified trophic web (KCal/M² • season) of epipelagic zone of the Sea of Okhotsk ecosystem in autumn 1985 (p=production). (Shuntov and Dulepova, In press)

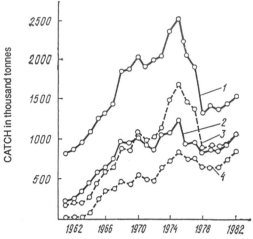

Figure 7. Fish catch in the Sea of Okhotsk, 1962 – 1982 (Shuntov, 1985): (1) Total catch, all countries; (2) USSR catch; (3) Total walleye pollock catch, all countries; (4) USSR walleye pollock catch.

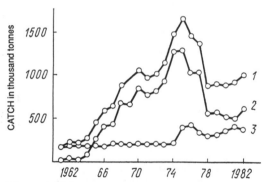

Figure 8. Walleye pollock catch by region in the Sea of Okhotsk, 1960 – 1980 (Shuntov, 1985): (1) The entire sea, (2) West Kamchatka region, (3) Other regions.

spawning grounds, which affects reproductive conditions. In general, the reasons for abundance fluctuations are not entirely clear.

Pacific salmon

In the late 1930s, the total catch of salmon (*Oncorhynchus*) in the Northwest Pacific was 400,000 – 600,000 tonnes (Semko, 1964; Moiseyev, 1967). Estimated total biomass of these fish in this region was 1.0 – 1.5 million tonnes, with about 0.8 – 1.2 million tons attributable to the Sea of Okhotsk stocks. The abun-

dance level in the 1920s – 1930s was probably close to maximum carrying capacity. A favorable period for salmon followed in the 1940s (Fig. 11). Up to the mid-1950s, the Soviet catch of all salmonid species in the Sea of Okhotsk ranged from 90,000 to 225,000 tonnes, which was not considered excessive. The mid-1950s were characterized by the beginning of the salmon stock depletion. Drop in abundance is attributed to deterioration of reproductive conditions in fresh water, in addition to the impact of a developing Japanese offshore fishery. In the second half of the 1950s and throughout the 1960s, salmon catches in most coastal areas were very low. In the 1970s, total catch in the sea increased slightly. The Japanese oceanic fishery was considerably limited in the late 1970s. At this time, there were increasing runs of salmon to the rivers, increases in catch, and higher stock density of spawning grounds.

The pattern of variability in abundance is different for various species and stocks of salmon. Fluctuations in abundance in the northern Sea of Okhotsk are often opposite to those in the Sakhalin-Kuril area. However, as a whole, the salmon resources remain low. With the existing fishing intensity, the catch accounts for no more than 25% of the total abundance, whereas biomass of salmon stocks passing through the Sea of Okhotsk is estimated at 200,000 – 320,000 tonnes, which is four times lower than in 1920 – 1940. An increase in the abundance of some large salmon stocks in the northern Sea of Okhotsk is expected, given the expansion of artificial propagation and limitations on the oceanic fishery.

Other fish species

A number of fish species, characterized by high abundance, occur and are not fully exploited in the Sea of Okhotsk, including cod (*Gadus morhua macrocephalus*), capelin (*Mallotus villosus socialis*), sand eel, and mesopelagic fish. Total cod biomass in the sea is estimated at 300,000 tonnes. Cod does not form dense or large concentrations and is caught mainly as bycatch (about 20,000 tonnes). Cod stocks have increased considerably in the 1980s. The catch of this species could be as high as 40,000 – 50,000 tonnes (Shuntov, 1985).

Biomass of capelin in the Sea of Okhotsk is estimated at 600 – 1,000 thousand tonnes, but a fishery for this species has not developed. Short-term spawning runs present difficulties for commercial utilization, and the potential for a fishery during other periods has not been investigated. Sand eel is also abundant in the Sea of Okhotsk. Estimates of its biomass are close to those of capelin. Smelts are highly abundant

Table 1. Relative abundance by species and biomass (in thousand tonnes) of flounder, Sea of Okhotsk shelf, 1981–1982*

	Sea areas			
	SE shelf of Sakhalin I	NE shelf of Sakhalin I	Northern shelf	West Kamchatka shelf
Species	Relative abundance (%)			
Greenland halibut				
(Reinchardtius hyppoglossoides)	16.43	17.07	31.54	10.85
Hyppoglossoides spp.	40.17	8.72	11.19	13.69
Yellowfin sole				
(Limanda aspera)	8.01	0.64	0.61	38.50
L. sakhalinensis	6.92	10.17	54.89	28.36
Alaska plaice				
(Platessa quadrituberculatus)	6.28	0.92	1.74	4.62
Starry flounder				
(Platichthys stellatus)	1.36	51.05	—	0.29
Glyptocephalus				
stelleri	14.39	9.70	—	0.02
Others	6.44	1.73	0.03	3.67
All species	Total biomass, thousand tonnes			
	22.2	40.4	635.7	364.2
	Biomass per km^2, tonnes			
	0.62	0.90	5.31	7.95

*(Borets, 1985)

along the entire coastal area of the Sea of Okhotsk, but these resources are not exploited.

In the southern area of the sea, especially off the Kuril Islands, there are concentrations of *Podonema longipes* with a biomass of several hundred thousand tonnes. It is probably a species whose abundance and migration patterns are highly variable.

Total biomass of fish from the continental slope, combined with mesopelagic fish, of which argentine is the most abundant, is assumed to constitute 2–3 million tonnes and possibly more. (This estimate excludes mesopelagic fish over deep water.)

Runs of Pacific saury (*Cololabis saira*) and west Pacific sardine (*Sardinops sagax melanosticta*) to the Sea of Okhotsk occur in summer. These species migrate through both the Kuril Straits and La Perouse Strait. Catch of Pacific saury by the Japanese fleet in the Sea of Okhotsk amounts to 40,000–60,000 tonnes. At present, sardine abundance is higher than that of saury, and its biomass is probably several hundred thousand tonnes (Shuntov, 1985).

With regard to exploitation of the Sea of Okhotsk living marine resources, expectations of further growth in the catch rate of such traditional species as walleye pollock, herring, flounder, salmon, and Greenland halibut are not realistic. Some aquatic organisms that do not form stable aggregations conducive to fishing, such as mesopelagic fish and squid and other species such as capelin and char, are in great part unexploited. Other marine flora and fauna, such as brown algae, crabs, shrimp,

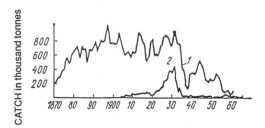

Figure 9. Herring catch of Sakhalin-Hokkaido population (Ayushin, 1963): (1) Catch in all regions, (2) Catch in the Sea of Okhotsk waters off Sakhalin.

echinoderms, and marine mammals are not fully utilized at present.

Variability in Biomass Yields

The historical record shows changes in the status of living marine resources in the Sea of Okhotsk ecosystem as a result of the effects of fishing (Shuntov, 1985). The stocks of demersal fish (mainly flounder) and crustaceans decreased considerably as a result of nonregulated fishing during 1950–1970. In the second half of the 1970s, after the introduction of a fishery conservation and management zone, the protection of resources was strengthened, which resulted in a considerable increase of demersal fish total biomass and high stability of demersal communities in the 1970s. The pelagic communities are characterized by lower stability. The dominant species in the pelagic ichthyocoenoses fluctuate (herring, walleye pollock, salmon, and Pacific sardine). Environmental factors are assumed to be the main reason for fluctuations in abundance. Variations in pelagic fish abundance exhibit cyclic patterns, which some scientists often associate with cyclical hydrographic processes. However,

attempts at long-term prediction of such variations are usually unsuccessful.

At the close of the 1980's there was a high level of biomass of walleye pollock. Under relatively stable climatic and oceanographic conditions and at moderate fishing intensity, a high level of walleye pollock resources can be maintained for a rather long period. Long-term unfavorable climatic or oceanographic anomalies and/or overexploitation constitute the possible reasons for future decrease of walleye pollock resources. A drop in abundance of walleye pollock and West Pacific sardine will result in a considerable and long-term decrease of fish productivity of the Far East seas, including the Sea of Okhotsk.

Recently, Sokolovskij and Glebova (1985a, 1985b) observed that sometimes a fishery can produce an increase in resources by rejuvenating the population and decreasing cannibalism, as is the case in walleye pollock. However, it has also been shown that with reference to West Pacific sardine, rejuvenation of a population can be accompanied by narrowed ranges of age classes, sizes of fish, and other characteristics, which inevitably decreases the overall stability of the population, leaving it more vulnerable to unfavorable environmental effects (Kuznetsov

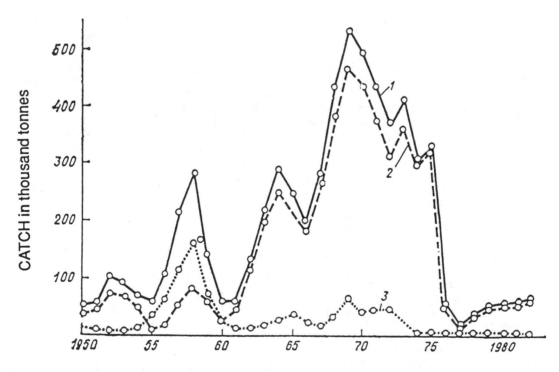

Figure 10. Catch of herring in the Sea of Okhotsk in the 1950s–1980s (Shuntov, 1985): (1) All populations, (2) Okhotsk populations, (3) Gizhiginsk-Kamchatsk population.

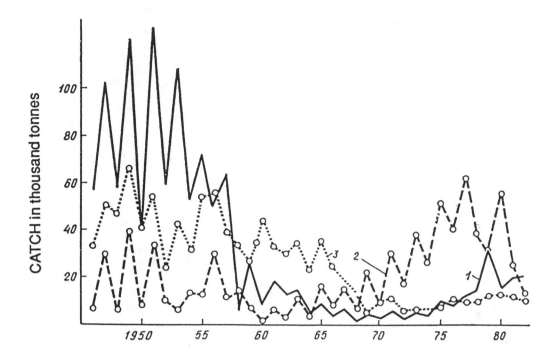

Figure 11. Catch (thousand mt) by region of Pacific salmon in the Sea of Okhotsk basin in the 1940s – 1980s (Shuntov, 1985). Regions: (1) West Kamchatka, (2) Sakhalin-Kuril, (3) Northern Sea of Okhotsk.

and Kuznetsova, 1988; Kuznetsov, 1989a, 1989b). In view of the above, it does not seem prudent to increase the catch of commercially valuable pelagic fish to the estimated maximum sustainable yield. It is necessary to allow a reserve to provide for stability of the population.

References

Ayushin, B. N. 1963. Abundance dynamics of herring population in the seas of Far East and reasons for the introduction of fishery regulation. Rapp. P.-v. Reun. Cons. int. Explor. Mer 154:262 – 269.

Borets, L. A. 1986. Composition of demersal fish communities at the Sea of Okhotsk shelf. Mar. Biol. 4:54 – 59. In Russian.

Chernyavsky, V. I. 1973. On the possibility of prediction of thermal conditions in the Sea of Okhotsk. Izv. TINRO 86:49 – 55. In Russian.

Chernyavsky, V. I. 1981. Circulation systems of the Sea of Okhotsk. Izv. TINRO 101:18 – 24. In Russian.

Davydov, I. V. 1984. On correlation of development of oceanological conditions in main fishing grounds of the Far East seas. Izv. TINRO 109. In Russian.

Dobrovolsky, A. D., and Zalogin, B. S. 1982. Seas of the USSR. M., Publ. by MSU. 192 pp. In Russian.

Fadeyev, N. S. 1980. Was it an "outbreak" of walleye pollock abundance in the North Pacific? Mar. Biol. 5:66 – 71. In Russian.

Iljinskij, E. N. 1988. Composition and qualitative distribution of mesopelagic fishes of the Sea of Okhotsk. Estimation and exploitation of biological resources of the ocean. Vladivostok. TINRO. pp. 8 – 10. In Russian.

Kachina, T. F. 1979. On the abundance dynamics of herring and walleye pollock in the Far East seas. Ryb. Khoz 3:7 – 9. In Russian.

Kachina, T. F., and Sergeeva, N. P. 1981. Abundance dynamics of the eastern Okhotsk walleye pollock. In: Ecology, Resources of and Fishery for Walleye Pollock. pp. 19 – 27. Vladivostok. In Russian.

Kuznetsov, V. V. 1989a. Retrospective analysis of large-scale variations in West Pacific sardine abundance. In: Biological background of abundance dynamics and catch prediction of fishes. pp. 84 – 97. Moscow. VNIRO. In Russian.

Kuznetsov, V. V. 1989b. Variations in abundance of coastal-pelagic fishes of the Kuroshio zone and successional processes in ichthyocoenosis. Scientific Papers of Higher School, Biological Sciences 4:25 – 31. In Russian.

Kuznetsov, V. V., and Kuznetsova, E. N. 1988. On

ecological interrelations and abundance dynamics of warm-water pelagic fishes in the zone of the Kuroshio influence. Rybn. Khoz 4:51 – 54. In Russian.

Leonov, A. K. 1960. Regional oceanography. Gidrometeoizdat, Leningrad. 765 pp. In Russian.

Markina, N. P., and Chernyavsky, V. I. 1984. Some data on quantitative distribution of plankton and benthos in the Sea of Okhotsk. Izv. TINRO 109:94 – 99. In Russian.

Moiseyev, P. A. 1967. Fisheries in Japan. Pishch. prom. Moscow. 200 pp.

Pushnikov, V. V. 1982. Population structure of walleye pollock from the Sea of Okhotsk and status of its stock. Synopsis of Ph.D. diss. Moscow State Univ. Moscow. 23 pp. In Russian.

Semko, P. S. 1964. Present state of the Pacific salmon, degree of their utilization and ways of reproduction. In: Salmon farming in the Far East. pp. 7 – 16. Moscow. In Russian.

Shuntov, V. P. 1985. Biological resources of the Sea of Okhotsk. Agropromizdat, Moscow. 224 pp. In Russian.

Shuntov, V. P., Borets, L. A., and Dulepova, E. P. In press. Ecosystems of the Far East seas and state-of-the-art review. In Russian.

Shuntov, V. P., and Dulepova, E. P. In press. Biological balance, present state, bio- and fish productivity of the Sea of Okhotsk ecosystem and elements of its functioning. In Russian.

Sokolovskij, A. S. and Glebova, S. Yu. 1985a. Long-term fluctuations of walleye pollock abundance in the Bering Sea. Izv. TINRO 110:38 – 42. In Russian.

Sokolovskij, A. S. and Glebova, S. Yu. 1985b. Population structure and productivity of the walleye pollock from the eastern Bearing Sea. In: Fish productivity of the Far East seas. pp. 30 – 37. Vladivostok. In Russian.

Sorokin, Yu. I. 1977. Microflora production. Biological productivity of the ocean. pp. 209 – 233. Nauka, Moscow. In Russian.

Tyurnin, V. B. 1980. On the reasons for the Okhotsk herring stock decline and measures for its restoration. Mar. Biol. 2:69 – 74. In Russian.

Udintsev, G. B. 1957. Bottom relief of the Sea of Okhotsk. Trudy AN SSSR 22:3 – 76. In Russian.

Chapter 10

Effects of Long-Term Physical and Biological Perturbations on the Contemporary Biomass Yields of the Yellow Sea Ecosystem

Qisheng Tang

Introduction

The Yellow Sea is a semi-enclosed body of water with unique hydrographic regime, submarine topography, productivity, and trophically dependent populations. It has well-developed multispecies and multinational fisheries. Over the past several decades, the resource populations in the sea have changed greatly, and, as compared with other continental shelf areas in the Northwest Pacific, the biomass yields have decreased markedly. At present, implementation of effective management strategies is an urgent and important task.

During the last several years, ocean scientists and managers have gained important new insights into the processes that govern the ecology of coastal and ocean species, which has led to a new conceptual framework called the large marine ecosystem (LME) approach. This approach provides the basis for developing new strategies for marine resource management and research from an ecosystem perspective (Sherman and Alexander, 1986, 1989).

The purpose of this chapter is to describe the Yellow Sea as an LME, emphasizing the effects of long-term physical and biological perturbations on the contemporary biomass yields. Detailed information on the physical setting, biological structure, and changes in biomass yields and their causes are reported. Suggestions for effective ecosystem conservation and management of the Yellow Sea are offered in the final section.

The Physical Setting

The semi-enclosed Yellow (Huanghai) Sea is

bounded by the Chinese mainland to the west, the Korean peninsula to the east, and a line running from north of the Changjiang (Yangtze) River mouth to Cheju Island. It covers an area of about 380,000 km^2, with a mean depth of 44 m. The central part of the sea, traditionally called the Yellow Sea Depression, ranges in depth from 70 m to a maximum of 140 m. This is the major overwintering ground for most fish and invertebrates in this ecosystem.

The seasonal mean circulation of the Yellow Sea is a basin-wide cyclonic gyre comprised of the Yellow Sea Coastal Current and the Yellow Sea Warm Current. The Yellow Sea Warm Current, which is a branch of the Tsushima Warm Current from the Kuroshio region in the East China Sea, carries water of relatively high salinity (>33%) and high temperature (>12°C) to the north along 124° E and then to the west, flowing into the Bo Hai Sea in winter. This current, together with the coastal current flowing southward, plays an important role in exchanging the waters in this semi-enclosed ecosystem (Fig. 1).

The Yellow Sea cold water mass is one of the most outstanding and important components in the hydrography of the sea (Ho *et al.*, 1959; Guan, 1963). A water mass characterized by low temperature and high salinity in the central part of its deep and bottom layers, it is believed to be the remnant of local water remaining from the previous winter (Fig. 2). In winter, the temperature distribution in the sea is vertically homogeneous, but has pronounced horizontal gradients with temperatures ranging from about 0°C in the northernmost part to about 10°–13°C along the boundary of the East China Sea. In contrast, during the summer, the horizontal surface temperature gradient is

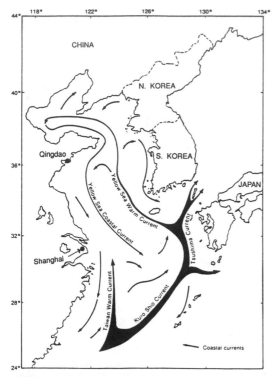

Figure 1. Schematic diagram of the major winter current system in the Yellow Sea (from Gu *et al.*, 1990).

Upward mixing of nutrient-rich bottom water at the boundary of the cold water mass is probably responsible for the higher silicate concentrations along the southern shores of both the eastern Liaodong and Shandong peninsulas (Fig. 3). The annual average phosphate concentration is about 0.4 mg/l, which is much lower than that found in other China seas (Yu *et al.*, 1990).

The Ecological Structure

The phytoplankton community is composed mainly of neritic diatoms. The dominant species are *Skeletonema costatum, Coscinodiscus, Melosira sulcata,* and *Chaetoceros.* The Yellow Sea ecosystem has relatively low primary productivity—about 60 g C $\cdot$ m^{-2} $\cdot$ yr^{-1}—compared with other shelf regions in the Northwest Pacific (Tang 1989). Primary production in the northern region is usually higher than that in the southern region, and the lowest level appears in the southwest coastal water of the sea (Liu, 1984). But a much higher level of primary production is found in the southeast coastal water of the sea, about 141 gC $\cdot$ m^{-2} $\cdot$ yr^{-1} (Choi *et al.*, 1988). The production of phytoplankton in the sea is estimated to be 0.52 $\cdot$ 10^9 tonnes $\cdot$ km^{-2} $\cdot$ yr^{-1}, which is lower than that of the East China and South China seas (Yang, 1985;

small and vertical stratification develops. The surface temperature in the central area of the northern Yellow Sea cold water mass may be as warm as 28°C or more; the lowest bottom (50 m) temperature may be 4°–5°C. The top of the thermocline lies 5 to 15 m below the surface with a vertical temperature gradient greater than 10°C/10 m; it is one of the sharpest thermoclines in the China Seas. Weng *et al.* (1989) have reported that the distribution range of the cold water mass has distinct secular interannual variations, with a "relative volume" in the stronger years that is 2.2 times greater than that in weaker years. Temperature and salinity varied by 7.7°C and 2.58%, respectively, during the period 1957–1985.

Concentrations of dissolved oxygen and nutrients have seasonal variations similar in some respects to those of the hydrographic variables. In winter, their distributions are vertically homogeneous, while in summer they are stratified. In general, the oxygen concentrations at the surface and bottom range from 6.4 to 8.0 ml/l and from 4.0 to 6.4 ml/l, respectively (Su, 1987). The higher nutrient concentrations in the nearshore shallow water zone are caused by river runoff. In deeper water, the nutrient concentrations are greater below the thermocline.

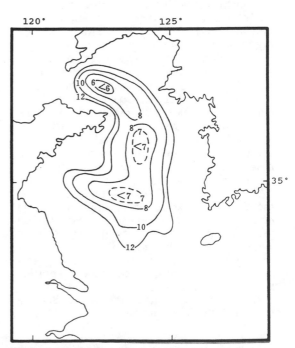

Figure 2. Distribution of bottom water temperatures (°C) of the Yellow Sea cold water mass in summer (from Sun *et al.*, 1981).

Phosphates, PO₄-P (mg/m³)

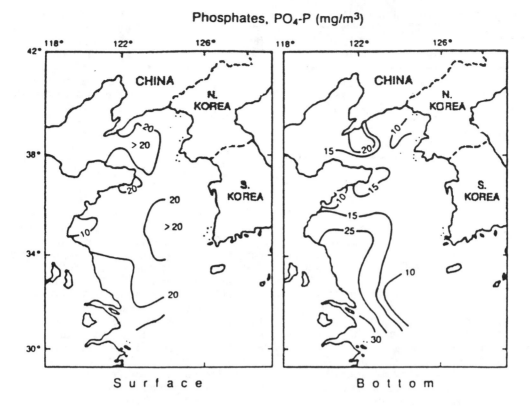

Silicates, SiO₃-Si (mg/m³)

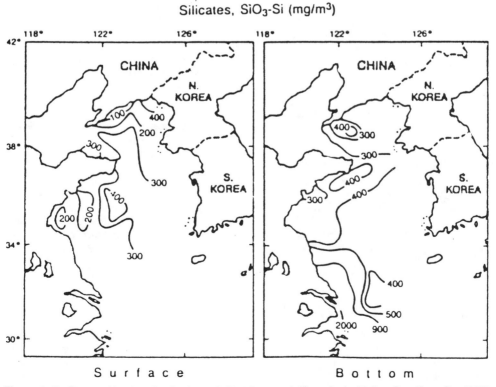

Figure 3. Surface and bottom distributions of phosphate and silicate in the Yellow Sea (from Su, 1987).

Table 1. Phytoplankton biomass in the mid-Yellow Sea (34°–37° 30'N; x 10^3 cells $\cdot$ m^{-3}).

Year	1961	1985
Spring	496	694
Summer	142	55
Autumn	818	1234
Winter	576	28
Average	508	550

Gu *et al.*, 1990). Over the past 30 years, the phytoplankton biomass has been relatively stable—about 508,000 cells $\cdot$ m^{-3} in 1961 and 550,000 cells m^{-3} in 1985 (Table 1). However, during this same period, the phytoplankton biomass in the Bo Hai Sea increased noticeably (Deng, 1988).

The biomass of zooplankton in the Yellow Sea is also lower than that of adjacent areas, ranging from 5 to 50 mg m^{-3} in the north to 25 to 100 mg m^{-3} in the south, because of the influence of the warm current. The dominant species, including *Sagitta crassa, Calanus sinicus, Euphausia pacifica,* and *Themisto gracilipes,* are all important food for pelagic and demersal fish and invertebrates. The seasonal changes in zooplankton production are con-

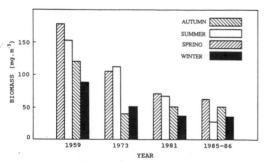

Figure 4. Seasonal trends of zooplankton biomass in the Yellow Sea for 1959, 1973, 1981, and 1985–1986.

sistent. However, the annual biomass yield in the sea has decreased noticeably since 1959 (Fig. 4); this is similar to the trend found in the East China Sea (Gu *et al.*, 1990).

The benthic biomass in the Yellow Sea is consistently low (Table 2). Of the groups comprising the benthic biomass, *Mollusca* is the most important group (about 37%), followed by *Polychaeta* (about 27%), *Echinodermata* (about 18%), *Crustacea* (about 3%), and others (about 15%). These bottom dwellers include important prey items and commercially important species (e.g., southern rough shrimp).

The fauna of the Yellow Sea are recognized as a sub-East Asia zoo-geographic province of the North Pacific Temperate Zone (Cheng, 1959; Liu, 1959, 1963; Dong, 1978; Zhao *et al.*, 1990). These resource populations are composed of species with various ecotypes. Warm temperate species comprise the majority, accounting for about 60% of the total biomass of the resource populations. Warm-water and cold temperate species account for about 15% and 25%, respectively. Demersal and semidemersal species account for about 58%, and pelagic species about 42% (Fig. 5). Because most of the species inhabit the Yellow Sea year-round, the faunal resource populations have formed an independent community. The diversity and abundance of this community are comparatively lower than are found in the East China and South China Seas. The Shannon-Wiener diversity index (H') and Simpson ecological dominance index (C) of the resource populations were determined to be 2.3 and 0.34, respectively (Tang, 1988).

The fishery resources in the Yellow Sea are multispecies in nature. Approximately 100 species are commercially harvested, including demersal fish (about 66%), pelagic fish (about 18%), cephalopods (about 7%), and *Crustacea* (about 9%). The annual catch of about 20 species exceeds 10,000 metric tonnes, accounting for less than 40% of the total catch. The remaining 60% includes about 80 commercially harvested species. However, in 1985–1986, about 20 major species accounted for 92% of the total biomass of the resource populations, and about 80 species accounted for the other 8% (Table 3).

Table 2. Benthic biomass in the mid-Yellow Sea (gm m^{-2}).

Year	Mollusca	Polychaeta	Echinodermata	Crustacea	Others	Total
1958–59	8.0	4.5	3.1	0.8	4.9	21.3
1975–76	8.8	7.9	4.8	0.7	1.8	24.0

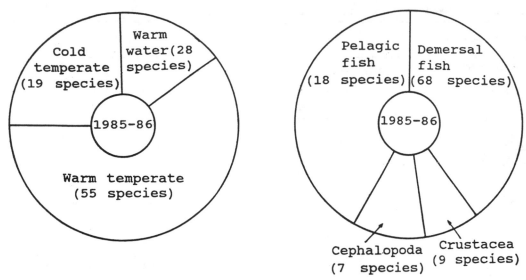

Figure 5. Proportion of the various ecotypes of the total biomass of resource populations based on research vessel bottom trawl surveys of the Yellow Sea in 1985–1986.

The Yellow Sea food web is relatively complex, with at least four trophic levels (Fig. 6). Japanese anchovy and macruran shrimp (e.g., *Crangon affinis* and southern rough shrimp) are keystone species. They occupy an intermediate position in the food web and are important food resources for higher trophic levels. About 40 species eat anchovy, including almost all of the higher carnivores of the pelagic and demersal fish, and the cephalopods. Japanese anchovies are abundant in the Yellow Sea, with an annual biomass estimated at 2.5–2.8 million metric tonnes in 1986–1988 (Zhu and Iversen, 1990). *Crangon affinis* and southern rough shrimp, which are eaten by most demersal predators (about 26 species), are also numerous and widespread in the Yellow Sea.

Changes in Biomass Yields and Their Causes

The Yellow Sea is one of the most intensively exploited areas in the world. Commercial utilization of the living resources in this ecosystem dates back several centuries. With a remarkable increase in fishing effort and its expansion to the entire Yellow Sea, nearly all the major stocks were fully fished by the mid-1960s, and by the end of that decade the resources in the ecosystem were being overfished (Fig. 7). Many studies have shown that biomass yields in the ecosystem have declined greatly (Xia, 1978; Liu, 1979; Chikuni, 1985; Zhao *et al.*, 1990). From the early 1960s to the 1980s, the fish and invertebrate biomass declined by about 40%, while fishing effort increased threefold. The proportion of the catch of traditional target species to the total catch decreased from 25% in the 1950s to about 10% in the late 1980s (Tang, 1989).

During this period, there was a marked shift in the dominant species—from small yellow croaker and hairtail in the 1950s and early 1960s to Pacific herring and chub mackerel in the 1970s. Small-sized, fast-growing, short-lived, low-valued species, such as Japanese anchovy, half-fin anchovy, and scaled sardine, increased markedly in abundance during the 1980s and assumed a prominent position in the ecosystem resources and food web thereafter. As a result, larger, higher trophic level, commercially important demersal species were replaced by smaller, lower trophic level, pelagic, less-valuable species. As a result, harvestable living resources in the ecosystem declined in quality. The major resource populations in 1958–1959 were small yellow croaker, flatfish, cod, hairtail, skate, sea robin, and angler, which accounted for 71% of the total biomass yield (Fig. 8). Herbivores represented 11%; benthophagic species, 46%; and ichthyophagic species, 43%. By 1985–1986, the major exploitable resources had shifted to Japanese anchovy, half-fin anchovy, squid, sea snail, flatfish, small yellow croaker, and scaled sardine. Of these, 59% were planktophagic species, 26% were benthophagic species, and 16% were ichthyophagic species. In addition, about 79% of the biomass yield in 1986 consisted of fish and invertebrates smaller than 20 cm. Their mean standard length was only 11 cm, with a

Table 3. Major resident species of resource populations in the Yellow Sea.

I. Pelagic species

Common name	Scientific name	Biomass[*] (%)	Harvest[†] (%)
Japanese anchovy	*Engraulis japonicus*	37.0	(2.8)
Half-fin anchovy	*Setipinna taty*	7.3	(3.0)
Scaled sardine	*Harengula zunasi*	3.7	(1.8)
Pacific herring	*Clupea pallasii*	2.0	0.2
Horse mackerel	*Trachurus japonicus*	2.9	
Chub mackerel	*Pneumatophorus japonicus*	1.7	2.2
Spanish mackerel	*Scomberomorus niphonius*	1.2	4.7
Butterfish	*Stromateides argenteus*	2.7	0.8

II. Demersal and semi-demersal species

Common name	Scientific name	Biomass[*] (%)	Harvest[†] (%)
Small yellow croaker	*Pseudosciaena polyactis*	4.2	1.2
Jewfish	*Johnius belengerii*	1.1	(2.8)
Croaker	*Collichthys niveatus*	1.0	(2.8)
Skates	*Rajidae*[‡]	3.2	(1.0)
Eelpout	*Zoarceselongatus*	1.7	1.4
Sea snail	*Liparis tanakae*	5.3	
Flatfish	—[§]	4.2	(1.8)
Angler	*Lophius litulon*	2.0	
Japanese squid	*Loligo japonicus*	3.5	(1.0)
Cuttlefishes	—[‖]	2.3	1.7
Fleshy prawn	*Penaeus orientalis*	1.6	
Southern rough shrimp	*Trachypenaeus curvirostris*	1.9	
Crangonid shrimp	*Crangon affinis*	1.6	
Blue crab	*Portunus trituberulatus*	3.5	(1.9)
	TOTAL	91.9	(34.6)

[*],[†] Based on research vessel bottom trawl surveys and fisheries statistics in 1985–1986.
[‡] Primarily *R. pulchra* and *R. chinensis*
[§] Primarily *Cleisthenes herzensteini*
[‖] Primarily *Sepiella maindroni*

mean weight of 20 g (Fig. 9), compared to a mean standard length in the 1950s–1960s of 20 cm. The biomass of less-valuable species increased by about 23% between the 1950s and 1980s (Fig. 10).

It is also possible that environmental factors may have had an influence on long-term changes in species abundance and composition. The catch of warm-water and temperate-water species tended to increase during the warm years, whereas the catch of boreal species such as herring tended to increase during the cold years (Fig. 11). However, the biomass flip of small pelagic species does not appear to be associated with temperature. There is no evidence of climatic or other large-scale environmental change during this period.

Ecosystem Effects of Overfishing

Small yellow croaker and hairtail were formerly the important commercial demersal species in the Yellow Sea, with catches in 1957 reaching a peak of about 200,000 metric tonnes (mt) and 64,000 mt, respectively. However, as

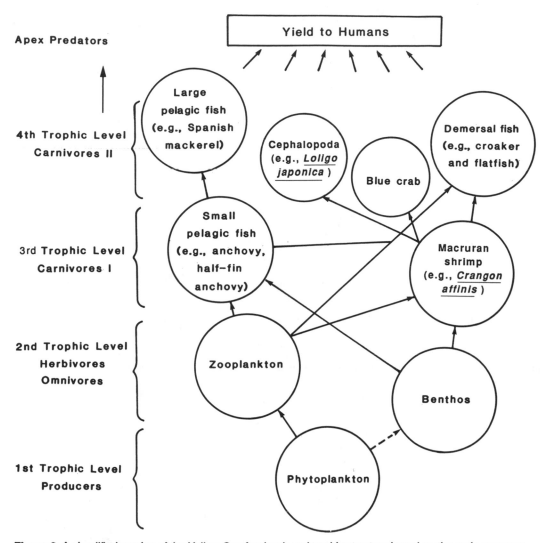

Figure 6. A simplified version of the Yellow Sea food web and trophic structure based on the main resource populations in 1985–1986.

young fish were heavily exploited in the over-wintering grounds and spawning stocks were intensively fished in their spawning grounds, the biomass of these two species declined sharply since the mid-1960s (Fig. 12). The Yellow Sea hairtail stock became a nontarget species in the 1970s, with a biomass estimated to be only 1/30 of previous levels (Lin, 1985). The present catch of small yellow croaker is only 30,000–40,000 mt, 80% of which are young fish aged 0 to 1 year. This decline in biomass was accompanied by a substantial reduction in its distribution, an increase in growth rate, earlier maturation, and a decrease in the mean age and body length of the spawning stock (Mio and Shinohara, 1975; Lee, 1977; Otaki and Shojima, 1978; Zhao *et al.*, 1990). Spawning

stock size in the early 1980s was thought to be 1/20 of its previous level (Mao, 1983).

After the resources of small yellow croaker off the Jiangsu coast in the southern Yellow Sea (about 33°N, 122°E) were depleted, the biomass yield of large yellow croaker increased, with the annual catch ranging from 40,000 to 50,000 mt during the period from 1965 to 1975. Because the overwintering stock was heavily exploited in the late 1970s, the biomass decreased sharply in the early 1980s and spawning stock size declined to about one-sixth of that in the 1960s. Conversely, the biomass yield of butterfish increased when small and large yellow croaker off Jiangsu decreased (Fig. 13). There was no evidence of climatic or large-scale environmental change during this period;

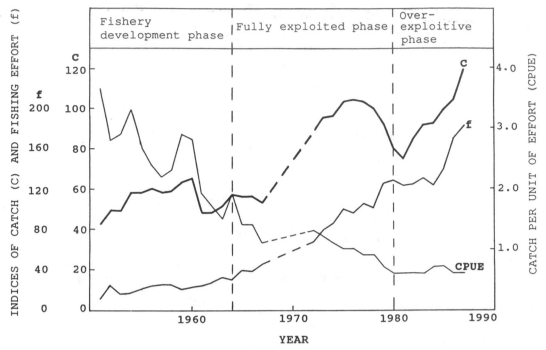

Figure 7. Trends in catch and effort for Yellow Sea fisheries, 1950s–1980s (generalized history of the fisheries in the Yellow Sea).

thus, the major cause of the fluctuations in biomass and shifts in species dominance in this area appears to be overexploitation. It is evident that overexploitation can be of sufficient magnitude to result in a species "flip" from a position of dominance to a subordinate position within an ecosystem. A biomass flip occurs when the population of a dominant species rapidly drops to a very low level and is replaced by another species (Sherman, 1989).

Flatfish and Pacific cod are important boreal species in the sea whose distribution and migration are related to the movement of the Yellow Sea cold water mass. The catches of flatfish and Pacific cod reached peaks of 30,000 mt in 1959 and 80,000 mt in 1972, respectively. As fishing efforts increased, young fish were heavily harvested, and the biomass yields of both species have decreased since the early 1970s (Fig. 12). The present catch is estimated at 20,000 mt for flatfish and about 3,000 mt for Pacific cod.

Overfishing has also caused a decline in biomass of several other demersal species, including sea robin, red sea bream, *Miichthys miiuy,* and *Nibea alibiflora.* However, under the same fishing pressure, the biomass yields of other exploited resources, including cephalopods, skate, white croaker, and daggertooth pike-conger, appear to be fairly stable (Fig. 12).

This stability may be a result of their scattered distributions.

Biomass of demersal species, such as fleshy prawn, may be affected by both natural and anthropogenic factors. Catch of fleshy prawn varies from year to year (Fig. 12), with an annual catch that ranged from 10,000 to 50,000 mt annually during the period from 1953 to 1988. Tagging and recapture results indicate that there are two geographically separated populations—a small one on the west coast of Korea and a large one on the coast of China. The main spawning grounds lie near the estuaries along the coast of the Bo Hai Sea. When water temperatures begin to drop significantly in autumn, the prawns migrate out of the sea to overwinter in the Yellow Sea Depression. We found that fluctuations in recruitment were correlated with both environmental factors and spawning stock size and that the relative importance of the two factors varied from year to year. Examples of the three possible combinations of dominant factors have been noted, namely, environment, spawning stock size, and environment plus spawning stock size. Similar trends in the indices of environmental conditions and recruitment have been documented for many years (Fig. 14), and spawning stock effects were apparent during 1981–1983 when low recruitment coincided with a low spawn-

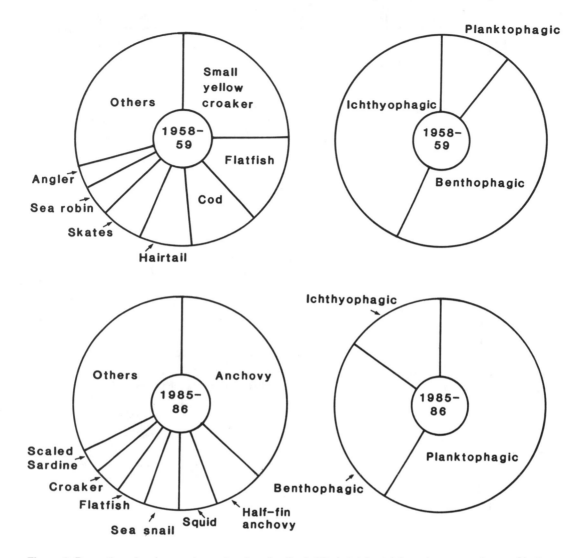

Figure 8. Proportion of major species and various feeding habits in total catch based on research vessel bottom trawl surveys of the Yellow Sea in 1958–1959 and 1985–1986.

ing stock (Tang *et al.*, 1989). Because of excessive fishing (the monthly coefficient of fishing mortality is about 0.7–0.9 [Deng *et al.*, 1982; Tang, 1987a]), fluctuations in recruitment have been aggravated and spawning stock has been reduced to below-normal levels in recent years.

Pacific herring, chub mackerel, Spanish mackerel, and butterfish are the major, larger-sized pelagic stocks in the sea. The annual catch from 1953 to 1988 fluctuated wildly, ranging from 30,000 to 300,000 mt per year. The causes of these fluctuations are more complicated. There may be two patterns of population dynamics. The variability is particularly significant for Pacific herring and chub mackerel stocks, whereas Spanish mackerel and butter-

fish stocks appear to be relatively constant (Figs. 15 and 16).

Yellow Sea Pacific herring has a long and dramatic history of exploitation. In the twentieth century, the commercial fishery experienced two peaks (in about 1900 and 1938), followed by periods of little or no catch. In 1967, a large number of young herring began appearing in the bottom trawl catches as the stocks recovered. After 1967, catches increased rapidly to a peak of 200,000 mt in 1972, but then declined to about 1,000 mt in the late 1980s. The fluctuations in recruitment of this stock are very large and directly affect fishable biomass; however, there is no strong relationship between spawning stock size and recruitment

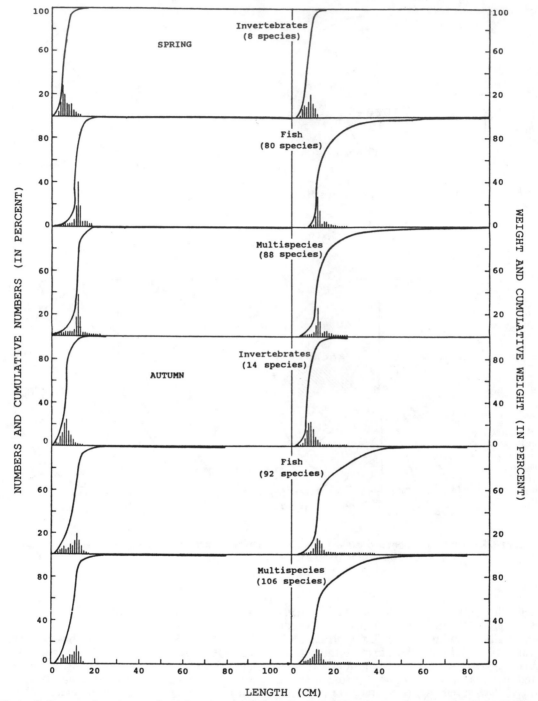

Figure 9. Percent of numbers and weight at length of all species caught from bottom trawl surveys of the Yellow Sea in 1986.

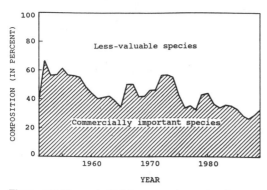

Figure 10. Percent of high- versus low-valued species in the annual catch from the Yellow Sea, 1951–1989.

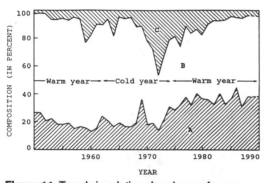

Figure 11. Trends in relative abundance of warm-water, temperate-water, and boreal species comprising the annual catch in the Yellow Sea from the 1950s to the 1980s, and long-term changes in environmental conditions. (A) Warm-water species; (B) Temperate-water species; (C) Boreal species.

(Tang, 1981, 1987b). When the herring stock increased in the 1970s, zooplankton biomass, the main food of herring, decreased (Fig. 4). Environmental conditions such as rainfall, wind, and daylight are the major factors affecting fluctuations in recruitment, and long-term changes in biomass yields may be correlated with a 36-year wetness oscillation cycle in eastern China (Fig. 17). It is quite likely that the high catch rate (F = 1.14–2.97 in 1973–1984) of the herring fishery has exacerbated the fluctuations in the stock size (Tang, 1987b).

Commercial utilization of Spanish mackerel stock began in the early 1960s, when both the catch and abundance of small yellow croaker and hairtail decreased. In 1964, the catch reached 20,000 mt. Since then, biomass yields of this stock have steadily increased. Catch has peaked at 60,000–70,000 mt in recent years, although the stock has borne excessive fishing predation. (The fishing mortality coefficient is about 1.0, and young fish, aged 0.3–1 year, account for 80% of the total catch.) The reason for this is not clear. Spanish mackerel feed on anchovy. As mentioned above, anchovy, a major prey for carnivorous species in the ecosystem, is very abundant in the sea. Perhaps this abundance is caused by an unusual combination of natural and anthropogenic conditions.

Conclusion

It is now accepted that both physical and biological factors can perturb the biomass yields of the resource populations in the Yellow Sea ecosystem. There may be two types of species shifts in the ecosystem resources: systematic replacement, in which one dominant species declines in abundance or is depleted by overexploitation while competing species use surplus food and vacant space to increase their populations; and ecological replacement, in which minor changes in the natural environment can have consequences for stock abundance, especially for pelagic species. Thus, in the long run, the effects of the two types of species shifts may be mingled, so that the causes of changes in biomass yields are extremely difficult to isolate. However, as discussed above, over the past 4 decades, human predation has been the principal cause for the biomass yield flips in quantity and quality in the Yellow Sea ecosystem, especially in demersal species.

Based on the above, effective ecosystem management in the Yellow Sea should be encouraged. A continuation of strict regulatory measures is essential, and intensified conservation of young fish and spawning stocks of commercially important species would be the most important task in ecosystem resource management. In order to realize this goal, it will be necessary to establish a multinational monitoring system of ecosystem resources in the Yellow Sea, including the following considerations: adequate data standardization, collection, and exchange (e.g., catch statistics by species, effort, and country); fishery-independent surveys of major resource populations; research on life history, stock assessment, and biological dynamics of migratory species; and process-oriented studies of ecosystem structure and function.

Since 1984, about 4 hundred million juvenile prawns (>3 cm) were introduced into the southern coastal waters of the Shandong Peninsula, and a total of 1,000 mt was caught 3 months later. About 13.2 billion juvenile prawns were released in the Bo Hai and Yellow Seas by

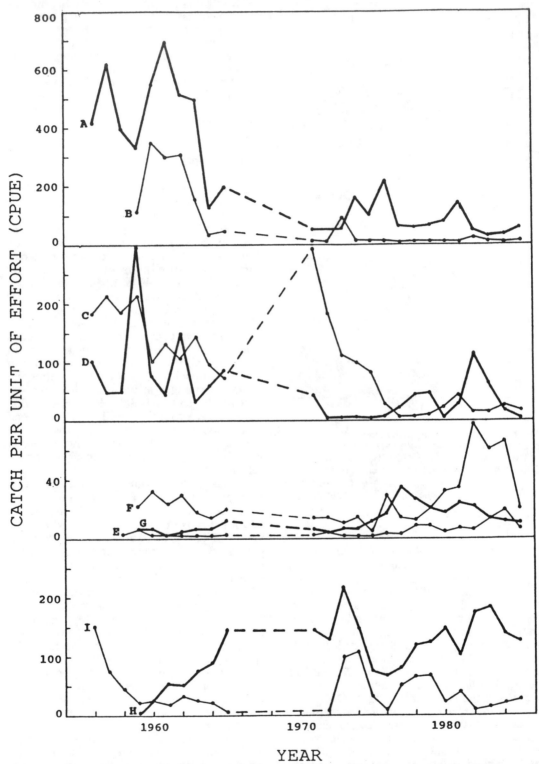

Figure 12. Catch per unit of effort (CPUE), expressed in kg caught per haul by paired trawlers, for (A) small yellow croaker, (B) largehead hairtail, (C) flatfish, (D) Pacific cod, (E) white croaker, (F) skate, (G) daggertooth pike-conger, (H) cephalopods, and (I) fleshy prawn.

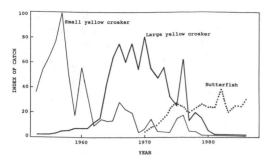

Figure 13. Changes in annual catch of dominant species of the fishery off Jiangsu in the southern Yellow Sea.

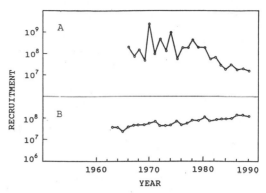

Figure 16. Recruitment of Pacific herring (A) and Spanish mackerel (B) in the Yellow Sea.

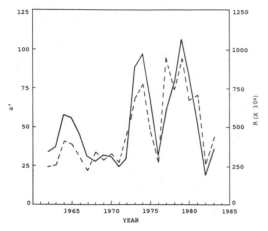

Figure 14. Indices of environmental conditions (a' = __) and recruitment (R = ---) for prawns (from Tang *et al.*, 1989).

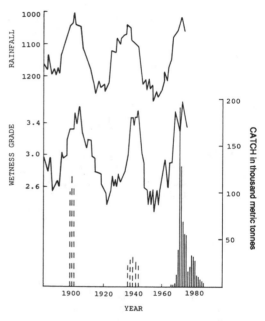

Figure 17. Relationship between the fluctuations in herring abundance of the Yellow Sea and the 36-year cycle of wetness oscillation in eastern China (adapted from Tang, 1981).

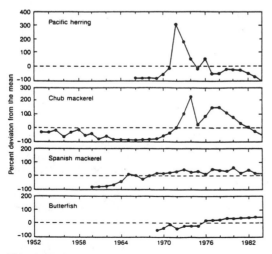

Figure 15. Annual catch of Pacific herring, chub mackerel, Spanish mackerel, and butterfish expressed as a percent deviation from the mean (from Tang, 1989).

1989, and the catch was about 25,000 mt. This success is encouraging. It not only promotes the development of artificial enhancement programs in the sea, but it also brings hope for recovery of ecosystem resources. Therefore, as an effective resource management strategy, artificial enhancement practices should be encouraged and expanded to an ecosystem level in the Yellow Sea.

References

Cheng, C. 1959. Notes on the economic fish fauna of the Yellow Sea and the East China Sea. Oceanol. Limnol. Sin. 2 (1):53–60. In Chinese.

Chikuni, S. 1985. The fish resources of the Northwest Pacific. FAO Fish. Tech. Paper 266. FAO, Rome.

Choi, J., et al. 1988. The study on the biological productivity of the fishing ground in the western coastal area of Korea, Yellow Sea. Bull. Nat. Fish. Res. Dev. Agency 42:143–168. In Korean with English abstract.

Deng, J. 1988. Ecological bases of marine ranching and management in the Bohai Sea. Mar. Fish. Res. 9:1–10. In Chinese with English abstract.

Deng, J., Han, G., and Ye, C. 1982. On the mortality of the prawn in the Bohai Sea. J. Fish. China 6(2):119–127. In Chinese with English abstract.

Dong, Z. 1978. The geographical distribution of cephalopods in Chinese waters. Oceanol. Limnol. Sin. 9(1):108–116. In Chinese with English abstract.

Gu, X. et al. (Editors). 1990. Marine fishery environment of China. Zhejiang Science and Technology Press, In Chinese.

Guan, B. 1963. A preliminary study of the temperature variations and the characteristics of the circulation of the cold water mass of the Yellow Sea. Oceanol. Limnol. Sin. 5(4): 255–284. In Chinese.

Ho, C., Wang, Y., Lei, Z., and Xu, S. 1959. A preliminary study of formation of Yellow Sea cold water mass and its properties. Oceanol. Limnol. Sin. 2:11–15. In Chinese.

Lee, J. 1977. Estimation of the age composition and survival rate of the yellow croaker in the Yellow Sea and East China Sea. Bull. Fish. Res. Dev. Agency 16:7–31. In Korean.

Lin, J. 1985. Hairtail. Agriculture Publishing House, Beijing. In Chinese.

Liu, B. 1984. Estimating the primary production of the Yellow Sea with satellite imagery. J. Fish. China 8(3):227-234. In Chinese with English abstract.

Liu, J. 1959. Economics of micrurus crustacean fauna of the Yellow Sea and the East China Sea. Oceanol. Limnol. Sin. 2(1):35–42. In Chinese.

Liu, J. 1963. Zoogeographical studies on the micrurus crustacean fauna of the Yellow Sea and the East China Sea. Oceanol. Limnol. Sin. 5(3):230–244. In Chinese.

Liu, X. 1979. Status of fishery resources in the Bohai and Yellow Seas. Mar. Fish. Res. Paper 26:1–17. In Chinese.

Mao, X. 1983. Status of yellow croaker stocks. In: Conditions of fisheries resources in the Yellow Sea and East China Sea. pp. 15–32. Ed. by East China Sea Fisheries Research Institute and Yellow Sea Fisheries Research Institute. In Chinese.

Mio, S., and Shinohara, F. 1975. The study on the annual fluctuation of growth and maturity of principal demersal fish in the East China and the Yellow Sea. Bull. Seikai Reg. Fish. Res. Lab. 47:51–95.

Otaki, H., and Shojima, S. 1978. On the reduction of distributional area of the yellow croaker resulting from decrease abundance. Bull. Seikai Reg. Fish. Res. Lab. 51:111–121.

Sherman, K. 1989. Biomass flips in large marine ecosystems. In: Biomass yields and geography of large marine ecosystems. pp. 327–331. Ed. by K. Sherman and L. M. Alexander. AAAS Selected Symposium 111. Westview Press, Inc., Boulder, CO.

Sherman, K., and Alexander, L. M. (Editors). 1986. Variability and management of large marine ecosystems. AAAS Selected Symposium 99. Westview Press, Inc., Boulder, CO.

Sherman, K., and Alexander, L. M. (Editors). 1989. Biomass yields and geography of large marine ecosystems. AAAS Selected Symposium 111. Westview Press, Inc., Boulder, CO.

Su, J. 1987. Physical oceanography of the Yellow Sea: With emphasis on its western part. Paper presented at the International Conference on the Yellow Sea, 23–26 June, East-West Center, Honolulu.

Sun, X., et al. (Editors). 1981. Marine hydrography and meteorology in the coastal water of China. Science Press, Beijing. In Chinese.

Tang, Q. 1981. A preliminary study on the causes of fluctuations in year class size of Pacific herring in the Yellow Sea. Trans. Oceanol. Limnol. 2:37–45. In Chinese.

Tang, Q. 1987a. Estimates of monthly mortality and optimum fishing mortality of Bohai prawn in North China. Collect. Oceanic Works 10:106–123.

Tang, Q. 1987b. Estimation of fishing mortality and abundance of Pacific herring in the Yellow Sea by cohort analysis (VPA). Acta Oceanol. Sin. 6(1):132–141.

Tang, Q. 1988. Ecological dominance and diversity of fishery resources in the Yellow Sea. J. Chinese Academy of Fishery Science 1(1):47–58. In Chinese with English abstract.

Tang, Q. 1989. Changes in the biomass of the Yellow Sea ecosystem. In: Biomass yields and geography of large marine ecosystems. pp. 7–35. Ed. by K. Sherman and L. M. Alexander. AAAS Selected Symposium 111. Westview Press, Inc., Boulder, CO.

Tang, Q., Deng, J., and Zhu, J. 1989. A family of Ricker SRR curves of the prawn under different environmental conditions and its enhancement potential in the Bohai Sea. Can. Spec. Publ. Fish. Aquat. Sci. 108:335–339.

Weng, X., Zhang, Y., Wang, C., and Zhang, Q. 1989. The variational characteristics of the Yellow Sea cold water mass. J. Ocean University of Qingdao 19(I–II):119–131.

Xia, S. 1978. An analysis of changes in fisheries resources of the Bohai Sea, Yellow Sea, and East China Sea. Mar. Fish. Res. Paper 25:1–13. In Chinese.

Yang, J. 1985. Estimates of exploitation potential of marine fishery resources in China. In: Proceedings of the strategy of ocean development in China. pp. 107–113. Ocean Press, Beijing. In Chinese.

Yu, M., *et al.* (Editors). 1990. Fishery resources of intertidal zone and shallow sea in China. Zhejiang Science and Technology Press, In Chinese.

Zhao, C., *et al.* (Editors). 1990. Marine fishery resources of China. Zhejiang Science and Technology Press, In Chinese.

Zhu, D., and Iverson, S. A. (Editors). 1990. Anchovy and other fish resources in the Yellow Sea and East China Sea, November 1984–April 1988. Mar. Fish. Res. 11:1–143. In Chinese with English abstract.

Chapter 11

Long-term Variability in the Food Chains, Biomass Yields, and Oceanography of the Canary Current Ecosystem

Carlos Bas

Introduction

The purpose of this paper is to characterize the Canary Current ecosystem. Historically, the continental shelf off Africa has been considered separately from the Canary Islands and as the only important area for primary production in the northwest African region (Bas and Cruzado, 1974, 1976; Cruzado, 1974; Cruzado and Manriquez, 1974; Fraga, 1973; and Font, 1977). This author thinks that it is necessary to con-

sider the Canary Islands and shelf as a whole. The intensity of the Canary Current is known to be very strong near the continental coast, becoming progressively weaker offshore. The acceleration of the current as it passes between the different islands is documented (Sangra, P. pers. com.; Fedoseev, 1970). In this paper, the phenomenon is examined from both a hydrographic and an ecological standpoint. Furthermore, the dynamic features of the Canary Current appear to be significant in explaining the functioning of the ecosystem.

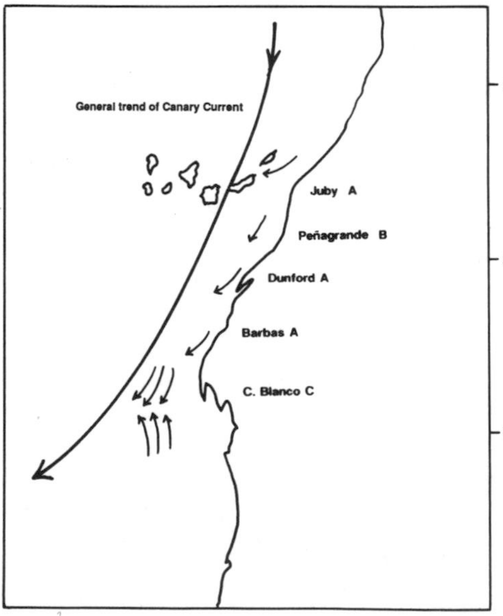

Physical Oceanography of the Area

Attention has been focused on the Canary Islands because of their geographic location between Gibraltar and Mauritania. The structure of the coast and the continental shelf are important in determining the position and features of major upwelling areas. The structure of the coastline at Cape Juby, Cape Bojador, Point Durnford, and Cape Barbas shifts the current oceanward and thus gives rise to upwelling (Fig. 1) (Anonymous, 1972).

The water at 50 m depth on the continental shelf also undergoes a major transverse shift at around 25° N off Peñagrande (Cruzado, 1974). This bottom structure is responsible for intense upwelling between 25° N and 23° N (Fig. 2).

In the southern part of the region, the Canary Current comes into contact with intertropical water over the shelf off Mauritania, producing the most active upwelling (Cape Blanco, 21° N).

The front that forms between the Canary Current and the intertropical water mass is not stationary but moves along the Mauritanian

Figure 1. General flow of the Canary Current upwelling areas: (A) coastal effects, (B) bottom influence, and (C) water mass interaction.

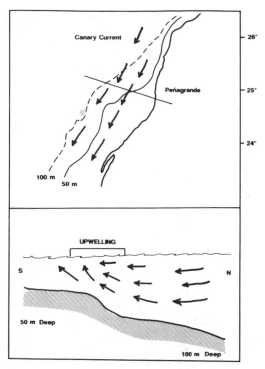

Figure 2. Upwelling mechanism and oceanographic dynamics in the Peñagrande area (Saharan coast, 25° N).

coast, particularly between Cape Blanco and Cape Timiris. It is situated near Cape Blanco during the summer. This is comparable to the situation found off the coast of Angola and Namibia in southern Africa (Fig. 3).

Variations in current intensity parallel this situation, with intensity strong in winter, with eddies at Bojador (27° N) and Cape Blanco (21° N) (Fig. 4). During summer, the situation is reversed (Fig. 5). The current is weak in the frontal zone and strong off Cape Blanco. This is clearly demonstrated by comparing the location of the north- and south-central water masses in the summer at different depth intervals (0–50–100–200 m) and clearly shows the front to be located around 21° N (Fig. 6) (Anderson, 1973; Fraga, 1973).

The influence of the Canary Islands is also quite distinct. First, the weak current intensity is boosted by the Venturi effect when it passes between the islands. Second, the island-mass effect produces a "shade" zone with warmer water on the southern side of the islands. This, in turn, leads to an accumulation of marine resource biomass (Hernandez-Leon, 1988). Lastly, contact and friction between the current and the shade zones give rise to two eddies, a cyclonic gyre in the west and an anticyclonic gyre in the east. The former has an effect similar to upwelling and acts to accumulate plankton (zooplankton) from the north.

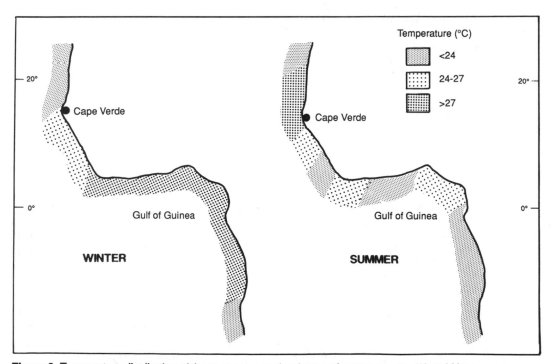

Figure 3. Temperature distribution of the water masses in winter and summer on the West African coast.

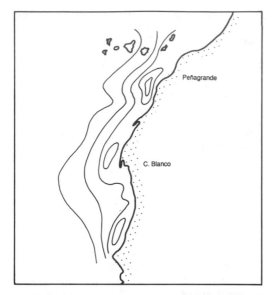

Figure 4. Marine currents in West Africa and Canary Islands. Winter: strong intensity. (From Fedoseev, 1970.)

Ecological Relationships

Past perceptions of the dynamic interrelationships in the region are not realistic; the true situation is more complex. There is great variability in the upwelling centers, and there also exists a shelf-slope front with numerous filaments (fingers) (Fig. 7). Water temperature variations also occur. In December, warm water masses occupy nearly the entire continental shelf, while cold inshore areas (upwelling centers) are small in number. The filament structure is important, because it promotes or hinders the movement of migratory species (Ruiz-Cañavate *et al.*, 1990).

Productivity

Productivity is high around Cape Juby, Cape Bojador, Peñagrande, Point Durnford, and Cape Barbas, and very high off Cape Blanco (Margalef, 1972, 1973; Fraga, 1973; Velasquez and Cruzado, 1974). In the northern part of the islands, the energy produced as the current collides with the coast supports very high plankton (particularly the small species) production. The biomass is not extremely large, but turnover is pronounced, resulting in high productivity (Fig. 8).

Zooplankton accumulates in the marginal southern areas, where it is heavily preyed upon by fish (*Scomber japonicus*) inhabiting the shade zone off the southern part of the

more important islands. Zooplankton (including amphipoda, copepoda, ostracoda, and mysidacea) biomass is quite important in the upwelling area off Peñagrande in the upper 50 m of the water column (Andreu, 1977). The normal community of zooplankton is composed of copepods, but, in general, mysidacea are very important in all regions (Fig. 9). Margalef (1972) considers the distribution of bacteria associated with phytoplankton to be of major importance. Bacterial abundance is high at the outer edge of the main phytoplankton concentration, especially in the Cape Blanco upwelling area.

Important Fisheries of the Canary Current Ecosystem

Yearly catch figures compiled by the Food and Agriculture Organization of the United Nations (FAO) are summarized in Tables 1–4 and yield the following information. For small pelagic species, peak catches occurred in 1970, 1977, 1981, and 1985; minimum catches occurred in 1971, 1982, and 1986 (Troadec and Garcia, 1979).

Catches of horse mackerel were highest in 1974, lowest in 1984, and fluctuations were generally small. The situation of chub mackerel was similar, with maximum catches in 1970 and 1988, and minimum catches in 1978.

The highest cephalopod catches were taken in 1974, 1983, and 1986; the lowest in 1970 and 1971 (Castro *et al.*, 1989a).

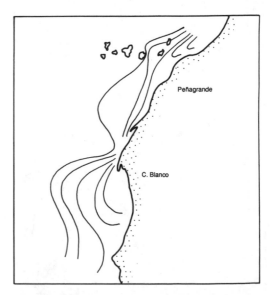

Figure 5. Marine currents in West Africa and Canary Islands. Summer: weak intensity. (From Fedoseev, 1970.)

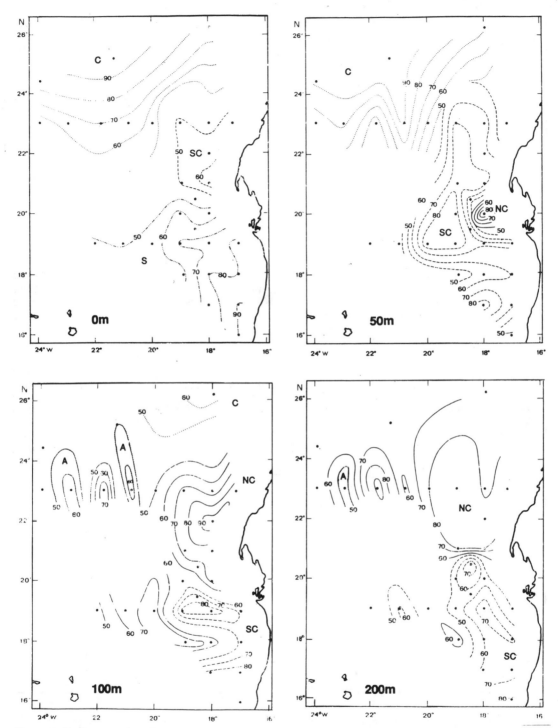

Figure 6. Percentage of north and south central water masses at 0, 50, 100, and 200 m off Cape Blanco during summer. South central [SC] water mass, north central [NC] water mass, Atlantic [A] zone, central [C] zone, southern [S] zone.)

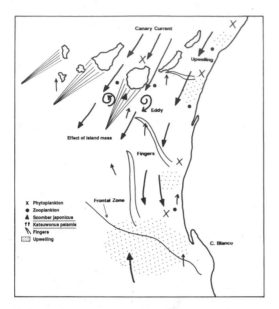

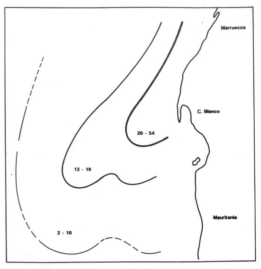

Figure 7. Interrelations between oceanographic dynamics, phytoplankton distribution, zooplankton areas, and *Katsuwonus pelamis* arrivals.

Figure 9. Distribution of the zooplankton (ml/50m³) in Cape Blanco area (summarized from *Atlor* cruises).

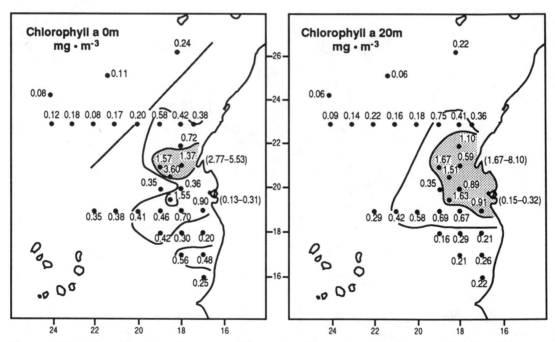

Figure 8. Distribution of the chlorophyll at 0 and 20 m in Cape Blanco area. Stippling indicates areas of high productivity. (From Margalef, 1972.)

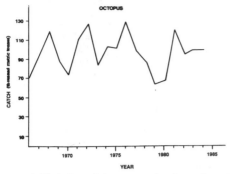

Figure 10. Variation of *Octopus vulgaris* catches (in thousand metric tonnes) in the Saharan shelf.

The influence of oceanographic conditions on the life span of small pelagic species, which normally live for a maximum of 5–7 years, is quite clear. The biomass levels of horse and chub mackerel are more stable because of their ability to prey on different species of zooplankton, their longer life span, and their marked horizontal and vertical mobility.

The behavior of octopus is quite similar to that of the small pelagic species, in that they feed on zooplankton and similar species and live in shallow water (upper 25 m) in the photic zone. Rapid growth and large numbers of eggs result in high turnover; thus, there is a marked resistance to heavy fishing pressure.

Maximum catches of *Sardina pilchardus* were taken in 1976 as a result of increased fishing effort; after that year, catch levels for this

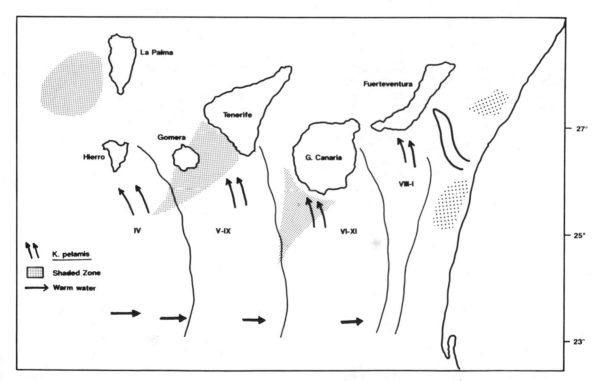

Figure 11. Several migration trends of *Katsuwonus pelamis* during the year, related to increasing temperature toward the coast.

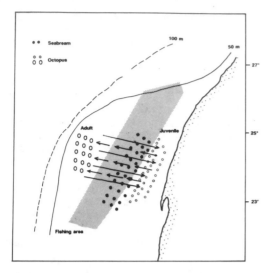

Figure 12. Fishery strategy for octopus on the Saharan shelf. Fishery vessels move in a direction parallel to the coast at 25–50 m deep. Seabream and octopus migrate for reproduction and feeding in a transversal direction across the shelf. Fishing pressure appears to decrease the seabream stocks.

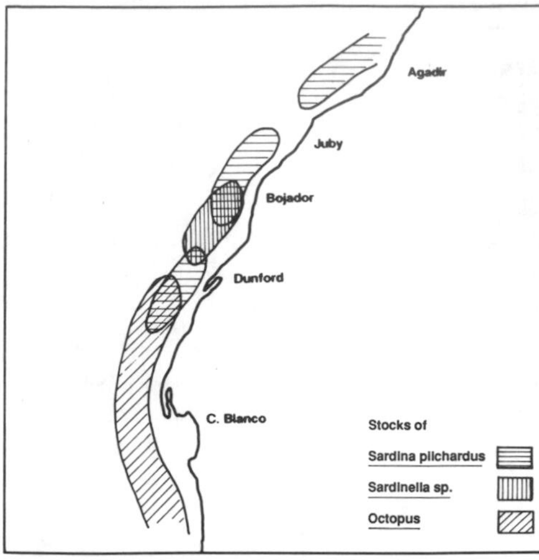

Figure 13. Distribution of *Sardina pilchardus*, *Sardinella aurita*, and *Octopus vulgaris* in the central east (CE) Atlantic zone.

stock were more regular. *Sardinella* sp. were heavily exploited in 1972, with another period of high catches from 1976 to 1981. The fishery for *Engraulis encrasicolus* has followed an upward trend. Catches of *Octopus vulgaris* were highest in 1968, 1972, 1976, 1982, and 1986; and lowest in 1970, 1973, 1979–1980, and 1984–1985 (Fig. 10). This 4-year cycle is brought about by the interaction of two factors: very high turnover, and a high sensitivity to environmental changes (Bas, 1974; Bas *et al.*, 1976).

Interactions between Several Pelagic Species

The most important interaction between pelagic species involves *Scomber japonicus* and *Katsuwonus pelamis* (Gonzalez and Bas, 1989). The early juvenile stages of *Scomber japonicus* live together with sardine stocks offshore and prey mainly on small fishes and zooplankton. Individuals between 1 and 4 years of age dwell in the shade zones on the southern sides of the islands, preying chiefly on zooplankton (copepoda and mysidacea) in the accumulation areas of the eddies. From the age of 4–5 years, they migrate and recruit to the large stocks in the open sea or to the stocks on the continental shelf off Africa.

There is evidence that skipjack (*Katsuwonus pelamis*) feed primarily on *Scomber japonicus*. They migrate in groups from the Gulf of Guinea to the Canary Islands during the year. The first group arrives at the western-most island (Hierro) in April; the last, at Lanzarote in December (Fig. 11). This migration pattern is closely related to the movement of warm water from west to east toward the African coast, until upwelling is confined to a narrow strip. *Katsuwonus pelamis* always migrate along the filaments from the Cape Blanco area. They may also use the filament from the Cape Bojador area to migrate to the eastern region of the islands (Castro *et al.*, 1989b).

The rapid growth and strong reproductive rates of *Octopus vulgaris* lead to high turnover. Fishing activity has collapsed the stocks of seabream, which have slower growth and reproductive rates (Fig. 12), leading to replacement of the seabream by octopus. On the other hand, fishing pressure on *Octopus vulgaris* stocks during reproductive and feeding migration is high, but the effects are mitigated by its rapid turnover rates.

Conclusions

- Hydrographic and climatic conditions of the Canary Current constitute the principal factor driving the dynamics of the Canary Current ecosystem. Oceanic conditions, in general, show both seasonal and longer-term period variations.

Table 1. Annual catches for major fisheries from 1970 to 1988 (in thousand metric tonnes).

Year	1970	1971	1972	1973	1974	1975	1976	1977
Sardine								
Anchovy	1813	945	1029	1108	1127	1268	1436	1443
Sardinella								
Horsemackerel	408	593	614	665	721	643	607	694
Mackerel	283	267	263	186	109	202	197	214
Cephalopoda	112	107	177	191	225	203	179	148

Year	1978	1979	1980	1981	1982	1983	1984	1985
Sardine								
Anchovy	1169	1004	1148	1146	856	1006	939	1836
Sardinella								
Horsemackerel	535	438	633	518	599	501	339	377
Mackerel	105	125	197	143	283	248	241	209
Cephalopoda	161	125	125	184	172	211	161	168

Year	1986	1987	1988
Sardine			
Anchovy	1052	1402	1607
Sardinella			
Horsemackerel	400	453	406
Mackerel	303	195	406
Cephalopoda	206	173	165

Table 2. Annual catches of the major small pelagic species from 1970 to 1988 (in thousand metric tonnes).

Year	1970	1971	1972	1973	1974	1975
Sardina pilchardus	196	211	220	421	341	361
Sardinella sp.	681	532	631	161	104	168
Engraulis encrasicolus	5	3	1	8	18	10

Year	1976	1977	1978	1979	1980	1981
Sardina pilchardus	983	778	435	390	500	521
Sardinella sp.	306	468	463	423	418	354
Engraulis encrasicolus	5	16	2	9	34	27

Year	1982	1983	1984	1985	1986	1987	1988
Sardina pilchardus	432	511	478	460	487	655	795
Sardinella sp.	178	263	211	289	263	298	296
Engraulis encrasicolus	73	51	84	73	61	161	222

Table 3. Annual catches of the major medium pelagic species from 1970 to 1988 (in thousand metric tonnes).

Year	1970	1971	1972	1973	1974	1975	1976
Trachurus sp.	297	488	477	486	501	437	430
Scomber japonicus	268	240	220	156	149	177	154
Year	1977	1978	1979	1980	1981	1982	1983
Trachurus sp.	492	330	248	477	359	405	33
Scomber japonicus	170	83	109	135	114	202	180
Year	1984	1985	1986	1987	1988		
Trachurus sp.	224	254	290	349	332		
Scomber japonicus	211	155	215	118	332		

Table 4. Annual catches of Octopus (*Octopus vulgaris*) from 1970 to 1988 (in thousand metric tonnes).

Year	1970	1971	1972	1973	1974	1975	
Octopus vulgaris	58	112	96	72	94	104	
Year	1976	1977	1978	1979	1980	1981	
Octopus vulgaris	91	69	71	48	65	97	
Year	1982	1983	1984	1985	1986	1987	1988
Octopus vulgaris	124	122	99	94	127	105	100

- There are three upwelling mechanisms: influence of the coastline, structure of the bottom, and interaction between water masses.
- The Canary Islands bring about enrichment of the oceanic zone through increased current intensity, formation of eddies in the southern coasts of the more important islands, and island-mass effects.
- This highly complex structure is complemented by movement of the frontal zone off Mauritania, filaments, and seasonal oceanic water temperature variations.
- The stocks with the highest biomass levels are *Sardina pilchardus*, *Sardinella* sp., *Trachurus* sp., and *Scomber japonicus*. *Octopus vulgaris*, benthic in nature, is regarded here as pelagic (Fig. 13).
- *Katsuwonus pelamis* is the most important predator of the pelagic species. Their migration routes are related to temperature and location.

- Small pelagic species exhibit the largest fluctuations in catches (7-to 8-year cycles). The catch levels for pelagic species are more regular for those that are moderate in size, longer-lived, more varied feeders, and more mobile (*Trachurus* sp. and *Scomber japonicus*).
- There is evidence of species interactions between *Scomber japonicus* and *Katsuwonus pelamis* in the southern area of the Canary Islands.
- Fishing activity is responsible for replacement of seabream by octopus.
- The cycle of maximum catches for octopus is 4–6 years and is closely related to environmental changes.
- Recommendations for scientific research and management include compilation of historical oceanographic data, analysis of mesoscale variations using remote sensing imagery, and investigation of the interactions between continental- and island-related phenomena and populations.

References

Anderson, J. J. 1973. Silicate water mass analysis off the NW coast of Africa. R. Exp. C. Cornide. 2:53–64.

Andreu, P. 1977. Valores de bimassa zooplanctónica de la zona costera de cabo Blanco, NW de Africa. R. Exp. C. Cornide. 6:205–210.

Anonymous. 1972. Hidrografía de la región de afloramiento del NW de Africa. Datos básicos de la campaña "Sahara II" del Cornide de Saavedra. R. Exp. C. Cornide. pp. 1–21.

Bas, C. 1974. Distribucion de especies demersales recogidas durante la expedición oceanográfica "Sahara I." R. Exp. C. Cornide. 3:189–247.

Bas, C., Arias, A., and Guerra, A. 1976. Pesca efectudas durante la campaña "Atlor V" (C. Bojador-C. Blanco, abril-mayo 1974). R. Exp. C. Cornide. 5:161–172.

Bas, C. and Cruzado, A. 1974. Campaña oceanográfica "Sahara I" (20 del VI a 3 del VIII, 1971). Información preliminar. R. Exp. C. Cornide. 3:1–52.

Bas, C. and Cruzado, A. 1976. Campaña oceanográfica "Atlor V" (C. Bojador-C. Blanco, abril-mayo 1974). Características y algunos resultados preliminares. R. Exp. C. Cornide. 5:113–122.

Castro, J. J., Lorenzo, J. M., and Bas, C. 1989a. The population of *Scomber japonicus* in Canary Islands; feeding and growth aspects. Poster presented at the International Symposium on Fish Population Biology, July 17–21, 1989. Aberdeen, Scotland, U.K.

Castro, J. J., Lorenzo, J. M., Gonzalez, A. J., and Bas, C. 1989b. Interaction between *Scomber japonicus* and *Katsuwonus pelamis* in the southern area of Canary Islands. CIEM. La Haja.

Cruzado, A. 1974. Resumen del análisis contínuo en Africa del NW entre 23° y 28° N. R. Exp. C. Cornide. 3:117–128.

Cruzado, A., and Manriquez, M. 1974. Datos hidrográficos de la campaña *Atlor III* en la región del afloramiento entre cabo Bojador y punta Dunford (Sahara español). R. Exp. C. Cornide.

3:89–115.

Fedoseev, A. 1970. Geostrophic circulation of surface waters on the shelf of NW Africa. AtlantNIRO. Rapp. P.-v. Reun. Cons. Int. Explor. Mer 119:32–37.

Font, J. 1977. Distribución superficial de variables oceanográficas en el NW de Africa (C. Bojador-C. Blanco, abril 1974), campaña oceanográfica *Atlor V.* R. Exp. C. Cornide 6:23–40.

Fraga, F. 1973. Oceanografía química de la región de afloramiento del NW de Africa. R. Exp. C. Cornide. 2:3–52.

Gonzalez, A. J., and Bas, C. 1989. The influence of environmental conditions of the skipjack tuna (*Katsuwonus pelamis*) in the Canary Island waters. Poster presented at the International Symposium on Fish Population Biology. July 17–21, 1989. Aberdeen, Scotland, U.K.

Hernandez-Leon, S. 1988. Gradients of mesozooplankton biomass and ETS activity in the wind shear area as evidence of an island mass effect in Canary Island waters. J. of Plankton Res. 10:1141–1154.

Margalef, R. 1972. Fitoplancton de la región de afloramiento del NW de Africa. Pigmentos y producción. (Campaña *Sahara II* de Cornide de Saavedra). R. Exp. C. Cornide. 1:23–51.

Margalef, R. 1973. Fitoplancton marino de la región de afloramiento del NW de Africa. R. Exp. C. Cornide. 2:65–94.

Ruiz-Cañavate, A., Villaneuva, P., Rico, J.A., Morales, F., and Villalobos, J.A. 1990. Campaña oceanográfica Canarias 88. Distribución de parametros físicos. Sección Oceanografía. Instituto Oceanográfico de la Marina. San Fernando (Cadiz-Spain).

Troadec, J. P., and Garcia, S. 1979. Les ressources halieutiques de l'Atlantique C. E. Premiere partie: Les ressources du Golfe de Guinee de l'Angola a la Mauritanie. FAO. Doc. Tech. 186.

Velasquez, Z. R., and Cruzado, A. 1974. Distribución de biomassa fitoplanctónica y asimilación de carbono en el no de Africa. R. Exp. C. Cornide. 3:147–168.

Chapter 12

The Large Marine Ecosystem of Shelf Areas in the Gulf of Guinea: Long-Term Variability Induced by Climatic Changes

Denis Binet and Emile Marchal

Introduction

The Gulf of Guinea ecosystem extends from the Bissagos Islands to Cape Lopez (Fig. 1). Its shelf waters are shared between subsystems; one is characterized by thermal stability of waters, the other by instability. In the first subsystem, nutrient input depends on land drainage, river flood, and turbulent diffusion through a stable pycnocline; in the second subsystem, periodic rising of mid-depth water raises nutrients to the euphotic layer (Binet, 1983a). This seasonal

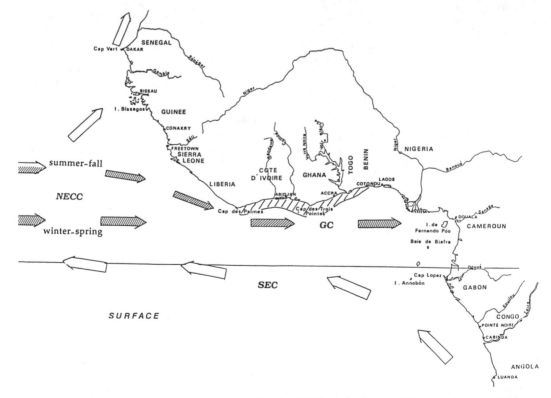

Figure 1. Surface circulation in the eastern tropical Atlantic. NECC: North Equatorial Counter Current; GC: Guinea Current; SEC: South Equatorial Current. Hatched belt along the coast: Seasonal upwelling and *Sardinella aurita* fishery area.

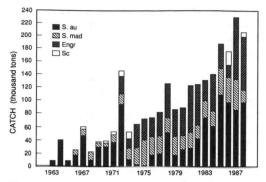

Figure 2. Pelagic catches of the Guinea upwelling LME (Côte d'Ivoire, Ghana, Togo, and Bénin) from 1963 to 1988. S.au: *Sardinella aurita*; S. mad: *Sardinella maderensis*; Engr: *Engraulis encrasicolus*; Sc: *Scomber japonicus*.

upwelling only concerns the part of the coast between Cape Palmas and Cotonou, from approximately 8°W to 2°E. The processes regulating the functioning of these two subsystems are quite different. The Gulf of Guinea upwelling ecosystem differs from eastern boundary current systems because of its proximity to the equator and its seasonal functioning.

Living resources of the Gulf of Guinea have supported increasing fishing effort since the end of the 1950s. Significant changes have occurred in the weight and taxonomic composition of landings since the start of this industrial fishery (Fig. 2). The most abundant pelagic species are *Sardinella aurita*, *S. maderensis*, *Engraulis encrasicolus*, *Brachydeuterus auritus*, and, in certain years, *Scomber japonicus*. During the last 2 decades, the most striking events were the dramatic increase of the trigger fish (*Balistes carolinensis*) during the 1970s, recently followed by a severe decrease (Caverivière, 1982, 1991), and the 1973 collapse of the *Sardinella aurita* fishery, followed by a recovery and unprecedented sustained landings during the 1980s (Binet *et al.*, 1991). Population dynamics alone do not explain these changes; thus, wise management of living resources requires, in addition, an understanding of the climatic forcings on the ecosystem to determine the level of fishing effort appropriate to these changing patterns.

Meteorology and Oceanography

The Inter-Tropical Convergence Zone (ITCZ) is the doldrums area separating northern and southern trade winds. It is a zone of instability, generating rains. On the northern coast of the Gulf of Guinea, this frontal zone separates dry, continental, heavy air masses from wet, maritime, lighter, southern trade wind air masses. Latitudinal migration of the ITCZ generates seasons over the tropical Atlantic Ocean and African land masses.

The Tropical Surface Waters (TSW) with high temperatures and varying salinity overlie a density discontinuity layer at the thermocline. Below this layer lies the South Atlantic Central Water (SACW) (Longhurst, 1962). The salinity of the TSW is lowered by input from rivers, especially during the wet monsoon. The first rains occur in June, when the ITCZ crosses the shoreline to begin its northward ascent. The short coastal rivers are in spate. The second, and most important flooding season, extends from September to November, when the ITCZ migrates southward and the floods of the great Sudanian rivers reach the sea (Fig. 3). Large amounts of warm water of low salinity (Guinean waters) originate also from the Bight of Biafra and off the Guinea coast (Berrit, 1961, 1962a, 1962b, 1966). These Guinean waters limit the seasonal upwelling area on one side at the front of Cape Palmas and on the other side in the Cotonou-Lagos region.

Off the Côte d'Ivoire and Ghana, the TSW are seasonally driven away by an upwelling. The underlying SACW takes their place from the end of June to late September. Does this cooling result from a true Ekman-type, wind-driven

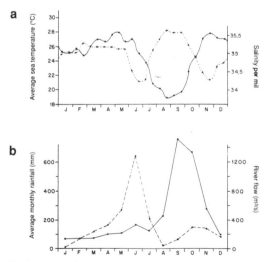

Figure 3. (a): Average sea temperature (solid line) and salinity (dashed line) at 10 m of the coastal station at Abidjan (data from Centre de Recherche Oceanographique [CRO], Abidjan). (b): Average monthly rainfall in Abidjan (dashed line) and flow of the rivers Comoé and Bandama (solid line), 1969–1975 (Office de la Recherche Scientific eu Technique Outre-Mer [ORSTOM] data).

upwelling, as Verstraete (1970) proposed? More recently, Colin (1988) defends this opinion, which has been criticized by Bakun (1978). Is the rise of the thermocline caused by the geostrophic adjustment of the Guinea Current (GC) (Ingham, 1970) or by the passage of an internal coastal trapped wave (Picaut, 1983)? Downstream eddies created in the GC can also have a part in this cooling (Marchal and Picaut, 1977). The answer may be composite and is unclear. Nevertheless, it is a biologist's concern because in a wind-driven upwelling the advective offshore transport, caused by wind stress, drives plankton away from the nutrient spring. In this case, upwelling intensification raises the nutrient input to the euphotic layer. But meanwhile, it also increases the loss of plankton from the coastal ecosystem.

An interesting feature of wind-driven upwellings is that the same Ekman offshore transport requires a weaker wind in lower rather than higher latitudes. Consequently, the same nutrient supply may be reached at a lower turbulence rate. Thus, in this near-equator upwelling, the primary production bloom can follow cool events with a shorter lag than in a subtropical region, such as the Canary Current (Roy, 1991).

The GC is an eastward, superficial flow that is fed by the North Equatorial Counter Current (NECC) off the Liberian coast (Fig. 1). The GC is not very deep—on average, it extends from the surface to 15 m near the coast and 25 m offshore. It overlays the Guinea Under Current (GUC) flowing westward. The GUC originates from the Bight of Biafra, as a return branch of the Equatorial Under Current (EUC) (Fig. 4). In the Bight of Biafra, the GUC may often be observed at the surface (Longhurst, 1964). As it runs westward, the GUC sinks progressively under the GC. Some inversions of the surface flow (i.e., westward circulation) occur from time to time; the percentage occurrences of these reversals decrease from the Niger River to Cape Three Points (Longhurst, 1962). In 1972, Lemasson and Rébert (1973a) located the dive of the GUC under the GC off the Ghana coast (Fig. 5).

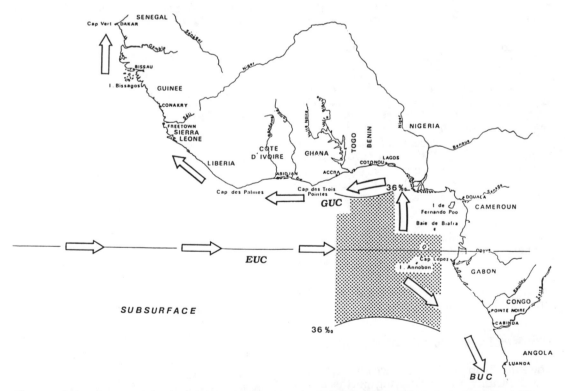

Figure 4. Subsurface circulation in the eastern tropical Atlantic. EUC: Equatorial Under Current; GUC: Guinea Under Current; BUC: Benguela Under Current. Dotted region is the area of the subsurface salinity maximum (>36 x 10⁻³) located around 50 m depth in May 1984 (from Piton and Wacongne, 1985).

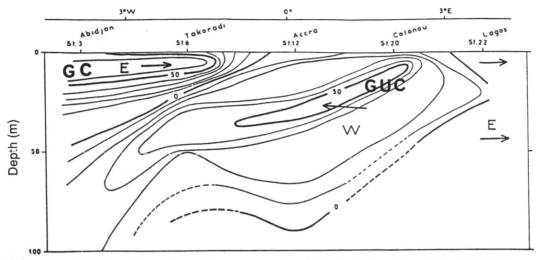

Figure 5. Longitudinal section of the zonal components of circulation off the Ivory Coast to Nigeria; 0–100 m profile on the continental shelf at 10 to 23 nautical miles from the shore, May 1972. The zonal velocities are indicated in cm/s; each contour line corresponds to a 10 cm/s variation. E and W indicate eastward and westward flows, respectively. The 0-contour is the shear layer separating the Guinea Current (GC) from the Guinea Under Current (GUC) (from Lemasson and Rebért, 1973a).

Plankton

Seasonal variations

Dandonneau (1971), Reyssac and Roux (1972), Bainbridge (1972), and Binet (1977, 1978, 1979, 1983b) showed that the taxonomic composition of phytoplankton and zooplankton was strongly related to the different hydrological seasons. Multivariate analyses show the coincidence between species assemblages and hydrological seasons: stability or upwelling, coastal rains, or flood season (Fig. 6). Thus, during every season, the specificity of nutrient supply, of vertical stratification or mixing, and perhaps coastal stream patterns, generate different states of the ecosystem.

Primary production is stimulated by the nutrient input carried to the surface of the sea by the runoff of the first rains over the land (June), then by the major upwelling, and finally by the flood of the larger rivers (September-October). Thus, the number of phytoplanktonic cells is at a peak during the months from June to September (Fig. 7). The depletion of external nutrients then leads to a progressive decrease in phytoplankton biomass. The production then depends only on regenerated nutrients, except during the minor upwelling

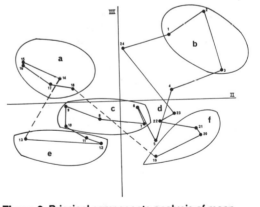

Figure 6. Principal components analysis of mean fortnightly abundances of zooplankton taxa during a composite year: projections of the average fortnights (points 1 to 24) in the factorial plane II, III (from Binet, 1977). The position of the points in the fractional space depends on the taxonomic composition of the corresponding fortnights; the nearer the points, the more similar the taxonomic compositions. The vicinity of points belonging to the same hydrological set show the similarity between hydrological and ecological seasons: (a) great cool season (July to September); (b) little cool season (January to mid-February); (c) great warm season (March to May); (d) little warm season (mid-November to mid-December); (e) first flood season (June); (f) second flood season (October to mid-November). (See Figure 3).

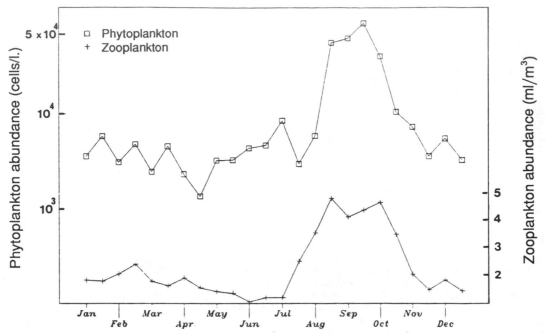

Figure 7. Average seasonal variation in abundance of phytoplankton (number of cells/l, log scale) and zooplankton (settled volumes in ml/m³). Averages calculated from data collected during the period 1969–1975 (data from Dandonneau and Binet, unpublished).

season (January to March), when small-intensity, short-lived upwellings take place—especially on the eastern sides of Cape Palmas and Cape Three Points—in the wake these headlands create in the GC (Dandonneau, 1973).

The zooplankton biomass follows the same seasonal pattern (Fig. 7). The number of copepods is also correlated with the upwelling cooling, but with a 2-week lag time (Binet, 1976).

Relationships between upwellings and river flows

A detailed analysis of the relationships between upwellings and river flows shows that the biomass of zooplankton is not linearly correlated to the input of deep-origin nutrients in the euphotic layer. At a coastal station near Abidjan, the correlation found between the upwelling intensity and the increase of zooplankton biomass is not valid for the months of August, September, and October, namely, the maximum cooling period (Binet, 1976). The upwellings benefit the zooplankton biomass especially when they break a period of thermal stability; but during the main cold season, after some weeks of continuous upwelling, the zooplankton biomass is no longer related to upwelling intensity.

Two explanations may be proposed. One possibility is that the input of deep-origin nutrients is linked to some turbulence that delays the phytoplankton growth and destroys patches of zooplankton. That would be the case in a wind-driven upwelling, but Cury and Roy (1989) do not accept this hypothesis for Côte d'Ivoire and Ghana. The second possibility is that nutrients may become a limiting factor. In upwellings, the growth of large diatoms consumes heavy quantities of silica, the regeneration of which is rather long. The flood from rivers enters the sea during these months and can alter the silica deficiency. Indeed, the biomass of zooplankton is correlated with the inflow of fresh water, especially during these 3 months (Binet, 1976, 1983b).

Finally, it appears that the planktonic biomass production depends on nutrient input from deep water or from the rain runoff from land. However, the relationship between the physical factors accountable for this enrichment and the planktonic production cannot be linear, except in the short range. In a wind-driven upwelling, the turbulence or deficiency of a given element slows down biological growth. In river plumes, the speed of the stream and the opacity of water caused by suspended materials inhibits primary production in the proximate lagoons and river mouths. To be optimal with respect to other factors, the transfer of organic matter along the food web needs a long residence time of grazers near their forage. An algal cell takes on the order of hours to divide, but the reproductive cycle of copepods, even in tropical waters, needs about 2 weeks to be achieved.

Zooplankton retention behavior

The quasi-coincidence of phytoplankton and zooplankton seasonal maxima must be pointed out. It may be explained by a behavioral particularity of some copepods. Many zooplankton species migrate between surface and subsurface waters. Diel migration is the most well known of these vertical movements. The progressive sinking of the core of a species distribution as it is growing older (i.e., ontogenetic migration) is another interesting vertical movement. As a matter of fact, *Calanoides carinatus*, the most abundant copepod species in the upwelling season, spends the first stages of its life in surface waters and then dwells in deeper layers (Binet and Suisse de Sainte Claire, 1975). The sinking of this population means a change from an eastward- to a westward-moving one; thus, horizontal advection of the total population is reduced (Fig. 8). This species and others exhibiting the same vertical migration pattern considerably reduce the numerical loss they would be subject to if they were to spend all of their lifetime in the same current layer. In addition, this behavior allows a more favorable transfer from primary to secondary production because the secondary consumers remain in the vicinity of the upwelling phytoplankton bloom.

The shallowness of the undercurrent and this pattern of migration are probably the reasons why phytoplankton and zooplankton seasonal cycles are in phase in the Guinea upwelling LME, an occurrence that is not observed in other eastern boundary current ecosystems. In the Canary Current, the undercurrent is very deep (500 to 1000 m) and the Ekman offshore transport is very fast. Thus, larger zooplankton are found off the shelf and the biomass peak lags several weeks behind that of plankton (Binet, 1988).

Pelagic tunicates and Cladocera are interesting exceptions. By their opportunistic strategy, they can harvest a plankton bloom and reproduce quickly by scissiparity or parthenogenesis, resulting in crowding populations. They live at the surface and their multiplicative forms can be found in great numbers in

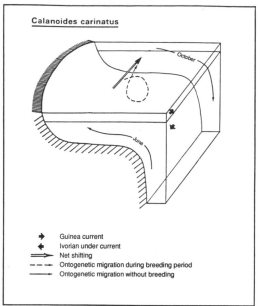

Figure 8. Diagram of the ontogenetic migration of the copepod *Calanoides carinatus* between the Guinea Current and undercurrent (Binet, 1979). At the beginning of the cool season (June), stage 5 copepodites rise from deep water and give way to adults who reproduce. During the great cool season, the development of copepodid stages occur over the shelf between the two current layers with a consequent reduction of eastward net shifting. As upwelling weakens (October), stage 5 copepodites from the last generation dive into deep water for the entire warm season.

the vicinity of upwellings in the Canary Current (Furnestin, 1957) and in the GC (Binet, 1977). When the period and location of food abundance is past, they generate resting forms.

Long-Term Environmental Changes

The drought

The Sahelian countries have been suffering from a severe drought for the past 2 decades. The mean annual flow of the great Ivorian rivers has been reduced to half the volume of the 1950s–1960s. The most important changes have occurred during the flooding season. Short coastal rivers did not undergo this calamity, because their basins received sufficient water during the first monsoon rains. Rivers of Ghana suffered the same flow reduction. The case of the Volta was aggravated by the erecting of a dam and the filling of Volta Lake. In Côte d'Ivoire, two dams on the Bandama were built

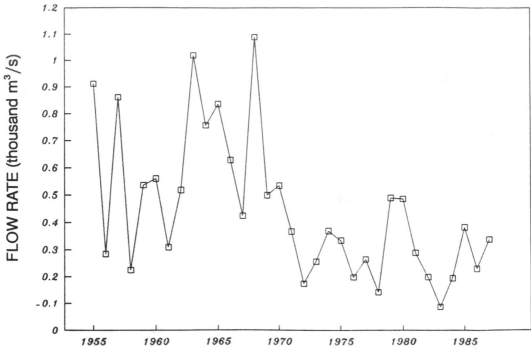

Figure 9. Annual flows (1955–1986) in thousand m³/s for the Bandama and Comoé rivers (data from ORSTOM and DRES Côte d'Ivoire).

during the last 2 decades. When the filling was achieved, these dams regulated the flow, especially during the floods, thereby decreasing the load of particles in downstream waters.

Figure 9 shows the marked reduction in discharge from the Bandama and Comoé Ivorian Rivers since 1972. They have not recovered their former values, despite some amelioration in recent years.

As described previously, floods have a favorable effect on planktonic production; thus, biomass may decrease as a direct result of reduced river flow. Indeed, after 1972, plankton biomass maxima at the end of the cool season have not reached the values observed between 1969 and 1971 (Fig. 10).

The upwelling

In the usually stratified coastal waters, the temperature decrease at 0 or 10 m depth is a good index of upwelling intensity. Annual indices for Tema (Ghana) and Abidjan (Côte d'Ivoire) show strong interannual variations but no significant long-term trend. Nevertheless, coastal monitoring at several shore stations shows interesting changes in the zonal gradient of sea-surface temperature (Arfi *et al.*, 1991).

From 1969 to 1980, the spatial pattern of surface temperature variations agreed with the concept of two upwellings, the cores of which would be on the eastern sides of Cape Palmas and Cape Three Points. The temperature dropped abruptly on the downstream side of each cape, then increased progressively eastward as the waters were carried by the GC (Morlière and Rébert, 1972).

In 1982, the decrease in temperature was almost the same along the coast. From 1983 to 1986 the pattern was reversed: The coldest waters were found in the western side of Cape Three Points. In 1987, upwelling was weak at all stations (Herbland and Marchal, 1991).

The sea-surface temperature (SST) observed by merchant ships in the area between the two capes (Servain and Lukas, 1990) shows a slight increasing trend after 1976 (Fig. 11).

The currents

The eastward-flowing NECC crosses a part of the Atlantic Ocean against the tradewind direction (Fig. 1). This flow migrates between a northern (summer and fall) and a southern (winter and spring) latitude, parallel to the movement of the ITCZ, and is controlled by changes in the curl of the wind stress. The seasonal variation of the NECC is part of the equatorial response to seasonal wind-forcing (Richardson and Reverdin, 1987). According to Philander (1986, p. 238): "Interannual variations in the Atlantic Ocean can be viewed as perturbations to the seasonal cycle. In 1984, the seasonal migration of the ITCZ took it further south than normal and it remained in a southerly position longer than normal."

We suppose that an interannual change in the migration pattern of the ITCZ will in-

duce a similar deviation in the latitude of the NECC.

During the 1970s, the ITCZ remained usually northward of 2° or 3°N in winter and spring at the most southerly latitude of its seasonal migration. During 1984, 1985, and 1986, the ITCZ moved southward as far as the equator (Citeau *et al.*, 1989). As Lamb (1978) observed previously, a southward shift of the ITCZ occurs during an Atlantic El Niño (warm event).

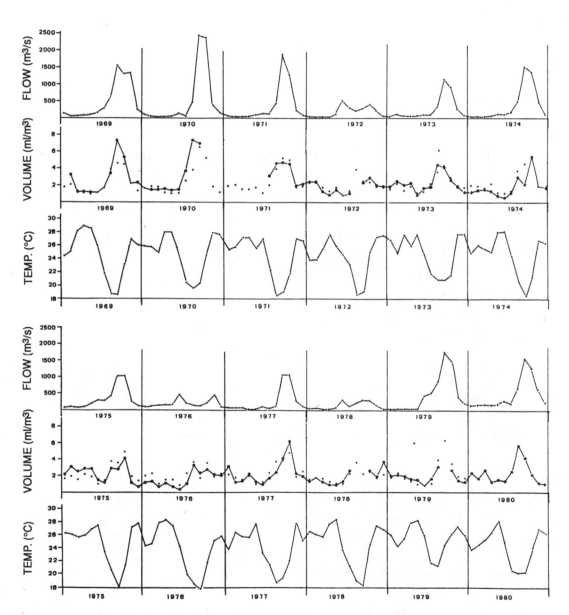

Figure 10. Coastal environmental variations, 1969–1980. From the top: monthly flow of the Bandama and Comoé rivers; settled volume of zooplankton at the coastal station at Abidjan; and temperature at 10 m the same station. Zooplankton settled volume: monthly averages of recorded zooplankton volumes (Large dots and solid lines). Predicted volumes of zooplankton using river-flow and sea-temperature multilinear regressions (Small dots) (from Binet, 1983b).

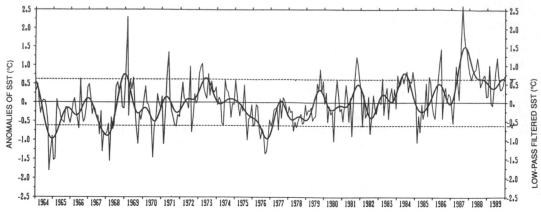

Figure 11. Sea-surface temperature anomalies for 1964–1990 from ship of opportunity data in the area 2°W–8° W, 4°N to the coast (from Servain and Lukas, 1990). Thin line: monthly anomalies (°C); Heavy line: filtered anomalies >12 months.

Atlantic El Niños occurred in 1968 and in 1984 (Hisard, 1988). The trade winds were reduced, the NECC intensified, and its axis shifted southward. Finally, anomalous southerly positions of the ITCZ and NECC might be related, especially during Atlantic El Niños.

Because the NECC directly feeds the GC, a southward shift of the former may induce an offshore displacement or a broadening and a flattening of the latter, along the Côte d'Ivoire and Ghana shelf (Fig. 12).

On the other hand, high-salinity waters (>36 *per mil*) of the EUC are usually upwelled along the equatorial divergence, and, in the subsurface waters of the Guinea Gulf, the high-salinity water belt is very narrow. However, in 1984 the relaxation of trade winds was followed by a weakening of the equatorial upwelling. The high-salinity waters of the EUC were not upwelled to the surface, and they spread in the Gulf of Guinea (Fig. 4).

This unusually important eastward transport of surface and subsurface waters reversed the sea-surface slope along the equator during the boreal winter, 1983–1984 (Hisard *et al.*, 1986). A strong rise in sea level occurred along the African coast (Katz *et al.*, 1986). A strengthening of the poleward (southward and westward) return-circulation may be expected from this unusual accumulation of water. Indeed, some of this water flowed southward along the African coast and suppressed coastal upwelling as far south as Angola and Namibia (Philander, 1986). A rather similar change is likely north of the equator, and the westward GUC should be intensified. Because of its increased velocity, the GUC undergoes higher Coriolis force, thus it should shift northward, closer to the coast, and sink under the GC to the west of the usual

longitude. Indeed, high-salinity waters were observed along the coasts of Bénin and Togo (Piton, 1987).

Upstream surface and subsurface circulation anomalies should change the regular pattern over the shelf (Figs. 5 and 12). The eastward surface flow (GC) will shift to the south and the westward flow (GUC) will remain near

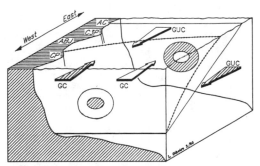

Figure 12. Diagram of the Ivory Coast and Ghana shelf illustrating hypothesized changes in the current pattern (arrows) and their consequences on *Sardinella* populations (circles). Hatched: situation prevailing in the 1960s and 1970s; White: situation in the 1980s. The oblique surfaces show the shear planes between two currents for 1972 (solid line) and 1984 (dotted line). CP: Cape Palmas; ABJ: Abidjan; C3P: Cape Three Points; AC: Accra; GC: Guinea Current; GUC: Guinea Under Current. The spawning areas of *Sardinella aurita* (circles) are situated eastward of Cape Palmas and Cape Three Points (see Fig. 14). In the 1960s–1970s, the Ghanaian population was larger than the Ivorian, but during the next decade, the new current pattern led to a rapid maturation of the Ivory Coast population.

the surface until it plunges under the GC, westward of the usual sinking region.

Measurements of superficial currents off Abidjan, Côte d'Ivoire (Table 1), made by Lemasson and Rébert (1973b) and Colin (1988), indicate westward transport in 7% and 9% of the observations during "normal" years (1969 and 1970, respectively), versus 23% and 35% during years of southward ITCZ shift (1968 and 1984, respectively). This supports the hypothesis of an increase in westward transport along the shelf, associated with a southward shift of the ITCZ. Similarly, the change observed in the coastal temperature zonal gradient sustains the idea of a shoreward and westward displacement of the GUC.

The *Sardinella* fishery

Two *Sardinella* species constitute the majority of the pelagic fishery of the coastal upwelling (Fig. 2). Until 1980, most of the catch from Ivorian purse seiners was comprised of *Sardinella maderensis*. Since that time, *Sardinella aurita* has ranked first in Côte d'Ivoire landings.

The main *Sardinella aurita* fishery was centered in Ghanaian waters during the 1960s and 1970s. Only a small part of the catch came from the Ivorian coast. That fishery took place during the major upwelling season (boreal summer) and fluctuated widely from year to year (Marchal, 1966, 1993), probably as a result of the influence of environmental factors, on one hand, and from stock abundance changes, on the other.

Influence of upwelling

Increased upwelling should improve plankton productivity and, consequently, increase the forage for plankton feeders, thus improving their survival rate in the early life stages. The fortnightly catch per unit effort (CPUE) for *Sardinella* spp. proved to be correlated to the upwelling intensity of the preceding fortnight

(Mendelssohn and Cury, 1987). This time lag corresponds to the mean life span of copepods, as if upwelled nutrients enhanced phytoplankton and zooplankton growth.

Based on these hypotheses, Cury and Roy (1987) modeled the annual *Sardinella* CPUE to assess the effects of upwelling intensity on fishing effort (the same year) and on recruitment (the next year). Their model worked for the years prior to 1981. It accounts for the total of both *Sardinella* species landings in Côte d'Ivoire only; but if we consider *S. aurita* separately, the sudden increase in landings in 1972 and the subsequent collapse remain unexplained. Bakun (1978) had observed that heavy rains were associated with poor *Sardinella* catches along this coast, and Binet (1982) hypothesized that the 1972 overfishing was favored by the drought. *S. aurita* is a stenohaline fish, avoiding turbid and low-salinity waters. During the dry year (1972), the salinity of coastal waters was not lowered by rivers, and juveniles frequented the coastline. Their proximity to the coast led to overfishing of juveniles, notably by canoe fishermen from Ghana. The stock suffered a 5-year collapse, but it had completely recovered by 1978. Fréon (1988) simulated these variations in a global production model in which two climatic variables—upwelling intensity and river flow—were incorporated to improve the fit of stock abundance and the catchability coefficient, respectively.

From the beginning of the 1980s, the *Sardinella aurita* fishery patterns changed again completely (Figs. 13 and 14):

- About half of the yield was caught on western Côte d'Ivoire, and not only in Ghana, as in past decades.
- The fishing season extended over the entire year, rather than being limited to the upwelling season only.
- The landings were high (over 100,000 metric tonnes [mt] in 1985) and were sustained over several consecutive years, even following a collapse in 1973 after a 90,000-mt catch.

Changes in the spatial distribution of pelagic biomass

Comparisons among acoustic surveys conducted in different years confirm that the change in the regional fishery pattern is not only caused by changes in fishermen's habits, but also results from a new geographical distribution of pelagic stocks.

Echo-integration surveys over the shelf (Marchal and Picaut, 1977) indicated the fol-

Table 1. Occurrences of Guinea Current (eastward) and Guinea Under Current (westward) in surface observations of Lemasson and Rebért (1973b) and Colin (1988).

	1968	1969	1970	!	1983	1984
% Eastward	77	93	91	!	71	65
% Westward	23	7	9	!	29	35

lowing fish biomass densities in 1974: Cape Palmas to Abidjan, 17.8 mt/nm²; Abidjan to Cape Three Points, 20.3 mt/nm²; Cape Three Points to Volta River mouth, 53.3 mt/nm². Because of the difference in the width of the continental shelf, the estimated fish biomass is greater eastward of Three Points (225,000 mt) than between the two capes (82,000 mt).

This pattern of biomass distribution was consistent until 1980. From 1980 to 1990, several acoustic surveys carried out in this area show a sudden increase in the fish density on the Ivorian shelf with equivalent values on both sides of Cape Three Points (Marchal, 1993).

One or two populations?

Because the preceding models cannot account for these changes, it seemed necessary to take a new look at stock and population localization.

Until recently, it was believed that a single stock supported the whole yield of the fishery. Based on tagging experiments, a population spawning in the vicinity of Cape Three Points during the major upwelling season has been identified. Its larvae are very abundant off Tema during July, August, and September (FRU *et al.*, 1976). Nevertheless, evidence from plankton sampled on surveys conducted between Cape Palmas and Cape Three Points from 1969 to 1972 indicated the likelihood of a second population (Marchal, 1991). The samples contained many larvae of *Sardinella aurita* that were found eastward of Cape Palmas during the minor upwelling (especially in March).

In addition, the lengths of *S. aurita* caught near Cape Palmas are notably greater than those in the Three Points area. Finally, the growth rates of these two groups also seem different (Pezennec *et al.*, 1993).

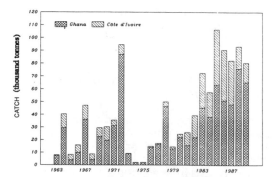

Figure 13. Catches of *Sardinella aurita* from the Ivory Coast and Ghana (in thousand tons) (from Pezennec *et al.*, 1993).

Thus, *Sardinella aurita* from this ecosystem may belong to two different populations (Fig. 14). The cores of these two populations are situated in the eddies created in the surface current downstream (i.e., eastward) of Cape Palmas and Cape Three Points. The two populations differ in their spawning times and locations. The fish of Cape Palmas grow faster and larger than those of Cape Three Points. They also differ in their seasonal availability, although the geographical ranges of adult distributions overlap slightly.

The changes in total landings for the last 3 decades may be understood as changes in the abundance and harvest of the two populations. The Ghanaian population of *S. aurita* was flourishing and supported the yield of the fishery in the 1960s; but in 1972, the drought caused an exceptional availability to coastal fishermen, which resulted in overfishing and a 5-year collapse, followed by a subsequent recovery, which presently yields variable catches.

In the beginning of the 1980s, an oceanographic event occurred that may have favored the increase of the Ivorian population. The fishery now harvests this population at the same time of year as formerly.

Improved retention of larvae by a shift of the currents

According to Sinclair's hypothesis (1988) on marine population regulation (member/vagrant), we think that the abundance of a population is partly controlled by the advective processes occurring before recruitment, as has been seen in copepods, where the vertical distribution of pelagic larvae ranging over two currents minimizes advective losses. Thus, a change in the current system may be responsible for the sudden increase in the Ivorian population.

The Ghanaian population spawns over the shelf, mainly between Cape Three Points and Tema (i.e., near the usual boundary of the two currents). Migration of larvae between these two layers ensures the retention of early stages over the shelf. Empirical results suggest that the strength of recruitment may be dependent on the food supply only, that is, on the upwelling intensity.

During the previous 2 decades, the larvae of the Ghanaian population distributed in the two currents were retained in this area and yielded good year classes. In the same time period, the larvae hatched in the western Ivory Coast were advected eastward by the GC; thus, Ivorian population recruitment was very poor.

Since the beginning of the 1980s, a possible change in the current pattern occurred

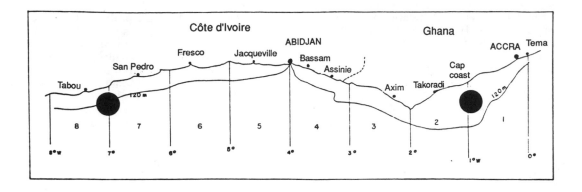

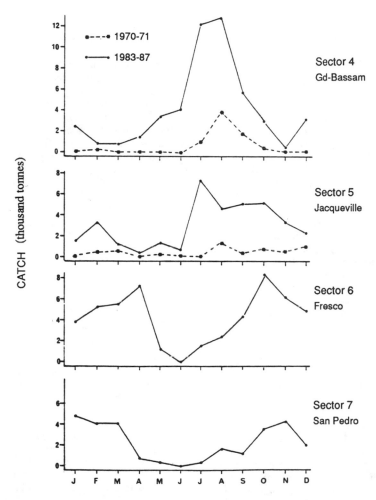

Figure 14. The map shows fishery divisions along Côte d'Ivoire and Ghana. Black dots correspond to the barycenters of the two populations of *Sardinella aurita*. The graphs show changes in the seasonal and regional patterns of Ivorian fisheries between 1970 and 1971 (before the collapse) and between 1983 and 1987 (during maximum yields) (from Binet *et al.*, 1991).

that may have reversed the abundance of the two populations. The shear surface between the two currents shifted westward, leaving the Ivorian and Ghanaian populations in reversed advective processes. The Ivory Coast population is now at the boundary of the two currents, and the advective losses of its larvae are minimized, whereas the Ghanaian population larvae may add to the recruitment of the other population.

Conclusions: Implications for Management

The northern coast of the Gulf of Guinea, cooled by a seasonal upwelling, may be considered a subsystem of the Gulf of Guinea LME. Hydrology and circulation features, nutrient enrichment processes, plankton, benthos, and fish species assemblages make it a different regime from its neighbors, who are situated in thermally stable regions. One commercial stock of the pelagic fish species *Sardinella aurita*, likely comprised of two populations, ranges all along the coast and is bounded by the zonal extent of the upwelling.

The management of coastal pelagic resources has never been easy. Their sudden increases or collapses seem to escape the classic rules of population dynamics. On one hand, it is difficult to correctly assess the fishing effort, because of the overdispersed distribution of fishes (schooling behavior) and the fact that rarity sometimes increases the catchability (contrary to randomly distributed species). On the other hand, availability and recruitment of these species seem very sensitive to environmental changes. Thus, approximate management rules may be proposed, provided that environmental changes remain in a narrow range. These conditions existed during the 1960s and probably late 1970s, after the stock had recovered from collapse; but if a climatic event such as the Sahelian drought or a modification to the current system leads to significant environmental changes, management rules must then be adapted to the new state of the ecosystem.

Practically speaking, how do we know if the Guinean upwelling ecosystem is undergoing a major change? Initially, the sea-surface temperatures were recorded as upwelling indices and the river flows were then monitored. Until recently, we thought that these parameters allowed the forecasting of plankton abundance, *Sardinella* recruitment, and availability. Now we think that the respective amounts of eastward and westward flows and the position of the boundary surface between GC and GUC may play a large part in the sudden change that occurred.

These alterations are probably linked to distant climatic anomalies, such as the ITCZ southward displacement during Atlantic El Niños. Improving ecological forecasts in the Gulf of Guinea requires a better understanding of oceanographic processes (what is the mechanism of the coastal upwelling?) and of remote ocean-atmosphere interactions (what determines the position and the flow of great oceanic currents?).

It is not clear how long the Ivory Coast population of *Sardinella aurita* will flourish. Perhaps the current system has already reversed to the prior situation and the high catches will quickly fall!

Acknowledgments

We wish to thanks scientists and assistants of the Centre de Recherches Oceanographiques of Abidjan, especially O. Pezennec, who provided us with fishery data. We are also indebted to P. Hisard, who suggested large-scale connections between different environmental changes, and to J. Servain, who computed the long-term SST trends (Fig. 11) from the data measured by ships of opportunity.

References

Arfi, R., Pezennec, O., Cissoko, S., and Mensah, M. 1991. Variations spatiale et temporelle de la résurgence ivoiro-ghanéenne. In: Pêcheries ouest-africaines. Variabilité, instabilité et changement. pp. 162–172. Ed. by P. Cury and C. Roy. Office de la Recherche Scientific en Technique Outre-Mer (ORSTOM), Paris.

Bainbridge, V. 1972. The zooplankton of the Gulf of Guinea. Bull. Mar. Ecol. 8:61–97.

Bakun, A. 1978. Guinea current upwelling. Nature 271:147–150.

Berrit, G. R. 1961. Contribution à la connaissance des variations saisonnières dans le golfe de Guinée. Observations de surface le long des lignes de navigation. Cah. Océanogr. C.C.O.E.C. 13:715–727.

Berrit, G. R. 1962a. Contribution à la connaissance des variations saisonnières dans le golfe de Guinée. Observations de surface le long des lignes de navigation. 2ème partie: Étude régionale. Cah. Océanogr. C.C.O.E.C. 14:633–643.

Berrit, G. R. 1962b. Contribution à la connaissance des variations saisonnières dans le golfe de Guinée. Observations de surface le long des lignes de navigation. 2ème partie: Étude régionale, suite et fin. Cah. Océanogr. C.C.O.E.C. 14:719–729.

Berrit, G. R. 1966. Les eaux dessalées du golfe de Guinée. Doc. sci. prov. C.R.O. Abidjan 9:1–15.

Binet, D. 1976. Biovolumes et poids secs zooplanctoniques en relation avec le milieu pélagique au-dessus du plateau ivoirien. Cah. ORSTOM Sér. Océanogr. 14:301–326.

Binet, D. 1977. Grands traits de l'Écologie des principaux taxons du zooplancton ivoirien. Cah. ORSTOM Sér. Océanogr. 15:89–109.

Binet, D. 1978. Analyse globale des populations de copépodes pélagiques du plateau continental ivoirien. Cah. ORSTOM Sér. Océanogr. 16:19–61.

Binet, D. 1979. Le zooplancton du plateau continental ivoirien. Essai de synthèse écologique. Oceanol. Acta 2:397–410.

Binet, D. 1982. Influence des variations climatiques sur la pêcherie des *Sardinella aurita* ivoiro-ghanéennes: Relation sécheresse-surpêche. Oceanol. Acta 5:443–452.

Binet, D. 1983a. Phytoplancton et production primaire des regions côtiéres à upwellings saissoniers dans le Golfe de Guinée. Océanogr. Trop. 18:331–355.

Binet, D. 1983b. Zooplancton des régions côtiéres à upwellings saisonniers du Golfe de Guinée. Océanogr. Trop. 18:357–380.

Binet, D. 1988. Rôle possible d'une intensification des alizés sur le changement de répartition des sardines et sardinelles le long de la côte ouest africaine. Aquat. Living Resour. 1:115–132.

Binet, D., Marchal, E., and Pezennec, O. 1991. *Sardinella aurita* de Côte d'Ivoire et du Ghana: Fluctuations halieutiques et changements climatiques. In: Pêcheries ouest-africaines. Variabilité, instabilité et changement. pp. 320–342. Ed. by P. Cury and C. Roy. ORSTOM, Paris.

Binet D., and Suisse de Sainte Claire, E. 1975. Contribution à l'Étude du copépode planctonique *Calanoides carinatus*. Répartition et cycle biologique au large de la Côte d'Ivoire. Cah. ORSTOM Sér. Océanogr. 13:15–30.

Caverivière, A. 1982. Le baliste des côtes africaines (*Balistes carolinensis*). Biologie, prolifération et possibilité d'exploitation. Oceanol. Acta 5:453–459.

Caverivière, A. 1991. L'explosion démographique du baliste (*Balistes carolinensis*) en Afrique de l'ouest et son évolution en relation avec les tendances climatiques. In: Pêcheries ouest-africaines. Variabilité, instabilité et changement. pp. 354–367. Ed. by P. Cury and C. Roy. ORSTOM, Paris.

Citeau, J., Finaud, L., Cammas, J. P., and Demarcq, H. 1989. Questions relative to ITCZ migrations over the tropical Atlantic Ocean, sea surface temperature and Senegal River runoff. Meteorol. Atmos. Phys. 41:181–190.

Colin, C. 1988. Coastal upwelling events in front of the Ivory Coast during the FOCAL program. Oceanol. Acta 11:125–138.

Cury, P., and Roy, C. 1987. Upwelling et pêche des espèces pélagiques côtieres de Côte d'Ivoire: Une approche globale. Oceanol. Acta 10:347–357.

Cury, P., and Roy, C. 1989. Optimal environmental window and pelagic fish recruitment success in upwelling areas. Can. J. Fish. Aquat. Sci. 46:670–680.

Dandonneau, Y. 1971. Étude du phytoplancton sur le plateau continental de Côte d'Ivoire. I.

Groupes d'espèces associées. Cah. ORSTOM Sér. Océanogr. 9:247–266.

Dandonneau, Y. 1973. Étude du phytoplancton sur le plateau continental de Côte d'Ivoire. III. Facteurs dynamiques et variations spatio-temporelles. Cah. ORSTOM Sér. Océanogr. 11:431–454.

Fisheries Research Unit (FRU), Tema; Centre de Recherche Oceanographique (CRO), Abidjan; ORSTOM. 1976. Rapport du groupe de travail sur la sardinelle (*S. aurita*) des côtes ivoiro-ghanéennes. 26 June–3 July, Abidjan. Ed. by ORSTOM.

Fréon, P. 1988. Réponses et adaptations des stocks de Clupéidés d'Afrique de l'ouest à la variabilité du milieu et de l'exploitation. Analyse et réflexion à partir de l'exemple du Sénégal. Études et Thèses. ORSTOM.

Furnestin, M. L. 1957. Chaetognathes et zooplancton du secteur atlantique marocain. Rev. Trav. Inst. Pêches Marit. 21.

Herbland, A., and Marchal, E. 1991. Variations locales de l'upwelling, répartition et abondance des sardinelles en Côte d'Ivoire. In: Pêcheries ouest-africaines. Variabilité, instabilité et changement. pp. 343–353. Ed. by P. Cury and C. Roy. ORSTOM, Paris.

Hisard, P. 1988. El Niño response of the tropical Atlantic Ocean during the 1984 year. Int. Symp. Long-Term Changes Mar. Fish Pop. 1986. Vigo, Spain pp. 273–290.

Hisard, P., Hénin, C., Houghton, R., Piton, B., and Rual, P. 1986. Oceanic conditions in the tropical Atlantic during 1983 and 1984. Nature 322:243–245.

Ingham, M. C. 1970. Coastal upwelling in the northwestern Gulf of Guinea. Bull. Mar. Sci. 20:1–34.

Katz, E. J., Hisard, P., Verstraete, J. M., and Garzoli, S. 1986. Annual change of sea surface slope along the equator of the Atlantic Ocean in 1983 and 1984. Nature 322:245–247.

Lamb, P. 1978. Case studies of the tropical Atlantic surface circulations pattern during recent sub-Saharan weather anomalies in 1967 and 1968. Mon. Weather Rev. 106:462–491.

Lemasson L., and Rébert, J. P. 1973a. Circulation dans le Golfe de Guinée. Étude de la région d'origine du sous-courant ivoirien. Cah. ORSTOM Sér. Océanogr. 11:303–316.

Lemasson, L., and Rébert, J. P. 1973b. Les courants marins dans le golfe ivoirien. Cah. ORSTOM Sér. Océanogr. 11:67–96.

Longhurst, A. R. 1962. A review of the oceanography of the Gulf of Guinea. Bull. IFAN 24(A)3:633–663.

Longhurst, A. R. 1964. The coastal oceanography of western Nigeria. Bull. IFAN 26(A)2:337–402.

Marchal, E. 1966. Fluctuations de la pêche des sardinelles en Côte d'Ivoire. Doc. Sci. Prov. C.R.O. Abidjan 1.

Marchal, E. 1991. Un essai de caractérisation des populations de poissons pélagiques côtiers cas de *Sardinella aurita* des côtes ouest-africaines. In: Pêcheries ouest-africaines. Variabilité, instabilité et changement. pp. 192–200. Ed. by P. Cury and C. Roy. ORSTOM, Paris.

Marchal, E. 1993. Biologie et écologie des poissons pélagiques côtiers. In: Environnement et ressource aquatiques de Côte d'Ivoire. I. Le milieu marin. Ed. by P. LeLoeuff, E. Marchal, and J.-B Amon Kothias. ORSTOM, Paris.

Marchal, E., and Picaut, J. 1977. Répartition et abondance évaluées par échointégration des poissons du plateau ivoiro-ghanéen, en relation avec les upwellings locaux. J. Rech. Océanogr. 2:39–57.

Mendelssohn, R., and Cury, P. 1987. Fluctuations of a fortnightly abundance index of the Ivoirian coastal pelagic species and associated environmental conditions. Can. J. Fish. Aquat. Sci. 44:408–421.

Morlière A., and Rébert, J. P. 1972. Étude hydrologique du plateau continental ivoirien. Doc. Scient. C.R.O. Abidjan 3:1–30.

Pezennec, O., Marchal, E., and Bard, F.-X. 1993. La pêche des petites espèces pélagiques en Côte d'Ivoire. In: Environnement et ressources aquatiques de Côte d'Ivoire. I. Le milieu marin. Ed. by P. LeLoeuff, E. Marchal, J.-B. Amon Kothias. ORSTOM, Paris.

Philander, S. G. H. 1986. Unusual conditions in the tropical Atlantic Ocean in 1984. Nature 322:236–238.

Picaut, J. 1983. Propagation of the seasonal upwelling in the eastern equatorial Atlantic. J. Phys. Oceanogr. 13:18–37.

Piton, B. 1987. Caractéristiques hydroclimatiques des eaux côtières du Togo (Golfe de Guinée). Doc. Scient. Brest 42.

Piton, B., and Wacongne, S. 1985. Unusual amounts of very saline subsurface water in the eastern Gulf of Guinea in May 1984. Tropical Ocean-Atmosphere Newsletter 32:5–8.

Reyssac, J., and Roux, M. 1972. Communautés phytoplanctoniques dans les eaux de Côte d'Ivoire. Groupes d'espèces associés. Mar. Biol. 13:14–33.

Richardson, P. L., and Reverdin, G. 1987. Seasonal cycle of velocity in the Atlantic north equatorial countercurrent as measured by surface drifters, current meters and ship drifts. J. Geophys. Res. 92:3691–3708.

Roy, C. 1991. Les upwellings: Le cadre physique des pêcheries côtières ouest-africaines. In: Pêcheries ouest-africaines. Variabilité, instabilité et changement. pp. 38–66. Ed. by P. Cury and C. Roy. ORSTOM, Paris.

Servain, J., and Lukas, S. 1990. Climatic atlas of the tropical Atlantic wind stress and sea surface temperature 1985–1989. Océans tropicaux-Atmosphère globale. IFREMER, ORSTOM, JIMAR. ORSTOM, Paris.

Sinclair, M. 1988. Marine populations: An essay on population regulation and speciation. Washington Press, Seattle.

Verstraete, J. M. 1970. Étude quantitative de l'upwelling sur le plateau continental ivoirien. Doc. Scient. C. R. O. Abidjan 1(3):1–17.

Chapter 13

Ecological and Fishing Features of the Adriatic Sea

Giovanni Bombace

Origin and History of the Mediterranean and Adriatic Seas

For most of the Tertiary age, the Mediterranean Sea was part of the Tethys Sea—that vast primeval ocean. The current configuration of the Mediterranean was produced by the tectonic events that followed during the upper Miocene, Pliocene, and the Quaternary periods (Ruggieri, 1967). According to the most widely accepted theory, this intermediate ocean was largely open on the Indo-Pacific region and populated by a tropical-type fauna (Paleomediterranean Element). After the salinity crisis at the end of the Miocene period (during the so-called Messinian period), resulting mainly from reduced contact with the Atlantic and the closing of intercommunication with the Indo-Pacific (previously via Syria), the Mediterranean consisted of a series of lagoons.

A certain degree of irregular intercommunication was maintained with the Indo-Pacific through the North Sea of Paratethys, populated by a distinctly Sarmatian fauna, while the Paleomediterranean Faunal Element was disappearing or being transformed as part of the Endemic Element, consisting of typical species of the Mediterranean. This geomorphological transformation and the process of speciation were completed during the Pliocene period. In the meantime, the Mediterranean was gradually becoming a temperate sea, which it remains today. Its intercommunications with the Indo-Pacific were closed permanently, while those with the Atlantic were reopened. In this phase, the Atlanto-Mediterranean Element was introduced, which remained constant and constitutes the main element of current populations.

During the events of the Quaternary period, with its alternating regressions (cold periods) and transgressions (hot periods), the Northern Element alternated with the Senegalian Element, which can be considered as superimposed on the Endemic and Atlanto-Mediterranean Elements (Peres and Picard, 1964). After the great post-Wurmian deglaciation, the situation became fundamentally as it is today. The gradual increase in salinity led to the extinction of almost all the northern species, resulting in the following faunal composition: a majority comprised of Atlantic-Mediterranean species (immigrated during the Pliocene) and endemic species (formed during the Calabrian and the Sicilian periods); plus the survivors of the Senegalian Element (which entered the Mediterranean during the Tyrrhenian period); and a number of survivors of the Northern Element (Peres and Picard, 1964).

It is important to emphasize that the alternation of warm and cold fauna in the Mediterranean was not caused by temperature variations, but by modifications in the flow of the currents in the Straits of Gibraltar and Bosporus (Mars, 1963), which were in turn related to the climatic fluctuations of the glacial and interglacial periods that characterized the Quaternary period.

Regarding the Adriatic Sea, several biogeographers and paleontologists maintain that the last contact of the Mediterranean with the Indo-Pacific occurred during the last Miocene and Pliocene periods through a corridor that connected the Northern Adriatic with the Sarmatic Sea. The Black, Caspian, and Aral Seas, which were once part of the Tethys Sea, drained into a vast basin that formed the Sarmatic Sea.

Biogeography of the Mediterranean and Adriatic Seas

The current biogeographical situation of the Mediterranean developed from the historical and paleogeographical events that occurred during the Miocene and Pliocene, but above all from climatic fluctuations during the Quaternary period.

As far as the benthos is concerned, the following distinct biogeographical districts can be identified: (i) the Alboran Sea, (ii) the Western Mediterranean, (iii) the Eastern Mediterranean, (iv) the Adriatic, and (v) the Black Sea, which is a sea in itself. With the exception of the Alboran Sea, these biogeographical areas can be further subdivided. The Western Mediterranean includes a northern area (the Ligurian Sea, the Lion Gulf) where, in winter, the deep Tyrrhenian waters form as a result of surface cooling and cascading; a central area; and a southern area tied to the North African coast. The Eastern Mediterranean, which is the richest in endemic elements and which has certain affinities with the Pacific region from a historical perspective, can be subdivided into a northern area that includes the Greek coasts and those of Asia Minor, and a southern area. The latter is also characterized by the so-called Lessepsian species, those entering the Mediterranean from the Red Sea, and consequently, from the Indo-Pacific region via the Suez Canal. Finally, we have the Adriatic, which is the most different sea of all and washes the Italian coasts. It can be divided into three subareas: the Northern and Central Adriatic, which are more similar ecologically and oceanographically, and the Southern Adriatic (Fig. 1).

The Northern Adriatic is also unique in terms of fish fauna, where several "sensu strictu" endemic species can be found (Tortonese, 1983), such as *Acipenser naccarii* and *Syngnathus taenionotus*, and several Gobides, such as *Knipowitschia panizzae* and *Pomatoschistus canestrini*.

In the Northern Adriatic and partially in the Central Adriatic, there are a number of fish species of pontic origin living in the colder areas of the Mediterranean, whereas the Atlantic species are very northern. For example, *Merlangius merlangus* (whiting) travels as far as the Norwegian and Icelandic coasts in the Atlantic, whereas in the Mediterranean it can be found in the Lion Gulf, the Northern and Central Adriatic, the Northern Aegean, the Sea of Marmara, and the Black Sea (*M. merlangus euxinus*). A second example is sprat (*Sprattus sprattus*), which in the Northern and Central Adriatic is an important stock, even if it is not fully exploited. It can be distinguished in an Atlantic form, which never travels any farther

than Gibraltar; a Baltic form; and a Mediterranean form, which lives in the cooler areas. Other pontic or ponto-caspian species present in the Northern Adriatic are *Huso huso*, *Syngnathus tenuirostris*, *Knipowitschia caucasica*, and *Platyichthys flesus luscus*.

Knipowitschia, a ponto-caspian genus, is well represented in the Northern Adriatic, because it is typically found in lagoon and estuarine brackish-water environments. In the case of these pontic-Mediterranean species, it is difficult to establish whether they are fish species passing through the Bosporus during the alternating events of the Plio-Quaternary periods or survivors of more ancient periods (Sarmatic Sea), which were trapped in particular Mediterranean biota.

Ecology and Fisheries of the Adriatic Sea

Several ecological parameters and aspects related to fishing will be described for the Northern and Central Adriatic (the two most-similar basins), apart from the Southern Adriatic, which, with respect to biogeography, ecology, and fisheries, is very similar to the Northern Ionian Sea (Ionic coasts of Calabria and Sicily), where there is a combination of "western" and "eastern" elements.

The coast, sea bottom, and inflow of fresh water (Fig. 1)

The Adriatic extends in a northwest/southeast direction for more than 700 km. The width between Pescara and Split is about 200 km. The major fetch is consequently by the southeast wind. The Adriatic has a surface area of 138,000 km^2, equal to $1/20$ of the entire Mediterranean. Its volume is 35,000 km^3, equal to 1/125th of the Mediterranean. If the fishing resources were distributed uniformly, considering the total production in the Mediterranean of 1 million tonnes per year, the Adriatic should produce 50 thousand tonnes of catch per year. On the contrary, it produces five times this amount. The coastal morphology of the Yugoslavian-Albanese and Italian coasts are quite different. The former is high, rocky, articulated, indented, and dotted with small and large islands, with little freshwater inflow. The Italian coast is generally a flat, alluvial plain to the north and south of the Po River, with lagoon facies in the north and near Gargano Lagodi (Lesina, Lagodi Varano, and Golfo di Manfredonia), or with raised terraces (plateau of the Puglie region, between Ancona and Pescara in the midcoast region).

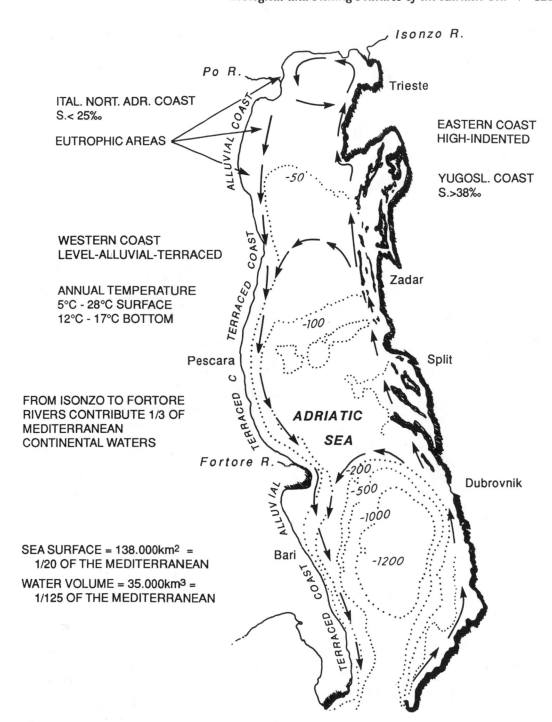

Figure 1. Selected ecological characteristics of the Adriatic Sea.

About one-third of all the Mediterranean continental waters flow into the Northern and Central Adriatic. Considering the number of rivers that flow into the Adriatic in the region between the Isonzo River near Trieste to the Fortore (Gargano), the volume of water per second which flows into the sea can vary from 3,000 to 10,000 m^3 or more, half of which comes from the Po River.

The sea bottom of the Northern and Cen-

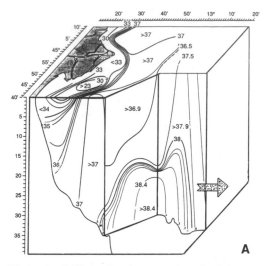

Figure 2a. Tridimensional perspective of salinity (per mil) in the northern Adriatic Sea (Po Delta) under autumnal-winter conditions. (By Franco, 1983, modified.) Arrow indicates migration of most demersal species in the offshore direction.

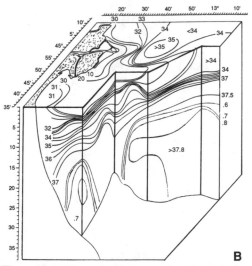

Figure 2b. Tridimensional perspective of salinity (per mil) distribution in the northern Adriatic Sea (Po Delta) under summer stratification conditions. (By Franco, 1983.)

tral Adriatic consists mainly of recent or ancient sediments (such as sandy, muddy, and silty) and constitutes a large shelf for trawling. This favors access to fishing resources with the use of "grazing" gear (such as trawling, midwater trawling, dredging for bivalve molluscs, and beam-trawling for flatfish), prima-

rily in international waters and along the Italian coast. The bottom topography on the Yugoslavian coast is rather uneven and trawling gear can be used only along the channels between the islands.

Thermohaline fronts and fisheries resources

The Northern and Central Adriatic are subject to strong annual thermal variation, particularly in the superficial layers of water (for example, 5°–28°C at the surface and 12°–17°C at the bottom).

During winter, the coastal water (within 6–7 miles from the coast) is uniform from the surface to the bottom, with low temperatures of 5°–6°C and a salinity below 37 per mil (Figs. 2a and 3). In the open sea, temperatures range from about 10°–12°C, and the salinity exceeds 38 per mil (Franco, 1970, 1972, 1983; Artegiani, 1984, 1987).

A thermohaline density front, which involves the entire water mass vertically, separates the coastal from open-sea waters.

During autumn and early winter, when thermal conditions are more favorable in the open sea, there is a seaward migration of exploitable demersal species from the coast. Therefore, during winter only a number of small eurythermic and euryhaline species (such as some Gobiidae and Atherinidae) remain inshore, where they constitute the only fishing resources available. For this reason and despite the national law, the sectors of the Northern Adriatic only are allowed to trawl within 3 miles of shore in the period October to March. This derogation involves the distance from the coast and the mesh size. From the ecological point of view, the vertical thermohaline front, which runs parallel to the coast, inhibits the flow of terrigenous materials (nutrients and other substances that flow from the rivers and waterways) out of the coastal area, resulting in favorable conditions for spring blooms. In summer (Figs. 2b and 4), stratification may occur that vertically separates (horizontal thermohaline front) the warmer superficial water of lower salinity from the deeper and colder water of higher salinity. The thermal difference between the two water layers is 11°–13°C. This stratification occurs only if certain climatic and hydrographic conditions are present (extended period of calm sea, strong insolation, high temperatures, and inflow of fresh water). These are ideal conditions for autumnal algal blooms and extended hypoxia or anoxia, although the two phenomena can occur separately as shown by the events of the last 20 years. Damage to demersal (both during the prerecruitment and

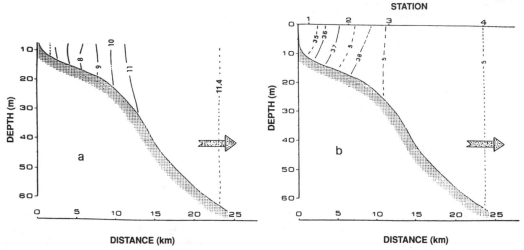

Figure 3. Central Adriatic Sea. Temperature (a) and salinity (b) distributions on Portonovo transect under winter conditions. (By Artegiani, 1987, modified.) Arrow indicates migration of most demersal species in the offshore direction.

subsequent phases) as well as to the sedentary, benthic resources (e.g., bivalves such as baby clams) may be considerable.

Sea depth

The Northern and Central Adriatic basins are not extremely deep. In the Northern Adriatic, the maximum depth does not exceed 75 m, whereas in the Central Adriatic it is about 100 m, with the exception of the so-called Pomo Pit ("Jabuka" to the Yugoslavians), where the depth is about 300 m.

It is interesting to note how "shallow-depth" plays an ambiguous role, according to the conditions and events resulting from climatic and hydrographic conditions.

Positive role for the environment

The shallow depth and slight inclination of the bottom reduce the time required for exchange of water masses. In the Northern Adriatic, the exchange of water masses occurs over a period of a few months.

Stormy seas, particularly those under the force of the northeast winds, resuspend sediment from the sea bottom with low inclination and recirculate the sedimented nutrients. This effect extends 3–4 miles offshore, and down to a depth of 10–15 m in the Northern Adriatic. In deeper seas, these salts are lost.

The shallow depth creates ecological

mechanisms that produce high productivity. The benthic and demersal species are in close proximity to their pelagic prey, thus reducing energy loss. For example, in Adriatic hake (Froglia, 1973), full stomachs sampled contained 50% clupeiforms in specimens longer than 14 cm and 85% in specimens longer than 16 cm (Fig. 5). This would be far less likely to occur in hake living in deeper seas.

Negative role for the environment

When thermohaline fronts involving the entire water column form, the terrigenous flows stagnate and accumulate along the coast during winter. However, during summer, there is no vertical mixing, with a consequent lack of oxygen on the sea bottom.

The ideal conditions for a basin characterized by the above ecological factors would involve periodic turbulence and remixing of water masses (wind and stormy seas), so as to distribute terrigenous flows and oxygenate bottom water. It is not by chance that some authors have related the frequency of algal blooms—as well as stratification and anoxia— in the Adriatic with the reduced frequency of northeast winds over the last 20 years.

Eutrophication and productivity of the sea

Because of the large flow of water and consequent influx of nutrients, productivity is high in spring (from several thousands of cells per liter to 600,000–700,000), especially in the northern and central Adriatic (Boni, 1985). Obviously, in the case of algal blooms, concentrations of millions of cells per liter of the dominant species occur. Together with high primary productivity in the Adriatic, there is considerable productivity contributed by particles of suspended organic material, which is the primary source of nutrients for the large biomass of sestonophagous and detritivorous molluscs (including razor shells, clams, striped venus, mussels, and oysters), which are of great economic importance in the Adriatic.

The process of eutrophication, which primarily involves the northern and central Adriatic coast, affects all of the links of the food chain and permits high levels of productivity in the fishing sector. The regions of major eutrophication are also those in which the catch is greater (Figs. 6 and 7). There is not only a clear correlation between eutrophication and marine resource catch, but also a link between the type of energetic material available, the nature of the bottom, and the type of catch. The large extent of soft substrate and the richness of both particulate matter and phytoplankton and zooplankton favor plankton, seston, and detritus feeders.

Fishing Features and the State of Resources in the Adriatic

Fifty-five percent of the total Italian catch of marine fish comes from the Adriatic. The Northern and Central Adriatic contribute 40% of this amount (Table 1). The portion of the contribution from the Adriatic composed of "small pelagic" groups and molluscs, particularly bivalves, is very important. The fishing features of the Adriatic basin can be obtained from the composition of the catch per group of species, both within the Adriatic production and in comparison to other Italian seas (Tables 2 and 3). The importance of the "small pelagic" group, primarily "other molluscs" (mainly represented by bivalves), is evident.

Status of resources

Fisheries in the Mediterranean are generally characterized by (i) a short average life span, permitting a renewal of resources within the space of a few years; (ii) multispecies fisheries—with various species in the fishery exhibiting different biological characteristics (e.g., growth, fertility index, mortality, and selectivity); various variations in compensation and biological alteration occur; this complexity leads to difficulty in using selective gear types and thus bycatch problems; (iii) excessive fishing effort out of proportion to available resources—fishing effort has increased considerably in the Mediterranean, primarily with the use of engine power; although the global catch has increased, the catch per unit effort (CPUE) and size at recruitment have decreased; generally, for demersal resources, it can be said that today fishing involves young age classes (young-of-the-year and 1 year), so that the fish-

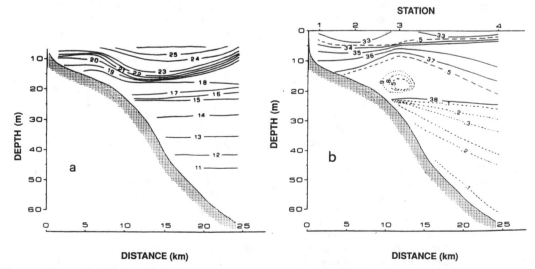

Figure 4. Central Adriatic Sea. Temperature (a) and salinity (b) distributions on Portonovo transect under summer stratification conditions. (By Artegiani, 1987.)

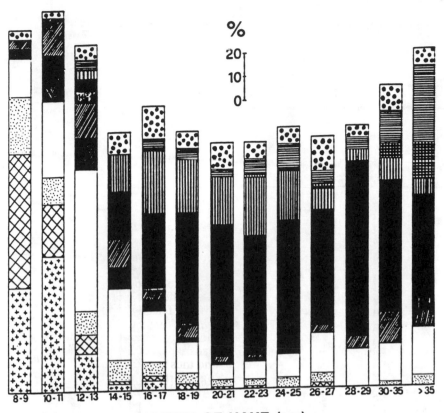

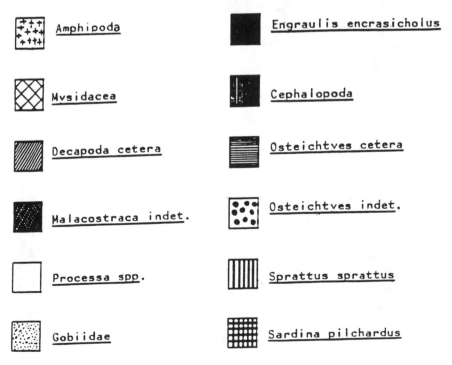

Figure 5. Percentage composition of prey consumed by hake in the central Adriatic Sea. (After Froglia, 1973.)

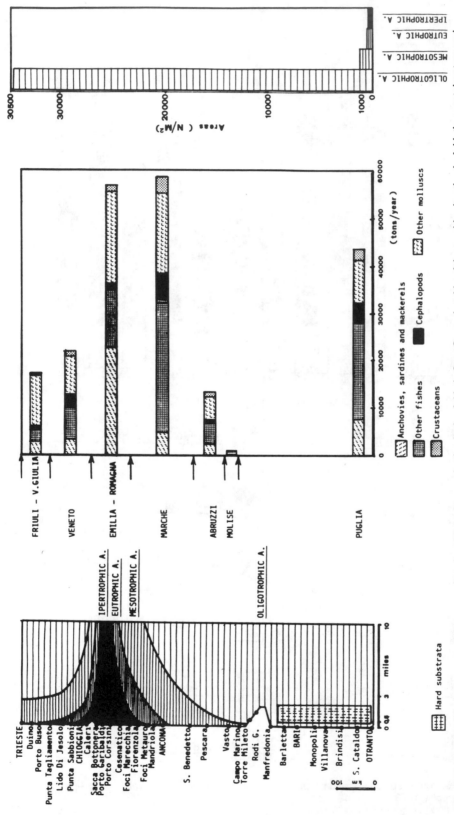

Figure 6. Relationship between trophic type and regional fisheries production in the Adriatic Sea. Synoptic table of the trophic situation in Adriatic coastal waters and regional fisheries production. (After Olivotti, 1989, modified.) The large fisheries production for the Apulia region is given by extension of hard substrata.

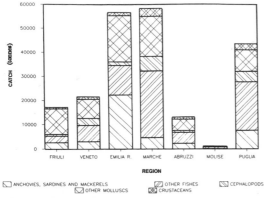

Figure 7. Total Italian catches in the Adriatic Sea per region. (Source: Italian Institute of Statistics [ISTAT] 1987.)

Table 1. Percent contribution of the Adriatic Sea to the global Italian catches (average for years 1982–1987)*

Resources	Adriatic	North. & Cent. Adr. Sea
Small pelagic/species	63	57
Demersal species	44	36
Molluscs	60	48
Striped venus	97	96
Crustaceans	30	25
Total Ital. catch (100%)	55	48

Source: Italian Institute of Statistics.
*Even if the absolute values are not precise, the relative values remain significant.

ing yield now largely depends on the degree of recruitment success; (iv) use of a variety of fishing gear on a single boat depending on the season and opportunities; and (v) a large number of ports for landing catch spread over an extensive coastline, causing difficulty in obtaining accurate fishery statistics.

As far as the Adriatic is concerned, the following observations can be made.

- All the Species:
  The contribution of the Adriatic to the Italian total catch dropped from 62.5% in 1982 to 52% in 1987, a decrease of 10% (Fig. 8). This reduction was primarily caused by a decrease in CPUE, predominantly in the Northern Adriatic. In 1982, a boat equipped with a 1-horsepower (HP) motor used in the Northern Adriatic caught 510 kg of biomass (different species); in 1987,

the same boat with a 1-HP motor produced a biomass of 258 kg. The CPUE decreased by one-half (Fig. 9).

- Common Demersal Species:
  There has been a large decrease in the catch of demersal species. The contribution of the Adriatic to the total national harvest dropped from 55% (1982) to 39% (1987), a decrease of 16% (Fig. 10). This decrease, from almost 20% (1982) to about 10% (1987), occurred mainly in the Northern Adriatic. Because the fishing effort in the Northern Adriatic has not increased significantly in this last decade (although it had reached high levels prior to the last decade), the reduction in catch could probably be attributed to a number of factors, among which natural mortality plays a particularly important role. One source of mortality in the prerecruitment stages (eggs and larvae), as well as during the adult life of sedentary or benthic species (e.g., baby clams) is anoxia or hypoxia on the seabed, caused by prolonged stratification (with or without algal blooms). In past decades, there have been intense and vast occurrences of anoxia on the Northern Adriatic bottom. Intensity and duration of the stratification, in turn, are affected by climatic and hydrographic conditions.

Baby clams or striped venus (Chamelea gallina L.)

Certainly environmental stress (anoxia in the coastal zone) and overfishing have contributed to a general decrease of the "immature clam" stock, an important resource. The baby clam stocks are subject to varying degrees of exploitation within each of the sectors of the Adriatic.

The Northern Adriatic contribution to global catch increased from 10.57% (1982) to 28.9% (1987), with a peak of 38.45% (1986) (Fig. 11). Populations sustain maximum exploitation at 1 year of age. The minimum size of 2.5 cm (2 years old), as provided by law, is rarely respected. The temporary increase in catch therefore results from the recruitment of juveniles, which can be very dangerous for a resource whose stock of reproducers may be decimated by extended anoxia.

In the Central Adriatic, there has been a decrease in the stock (Fig. 12) such that the percent contribution to global production dropped from 88.28% (1982) to 63.45 (1987), a

Table 2. Catch composition in Adriatic Sea[*]

1982–1987 (Average)	Adriatic		Contribution of Northern and Central Adriatic	
Resources	Tonnes	%	Tonnes	%
Small pelagic species	63.403	30	57.361	90
Demersal species	80.076	35	65.758	82
Molluscs	41.584	18	33.147	80
Striped venus	30.701	13	30.324	99
Crustaceans	8.506	4	6.697	79
	224.270	100	193.287	86

Source: Italian Institute of Statistics.
[*]ISTAT data underestimate fisheries production; nevertheless, the percent relationships remain significant.

Table 3. Catch percent composition per groups of species in the different Italian seas (average for years 1985–87)

Resources	Adriatic	Ionian	Sicilian	Sardin.	Tyrrhen.	Ligur.
Pelagic	27	13	10	6	28	32
Demersal	35	57	62	69	49	48
Cephalopods	8	9	9	10	9	7
O.[*] Molluscs	26	13	6	10	6	9
Crustaceans	4	8	13	5	8	4
	100	100	100	100	100	100

Source: Italian Institute of Statistics.
[*]O. = other

reduction of about 25%. On the other hand, fishing effort cannot be controlled and has nearly doubled (from 400 to 760 dredge vessels) in the past decade. Initiatives such as gradually reducing fishing quotas, closing the fishing season for 2 months (rather than 1), and reducing weekly fishing activities, have not managed to slow the impact on this dwindling resource. In recent years, fishing boats have integrated their catch by fishing Golden carpet shells *(Tapes aureus)*, which live in the baby clam boundary area. The processing industry, which could not compete with the baby clam market prices, started purchasing Golden carpet shell supplies, but even this resource will not be able to sustain the growing fishing effort and demand.

Other molluscs

Catch of cephalopods has been generally constant (Fig. 12), although there has been an increase in mussels and oysters as a result of ex-panded offshore mariculture, as will be explained further.

Crustaceans

Following a decrease in 1983–1985, the contribution from the Adriatic to the total production of crustaceans has returned to the 1982 values of about 35% (Fig. 13).

Small pelagic species

The percent contribution from the Adriatic to the total Italian production of anchovy, sardine, and mackerel remains very high, on average about 63% (Fig. 14).

In addition to the fluctuations that characterize small pelagic species, something happened after 1986 that has caused a decrease in pelagic biomass and corresponding catch, especially in the northern Adriatic, where the contribution dropped from 38.56% (1982) to

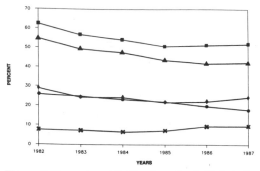

Figure 8. Percent contribution from the Adriatic to the total Italian catch (all species) for the entire Adriatic (■), the northern plus central Adriatic (▲), the northern Adriatic (•), the central Adriatic (♦), and the southern Adriatic (x).

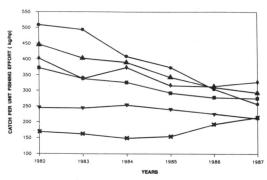

Figure 9. Catch per unit fishing effort (all species, all types of gear) expressed as annual kilograms biomass harvested per unit horsepower of engine, for the entire Adriatic (■), the northern plus central Adriatic (▲), the northern Adriatic (•), the central Adriatic (♦), the southern Adriatic (x), and for Italy (▼).

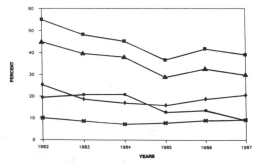

Figure 10. Percent contribution from the Adriatic to the total Italian catch (mainly demersal species) for the entire Adriatic (■), the northern plus central Adriatic (▲), the northern Adriatic (•), the central Adriatic (♦), and the southern Adriatic (x).

30% (1987). In the southern and central Adriatic, the percent reduction in 7 years (1982–1987) was 10% of the contribution to the total catch for the same area.

A more precise idea of what has happened to the small pelagic species in the Adriatic ecosystem can be obtained by examining the results of acoustic surveys conducted by Instituto Richerche Pesca Marittima (I.R.PE.M.) in the last 13 years. In the Northern Adriatic, the total 1988 pelagic biomass was about 283,000 tonnes (± 35%) over an area of about 8,000 square nautical miles. For 1987, the pelagic biomass estimated was more than double this amount.

In terms of geographical distribution (Fig. 15 a,b), certain observations have been made, such as the disappearance of the high-density concentration from the Gulf of Trieste, a certain concentration adjacent to the Istria peninsula, and the breaking down of the concentration located formerly near the Po delta and along the coasts between the regions of Emilia-Romagna and Marche. This leads to a number of considerations. The trend in total pelagic biomass indicates a decrease in 1988 to almost the same level as found in 1985 and 1980.

The average density of pelagic biomass is very high, namely 77 metric tonnes (mt)/nm^2, but with large oscillations (72% to 58%). The average biomass per year is 596,312 tonnes, with a maximum of 1,025,187 and a minimum of 249,096 mt/year.

Anchovies

The anchovy stock has collapsed within a 10-year period (Fig. 16). From a maximum biomass of more than 640,000 mt in 1978, it has dropped to a biomass of just several tonnes. Many midwater pair-trawlers have had to stop their activities, resulting in a 10-fold increase in prices. The overall fishing effort in the Northern and Central Adriatic for pelagic stocks also decreased from 1981–1988 (Fig. 17). The actual average density of anchovy is 24 mt/(nautical mile [nm])^2.

The stock decreased from 1980 to 1985 to an average biomass of about 200,000 tonnes/year, and totally collapsed in 1986.

Sardines

From 1981 to 1984, there was an increase in sardine production (Fig. 18) concomitant with the decrease in anchovy production (Fig. 16). A maximum peak was sustained in 1983. Another increase occurred in 1987, which was once again the opposite of the anchovy trend

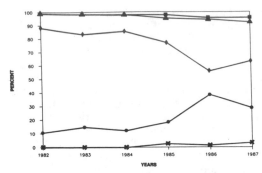

Figure 11. Percent contribution from the Adriatic to the total Italian catch of striped venus for the entire Adriatic (■), the northern plus central Adriatic (▲), the northern Adriatic (•), the central Adriatic (◆), and the southern Adriatic (x).

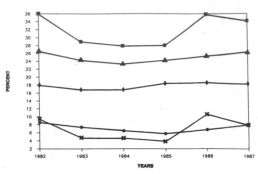

Figure 13. Percent contribution from the Adriatic to the total Italian catch of crustaceans for the entire Adriatic (■), the northern plus central Adriatic (▲), the northern Adriatic (•), the central Adriatic (◆), and the southern Adriatic (x).

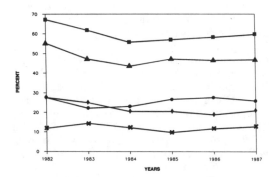

Figure 12. Percent contribution from the Adriatic to the total Italian catch of molluscs (not including striped venus) for the entire Adriatic (■), the northern plus central Adriatic (▲), the northern Adriatic (•), the central Adriatic (◆), and the southern Adriatic (x).

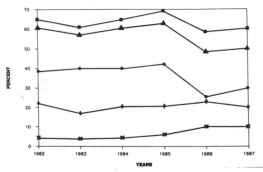

Figure 14. Percent contribution from the Adriatic to the total Italian catch of anchovies, mackerel, and sardines for the entire Adriatic (■), the northern plus central Adriatic (▲), the northern Adriatic (•), the central Adriatic (◆), and the southern Adriatic (x).

(partial biological compensation). This increase was followed by a decrease in 1988.

Sprat

Fluctuations are more regular in sprat than in sardines and anchovies (Fig. 19). The maximum peak occurred in 1987 with a density of about 32 mt/nm^2. In 1988, a decrease to just a few mt/nm^2 was reported. Average density is currently 14 mt/nm^2.

Other pelagic species

The major commercially important pelagic species are the horse mackerels (*Trachurus* spp.). Their trend from 1981 through 1988 (Fig.

20) seems to repeat that of the sardine (Fig. 18). Maximum abundance was reported at about 28 mt/nm^2 in 1983, with a minimum of just several mt/nm^2 in 1986. Average density is now 8 mt/nm^2. Even in this group, 1988 proved to be a year in which density dropped.

Observations

Anchovy and sardine catches have followed the biomass fluctuations, with prices compensating for falling catches. In particular, the collapse of the anchovy stock has been felt considerably. Sprat is a largely unexploited resource, and sardines are difficult to market because of economic problems. The question of the causes for the depletion of the pelagic resources must be addressed.

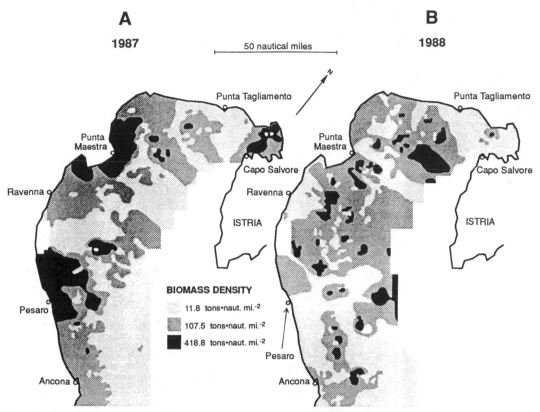

Figure 15. Distribution of pelagic fish biomass from 1987 (A) and 1988 (B) echo-surveys of the northern Adriatic Sea.

From an ecological viewpoint, the major factors that apparently contributed to the depletion are the following:

- Break-up of the large density area that previously existed around the Po delta.
- Below-average biomass trend in the period 1981–1985 and the collapse of the anchovy stock from 1986 to 1988.
- Decrease of the sardine stock in just 1 year (1989) during which the competitor species (anchovy) was at a minimum. It seems that the mechanism of biological compensation no longer works as it did in the period 1981–1984.
- Rapid decrease of the sprat stock from a maximum to an absolute minimum in just 2 years (1987–1988), preceded by a more gradual decrease from 1983 to 1986.
- The fact that fishing effort, in terms of number of boats (paired midwater trawlers), has decreased in recent years should have allowed the renewal of the pelagic biomass; however, the opposite has occurred.

Probably one can attribute the above to anthropogenic, ecological, and climatic factors. Most certainly, there has been a considerable depletion of pelagic stocks in the Adriatic during the 1970s and 1980s, resulting from overfishing, the complicating factor of refunds for unsold products permitted by European Economic Community (EEC) regulations, and finally, authorized fishing for juveniles, which still occurs today.

However, fisheries-related issues alone would not have led to such radical stock decreases without the influence of ecological and climatic factors. A hypothesis, which seems to be feasible, but has yet to be substantiated by data, is that of a reduced flow of fluvial waters resulting from drought conditions and higher average temperatures, both on land and at sea. In fact, a reduced amount of water means lower nutrient levels and less particulate organic matter, and consequently, reduced productivity. For example, a reduction of 12% in the annual average flow of the Po and lowered organic load have been recorded recently (Grego, 1990; Sansebastiano and Sansebastiano, 1990). This naturally leads to the decline in the pelagic fish concentration around the Po delta. Higher average temperatures, especially on the sea bottom, produced unfavorable conditions for the sprat stocks that normally live in the colder part of the water mass, the layer in direct contact with the sea bottom. This species is in fact typically northern and lives only in the Northern Adriatic. In addition, the composition of the phytoplankton community seems to have changed in the last decade in favor of microflagellates, and we are not yet certain how this will affect the diet of the planktivorous organisms. Research on climatic and hydrographic data is underway to test this hypothesis.

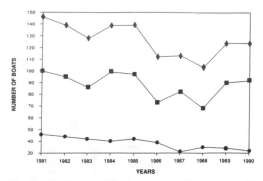

Figure 17. Fishing effort in number of boats operating on pelagic stocks (northern [■] Adriatic, central [●] Adriatic, and northern plus central Adriatic [◆]).

Management and Exploitation of the Coastal Zone

Managing the coastal zone from an ecosystem perspective can resolve a number of fishing problems and can facilitate the renewal of impoverished resources.

It is necessary to regulate the fishing effort on marine resources, and if necessary, close the fisheries. Fishing season and areal regulations must be respected, and all biotechnological mechanisms that are required to reduce fishing mortality during the prerecruitment phase must be implemented (for example, gear selectivity and protection of the juveniles).

Artificial Reef Experiments

It is important ecologically to use the biochemical energy that accumulates inshore from recycling systems. Eutrophication should be channeled toward halieutic production. This could possibly be achieved by introducing multipurpose artificial reefs and mariculture structures in the open sea, as demonstrated by the experiments conducted by the I.R.PE.M. in 1974

and which are still under way (Fabi and Fiorentini, 1989). There are currently a total of 16 experimental artificial reefs in Italy, of which 8 are in the Adriatic, 4 are in the Ligurian Sea, 2 are in the Tyrrhenian, and the 2 remaining are in Sicily.

Results include effects of protection and production. As far as protection is concerned, the following have been obtained:

- Protection of the sea bottom in areas affected by illegal trawling and consequent growth of the juveniles of native species.
- Protection and refuge for many organisms, such as eggs, juveniles, and adults of pelagic, nekto- and benthic fish, crustaceans during ecdysis, and cephalopods (cuttlefish, octopus, and squid).
- Protection of small-scale fishing gear from the impact of trawling. It is difficult to quantify the extent of these benefits.

As far as fishing and ecological results are concerned, the following data are available:

- Total catch data for the various years (Porto Recanati) (Bombace, 1982; Bombace and Rossi, 1986).
- Fishing yields for fish, molluscs, and crustaceans obtained using standard experimental trammel nets for many of the artificial reefs (Bombace *et al.*, 1989).
- Yields for the gastropods, *Nassarius mutabilis* and *Hinia reticulata*, based on trap sampling.
- Estimated biomass for mussels settled both on substrate and artificial modules, as well as for the production of mussels on suspended structures associated with the reefs (Fabi and Fiorentini, 1989).

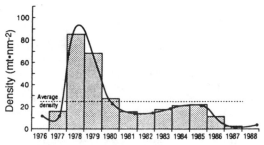

Figure 16. Anchovy biomass from echo-surveys of the northern and central Adriatic Sea during 1976–1988. (From Azzali *et al.*, 1990.)

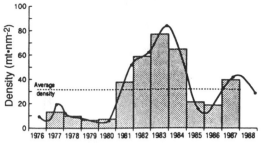

Figure 18. Sardine biomass from echo-surveys of the northern and central Adriatic Sea during 1976–1988. (From Azzali *et al.*, 1990.)

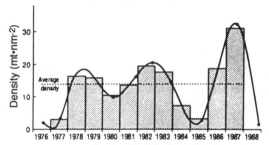

Figure 19. Sprat biomass from echo-surveys of the northern and central Adriatic Sea during 1976–1988. (From Azzali *et al.*, 1990.)

Biological data are available on growth, recruitment, and mortality of mussels, and limited data are available for oysters. Data on the feeding ecology of various fish species that frequent the artificial reefs and biological parameters for the two species of gastropods mentioned above are also available. Observations on sessile settlements, changes in the nature of the substrate, interspecific relationships, environmental and technological data concerning both the substrate and the structural supports for mariculture, other methods of immersion, and socioeconomic considerations are also available.

An evaluation of these catch data and yield indexes has led to the following conclusions:

- The unit fishing yield using standard nets increases progressively as the sampling stations are moved out to sea toward the heart of a protected area (artificial reefs).
- The unit fishing yield using standard nets or the capture of organisms is always greater (about 3:1) in a reef area compared to other areas without reefs.
- The fishing yield in an area is considerably higher after the construction of artificial reefs, especially for prized species (e.g., *Sciaenidae* and *Sparidae*)

belonging to the nekto-benthic group (from three to eight times higher, and much more in the case of the *Sciaenidae*) (Fig. 21).

- The capture of gastropods is 2 1/2 to 6 times greater after the installation of a reef.
- The mussel and oyster biomass is a total gain because they were introduced via mariculture to locations they had not inhabited previously. Today, in 1990, in the reef and mussel-culture installations of the protected marine areas of the Adriatic, the biomass averages several hundreds of tonnes per installation. Mussel and oyster cultures, as well as breeding of other bivalves (such as *Tapes* spp.), can be achieved with suspended or submerged installations in the open sea, whether or not they are associated with artificial reefs.

The above results, in addition to other reports, point to artificial reefs as being useful from a bioecological and halieutic point of view. From the economic viewpoint, analysis of data obtained from the experiments carried out in Porto Recanati indicated that the return on the investment tripled from 1977 to 1980 (Bombace, 1982) and the gross income doubled for a reef fisherman in 1982–1984 (Bombace and Rossi, 1986).

Finally, (i) if there is to be a recovery of the small-scale fishing industry equipped with fixed gear (about 7,000 boats in Italy), (ii) if the small trawlers are to be converted for use in the management of reefs and mariculture (about 2,000 boats in Italy), and (iii) if there is to be a general policy for protecting nurseries and damaged stocks, rebuilding resources, energy recycling, and appropriately managing the coastal environment, we must aim at integrated initiatives that provide protected marine areas

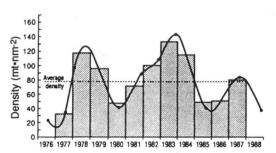

Figure 20. Total pelagic fish biomass from echo-surveys of the northern and central Adriatic Sea during 1976–1988. (From Azzali *et al.*, 1990.)

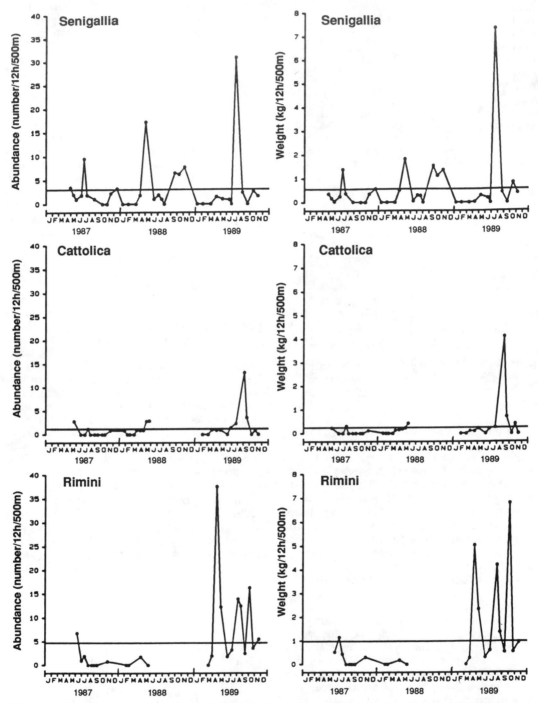

Figure 21. Number and biomass of nekto-benthic fish in catches at Senigallia, Cattolica, and Rimini before and after reef deployment. The horizontal lines represent the average values estimated for the whole survey period. (From Bombace *et al.*, 1989.)

enhanced by artificial reef habitat as the first priority. These must be planned, implemented, and monitored. The water parcels involved would be allocated based on management decisions involving various species and user groups. A comprehensive plan for the coastal areas in question is needed that considers an integrated system of artificial reefs, hatcheries, mariculture, research stations, and so forth. Cooperation among management, research, and harvesting communities will be required.

An International Technology Conference sponsored by the Food and Agriculture Organization of the United Nations (FAO) addressing this subject was held in Ancona in 1986 (FAO, 1986). Another international conference on the same topic was convened in Ancona in November 1989 by the Conseil Général Pêche Méditerranée (CGPM)-FAO, and a working group was formed by CGPM and by the Societa' Italiana Biologia Marina (SIBM). Artificial reefs, in fact, provide an excellent opportunity for interdisciplinary research. Research on the theoretic bioecological and fishing aspects and the trophic relations within the reef and those between the reef and the surrounding area should be continued and expanded. Sampling methods need further development and integration with visual methods of evaluation.

The technological aspects of implementing artificial reefs including the relationships between forms, architecture, and geometry of the habitats, density, and distribution of fishing resources must be examined more thoroughly. The casting and immersion techniques used for the modules as well as the structures for reducing production costs need refining.

Legal, administrative, financial, and institutional requirements to expedite the installation of such systems include simplification of the procedure for state concession of coastal marine waters; consideration of an institutional liaison with the coastal regions; consideration of the "coastal area management and marine culture" as a distinct problem when revising the law on fishing; modification, at the EEC level, of several administrative laws regarding initiatives financed with community funds; and the introduction of a new chapter for planned general initiatives.

Experiments with artificial reefs are expensive and the small, private fishing sector cannot sustain these costs. Substantial public intervention, both by the EEC and the member state, is necessary at least for the first system.

Acknowledgments

Special thanks go to G. Fabi, F. Fiorentini, S. Bolognini, N. Cingolani, M. Luna, M. G. Angeli Temperoni, and A. Sala.

References

Artegiani, A. 1984. Seasonal flow rates of the Italian rivers having outlets in the Northern and the Central Adriatic. FAO Fish. Rep. 290:81–83.

Artegiani, A. 1987. Parametri ambientali, dinamica delle acque costiere, della produtt. primaria, delle caratt. geomorfol. e fisionomia della fascia costiera marchigiana. Report for the Marche Region.

Azzali, M., Cosimi, G., and Luna, M. 1990. Rapporto sulle risorse pelagiche dei mari italiani stimate con metodi acustici. Ancona Rapporto per il Min. Marini Mercantile:1–12.

Bombace, G. 1982. Il punto sulle barriere artificiali: problemi e prospettive. Naturalista Sicil., Ser. 4, 6 (Suppl. 3):573–591.

Bombace, G., and Rossi, V. 1986. Effects socioeconomiques consecutifs à la réalisation d'une zone marine protegée par des recifs artificiels dans la zone de Porto Recanati. C.G.P.M. F.A.O. Rapport sur les pêches 357:157–164, C.G.P.M.

Bombace, G., Fabi, G., and Fiorentini, L. 1989. Preliminary analysis of catch data from artificial reefs in the Central Adriatic. FAO Fish. Rep. 428:86–98.

Boni, L. 1985. Problematiche del fitoplancton in Adriatico, in "Eutrofizzazione auali interventi"? Conv. Nazion. Ancon—Fiera di Ancona: 27–30.

Fabi, G., and Fiorentini, L. 1989. Shellfish culture associated with artificial reefs. FAO Fish. Rep.428:99–107.

FAO (Food and Agriculture Organization of the United Nations). 1986. Technical consultation on open sea shellfish culture in association with artificial reefs. FAO Fish. Rep. 357:1–175.

Franco, P. 1970. Oceanography of the Northern Adriatic Sea. Cruises July, August, October, and November 1965. Arch. Oceanogr. Limnol. 16 Suppl.:1–93.

Franco, P. 1972. Oceanography of the Northern Adriatic Sea. Cruises January–February and April–May 1966. Arch. Oceanogr. Limnol. 17 Suppl.:1–97.

Franco, P. 1983. L'Adriatico settentrionale: caraterri oceanografici e problemi. Atti 5 Congresso AIOL, (Stresa):1–29.

Froglia, C. 1973. Osservazioni sull'alimentazione del Medio Adriatico. Atti V Congresso S.I.B.M. Ed. Salent.:327–341.

Grego, G. Idrologia del Delta del Po— Evoluz. nell'ultimo ventennio. Conv. Ecol. Delta Po. Albarella. In Press.

Mars, P. 1963. Les faunes et la stratigraphie du Quaternaire mediterraneen. Rec. Trav. St. Mar. Endoume 28:61.

Peres, J. M., and Picard, J. 1964. Nouveau manuel de bionomie benthique de la mer Méditerranée. Rec. Trav. St. Mar. Endoume. Bull. 31, Fasc. 47.

Olivotti, R. 1989. Tutela delle acque costiere. Cosa e cambiato negli ultimi venti anni. Atti Conv. Nazion. Ancona 4 Aprile:116–142.

Ruggieri, G. 1967. The miocene and later evolution of the Mediterranean Sea. In: Aspects of Tethyan biogeography. Ed. by C. G. Adams and D. V. Ages. Syst. Ass. Publ., London.

Sansebastiano, G., and Sansebastiano, A. In Press. Qualita delle acque in rapporto al funzionamento della Centrale di Porto Tolle. P. II Aspetti microbiologici. Conv. Ecol. Delta Po. Albarella.

Tortonese, E. 1983. Distribution and ecology of endemic elements in the Mediterranean fauna (fishes and echinoderms). Medit. Mar. Ecosyst. NATO Conf. Ser.:57–83.

Suggested Further Reading

Arneri, E., and Jukic, S. 1985. Some preliminary observations on the biology and dynmamics of *Mullus barbatus* L. in the Adriatic Sea. FAO Fish. Rep. 345:79–86.

Artegiani, A., Azzolini, R., Marzocchi, M., Morbidoni, M., Solazzi, A., and Cavolo, F. 1985. Prime osservazioni su un bloom fitoplanctonico lungo la costa marchigiana nell'anno 1984 Nova Thalassia 7, Suppl. 3:137–142.

Azzali, M., and Luna, M. 1988. Nota preliminare sulle risorse pelagiche nei mari italiani, stimate con tecniche acustiche. Atti Semin. Un. Operat. Min. Mar. Mercantile C.N.R. Vol. I:99–174.

Bombace, G. 1981. Note sur les experiences de creation de recifs artificiels en Italie. C.G.P.M. FAO Études et Revues 58:231–337.

Bombace, G. 1985. Eutrofizzazione e produzione di pesca. Nova Thalassia 7, Suppl. 3:277–295.

Bombace, G., Cingolani, N., Coppola, S. R., and Mortera, J. 1987. Summary report on the quality check sample survey of fisheries catch and effort statistics. Adriatic Area. FAO Fish. Rep. 394:245–259.

Bombace, G. 1987. Iniziative di protezione e valorizzazione della fascia cosiera mediante barriere artificiali a fini multipli. Atti LIX Riun. SIPS:201–233.

Chapter 14

Contrast Between Recent Fishery Trends and Evidence for Nutrient Enrichment in Two Large Marine Ecosystems: The Mediterranean and the Black Seas

John F. Caddy

Introduction

The Food and Agriculture Organization has been active in promoting sustainable development and sound environmental management in both developed and developing countries. Its program has evolved over the years specifically to address environmental problems arising from growing population pressures caused by agricultural practices. These considerations also apply to aquatic systems, where downstream effects of human activities on the fish fauna of watersheds and coastal lagoons are well documented (Kapetsky, 1984; Welcomme, 1985).

Until recently, however, there has been very little in the way of documentation of the impacts of various terrestrial inputs to the fisheries of open-sea areas. With the exception of studies such as those of Steemann Nielsen (1971) and Sutcliffe (1972), it has generally been assumed (wrongly as it turns out) that these inputs are either of little importance, or, if measurable in impact, uniformly negative. It is clear also that the definition of "optimal" may not be the same for all users of the marine ecosystem, which adds a socioeconomic dimension to the question of marine pollution. Of particular interest here is that type of pollution caused by nutrient enrichment—a phenomenon referred to as "cultural eutrophication" when it results from human activities (UNEP/UNESCO/FAO, 1988).

The first impacts of human activities on marine systems are expected to be found in those closed and semi-enclosed seas where dilution effects are likely to be reduced. Despite certain favorable hydrographic characteristics, this is the case for the Mediterranean, with a turnover period of roughly 80 years for Atlantic water entering through the Straits of Gibraltar (UNEP, 1989). Problems of coastal pollution have been a focus of the Regional Seas Program of the United Nations Environmental Program (UNEP) since 1974 (UNEP, 1989). The Mediterranean, with a coastal population now estimated at 132 million inhabitants (UNEP, 1989) plus a very large transient population, was in fact the first regional sea to move to an Action Program in 1975. The Blue Plan, a component of the Mediterranean Action Plan (Anonymous, mid-1980s), has since focused on a broadbrush description of the effects of pollution on the Mediterranean system; the reports of these initiatives should be consulted for further details. However, the least studied aspect of cultural eutrophication may be its effect on the fisheries sector. This was perhaps because it had been assumed, until recently, that fishing effort is the key controlling variable, and perhaps also because fisheries statistical systems in some coastal countries of the Mediterranean have serious weaknesses.

Some relevant studies that at least point to cultural eutrophication as having an influence on Mediterranean fisheries include the work of the General Fisheries Council—for the Mediterranean as summarized in Caddy (1990); for the Black Sea in Ivanov and Beverton (1985), Balkas *et al.*, (1990), and in Caddy and Griffiths (1990); and the work carried out under the Mediterranean Action Plan.

Oceanographic Overview of the Black and Mediterranean Seas and Their Hydrographic Inputs

This brief account is extracted from reviews of the main oceanographic features of the Mediterranean found in Murdoch and Onuf (1972) and Miller (1983), and from Sorokin's description of the Black Sea (1983). The differences in responses of these two seas to nutrient enrichment appear to be related predominantly to three main physical features of these basins.

Inputs

Although inflows make up a relatively small proportion of the volume of both seas, the Black Sea (Fig. 1) is a largely closed body of water that receives a significant proportion of its inputs at the surface as fresh water from the Danube, Dnepr, Dnestr, Don, and Kuban rivers, and forms a large catchment basin (Balkas et al., 1990). Inflows of Mediterranean water and migratory incursions of marine fauna enter through the narrow Bosporus.

In the case of the Mediterranean, freshwater inflows are relatively lower, and only about 10% of the total, including the Nile (UNEP, 1989), enters along the southern shores. Inputs are roughly compensated for by evaporation, especially in the Levant Basin. For both seas, inputs of fresh water, especially from the Nile and larger Black Sea rivers, have been re-

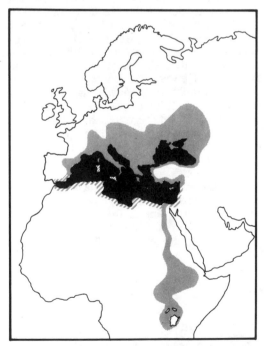

Figure 1. Schematic view of catchment areas drained by permanent rivers (grey shading) and from largely seasonal runoff (striped shading) into the Mediterranean and Black seas. (Note: Nile catchment is largely excluded since construction of the Aswan Dam.)

duced over past decades by increased freshwater usage, and nutrient levels of the remaining flow have generally increased.

Stratification

The Mediterranean, bordered by arid or semiarid lands to the south and east, receives relatively minor freshwater discharges, and its major inflow is nutrient-poor, oxygenated Atlantic surface water through the Straits of Gibraltar. As a consequence, bottom waters are generally well oxygenated. Earlier investigations of biological oceanography, as mentioned in Murdoch and Onuf (1972), described nutrient levels in the Mediterranean as very low, generally decreasing from west to east and from north to south. Nutrient levels serve as indices of production which also, broadly speaking, are reflected in the abundance of fish stocks. Nutrient levels in the Levant area are very low, and Azov (1990) reported chlorophyll concentrations off the coast of Israel at between 1/2 and 1/10 of those for the Sargasso Sea, an area of very low primary productivity. They noted that phosphates and nitrogen levels also remain very low in subsurface waters, down to at least 3,000 m.

The outflow of highly saline bottom water through the Straits of Gibraltar is richer in nutrients and provides a potential feedback mechanism that, in the very long-term, may reduce accumulation of nutrients and pollutants. Over recent decades, freshwater inputs to both seas have been reduced because of diversion of river outflows into irrigation, and this has had significant impacts on estuarine fauna, especially for the Black Sea (Ivanov and Beverton, 1985). In the case of the marked reduction in Nile outflow since the late 1960s—apart from the consequent decline in enrichment of the nutrient-poor eastern Mediterranean—one suggestion is that the removal of a seasonal lower salinity barrier caused by Nile outflows plus higher water temperatures in the Levant area have facilitated colonization of the eastern Mediterranean by species adapted to higher salinities. Some 500 species have been recorded in the Levant area, entering from the Red Sea and Indian Ocean through the Suez Canal, described by Por (1978) as Lessepsian migration.

Bathymetry

Both the Black and Mediterranean Seas are characterized by narrow shelves; the exceptional areas of wider shelf (also areas of higher than average biological productivity) are the

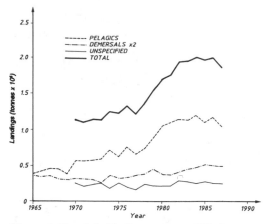

Figure 2. Trends in Mediterranean and Black seas combined landings from FAO statistical sources (demersal landings doubled for ease of illustration) (from Caddy, 1990).

Adriatic, the gulfs of Lions and Gabes for the Mediterranean, and the Sea of Azov for the Black Sea. In contrast, because of bathymetry and stratification (Balkus *et al.*, 1990), the lower (and major) part of the water column of the Black Sea has in recent times consisted of unoxygenated water with well-developed stratification, and there have been suggestions that the upper shelf water has also suffered low-oxygen episodes in recent years.

Recent Effects of Increased Nutrient Discharges on the Mediterranean and Black Sea Systems

The impacts of discharges of nutrients on aquatic systems are usually considered synonymous with eutrophication. Although the overall effects of high levels of enrichment are likely to be negative from the perspective of fisheries, a moderate degree of enrichment will certainly increase the biological production in oligotrophic systems, such as the Mediterranean (UNEP, 1989) and the Black Sea (Ivanov and Beverton, 1985) were considered to be earlier in the twentieth century. Recent evidence (UNEP, 1989) suggests that roughly half of the organic inputs to the Mediterranean are industrial in origin. The remaining half is derived equally from human sewage and agricultural sources. Of these totals, some 60% to 65% of organics entering the Mediterranean come from coastal runoff and discharge, and the rest of the load is transported by rivers. Direct atmospheric inputs are also important, particularly with respect to nitrogenous compounds, contaminants, and heavy metals.

The proportion of the nutrient-enriched production that is translated into enhanced fisheries yield has not been considered until recently, and it is still impossible to provide estimates. It is more certain, and more easily observed, that nutrient discharges have a locally deleterious impact on the high diversity of Mediterranean fauna and flora close to the points of discharge. Nonetheless, there is strong circumstantial evidence (Caddy, 1990) that Mediterranean and Black Sea fishery yields (Fig. 2) rose in the 1990s, at a time when fishing effort would appear to have been in excess of the maximum sustainable yield, so that these increases are not easy to attribute to fishing effort alone.

At the same time, given that effects of nutrient discharges from point sources, rivers and lagoons, and direct runoff are most pronounced in coastal areas (Fig. 3) and are enhanced in more enclosed systems such as the northern Adriatic (Degobbis, 1989), it is clear that their negative effects will be more readily apparent to land-based users, well before the productivity of the entire system is affected.

The Fisheries of the Mediterranean and Black Seas: A Brief Historical Perspective

In general, the concept that a steady-state production is possible in marine fisheries is one that has come into serious question over the last 2 decades. At the global level, the world marine production, documented in FAO Annual Yearbooks of Fisheries Statistics, has risen slowly by some 3% per year since 1980. Technological improvements in boats and gear have continued unabated, and the rate of exploitation has increased through global expansion of industrial fleets (particularly of distant water fleets in the 1960s–1970s) and by a more diversified national exploitation of Exclusive Economic Zones (EEZs) in the 1980s–1990s. For many of these renewable natural resources, the emphasis has already shifted from development to management, with efforts being made to estimate their potential long-term yields and to begin extracting improved resource rents.

One of the axioms underlying most earlier approaches to marine resource management has been that the only factors whereby yields could be affected (positively or negatively) were manipulation of areas and seasons of fishing; control of the amount of effort; and modification to, or restrictions on, the type of fishing gear. As noted later, it may be necessary to add other anthropogenic effects to these "control mechanisms," in particular, man's influence on the nutrient balance of inshore waters. In parallel with the impacts of nutrient

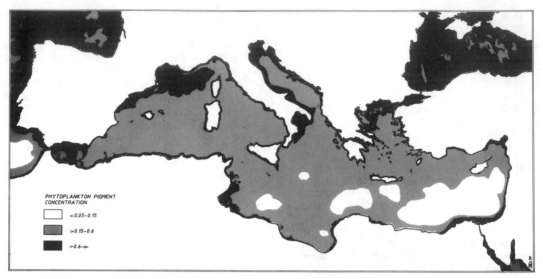

Figure 3. Distribution of phytoplankton concentrations (mg/m³) in the Mediterranean and Black seas, drawn from 30 National Science Foundation/National Aeronautic and Space Administration images taken during May 1980.

enrichment on lagoon and lake fisheries that are already documented, inland or enclosed seas are likely to be the first areas where we should look for such anthropogenic effects on fisheries.

Fisheries statistical data, although providing one of the few long-term information series for oceanic systems, have serious disadvantages from the perspective of monitoring environmental changes. An examination of landing trends on a stock-by-stock basis (Caddy and Gulland, 1983) showed that for most marine resources, it is difficult to say whether or not stocks are being exploited at a steady-state level. It is even difficult to determine the relative contributions of changes in fishing effort versus major fluctuations in annual recruitment caused by variation in climatic or other factors. In addition to natural fluctuations, concern has been expressed in recent years about global trends of a more unidirectional type, such as the sierra effect, and, more immediately, about other anthropogenic influences such as those discussed in this chapter.

The inshore fishery of the Mediterranean has made an important contribution to the nutrition of the peoples along its littoral for millennia, and the wide diversity in types of fishing gear in use is unparalleled elsewhere in Europe. The motorization of fishing boats began as early as the 1920s in the northern Mediterranean, and by 1935 was well under way, accompanied by increases in vessel size, iceholds, mechanical haulers, and a wider range of action of the trawlers (Doumenge, 1968). A doubling or tripling of the range of action of

these improved fishing vessels took only several years to accomplish and was paralleled by a corresponding decline in catch rate. For example, the catch per boat of one important species group, the red mullets (*Mullus* spp.), for a selected port, Castellon, declined from some 75 kg/boat/day in 1943 to roughly 10 kg by 1961 (Doumenge, 1968). The corresponding wave of motorization and mechanization of fishing activities along the southern and eastern shores of the Mediterranean occurred somewhat later.

Probably by the 1970s, a significant proportion of the generally less-productive southern shelves was already being harvested, even by trawlers fishing distant from their ports of origin. It is thus reasonable to deduce, given the relatively limited areas of shelf for trawling, that by the early 1970s most of the demersal resources of Mediterranean shelves (and this applies also for Black Sea demersals) were being intensively exploited at close to an overall maximum sustainable yield (MSY) and that stocks in the northern Mediterranean were being overfished. In a sense, the expansion of the trawl fishery in the northern Mediterranean to some areas of the southern shelf, and to the less easily fished deeper slope areas, indirectly confirms this supposition. The sessions of the General Fisheries Council for the Mediterranean (GFCM) in the late 1970s and 1980s also reflected the council's concern for an actual or potential state of overexploitation of demersal resources.

The situation for small and large pelagic fish stocks is somewhat different. The rising prices paid for Mediterranean demersal fish, crustaceans, and molluscs are among the highest prices in the world for these species (Josupeit, 1987). In comparison, a relatively lower demand for small pelagics, with the exception of anchovy, has meant that exploitation rates for small pelagic fish have generally remained at moderate levels (Caddy, 1990). In contrast, the development of intensive fisheries for small pelagics in the Black Sea was pursued on an industrial scale by the former eastern block countries and later by Turkey. The fisheries for large pelagics, notably swordfish, bluefin tuna, and albacore, have long histories of exploitation by harpoon, hook-and-line, and fixed traps. With the introduction of purse seining in the 1960s and 1970s, and large-scale surface longline and gillnet fisheries in the 1980s, these species have also become heavily exploited.

In the Black Sea, industrial fisheries also began relatively early, particularly for small pelagics; the landings in the demersal trawl fishery increased steadily since about 1960 (Ivanov and Beverton, 1985). In recent decades, landings of turbot, migratory pelagics, and anadromous species, especially sturgeon, declined to low levels in the northern and western Black Sea, and in the last few years the major small pelagic fishery for anchovy apparently collapsed. Therefore, from preliminary evidence, we may conclude that this is the first semi-enclosed sea whose productivity has been decimated as a result of human activities, with the consequent nutrient enrichment having played a major part.

A Broad Overview of Recent Fishery Trends in the Mediterranean and Black Seas

Given the relatively early expansion in fleet size and fishing power in both seas, it might be expected that reported landings would have reached a plateau by the early 1970s. Prior to a review of fisheries statistics for the whole region, a plateau or decline in production was expected to have occurred as a result of the high levels of fishing intensity, which were prompted by the high prices for Mediterranean demersals. In fact, a review of national landing data summarized for the Mediterranean and the Black Sea (treated together in the FAO Yearbook of Statistics as Statistical Area 37) shows that the trend in reported landings for both demersals and pelagics from the two seas is a steady rise, beginning around the mid-1970s (see Fig. 2). When broken down by major subarea for the period 1973–1974 to 1984–1985, the rates of increase for all species combined are shown in Table 1.

One area, the Levant in the eastern Mediterranean, has shown a zero or slightly negative trend in reported landings over the period, but other subareas of the Mediterranean and the Black Sea have all shown increases, averaging between about 2% and 6% per year until 1984–1985 (Fig. 4). These increases have been noted for a wide range of demersal and small and large pelagics, with the exception of estuarine species.

Uncertainties were expressed in Caddy (1990) as to whether some of these trends could be caused by improvements in fishery statistics. However, a contrary effect may also have been important; namely, the more recent motorization of small-scale fisheries whose catches often are poorly represented in official statistics reported to FAO and whose contributions to the total catch may have been progressively underestimated.

An Alternative Hypothesis to Account for Rises in Production

The tentative conclusion arrived at in Caddy (1990) was that, in the absence of effective control in most countries, fishing effort in the Mediterranean and Black seas has almost certainly increased over the period in question in response to a real rise in fish prices, especially for demersals in the Mediterranean (Josupeit, 1987), or to an increase in demand for cheap protein, as in the case of the Black Sea fisheries. However, because most fisheries were probably already operating at or close to MSY in the early 1970s, it seems unlikely that this explains the significant increases in production observed subsequently. Given that the areas showing highest rates of increase are those semi-enclosed basins of the Mediterranean system where significant impacts of nutrient enrichment have already been reported, it was suggested that the increase in landings might, in part, reflect a transition from the original oligotrophic to a more highly nutrified or even eutrophic situation. Some evidence in support of this general hypothesis has subsequently been provided by remote sensing imagery and is the main focus of this chapter.

The coastal zone color scanner (CZCS) on the National Aeronautic and Space Administration's (NASA's) *Nimbus-7* satellite (NSF/NASA, 1989) was in operation from 1978 to 1986. A series of images taken in May 1980, outside the spring period of phytoplankton bloom, revealed the distribution and concentration of phytoplankton over the Mediterranean and Black Sea areas. Figure 3 provides an impression of the major features of this com-

Table 1. A comparison of subregional landings within the Mediterranean/Black Sea Basin between two pairs of reference years, 1973–1974 and 1984–1985.

Subarea	Mean landing (mt) 1973–1974	Mean % per year 1984–1985	(all species combined)
Gulf of Lions	45,061	42,517	-0.5 (*)
Levant	23,371	21,998	-0.6
Sardinia	97,321	98,646	+0.1
Balearic	170,581	241,972	+3.2
Adriatic Sea	139,213	230,331	+4.7
Ionian Sea	130,434	226,975	+5.2
Aegean Sea	55,503	102,113	+5.7
Black Sea	446,972	881,218	+6.4

Based on GFCM Statistical Bulletin No. 6.

*Improved statistics made available in a recent publication (Anonymous, in press) suggest that for the predominantly French fleets fishing the Gulf of Lions, landings have increased by some +2% over the period.

posite image, which is noteworthy in confirming that over the greater part of its offshore surface area the Mediterranean must still be regarded as a nutrient-limited system, compared with Atlantic coastal waters. Remote sensing imagery of this key inland sea provides uniquely valuable information on coastal enrichment processes in general, as well as those resulting from hydrodynamic effects, without the higher level of natural background production typical of open-ocean areas obscuring anthropogenic effects. In the following account, this particular set of imagery is contrasted with existing information on marine fisheries, principally obtained through reports of the GFCM and summarized in Caddy (1990).

Primary Production and Fisheries: A Brief Overview by Subregion

In contrast to oceanic production, many scientists consider phosphorus, rather than nitrogen, the main limiting factor in Mediterranean fisheries. This agrees well with fisheries-modeling approaches, notably by Andersen and Ursin (1977) and Tatara (1990), who both modeled fisheries production theoretically and empirically in units of phosphorus, although it is clear that at times other nutrients or growth factors can be limiting.

Shallow-water production

Shallow-water production: Boundary effects leading to turbulence and release of nutrients from sediments in shallow water probably ex-

plain higher production close to the coast, even in relatively unpopulated areas of the south-central Mediterranean littoral. This only partly explains, however, the areas of high primary production associated with the shelf area of the upper Adriatic, the Gulf of Lions, and the Gulf of Gabes in Tunisia (Fig. 3). These last three areas support some of the more productive fisheries in the Mediterranean, and at least the first two are subject to nutrient enrichment by dense coastal populations and river discharge, and the third by industrial effluents, particularly from the phosphate industry (Darmouli, 1988).

When compared with the distribution of major urban concentrations, the imagery seems to show very localized high levels of primary production immediately offshore from some coastal population centers (e.g., Genoa, Rome, Tripoli, Benghazi, and Alexandria), where this presumed effect of cultural eutrophication is not obscured by background levels of "natural" production. Anomalies such as the high levels of production in the Gulf of Taranto seem to imply that it is receiving some outflow of nutrient-rich Adriatic water that is in accord with prevailing southerly current patterns in the western Adriatic, but point to the need for a more careful documentation of these effects through a longer time series of information.

Generally speaking, judging from this set of imagery, levels of production attributable to nonriverine sources drop off rapidly with distance offshore and are very low in the southeastern Mediterranean and Levant region. Direct runoff seems, at first glance, to have relatively less impact on primary production than do those effects caused by larger river plumes, even though other sources suggest that inputs of organics to the Mediterranean by routes other than via estuaries are more important.

Gyres and upwellings

The high levels of primary production noted in the Sea of Alboran are associated with the gyre resulting from the jet of Atlantic surface water entering the Mediterranean. Some other areas of seasonally high phytoplankton production, such as in the Gulf of Sirte and in the eastern Aegean, are also believed to be associated with upwellings. A complex series of gyres described by the cooperative oceanographic program, POEM (Physical Oceanography of the Eastern Mediterranean), in the eastern Mediterranean also contributes to local productivity. Evidence suggests that local populations of sardine and other small pelagic species are associated with these natural hydrographic features.

The western Mediterranean

This region is particularly dominated by inflows of Atlantic surface water, which enters along the Moroccan and Algerian littoral, before sweeping north toward Sardinia. Upwellings along boundaries between water masses dominate biological production in the southern part of this region.

Outflows from the Rhône and the rivers of the Spanish littoral include pollution loads from some 40% of the coastal population of the Mediterranean, and this is likely to have an impact on marine production along the northern littoral.

River plumes and influence of inflows

Other areas of intense phytoplankton production are associated with river plumes, notably, the River Rhône, the River Po, and the rivers along the Catalonian coast. The areas around these outflows have shown increases in fisheries production over the study period.

In addition to sustained landings of demersal fish in the face of heavy fishing effort, mussel culture, which was at a relatively low level in the early-to-mid-1970s, had reached reported levels of 92,000 metric tonnes (mt) in 1986, particularly in the Gulf of Lions. Figure 3 appears to support the conclusion that detrital and phytoplanktonic pathways to filter and suspension feeders are being promoted by nutrient enrichment from the Rhône outflow.

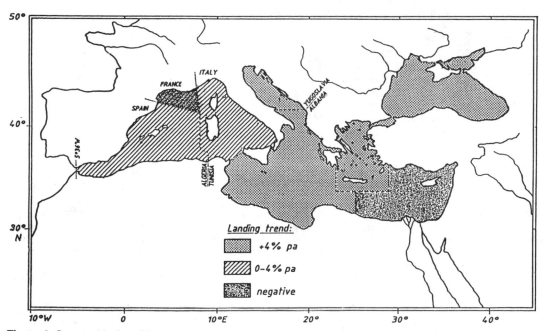

Figure 4. Geographical partition of landing trends (per annum) for Mediterranean and Black seas fisheries between 1973–1974 and 1984–1985 by major statistical area in the 1989 GFCM Statistical Bulletin No. 6. (Modified, after Caddy, 1990).

The Black Sea and Sea of Azov

River inflows from the Danube, Dnepr, Dnestr, Don, and Kuban rivers supposedly account in part for the high phytoplankton production in the Black Sea (Balkas *et al.*, 1990). The Black Sea, in contrast to the Mediterranean, but similar to the Bay of Biscay, in this imagery shows high levels of primary production widely distributed throughout the surface waters and little evidence of nutrient limitation at this time of the year. There is circumstantial evidence (summarized in Ivanov and Beverton [1985] and Caddy and Griffiths [1990]) that, despite a growing fishery, small pelagic fish, especially sprat, increased in abundance in the 1970s, when enrichment effects were becoming notable. There has also apparently been a decline in many demersal fish and benthic invertebrates over the period, a rather drastic drop in diversity of demersal species, as well as reduced entry of migratory species (bonito, bluefish, and mackerel) into the northern Black Sea (Ivanov and Beverton, 1985).

Benthic systems of the shelf and nearshore areas have become dominated by species such as *Mya arenaria*, introduced into the Black Sea, which are better adapted to low-oxygen conditions than many of the native species that have declined in abundance. Swarms of jellyfish (*Aurelia aurita*) were replaced in 1988–1989 by very high densities of ctenophores (predators on fish eggs and larvae). Blooms and red tides of various phytoplanktonic organisms have been reported over the last decade in the north and west of the basin, suggesting that dramatic changes in the pelagic ecosystem have been occurring. Hydrographic changes caused by increased use of fresh water from rivers for domestic, industrial, and agricultural purposes have also significantly increased the salinity of the Sea of Azov in recent times (Balkas *et al.*, 1990) and have changed the dominant species of its ichthyofauna.

The Nile

The effects of the Aswan Dam on Nile outflow changed the hydrography of the Levant area in the late 1960s and was considered responsible for the decline in *Sardinella* production. This may be one reason for the apparent decline in landings in the Levant area over the period of statistical review mentioned above.

It is significant that there is no evidence in Fig. 3 for a Nile plume, although strong and diffused nutrient enrichment is shown offshore from the whole area of the delta. This is attributed to the high level of nutrient enrichment in drainage water from agriculture and high population densities, often discharged via coastal lagoons, whose fish production has been considerably augmented over the last few decades (Kapetsky, 1984). Egypt's marine fish catch dropped significantly following the opening of the Aswan Dam in the mid-to-late 1960s, and their annual Mediterranean catch of small pelagics declined from about 25,000 tonnes in 1964–1965 to around 3,000 tonnes in the mid-1970s (GFCM, 1989). Production of small pelagics and demersals offshore from the delta has grown in recent years (Savini and Caddy, 1989) and appears to reflect an increase in concentration of nutrient outflow in drainage waters from increased urban growth (Fig. 3).

The Adriatic

The high levels of primary production in the northern Adriatic are not surprising, and serious anthropogenic effects of nutrient enrichment on the upper Adriatic are a current cause for concern, principally caused not only by runoff from the polluted discharges of the River Po, but also from extensive littoralization of human populations. This is significantly augmented in summer months by a heavy tourist influx. (About 0.4 tourists per linear meter of coastline have been estimated to visit the Mediterranean annually according to Blue Plan publications.) Blooms of phytoplankton (UNEP/FAO, 1990) and benthic diatoms and the mucous rafts they produce have adversely affected the tourist industry and have led to local "die-offs" of fish and invertebrates as a result of anoxic conditions. Planktonic blooms and sewage contamination of coastal waters have also caused health problems associated with ingestion of contaminated shellfish (UNEP, 1989).

The effects of nutrient enrichment have been documented in the central Adriatic by Yugoslavian researchers (Vucetic, 1988) and seem to have increased over the last 40 years. A correlation between enhanced phytoplankton production in the central Adriatic and sardine production has also been established (Marasovic *et al.*, 1988). Based on a long historical series of Yugoslavian sardine catches, Regner and Gacic (1974) showed cycles in production associated over the early period with a roughly 11-year periodicity of increased inflow of less oligotrophic Mediterranean water masses into the (then) nutrient-poor Adriatic. More recent work suggests that with the massive increase in sources of anthropogenic nutrient inputs to the Adriatic over recent decades, the enhancing effect of Mediterranean water inflow on biological production of pelagic food webs is less evident, or may even be reversed, because Mediterranean water is now less nutrient-rich. The band of high production along the western Adriatic and extending into the Gulf of Taranto (Fig. 3) may be caused, in part, by low coastal runoff, but also presumably reflects the southerly drift of enriched water from the upper Adriatic, especially that resulting from discharges from the River Po.

The effects of anthropogenic sources of enrichment on Adriatic fisheries have not been wholly negative. From informal reports, recruitment and production of some demersal as well as small pelagic fish may have increased, and following imposition of a new closed season for trawling by Italian authorities in 1988–1989, recruitment of mullet (*Mullus* spp.) to the fishery appears to have occurred at above-average levels. The net economic impact of these factors is almost certainly negative, however. Phytoplanktonic blooms and the dense rafts of mucilage produced by benthic diatoms cause local anoxia, foul fishing nets, and have still more serious economic repercussions along tourist beaches, especially in the western and northwestern Adriatic as seen in the summer of 1989. Concern has been expressed, especially in the northern Mediterranean, over declines in other species or losses in species diversity. To what extent these can be quantified, and whether these effects are principally caused by overfishing or by eutrophication, is less clear.

The northern Aegean

In the northern Aegean, cultural eutrophication has been documented in coastal bays (Friligos, 1989) and is evident in the imagery in Fig. 3. At the same time, Friligos points to nutrient-rich discharges of surface water from the Dardanelles as a contributor to high primary production in the northern Aegean. Figure 3 illustrates that high primary production levels in this area are continuous with those in the Sea of Maramara and supports his deduction.

Although caution is necessary in comparing disparate phenomena observed over different time durations, the overall picture of phytoplankton production offered by Fig. 3, an integration of 30 images during May 1980, corresponds rather well with that determined from fishery statistics for the Mediterranean during the overlapping period, 1973–1985 (Fig. 4). There seems no reason to reject the tentative conclusion that areas of high nutrient enrichment are geographically associated with increased fisheries yield, at least in the early stages of enrichment. Because there is circumstantial evidence of growing negative impacts of overfishing and pollution on productive coastal systems, extrapolating conclusions from

fisheries landings data and from nutrient levels used as an indicator of primary production to beyond 1985 is risky. These impacts need to be documented in a quantitative fashion to better understand the response of this complex system, and, in addition, parallel, more-intensive efforts need to be given higher priority so that accurate fisheries and environmental data series for "ground truthing" of images can be gathered.

Such studies may even be overdue in the case of the Black Sea, where the time span for corrective action to avert serious adverse changes to the fishery ecosystem, if this is feasible, may be very short. The importance of a closely coordinated approach to impact evaluation and an early search for remedies and their application seems indicated.

Discussion

In freshwater systems, a generally linear relationship has been established between fisheries yield and total phosphorus (Hansen and Leggett, 1982) after correcting for flushing (Lee and Jones, 1981), although the extent to which this relationship holds for marine systems is not yet elucidated and attention to date has been largely directed at negative impacts. Eutrophication undoubtedly has an impact on marine ecosystems throughout the Mediterranean and Black Sea basins, and, from many points of view—notably, preserving species diversity and conserving the integrity of the coastal, benthic, and demersal components of ecosystems—the situation is one rightly provoking alarm. This is also because of the impact of coastal degradation on human activities, at least one of which, tourism, is generally conceded to be of higher economic value than fisheries in many Mediterranean countries.

> The application of environmentally sound management practices in coastal maritime activities is now accepted as the key to safeguarding the marine environment. (UNEP, 1982).

The above statement underlies the philosophy of the Regional Seas Program and reflects a general realization that land-based activities can affect the marine environment in ways that are potentially or actually harmful to a variety of human uses of the marine ecosystem. At the same time, this statement implicitly modifies the original concept that the oceans must be preserved unchanged, allowing for a concept where the oceans may be used rationally for a variety of purposes.

What has perhaps not been so well appreciated until recently is that a moderate level of nutrient enrichment in naturally oligotrophic systems, if controlled, may increase production of some economically important species and contribute to trophic pathways leading to detritus feeding organisms, and hence, to demersal food chains of a mesotrophic type. Of course, at the same time, a variety of negative factors, namely, the discharge of nonbiodegradable contaminants in wastewater, is of concern, and blooms of toxic microorganisms, caused by such things as nutrient discharges and local anoxic events, may occur close to the point of discharge. The correct siting of effluent discharges in relation to oceanographic and physical criteria, as a means of reducing some of these negative impacts, does not appear to have received the attention it merits in the Mediterranean.

Recent events in the Mediterranean and Black seas demand high priority for coordinated action, both with respect to the control of fishing effort and—probably more seriously—with respect to the control of eutrophication and pollution by nonbiodegradable contaminants. The situation for the Black Sea, a stratified body of water with a low level of flushing, must already be regarded as critical, or even disastrous, from both environmental and fisheries perspectives.

An intensification and greater coordination of scientific studies of fish resources and the biological and oceanographic environment is undoubtedly called for in both the Mediterranean and the Black Sea. Such research is necessary not only because of the historical, cultural, and human significance of these seas, but also because of the advantages that studies of the eutrophication of basically oligotrophic systems offer, that is, a more ready separation of natural from anthropogenic effects.

Could increased nutrient discharges have played a role in increasing biological production and fishery yield in both the Mediterranean and Black seas? The answer to this question seems unambiguously positive, and not only from inspection of Figs. 3 and 4.

Has nutrient enrichment increased first, the amount, and second, the value, of Mediterranean fishery landings? This question requires a much more provisional and circumspect answer. There is evidence that a moderate level of enrichment can relax the nutrient limitation that has applied historically in the Mediterranean ecosystem and can increase some fisheries yields. At the same time, considering the Mediterranean as the world center for international tourism, any such increases must be relatively modest in comparison with the potential and actual losses in revenue that must occur as a result of cultural eutrophication of coastal waters.

Even taking a limited perspective, and considering fisheries separately from other economic sectors of the littoral region, makes it doubtful that nutrient enrichment has resulted in any significant economic benefit, if only because other factors, principally the lack of measures for effort control in most countries, have led to overcapitalization of the fishing industry and dissipation of economic rent (Hannesson, 1989). Other clearly negative aspects—including health hazards from untreated sewage, the greater risk of contamination of shellfish, and increased incidence of toxic blooms and their undesirable side effects—have increased in recent years. All of these factors undoubtedly impose extra costs. The fact that levels of mercury in Mediterranean fish tend to be higher than in Atlantic fish (UNEP, 1989) reflects to a large extent natural sources of this metal in a volcanic region. The risks of analogous accumulation of other nonbiodegradable chemicals in this marine system must evidently be viewed with concern.

What is the long-term perspective for Mediterranean fisheries? At present, neither the nutrient budget nor accurate estimates of fishing effort and yield are well documented; more knowledge of the circulation, carrying capacity, and rate of flushing of the Mediterranean Sea is needed before we can make predictions about the long-term risks posed by a combination of overfishing, nutrient enrichment, and contamination by nonbiodegradable effluents. At the same time (though not discussed here), both the Mediterranean and Black Sea regions will be extremely sensitive to climatic change. There is little or no room for complacency and much scope for cooperative action of all countries and international agencies concerned with the long-term health of the Mediterranean basin and its watershed.

References

Andersen, K. P., and Ursin, E. 1977. A multispecies extension to the Beverton and Holt theory of fishing, with accounts of phosphorus and primary population. Medd. Danm. Fisk-og. Havunders. N.S. 7:319–435.

Anonymous. Mid-1980s. Overview of the Mediterranean Basin (development and environment). Brochure describing the Mediterranean Action Plan (Blue Plan). UNEP, Athens.

Anonymous. In press. Rapport IFREMER concernant l'exploitation et la recherche halieutique dans la sous-zone 37–1–2 (Golfe du Lion). Document presented to the Sixth Session of GFCM Technical Consultation in the Balearic and Gulf of Lions Statistical Divisions, 28 May–1 June 1990, Casablanca.

Azov, Y. 1990. Eastern Mediterranean: A marine desert? Mar. Pollut. Bull. 23:225–232.

Balkas, T., Decheo G., Mihnea, R., Serbanescu, O., and Unluata, U. 1990. Review of the state of the marine environment of the Black Sea. UNEP Reg. Seas Rep. Stud. No. 124. UNEP, Nairobi.

Caddy, J. F. (1990). Recent trends in Mediterranean fisheries. In: Recent trends in the fisheries and environment in the General Fisheries Council for the Mediterranean (GFCM) area. pp. 1–42. Ed. by J. F. Caddy and R. C. Griffiths. GFCM Studies and Reviews. No. 63. Food and Agriculture Organization of the United Nations (FAO), Rome.

Caddy, J. F., and Griffiths, R. C. (1990). A perspective on recent fishery-related events in the Black Sea. In: Recent trends in the fisheries and environment in the General Fisheries Council for the Mediterranean (GFCM) area. pp. 42–71. Ed. by J. F. Caddy and R. C. Griffiths. GFCM Studies and Reviews. No. 63. FAO, Rome.

Darmouli, B. 1988. Pollution dans le Golfe de Gabes (Tunisia). Bilan de six années de surveillance (1976–1981). Bull. Inst. Natn. Scient. Tech. Oceanogr. Pêche Salammbo 15:61–84.

Degobbis, D. 1989. Increased eutrophication of the Northern Adriatic Sea. Marine Pollution Bulletin 20(9):452–457.

Doumenge, F. 1968. Hydrologie, biologie pêche en Mediterranee occidentale. Soc. Languedocienne de Geographie; Faculte des Lettres; Route de Mende, Montpellier, France.

Por, F.D. 1978. Lessepsian migration. Ecological Studies 23. Springer-Verlag. Berlin, Fed. Rep. Ger.

Friligos, N. 1989. Nutrient status in the Aegean waters. Appendix V. In: FAO Fish. Rep. No. 412. pp. 190–194. Doc. FIPL/R412. FAO, Rome.

General Fisheries Council for the Mediterranean (GFCM). 1989. General Fisheries Council for the Mediterranean, Statistical Bulletin No. 6. FAO, Rome.

Hannesson, R. 1989. Optimum fishing effort and economic rent: A case study of Cyprus. FAO Fish. Tech. Paper No. 299. FAO, Rome.

Hansen, J. M., and Leggett, W. C. 1982. Empirical prediction of fish biomasses and yield. Can. J. Fish. Aquat. Sci. 39:257–263.

Ivanov, L., and Beverton, R. J. H. 1985. The fisheries resources of the Mediterranean. Part Two. Black Sea. GFCM Studies and Reviews. No. 60. FAO, Rome.

Josupeit, H. 1987. Prices and demand for small pelagic fish in Mediterranean countries. DOC.GFCM:SP/III/87/Inf. FAO, Rome.

Kapetsky, J. M. 1984. Coastal lagoon fisheries around the world. Some perspectives of fishery yields, and other comparative fishery characteristics. GFCM Studies and Reviews. No. 61.

Lee, F., and Jones, R. A. 1981. Effect of eutrophication on fisheries. In: Eutrophication management framework for the policy maker. Ed. by W. Rast, M. Holland, and S.O. Ryding 1989. MAB Digest Series 1. Am. Fish. Soc., Washington, DC.

Marasovic, I., Pucher-Petkovic, T., and Alegria, V. 1988. Relation between phytoplankton productivity and *Sardina pilchardus* in the Middle Adriatic. FAO Fish. Rep. No. 394 (FIPL/R394). FAO, Rome.

Miller, A. 1983. The Mediterranean Sea. A. Physical aspects. In: Estuaries and enclosed seas. pp. 219–238. Ed. by B. H. Ketchum. Elsevier Press, Amsterdam.

Murdoch, W. W., and Onuf, C. P. 1972. The Mediterranean: An ecological overview. In: The Mediterranean marine environment and development of the region. Pacem in Maribus III, 28–30 April, Split, Yugoslavia.

NSF/NASA (National Science Foundation/NASA). 1989. Ocean Color from Space. A folder of remote sensing imagery and text prepared by NSF/NASA-sponsored U.S. Global Ocean Flux Study Office, Woods Hole Oceanographic Institution, Woods Hole, MA.

Regner, S., and Gacic, M. 1974. The fluctuations of sardine catch along the eastern Adriatic coast and solar activity. Acta Adriat. 15(11).

Savini, M., and Caddy, J. F. (Editors). 1989. Report of the second technical consultation on stock assessment in the eastern Mediterranean. FAO Fish. Rep. 412 (FIPL/R412). FAO/Rome.

Sorokin, Y. I. 1983. The Black Sea. In: Ecosystems of the world. Vol. 26. Estuaries and enclosed seas. pp. 253–291. Ed. by B. H. Ketchum. Elsevier Press, Amsterdam.

Steemann Nielsen, E. 1971. Production in coastal areas of the sea. Thalassia Jugoslavica. 7:383–391.

Sutcliffe, W. H., Jr. 1972. Some relations of land drainage, nutrients, particulate material and fish catch in two eastern Canadian bays. J. Fish. Res. Board Can. 29(4):357–362.

Tatara, K. 1990. Utilization of the biological production in eutrophicated sea areas by commercial fisheries and the environmental quality standard for fishing ground. Mar. Pollut. Bull. 23:315–320.

UNEP (United Nations Environmental Program). 1982. Achievements and planned development of UNEP's Regional Seas Program and comparable programs sponsored by other bodies. UNEP Regional Seas Reports and Studies. No. 1. UNEP, Nairobi.

UNEP. 1989. State of the Mediterranean marine environment. The Mediterranean Action Plan. Technical Report. No. 28. UNEP, Nairobi.

UNEP/FAO. 1990. Final reports on research projects dealing with eutrophication and plankton blooms (Activity H). MAP Technical Reports Series. No. 37. UNEP, Nairobi.

UNEP/UNESCO (United Nations Educational, Scientific, and Cultural Organization)/FAO. 1988. Eutrophication in the Mediterranean Sea: Receiving capacity and monitoring of long-term effects. Mediterranean Action Plan. Technical Report. No. 21. UNEP, Nairobi.

Vucetic, T. 1988. Long-term (1960–1983) fluctuations of zooplankton biomass in the Palagruze-Gargano area. FAO Fish. Rep. No. 394 (FIPL/R394). FAO/Rome.

Welcomme, R. L. 1985. River fisheries. FAO Fish. Tech. Pap. No. 262. FAO/Rome.

Chapter 15

Stratified Models of Large Marine Ecosystems: A General Approach and an Application to the South China Sea

Daniel Pauly and Villy Christensen

Introduction

This contribution provides an approach for constructing models of large marine ecosystems (LMEs) as defined in Sherman (1990), Sherman and Alexander (1986, 1989), and Sherman and Gold (1990).

This contribution results from an attempt to follow up on some of the implications of the LME concept for ecological modeling, especially approaches that place emphasis on fish and other living resources, and hence on fisheries management. Conversely, we shall neglect models that emphasize only the lower part of food webs.

Modeling of LMEs: The Need, the Constraints, and a Resolution

Given our inability to conduct controlled experiments[1] at the LME scale and the absence of a comprehensive theory that could predict interactions within LMEs and their evolution through time, modeling of such systems appears to be a necessary tool to link understanding of organism-level interactions with ecosystem dynamics (Toft and Mangel, 1991).

The ecological models that might be considered for describing LMEs can be grouped into two broad, nondistinct classes: (i) dynamic models, built of coupled differential equations describing major transfer and growth rates and integrated to provide time series of, for example, biomasses for key species/groups; and (ii) steady-state models, in which the species/groups compared are assumed to maintain their biomass (and related statistics) around some average level, valid for the period under consideration.

Andersen and Ursin (1977) and Laevastu and Favorite (1977) developed models of the first type to describe resource dynamics in the North and Bering seas, respectively, and Larkin and Gazey (1982) developed the first simulation model of a tropical LME, the Gulf of Thailand.[2]

The latter model was used to illustrate that relatively simple simulation models can be rapidly constructed, parameterized, and used to test various competing hypotheses on the interactions among the resources of an LME, and between fisheries and their resources.

An often encountered problem with more comprehensive dynamic models is that the complex interactions among the simulated processes often lead to invalidation, even when using input data well within observed ranges. There are various routes for overcoming this constraint. One, briefly sketched by Larkin and Gazey (1982), consists of drastically reducing the number of processes that are simulated and increasing the number of external inputs. An example of this approach is the reduction of the North Sea model of Andersen and Ursin (1977) to Multispecies Virtual Population Analysis (MSVPA), which, for the fish in the system, requires the input of sizes at-age (rather than simulating individual fish growth) and which combines (externally inputted) catch-at-age data with numbers of consumed prey items to estimate biomasses using VPA (Sparre, 1991).

Another approach for dealing with the problem is to abandon all pretenses of being able to model LMEs realistically in the time domain, and to turn to the steady-state models described above.

Thus, Polovina (1984) reduced a dynamic

model, the Bering Sea model of Laevastu and Favorite (1977), to a static system of linear equations in which, for each species/group,

Production = exports + mortality
due to predation + other mortality   ...1)

or in more detailed fashion, for any species/group (i),

$$P_i = Ex_i + \sum B_j(Q/B_j)(DC_{ji}) + B_i(P/B_i)(1-EE_i) \quad ...2)$$

where P_i is the production during any nominal period (here, 1 year) of group i; Ex_i represents the exports (fishery catches and emigration) of i; $\sum_j$ represents summation over all predators of i; B_j and B_i are the biomass of the predator j and group i, respectively; Q/B_j is the relative food consumption of j; DC_{ji} is the fraction that i constitutes of the diet of j; B_i is the biomass of i; and $(1-EE_i)$ is the other mortality of i, that is, the fraction of i's production that is not consumed within, or exported from, the system under consideration. (In the text below, we refer to EE as "ecotrophic efficiency"; its definition is inverse to that of "other mortality.")

Polovina and Ow (1983) implemented this approach in the form of a program called ECOPATH, which they used to estimate the biomasses of the major species/groups of French Frigate Shoals, a coral reef system north of Hawaii.

Since its original presentation, the ECOPATH approach has been extended to include estimation of not only biomasses, but also of other variables in equation (2), and description of the network of trophic flow between the "boxes" of a model using the theory of Ulanowicz (1986) and related concepts (Christensen and Pauly, in press, a).

These changes led to a much improved ECOPATH II software system (Christensen and Pauly, 1991). The ECOPATH II was applied to a wide variety of aquatic ecosystems, ranging from aquaculture ponds in China to the Antarctic Shelf (Christensen and Pauly, in press, b). This exercise allowed evaluation of various aspects of the approach. Notably, it led to the conclusion that steady-state models such as ECOPATH can be used to model systems that

are changing with time either: (i) by constructing models that apply to longer periods with no major changes in biomasses, and during which all rates and states can be averaged (see Walsh, 1981); or (ii) by constructing a model representing a "snapshot" of a rapidly changing situation, such as representing the midpoint of the growing period in an aquaculture pond, or a given month in a system subjected to strong seasonal oscillations (Jarre and Pauly, 1990[3]).

Our models are based on the first of these two approaches, with the bulk of the data used for model construction stemming from the decade from the mid-1970s to the mid-1980s.

The South China Sea: The Reality and the Models

Figure 1 defines the South China Sea (SCS) as discussed here. We see the SCS as bounded in

Figure 1. Map of the South China Sea, as defined in this paper (i.e., 3.5 x 10⁶ km²), with 50 and 200 m isobaths and major coralline areas (thin dotted lines).

---

(1) We are aware that various uncontrolled experiments have been and continue to be conducted at the scale of the LME, for example, through overfishing or massive pollution.

(2) Walsh (1975) developed another early simulation model of a tropical LME, the Peruvian upwelling system; this is not discussed further here because it dealt mainly with phytoplankton production and consumption (i.e., with the lower part of the food web) and hence could not be used to deal with fishery management issues (as opposed, for example, to the model of the same system documented in Jarre *et al.*, 1991).

(3) This paper will appear in revised form in Christensen and Pauly (in press, b), together with other contributions from the same meeting cited here.

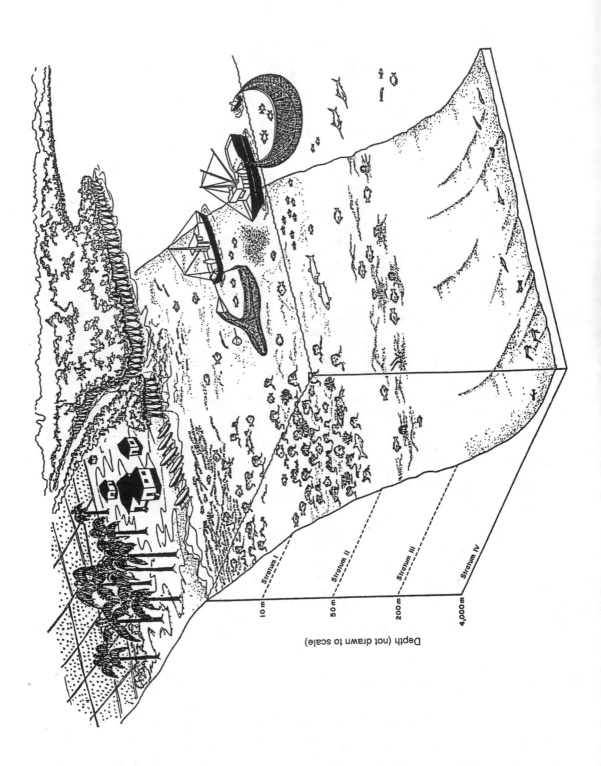

Depth (not drawn to scale)

Stratum I 10 m
Stratum II 50 m
Stratum III 200 m
Stratum IV 4,000 m

the north by the 25th parallel linking Taiwan and the Chinese mainland, and to the east by the Taiwanese coast, the 121° line between Taiwan and Luzon, by straight lines from Luzon to Mindoro and from Mindoro to Palawan, and by a line linking Palawan to northwestern Borneo (Shindo, 1973).

The southern limit is defined by a line crossing the Bangka and Karimata straits between Sumatra and Borneo at 3° S, while the western limit is the line crossing the Malacca Strait at 103° E, slightly west of Singapore.

The western border of the SCS includes a number of large subsystems, such as the Gulf of Thailand and the Gulf of Tonkin. These gulfs are completely open to the SCS proper. The SCS, as we have defined it, is an ecosystem bounded by rather narrow straits and sills. This system can be conceived as having negligible biological exchanges with other marine ecosystems. A point illustrating this is that the surface salinity in the SCS proper is relatively low because of large inflow from rivers and limited water exchange with open oceans (Wyrtki, 1961).

Overall, the SCS covers $3.5 \times 10^6 km^2$, or about 15 times the minimum size of an LME as conventionally defined (Sherman and Alexander, 1989). One implication of our choice of system is that we could not construct any single, manageable trophic box model reflecting the biological diversity of the subsystems (mangrove, coral reefs, soft-bottom communities, open seas) composing the SCS as a whole.

The approach we used was, therefore, to represent the overall system by a series of interlinked models representing subsystems (i.e., strata) as identified in Table 1. Here, the stratification ensures that the estimated biomasses of various species/groups and the extent of their trophic interactions remain biologically meaningful. Yet the interconnections between subsystems ensure that the overall system functions as an ecosystem, that is, with its various parts interacting (via export or import of production or detritus).

As a compromise between ecological reality and actual availability of data, we used the

Table 1. Summary statistics for 10 subareas of the South China Sea (SCS).

Model	Depth (m)	Area (10^3 km²)
A. Shallow waters	0 – 10	172
B. Reef-flats/seagrass	0 – 10	21
C. Gulf of Thailand	10 – 50	133
D. Vietnam/China	10 – 50	280
E. NW Philippines	10 – 50	28
F. Borneo	10 – 50	144
G. SW SCS	10 – 50	112
H. Coral reef	10 – 50	77
I. Deep shelf	50 – 200	928
J. Open Ocean	200 – 4,000	1,605
Total SCS	0 – 4,000	3,500

following strata for our overall SCS model (from inshore to offshore, see also Fig. 2):

(Ia) Estuarine, mangrove-lined, shallow waters down to 10-m depth

(Ib) Reef-flats/seagrass-dominated shallow waters down to 10-m depth

(IIa) Soft-bottom communities, from 10 to 50 m

(IIb) Coral reef communities, from 10 to 50 m

(III) Deep shelves, from 50 to 200 m

(IV) Oceanic waters, all areas deeper than 200 m.

The surface area of each stratum was determined by planimetry, whereas the separation of stratum I into Ia and Ib and of stratum II into IIa and IIb was done on the basis of a 9:1 ratio estimated by visual assessment of mangrove and coral maps in White (1983). This results, for the corals, in a total area for the SCS, which closely matches the estimate of Smith (1978) for the northern part of the "Southeast Asian Mediterranean."

We have further divided the most productive stratum (IIa) by describing soft-bottom communities from 10 to 50 m depth, into six substrata as follows (Table 1):

Figure 2. Schematic representation of a "slice" of the South China Sea (SCS) illustrating major elements considered in our 10 submodels of the SCS. Stratum I (0–10 m) comprises two subareas (mangrove-lined, "estuarinized" coasts, right, and reef-flats/seagrasses, left). Stratum I is exploited mainly by small-scale fishermen, including gleaning by women and children. Stratum II (10–50 m) also comprises two subareas (soft-bottom communities, right, and coral reefs, left). The former of these two subareas supports extensive trawl fisheries. Stratum III (50–200 m) represents the deep shelf, generally trawlable, but often unexploited because of technological or economic constraints. Stratum IV (200–4,000 m) represents the oceanic part of the SCS, in which only the large pelagics (tuna, billfishes) are exploited. (See text for details and quantitative estimates relating to this graph [kindly drawn by Mr. Chris Bunao, International Center For Living Aquatic Resources Management ICLARM Manila, Philippines]).

(i) Gulf of Thailand
(ii) Vietnam/China (the Vietnamese coast and southern China including Taiwan)
(iii) Northwest Philippines (entire Philippine SCS Coast)
(iv) Borneo (the northwestern coast of Borneo)
(v) Southwest SCS (representing the coast of eastern peninsular Malaysia, especially Kuala Terengganu, and southeastern Sumatra).

Each stratum or substratum is represented by a steady-state model constructed using the ECOPATH II software with the data documented briefly below. The following major biological interactions were assumed.

- The detritus (especially mangrove leaf litter), and the fish and invertebrate production not consumed inshore are exported from stratum I to the detritivores and carnivores of stratum II.
- The detritus, and the fish and invertebrate production not consumed in stratum II are exported to stratum III.
- The detritus, and the fish and invertebrate production not consumed in stratum III are exported to stratum IV.
- Only stratum IV exports detritus out of the South China ecosystem (for burial on the sediments covering bottoms deeper than 200 m).

The "real" SCS is characterized by far more interactions among its subsystems; however, we believe it appropriate at this stage to present a simplified implementation of our approach—one that would allow us to retain simplicity and ease of application.

Source of Data and Model Construction

For all consumer groups in all models, it is assumed that 20% of the consumption is not assimilated (Winberg, 1956). Throughout, wherever biomasses are not known, it is assumed that 95% of the production is eaten or caught (Ricker, 1968). A diskette with the 10 data sets is available from the authors, along with the ECOPATH II software and a user's manual (Christensen and Pauly, 1991).

Model A: Shallow waters
(0–10 m, all around SCS)

This model for shallow waters (Fig. 3) is based on data from the Gulf of Thailand. The catches from the Gulf of Thailand in 1979 (SEAFDEC,

1981) were separated into two depth ranges (0–10 m and 10–50 m), based on the assumption that the large-scale fishery (excluding bamboo stake traps) operates between 10- and 50-m depth, whereas the small-scale fishery operates in the shallower parts of the gulf, where the bamboo stake traps are also located.

The biomass of the apex predators (mainly tuna) is from Olson and Boggs (1986) and originally pertained to a stock of eastern Pacific tuna. The diet matrix and estimates of production and consumption were nearly identical to that of the Gulf of Thailand model (below) with only minor adjustment to reflect differences in abundances.

The estimates of production and consumption rates were also taken from model C.

Model B: Reef-flats/seagrasses
(0–10 m all around SCS)

The model for reef-flats/seagrasses (Fig. 4) is based on a model of areas near Bolinao, northwest Luzon, Philippines, described by Aliño et al. (1990).

The Bolinao model is characterized by extremely high primary production of seagrass and seaweeds, comparable to that of the most productive terrestrial ecosystems (Rodin et al., 1975). This type of ecosystem is common only in the Philippine part of the SCS (White, 1983). We have therefore reduced the production of benthic producers in our model of reef-flats/seagrasses to 20% of the Bolinao model, so that the resulting primary production becomes similar to that of the other shallow water area (model A).

Model C: Gulf of Thailand
(soft bottom, 10–50 m)

The groupings for this model of the Gulf of Thailand soft-bottom community (Fig. 5) are mainly based on information in Pauly (1979), assuming that only the tuna fishery operates in areas of the gulf that are deeper than 50 m. Zooplankton biomasses were adopted from Piyakarnchana (1989), assuming a mean water depth of 30 m and leading to a rather high estimate of $17.3 \text{ g} \cdot \text{m}^{-2}$. The benthic biomass is from Piyakarnchana (1989); those for the demersal fish groups are from Pauly (1979).

No reliable estimates of biomass were available for pelagics; the biomass of apex predators was based on information from Olson and Boggs (1986).

Only a few estimates of production/biomass ratios (P/B) are available from the Gulf of Thailand. For phytoplankton, the total produc-

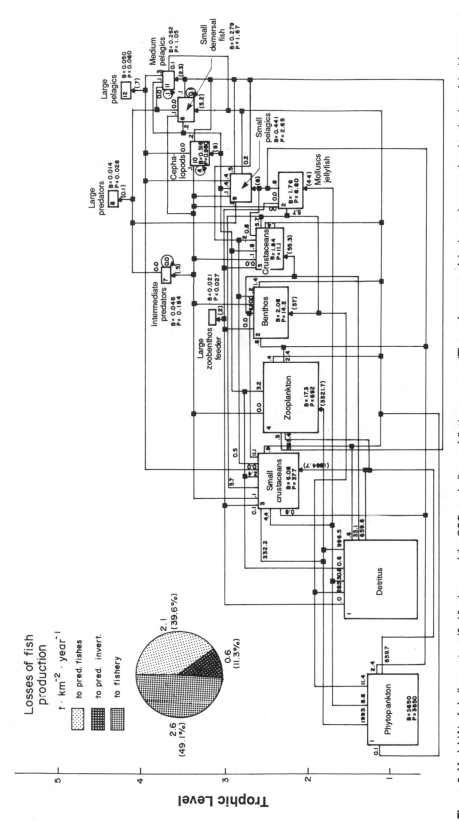

Figure 3. Model (A) of shallow waters (0–10 m) around the SCS, excluding reef-flat/seagrasses. (The surface area of the boxes is proportional to the log of the biomasses; all flows are in tonnes • km⁻² • year⁻¹; catches, respiration, and detrital backflows are omitted; see text for details on construction; based on miscellaneous data, including catches in SEAFDEC, [1981].)

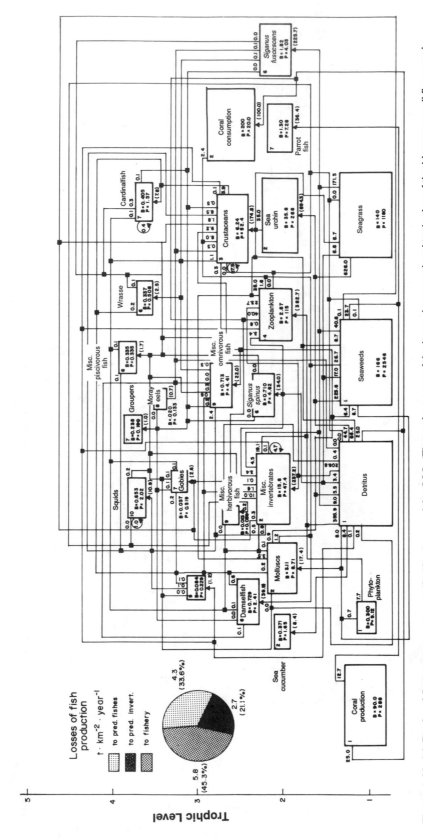

Figure 4. Model (B) of reef-flat/seagrasses areas (0–10 m) around the SCS. (The surface area of the boxes is proportional to the log of the biomasses; all flows are in tonnes • km⁻² • year⁻¹; catches, respiration, and detrital backflows are omitted; see text for details on construction; based on Aliño et al., [1990].)

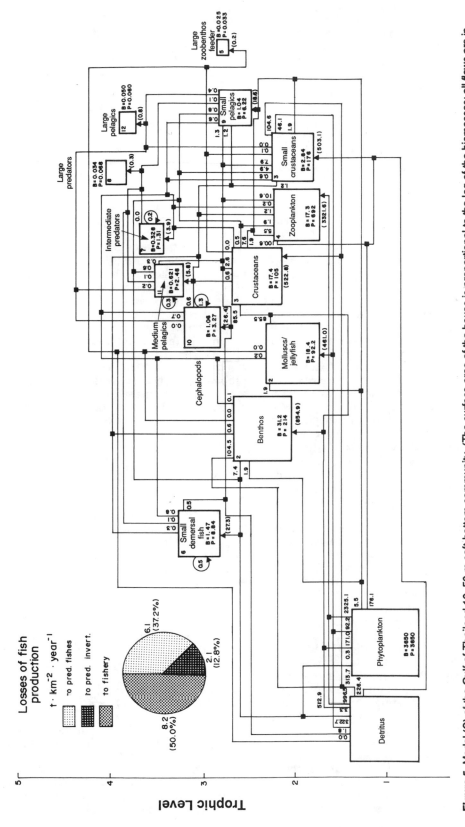

Figure 5. Model (C) of the Gulf of Thailand 10–50 m soft-bottom community. (The surface area of the boxes is proportional to the log of the biomasses; all flows are in tonnes • km⁻² • year⁻¹; catches, respiration, and detrital backflows are omitted; see text for details on construction; based on various information, with emphasis on data in Pauly [1979] and Piyakarnchana [1989].)

tion is approximately 1 g wet weight · m-2 · day-1 (Piyakarnchana, 1989). Jellyfish and molluscs were assumed to have a P/B ratio that is intermediate between the values reported by Silvestre et al. (1990) for heterotrophic benthos and the value for shrimps/crabs used here. The estimates for cephalopods and zooplankton are based on Buchnan and Smale (1981) and Polovina (1984), respectively. The P/B ratio for shrimp is based on Chullasorn and Martosubroto (1986), and the P/B ratio for benthos is from Liew and Chan (1987). The estimate for crustaceans is from Silvestre et al. (1990).

For the rays (i.e., our "large zoobenthos feeders") and for large predators, we have used the P/B estimates of Liew and Chan (1987). The P/B ratio for intermediate predators is within the range reported by Pauly (1980); the estimates for small pelagics and for small demersal fishes are based on Chullasorn and Martosubroto (1986).

The P/B value for medium-sized pelagics was assumed to be intermediate between those for large predators and for small pelagics. The P/B ratio for large pelagics was again from Olson and Boggs (1986).

For most fishes, consumption/biomass ratios (Q/B) were estimated from the regression of consumption as a function of temperature, weight, and feeding mode given by Pauly et al. (1990). Weights were estimated from mean lengths given by Pauly (1979). Mean Q/B values were estimated from the biomass-weighted means of Q/B values of the various species/groups.

For jellyfish, molluscs, and crustaceans, Q/B was estimated based on an assumption of a gross food conversion efficiency (production/consumption) of 0.2.

Q/B ratios of 29 and 16.6 were used for shrimp and cephalopods, respectively (Sambilay et al., 1990). For zooplankton, a Q/B estimate of 192 was adopted from Ikeda (1977). The Q/B value for large pelagics was adopted from Olson and Boggs (1986).

There are a number of sources for diet compositions of the abundant fishes in Gulf of Thailand waters. Menasveta (1980) provides qualitative but useful information. Quantitative information is available in Menasveta (1986) for cephalopods, in Browder (1990) for shrimp, and in Liew and Chan (1987) for large zoobenthos feeders and large predators. Their diet compositions have been adapted here, in slightly modified form, to reflect local conditions.

The diet composition of small demersal prey fish is based on Yamashita et al. (1987), who give quantified diets for seven species/groups in this category, and on Menasveta (1980), who reports on the diet of two species. Yamashita et al. (1987) describe the diet of six

intermediate predators. These, together with data in Siti and Taha (1986), Menasveta (1980), and Pauly (1979) were used to derive an average diet composition.

The sources of SCS diet compositions are as follows: small pelagics from Yamashita et al. (1987) and Menasveta (1980); medium-sized pelagics from Menasveta (1980); and large pelagics from Olson and Boggs (1986) and Tandog-Edralin et al. (In press).

Model D: Vietnam/China
(Cape Cambodia–China, 10–50 m)

This is a very productive area for which primary production and phytoplankton biomass estimates are given by Nguyen (1989).

From the mid-1970s to the mid-1980s great changes occurred in Vietnam; thus, information on the fisheries is limited. Menasveta et al. (1973) reported that a substantial fraction of Vietnamese catches were taken by artisanal, nonmechanized boats in coastal and estuarine areas. Therefore, Vietnamese catch data are not included in the present (more offshore) model (Fig. 6). Yeh (1981) reports that the demersal resources off southern Vietnam were exploited primarily by Taiwanese vessels, and gives catch and effort data. Based on this information, catches and biomasses for the demersal fish groups could be estimated. These data are assumed to be representative for the whole Vietnam/China area.

The biomass of planktivorous fish was estimated by Nguyen (1989) as 3 g · m⁻² for the whole Vietnamese shelf area and was separated into small and medium-sized pelagics based on an assumed 2:1 ratio. The catches for these groups were set at zero.

For the fish groups mentioned above, the P/B ratios were then estimated using the ECOPATH II program, assuming an ecotrophic efficiency of 0.95.

Zooplankton biomass and P/B were taken from Nguyen (1989) and the cephalopod biomass and catches from Yeh (1981).

For other groups, P/B and Q/B values are assumed to be similar to the values of model C.

Model E: Northwest Philippines
(all Philippine coast, 10–50 m)

This model of the soft-bottom community along the northwestern Phillipine coast (Fig. 7) is based on data recently assembled by Guarin (1991) for an ECOPATH II model of the Lingayen Gulf, northwestern Philippines. This is a soft-bottom area fished intensively, mainly by trawlers (Silvestre et al., 1989).

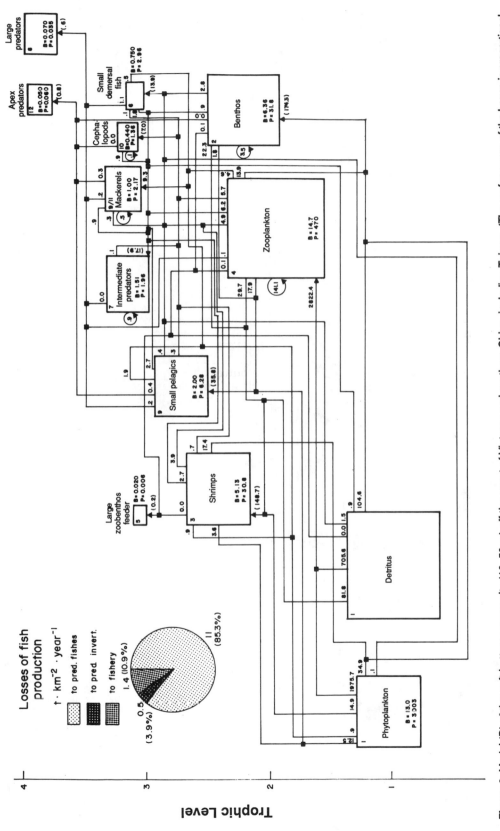

Figure 6. Model (D) of the soft-bottom community (10–50 m) off the coast of Vietnam and southern China, including Taiwan. (The surface area of the boxes is proportional to the log of the biomasses; all flows are in tonnes · km⁻² · year⁻¹; catches, respiration, and detrital backflows are omitted; see text for details on construction; based on various data sets, with emphasis on Nguyen [1989] and Yeh [1981]).

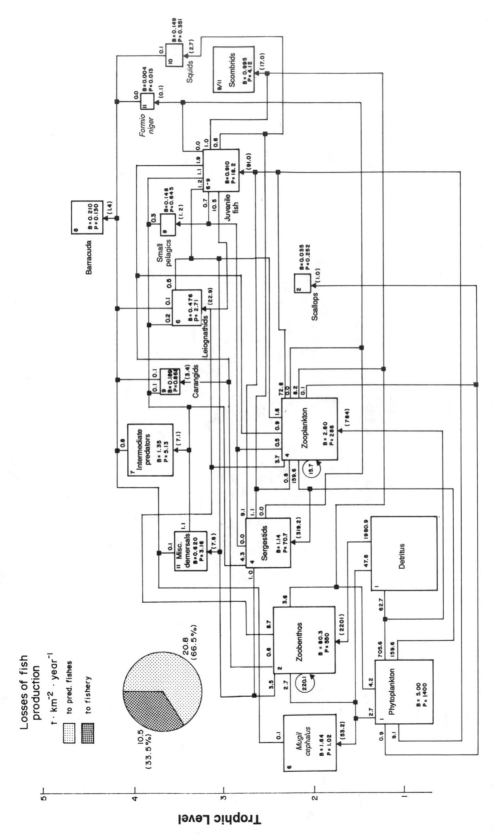

Figure 7. Model (E) of the 10–50 m soft-bottom community along the northwestern Philippine coast. (The surface area of the boxes is proportional to the log of the biomasses; all flows are in tonnes · km⁻² · year⁻¹; catches, respiration and detrital backflows are omitted; see text for details on construction; based on preliminary information from the Lingayen Gulf, assembled by Ms. F. Y. Guarin [pers. com.].)

Model F: Borneo (all Borneo coast, 10–50 m)

This model of the soft-bottom community (Fig. 8) is based on the model by *Silvestre et al.* (1990) of the moderately exploited Brunei Shelf and is also considered representative of the areas off Sarawak and Sabah, which have nearshore (0–10 m) fisheries, but where the more offshore areas only began to be exploited during the time period considered here.

Model G: Southwestern South China Sea (eastern peninsula of Malaysia and southeastern Sumatra, 10–50 m)

This model of the soft-bottom community (Fig. 9) is a modification of that of Liew and Chan (1987), who constructed a model of the area off Kuala Terengganu, on the eastern coast of western peninsular Malaysia. For this model, primary production is two-thirds of that in the Gulf of Thailand. This is supported by the pattern of primary production given by Lieth (1975) and the Food and Agriculture Organization of the United Nations (FAO) (1981).

Model H: Coral reefs (all around SCS, 10–50 m)

The model of coralline areas (Fig. 10) is based on Polovina's (1984) model of the French Frigate Shoals (FFS), north of Hawaii. This "import" of an entire model in the SCS appears legitimate because their latitudes are compatible. However, to render this model compatible with the other nine, we added detritus and detritivory by heterotrophic benthos.

FFS is an unfished ecosystem, whereas coral reefs in the SCS tended, in the 1970s, to be at least moderately exploited. To adjust for this, we deleted three apex predators (birds, monk seals, and tiger sharks) from the original model and treated their prey consumption as fishery catches. This adjustment resulted in a catch composition roughly similar to that observed from coral reef fisheries in the Philippines (Murdy and Ferraris, 1980).

Model I: Deep shelf (all around SCS, 50–200 m)

The deep shelf area utilized in this deep shelf model (Fig. 11) occupies more than one-quarter of the total SCS area. Yeh (1981) reported that the predominant fishing activity in this area is by Taiwanese vessels.

The primary production for this sub-system was estimated to be 0.2 g C $\cdot$ m^{-2} $\cdot$ day^{-1} or approximately 730 g wet weight $\cdot$ m^{-2} $\cdot$ year^{-1} (Nguyen, 1989). In line with this relatively low primary production, the biomass of zooplankton is assumed to be 25% of that off Vietnam/China (10–50 m), whereas its P/B and Q/B values were taken as equal to those in model C.

For shrimp and crabs there are no catch or biomass data, and other parameters are assumed to be identical to those in model C. The benthos parameters were adopted from model D.

Information on catches of demersal fish groups is sparse. Based on the South China Sea Programme (SCSP) (1978), the catches in 1975 in two deep offshore areas (Gulf of Thailand, depths over 50 m, and Natuna Islands–Central Sunda Shelf) were estimated as 45,100 tonnes from 318,000 km^2 (i.e., 0.11 tonnes $\cdot$ km^{-2}). This estimate reflects a low fishing pressure and is assumed to be representative of the whole area. The biomasses of the demersal groups and cephalopods were estimated using data in Yeh (1981); the Q/B estimates were from model C.

No information seems available on the pelagic stocks in this subsystem. We have therefore assumed that the biomasses of small and medium-sized pelagics are 50% of those in model D. For both groups, the Q/B values were assumed to be similar to those in model C. The large pelagics group was assumed to have the same parameters as the other models.

Model J: Oceanic waters (central SCS, 200–4000 m)

This stratum covers nearly one-half of the SCS (1.6 million km^2). The fisheries are limited to catching large pelagic fishes, mainly tunas.

The system represented by this open-ocean community model (Fig. 12) is divided into the following components based on Blackburn (1981), Rowe (1981), and Mann (1984):

- Apex predators (tuna, billfish, swordfish, sharks, and porpoise) occurring in the upper 200–300 m. Olson and Boggs (1986), based on studies conducted in the eastern Pacific, estimated the biomass of this group as approximately 0.05 g wet weight $\cdot$ m^{-2}; the corresponding P/B was 1.2 year^{-1} and the Q/B, 15 year^{-1}. Their paper also presents the diet composition of yellowfin tuna, used here as representative of the whole group. In the absence of reliable data for the SCS, the catch per area was also taken from this source.

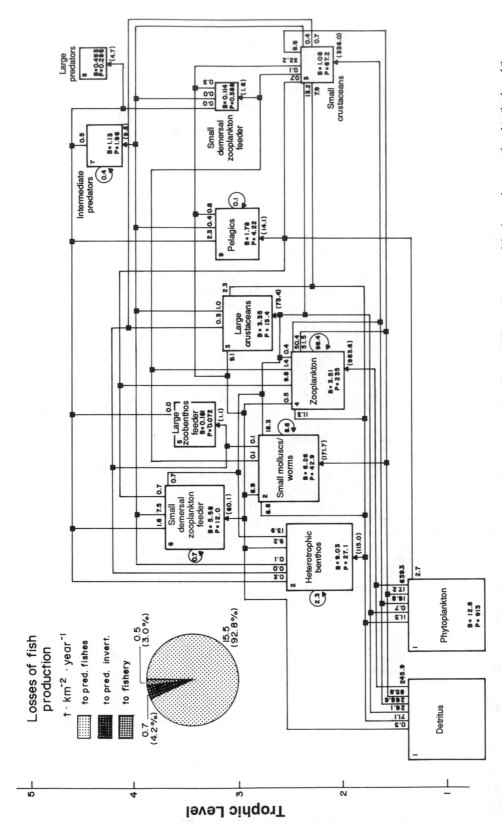

Figure 8. Model (F) of the 10–50 m soft-bottom community along the northwestern coast of Borneo. (The surface area of the boxes is proportional to the log of the biomasses; all flows are in tonnes • km⁻² • year⁻¹; catches, respiration, and detrital backflows are omitted; see also Table 1 and text for details on construction; adapted from a model of the Brunei Shelf assembled by Silvestre *et al.* [1990].)

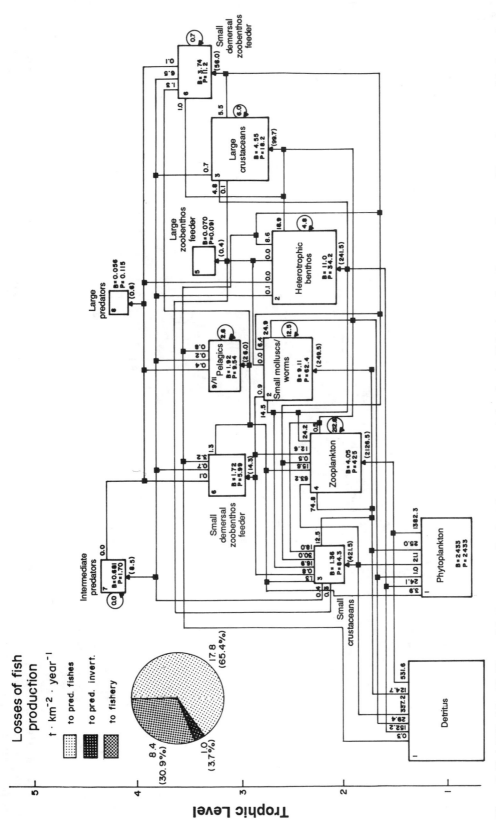

Figure 9. Model (G) of the 10–50 m soft-bottom community along the eastern coast of peninsular Malaysia and southeastern Sumatra. (The surface area of the boxes is proportional to the log of the biomasses; all flows are in tonnes • km^{-2} • year^{-1}; catches, respiration, and detrital backflows are omitted; see text for details on construction; modified from model of Kuala Terengganu, Malaysia, of Liew and Chan [1987].)

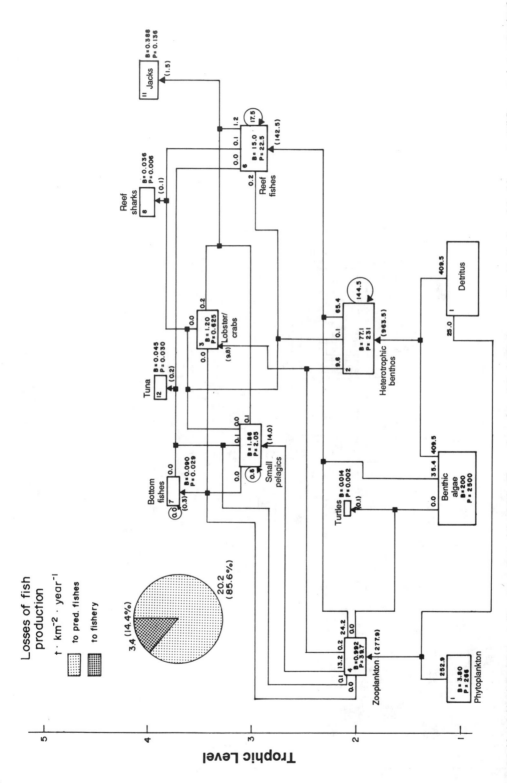

Figure 10. Model (H) of 10–50 m coralline areas around the SCS. (The surface area of the boxes is proportional to the log of the biomasses; all flows are in tonnes • km⁻² • year⁻¹; catches, respiration, and detrital backflows are omitted; see text for details on construction; adapted from Polovina [1984], through deletion of three top predators and their replacement by the fishery, and the addition of detritus and detritivory by heterotrophic benthos.)

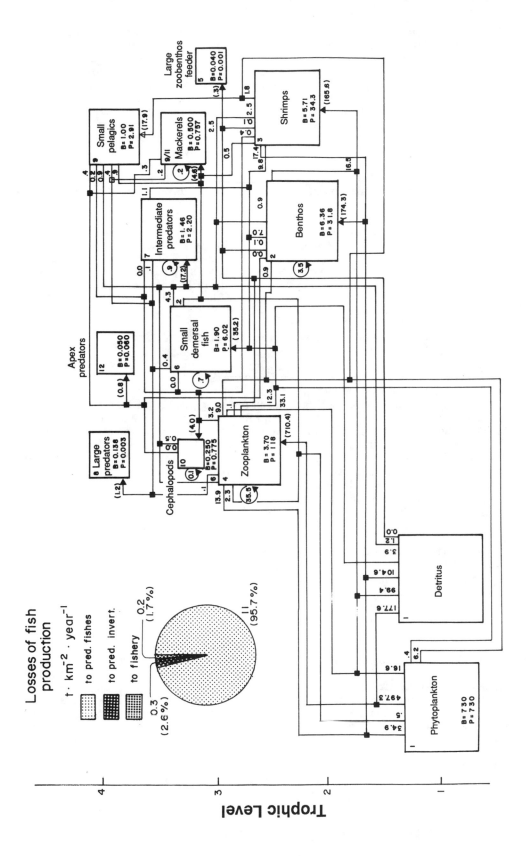

Figure 11. Model (I) of the 50–200 m soft-bottom, deeper shelf communities around the SCS. (The surface area of the boxes is proportional to the log of the biomasses; all flows are in tonnes · km⁻² · year⁻¹; catches, respiration, and detrital backflows are omitted; see text for data sources and details on construction.)

- Epipelagic nekton (mackerel, small tuna, nomeids, flyingfish, cephalopods) occurring in the upper 200–300 m. Mann (1984), considering oceanic areas in general, gives a biomass of 0.5 g · m^{-2} and a production of 0.5–1.3 g · m^{-2} · year^{-1} (i.e., a P/B of 1.0–2.6 year^{-1}). We adopted a P/B value of 2 year^{-1} and assumed a Q/B value of 9.3 year^{-1}, as for mackerel in model C. The diet composition is based mainly on Mann (1984).
- Mesopelagics (myctophids, gonostomatids, and sternoptychids) occurring between 200 and 1,000 m during daytime. At night, a large proportion of the mesopelagics migrate to the epipelagic zone to feed, mainly on zooplankton. The biomass of this group is assumed to be 2.6 g · m^{-2} based on data from the western central Pacific in Gjøsaeter and Kawaguchi (1980). Mann (1984) estimated the biomass to be in the range of 1.75–3.0 g · m^{-2}. As in Mann (1984), who used a bioenergetic model and derived a Q/B value of 2.9 year^{-1} for the mesopelagics, we set the P/B value to 0.6 year^{-1}. Hopkins and Baird (1977) estimated that more than 70% (by volume) of their food consists of crustaceans.
- Bathypelagics (anglerfish and Cyclotone) occurring at depths greater than 1000 m. These fish tend to minimize their energy expenditure and are capable of taking prey over a large size range. Mann (1984) reported a biomass of 0.02 g · m^{-2} and a P/B of 0.1 year^{-1}. We used a Q/B value of 0.4 year^{-1}.
- Benthic fish (Bathysaurus, Chlorophthalmidae, Macrouridae, Moridae, and Brotulidae) for which there is scarce quantitative information. According to Mann (1984), their joint biomasses range from 1.0–2.0 g · m^{-2} (we used 1.5), and their P/B values from 0.05 to 0.10 year^{-1} (we used 0.075); Q/B was assumed to be 0.3 year^{-1}. The diet composition was assumed based on scattered information in Mann (1984) and constrained the limited number of boxes used to describe the system.
- Benthos (amphipods, shrimp, and other decapods). Mann (1984) reported biomass as 5.0 g · m^{-2}, with a P/B value of 0.1 year^{-1}; Q/B was assumed as 0.4 year^{-1}.
- Zooplankton (larger copepods, euphausiids, and decapods). Blackburn (1981) reported biomasses as 8–13 g · m^{-2} (we used 10 g · m^{-2}), and Mann (1984) reported a P/B ratio of 0.5 year^{-1}. Q/B was assumed to be 2.5 year^{-1}.
- Phytoplankton. Blackburn (1981) reported primary production rates from oceanic areas of 0.1–0.5 g C · m^{-2} · day^{-1}. We adopted a value of 400 g wet weight · m^{-2} · year^{-1}, corresponding to a value in the lower part of the range given by Blackburn (1981).
- Microzooplankton. Blackburn (1981), in a review of low-latitude gyral regions, summarized information suggesting that the biomass of microzooplankton (which is usually not sampled) may be about 25% of that of net-caught zooplankton. The P/B and Q/B ratios were assumed to equal only half the P/B and Q/B values of inshore zooplankton, because of lower primary production.

Results and Discussion

The models

Each of the 10 models in this study has been drawn so that the area of a box is proportional to the logarithm of the biomass of the box. All boxes included the biomass (B) and production (P), in tonnes wet weight · km^{-2} and tonnes · km^{-2} · year^{-1}, respectively. To minimize the number of "wires" needed to draw the connections (i.e., energy flows) between groups, we used the following rules: (i) flows exiting a box do so from the top half of a box, whereas flows entering a box do it in the lower half, and (ii) flows exiting a box cannot branch, but they can be combined with flows from other boxes, if they all go to the same box.

Backflows to the detritus box, respiration, and fishery catches have been omitted on Figures 3–12 for the sake of clarity. Nonetheless, all boxes have been balanced by the ECOPATH II system so that inputs equal outputs.

Based on these models, some generalizations can be made. They are presented in Table 2, which includes the total primary production of each of the 10 areas. This varies between 4,000 g · m^{-2} · year^{-1} in the highly productive reef-flat/seagrass area down to 400 g · m^{-2} · year^{-1} in the open-ocean waters.

Primary production

The general pattern of primary production indicates high production in coastal and gulf areas and decreasing production with

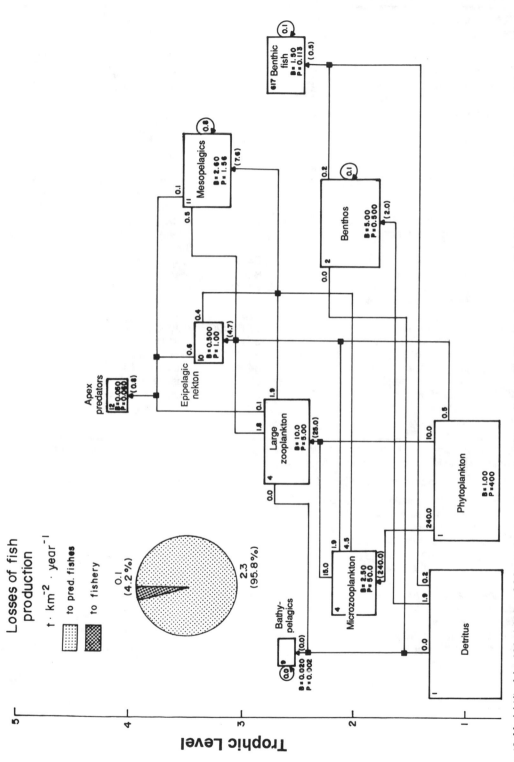

Figure 12. Model (J) of the 200–4,000 m SCS open-ocean community. (The surface area of the boxes is proportional to the log of the biomasses; all flows are in tonnes · km^{-2} · year^{-1}; catches, respiration, and detrital backflows are omitted; see text and Figs. 3–11 for details on construction.)

Table 2. ISSCAAP numbers and common names of fish and invertebrates caught in the SCS and their corresponding "boxes" in ECOPATH II models in Figs. 3 to 12.

ISSCAAP numbers*	Common names†	"Box" number‡	ISSCAAP numbers*	Common names†	"Box" number‡
921	sea weeds	1	3403	mullets	6
941	misc. plants	1	391/392	misc. fishes (+ 3320/3412)	6
831	sponges	2	3301/2	catfishes & eels	7
751	sea urchins	2	3303	lizardfish	7
752	sea cucumbers	2	3305	groupers	7
761	jellyfish	2	3307/8	misc. snappers	7
581	misc. mollusks	2	3313	drums & croakers	7
541	mussels	2	3316	bigeyes	7
531/2	oysters	2	3404	threadfins	7
551	scallops	2	381	sharks	8
561	cockles	2	3304	pike & conger eels	8
562	misc. bivalves	2	3402	barracuda	8
421	swimming crabs	3	351/2/3/4	clupeids	9
422	mangrove crabs	3	373	Indian mackerels	9
431	spiny lobsters	3	3309	fusilier	9
432	slipper lobsters	3	3405	round scads	9
451/2	penaeid shrimps	3	3408	hardtail scads	9
471	misc. crustaceans	3	571/2	cuttlefish/squids	10
—	sergestids/zooplankton	4	3410	black pomfret	11
382	rays	5	375	hairtail	11
311/3/4	flounders, soles	6	374	Indopacific mackerel	11
312	Indian halibut	6	371/2	king mackerel	11
3306	sillago whiting	6	3406/7	misc. jacks	11
3311	pony fishes	6	3411	white pomfret	11
3312	grunters/sweetlips	6	355	wolfherring	11
3314	goatfishes	6	368	sailfish & billfish	12
3317	breams	6	361–367	tuna & frigate mackerels	12
3319	rabbitfishes	6			

* These numbers refer to a coding system developed by FAO, the "International Standard Statistical Classification of Aquatic Animals and Plants."

† Some common names were adjusted to account for the pooling of ISSCAAP groups.

‡ Number-specific definition of the "boxes" (see text and Figs. 3–12, 14) are:
   1: benthic producers
   2: misc. invertebrates
   3: crustaceans (excl. plankton)
   4: sergestids/zooplankton
   5: large zoobenthos feeders
   6: small demersal prey fishes
   7: intermediate predators
   8: large predators
   9: small pelagics
  10: squids and cuttlefish
  11: medium pelagics
  12: large pelagics

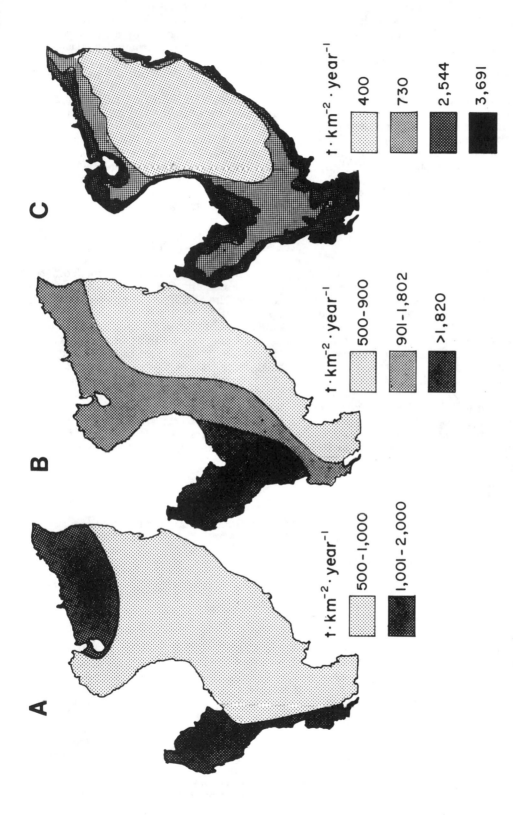

Figure 13. Mean annual distribution of primary production in the SCS. Map A, adapted from Lieth (1975), suggests a primary production of 3.36 · 10⁹ tonnes per year for the entire SCS. Map B, adapted from FAO (1981), suggests a total primary production of 4.2 · 10⁹ tonnes per year if one assumes, as in Lieth (1975), an upper limit of 2,000 tonnes · km⁻² · year⁻¹ for the Gulf of Thailand and adjacent areas. Map C presents our depth-based stratification; the corresponding estimate of total production is 4.0 · 10⁹ tonnes, within the range of the other two estimates.

depth. This does not match the distribution patterns given by FAO (1981) or by Lieth (1975), both of whom reported a northwest/southeast gradient in the primary production of the SCS (Fig. 13a, b, and c). Based on our 10 submodels, we estimated a mean primary production of the whole SCS ecosystem of 1,143 tonnes · km^{-2} · year^{-1}, corresponding to 4.0 x 10^9 tonnes wet weight · year^{-1}. From planimetry of the primary production maps given by Lieth (1975) and FAO (1981) (Figs. 13a and b), total primary production for the SCS system ranges from 336 to 418 million tonnes carbon · year^{-1}, which neatly brackets our estimate (if a carbon-to-weight conversion factor of 10 is used, as we have done throughout).

Transfer efficiencies

Table 2 also gives transfer efficiencies between trophic levels. For this analysis, the consumption of each group in each system has been split in discrete trophic levels (Christensen and Pauly, 1991). These transfer efficiencies depend on the structure of the food webs (and thus on a multitude of assumptions, many of which may not have been met), yet a general pattern emerged with an overall mean transfer efficiency of around 9%. This is in line with values generally assumed—but often not estimated. Table 2 also suggests that there is no correlation between the mean transfer efficiencies of various models, which vary from 4% to 15%,

and primary productivity in the areas represented. Thus, even the least productive offshore systems pass their energy up the food chain as efficiently as the more productive coastal systems.

Transfer efficiencies from primary producers to fishery catches can be viewed as measures of the efficiency of the various fisheries and are found to vary by two orders of magnitude between systems (Table 2). This indicates that the systems are harvested at different trophic levels. A difference of one trophic level between fisheries implies a difference of at least one order of magnitude in their catches—even if the systems these fisheries exploited are otherwise similar.

Catch estimates (Table 3)

The highest catches come from the coastal areas, the Gulf of Thailand, and the southwestern SCS. The estimated catches add up to nearly 5 million tonnes · year^{-1}. Small demersal fishes, small pelagics, and intermediate predators are the most important groups caught (Fig. 14a). In order to compare the annual catch data with those from SEAFDEC (1981) for the year 1979, the latter had to be adjusted to fit our definition of the SCS. Thus, we included 33% of the overall catch of Taiwan, the catches from eastern peninsular Malaysia (from Sarawak and Sabah), and from the Gulf of Thailand (totaling 1.96 million tonnes).

Table 3. Summary statistics for 10 models, representing different subareas of the SCS.

Model	Fig.	Primary production (t · km^{-2} · year^{-1})	Transfer efficiency (%) by trophic level*			Geometric mean	Transfer efficiency from primary production to fishery (%)
			II	III	IV		
A. Shallow waters	3	3,650	6.2	3.1	9.6	5.7	0.17
B. Reef-flats/seagrass	4	4,023	8.6	11.0	6.6	8.6	0.33
C. Gulf of Thailand	5	3,650	7.2	13.8	7.0	8.9	0.26
D. Vietnam/China	6	3,003	3.5	10.7	6.9	6.4	0.05
E. NW Philippines	7	913	9.3	8.9	9.3	9.2	1.23
F. Borneo	8	913	15.9	18.4	11.7	15.1	0.08
G. SW SCS	9	2,433	11.7	15.1	8.7	11.5	0.35
H. Coral reef	10	2,766	10.0	1.4	—	3.7	0.14
I. Deep shelf	11	730	8.0	13.0	8.1	9.4	0.03
J. Open Ocean	12	400	9.3	12.3	7.4	9.5	0.01
SCS Weighted means	2	1,143	8.3	10.4	7.0	9.2	0.12

* II refers to first consumer level, III to second, etc.; transfer efficiencies computed after removal of cycles.

We added Vietnamese catches (approximately 700,000 tonnes) (Nguyen, 1989), Chinese catches (e.g., 400,000 tonnes) (Shindo, 1973), and catches from Hong Kong, northwest Philippines, Cambodia, Brunei, northwest Indonesia, and Singapore. The total catch is about 4 million tonnes annually—a figure similar to our estimate and to the figure of 4.6 million tonnes derived by Marr (1976) for the SCS as defined here. We conclude that our models incorporate and/or lead to a reasonable estimate of total catches.

Potential catches

The notable differences in the efficiency of the fishery discussed above raise the question whether the catches can be increased by directing the fishery toward the lower parts of the food web. One way to consider this is to look at the fate of the fish production within the system. To facilitate this, pie charts were added to each submodel (Figs. 3–12) showing the fate of fish production. For all submodels, the bulk of the fish production is consumed by fish predators, while the fisheries and invertebrate predators take the rest.

Total fish production in the SCS area is estimated at about 30 million tonnes annually. About 13% is harvested by the fisheries and the rest is eaten by predators (Fig. 14 a, b).

If we assume that it is possible to harvest all systems as efficiently as the fully exploited coastal systems (i.e., models A, B, C, and G, but disregarding E, the northwest Philippines system, whose transfer efficiency may be biased upward because of a low estimate of primary production), we obtain a mean potential fishery efficiency of 0.275% of primary production. If the less-exploited systems could be harvested with this efficiency, the additional catches from the SCS would be about 5.8 million tonnes annually, more than doubling the catch; however, this potential may not be feasible in practice. The Vietnam/China system was only lightly exploited in the period covered here, the mid-1970s to mid-1980s. The potential for the area is estimated to be 1.86 million tonnes per year, corresponding to an increased catch rate of 1.6 to 8.2 tonnes $\cdot$ km^{-2} $\cdot$ year^{-1}, which is extremely high and probably unrealistic.

The Bornean coast, beyond 10 m deep, was not exploited intensively in the late 1970s, and this is reflected in the potential for increases in catch of some 260 thousand tonnes $\cdot$ year^{-1}. This corresponds to an increase in catch rate of 0.7-2.5 tonnes $\cdot$ km^{-2} $\cdot$ year^{-1}. Much of this potential has probably been realized, since the fishery in Sarawak and Sabah has increased considerably in the last decade.

The potential for the deeper coralline areas is estimated at about 300,000 tonnes $\cdot$ year^{-1}, doubling the catch and bringing the total catch rate to about 7.5 tonnes $\cdot$ km^{-2} $\cdot$ year^{-1}. There are large coralline areas in the central part of the SCS that are only lightly exploited. Because catch rates for intensively exploited coral areas can exceed 20 tonnes $\cdot$ km^{-2} $\cdot$ year^{-1} (Alcala, 1981; White, 1989), we conclude that there may be a basis for some increase. However, because the model is based on data from outside the region, we stress that one should consider these estimates with care.

The bulk of the additional potential 5.8 million tonnes $\cdot$ year^{-1} comes from the deeper areas of the SCS. This potential may not be realized because it is difficult to fish the deeper areas in a way that is economically viable.

The same problems also occur with regard to the abundant offshore resources of mesopelagic fish such as myctophids. It may well be that the only realistic way to harvest these resources is by catching their predators, the large pelagics and cephalopods.

FAO (1981) estimated the potential of the SCS at 3.3 and 2.8 million tonnes $\cdot$ year^{-1} for demersal and pelagic fishes, respectively. Our corresponding estimates are 2.6 and 1.5 million tonnes $\cdot$ year^{-1}.

The potential catches presented here are tentative, as were those of FAO (1981); however, this does not mean that the method we employed is not useful. Indeed, the problems associated with estimating potential catches in data-sparse areas make even indicative approaches worthwhile.

Many fisheries in the region suffer from high fishing pressure, use of small-meshed nets, and sometimes from destructive fishing methods (Pauly and Chua, 1988). Carefully designed, new fishing regimes could therefore, even in these cases, be expected to lead to increased catches.

Detritus flows within the South China Sea

The models we have presented here can be linked to show the flow of detritus within the SCS system (Table 5, Fig. 14c). Total sedimentation is estimated at 10^9 tonnes $\cdot$ year^{-1} or 25% of primary production. Assuming all of this to be deposited at depths in excess of 200 m, the total input to the bottom is about 650 tonnes $\cdot$ km^{-2} $\cdot$ year^{-1}. This estimate is one order of magnitude higher than the estimate of Rowe and Gardner (1979) for the deep North Atlantic. Expecting a higher estimate for the SCS, and bearing in mind our assumption that all detritus is deposited in the deepest stratum (which makes up only 50% of the SCS) and that we

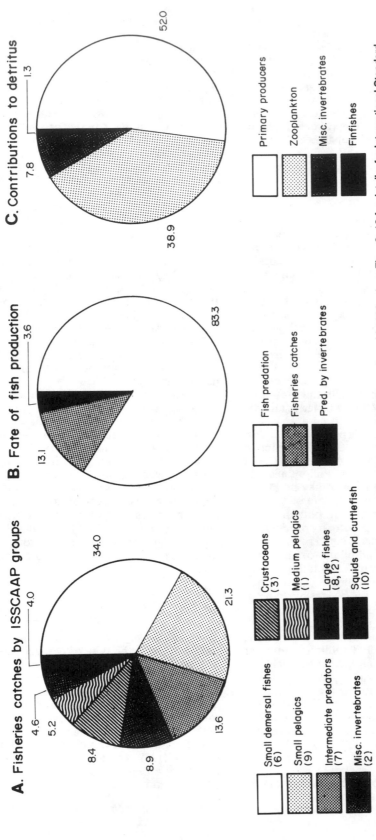

Figure 14. Fate of biological production for the entire SCS (especially finfishes), mid-1970s to the mid-1980s. (see Figs. 3–12 for details; for International Standard Statistical Classification of Aquatic Animals and Plants [ISSCAAP] groupings, see Table 2).

A- Percent composition of fisheries catches by ISSCAAP groups.

B- Percent of fish production consumed by fish predators, harvested as fisheries catches, and consumed by invertebrate predators.

C- Percent contribution to detritus of primary producers, zooplankton, miscellaneous invertebrates, and finfishes.

Table 4. Estimated catches (t · km⁻² · year⁻¹) in 10 submodels from the SCS, representative of the late 1970s.

Group/Model*	A	B	C	D	E	F	G	H	I	J	Total (10³ t · year⁻¹)
1. Primary producers	—†	0.988	—	—	—	—	—	—	—	—	21
2. Miscellaneous invertebrates	1.840	5.210	0.053	—	0.210	0.003	—	—	—	—	439
3. Crustaceans (excl. plankton)	1.455	0.276	0.594	—	—	0.174	0.174	0.404	—	—	411
4. Sergestids/zooplankton	—	—	—	—	0.350	—	—	—	—	—	10
5. Large zoobenthos feeders	0.026	—	0.022	0.006	—	0.012	0.086	—	0.001	—	21
6. Small demersal predators	1.075	5.409	5.449	0.408	4.514	0.136	1.541	2.375	0.036	—	1,672
7. Intermediate predators	0.140	0.440	0.251	0.867	3.220	0.060	1.556	0.129	0.077	—	664
8. Large predators	0.260	—	0.058	0.033	0.130	0.058	0.110	0.006	0.003	—	49
9. Small pelagics	0.925	0.006	1.497	—	0.340	0.283	5.126	0.831	—	—	1,047
10. Squids and cuttlefish	0.152	0.750	0.440	0.255	0.176	—	—	—	0.023	—	198
11. Medium pelagics	0.402	—	0.921	—	0.320	—	—	—	—	—	257
12. Large pelagics	0.042	—	0.050	0.050	—	—	—	0.028	0.050	0.050	157
Total	6.083	13.079	9.335	1.619	11.260	0.726	8.593	3.773	0.190	0.050	4,945
Total catch‡ (10³t · year⁻¹)	1,046	275	1,242	453	315	105	962	291	176	80	4,945
Potential catch§ (10³t · year⁻¹)	—	—	—	1,860	—	257	—	295	1,688	1,686	5,786

* Refer to Table 1 for the stratification of the SCS used in this analysis.

† Dashes indicate missing information or zero catches.

‡ From SEAFDEC (1981), referring to the year 1979.

§ This "potential" may not be realizable (see text).

Table 5. Estimated flows (10^3t · year^{-1}) of excess fish and detritus production for the SCS.

	Fish		Detritus	
Model*	Import	Export	Import	Export
A	0	60	0	30,000
B	0	280	0	56,000
C	60	410	15,000	22,000
D	120	180	31,000	260,000
E	10	23	3,100	3,800
F	60	330	16,000	30,000
G	50	210	12,000	64,000
H	30	130	8,500	150,000
I	1,300	4,300	530,000	700,000
J	4,300	0	700,000	1,000,000

* See Table 1 and text for definition and construction.

have no independent estimate of flux to the deep bottom layer from other highly productive tropical areas, we conclude that our estimate, although high, is not unrealistic.

We initiated this exercise in response to the challenge represented by the LME concept and are surprised and pleased to see that some sense has come out of our rather crude modeling approach. We view our results as an indication of the robustness of the approach incorporated in ECOPATH II and are confident that the steadily increasing number of aquatic ecosystems that are being quantified using steady-state models will help to improve our understanding of the ecology of LMEs.

Acknowledgments

We wish to take this opportunity to express our heartfelt thanks to Ms. Lou Arenas and Messrs. Chris Bunao, Igy Castrillo, Albert Contemprate, and Jun Torres for their dedicated after-hour support toward the completion of this paper. We also wish to thank Ms. Poyie Guarin for permission to use her unpublished data for the Lingayen Gulf. Danida, the Danish International Development Agency, provides the funding for the ECOPATH II project.

References

Alcala, A. C. 1981. Fish yields of coral reefs of Sumilon Island, Central Philippines. Nat. Resource Counc. Philipp. Res. Bull. 36(1):1–7.

Aliño, P. M., McManus, L. T., McManus, J. W., Manola, C., Fortes, M. D., Trono, G. C., Jr., and Jacinto, G. S. 1990. Initial parameter estimation of a coral reef flat ecosystem in Bolinao, Pangasinan, northwestern Philippines. ICES CM 1990/L:55.

Andersen, K. and Ursin, E. 1977. A multispecies extension to the Beverton and Holt theory of fishing, with accounts of phosphorus circulation and primary production. Medd. Danm. Fisk=og. Havunders. N.S. 7:319–435.

Blackburn, M. 1981. Low latitude gyral regions. In: Analysis of marine ecosystems. pp. 3–29. Ed. by A. R. Longhurst. Academic Press, San Diego.

Browder, J. A. 1990. Trophic flows and organic matter budget, Gulf of Mexico continental shelf. ICES CM 1990/L:116. Sess. Q.

Buchnan, P. R., and Smale, M. J. 1981. Estimates of biomass, consumption and production of *Octopus vulgaris* Cuvier off the east coast of South Africa. Invest. Rep. Oceanogr. Res. Inst. (50). Durban.

Christensen, V., and Pauly, D. 1991. A guide to the ECOPATH II program (Ver. 2.0). ICLARM software 6 (Rev. 1). International Center for Living Aquatic Resources Management, Manila, Philippines.

Christensen, V., and Pauly, D. (In press, a). The ECOPATH II—a software for balancing steady-state ecosystem models and calculating network characteristics. Ecol. Modelling.

Christensen, V., and Pauly, D. (Editors). (In press, b). Trophic models of aquatic ecosystems. ICLARM Conference Proceedings 26.

Chullasorn, S., and Martosubroto, P. 1986. Distribution and important biological features of coastal fish resources in Southeast Asia. FAO Fish. Tech. Pap. (278). FAO, Rome.

Food and Agriculture Organization of the United Nations (FAO). 1981. Atlas of the living resources of the seas. FAO Fisheries Series. Rome.

Gjøsaeter, J., and Kawaguchi, K. 1980. A review of world resources of mesopelagic fishes. FAO Fish. Tech. Pap. 193. FAO, Rome.

Guarin, F. Y. 1991. A model of the trophic structure of the soft-bottom community in Lingayen Gulf, Philippines: An application of the ECOPATH II software and modeling approach. Ph.D. diss., Institute of Biology, College of Science, University of the Philippines, Quezon City.

Hopkins, T. L., and Baird, R. C. 1977. Aspects of the feeding ecology of oceanic midwater fishes. In: Oceanic sound scattering prediction. pp. 325–360. Ed. by N. R. Andersen and B. J. Zaruranec. Plenum Press, New York.

Ikeda, T. 1977. Feeding rates of planktonic copepods from a tropical sea. J. Exp. Mar. Biol. Ecol. 29:263–277.

Jarre, A., Muck, P., and Pauly, D. 1991. Two approaches for modelling fish stock interactions in the Peruvian upwelling ecosystem. ICES Mar. Sci. Symp. 193:178–184.

Jarre, A., and Pauly, D. 1990. Seasonal changes in the Peruvian upwelling ecosystem. ICES CM 1990/L:50.

Laevastu, T., and Favorite, F. 1977. Preliminary report on dynamical numerical marine ecosystem model (DYNUMES II) for Eastern Bering Sea. U.S. Natl. Mar. Fish. Serv. Northwest and Alaska Fish. Center, Seattle. (mimeo).

Larkin, P. A., and Gazey, W. 1982. Applications of ecological simulation models to management of tropical multispecies fisheries. In: Theory and management of tropical fisheries. pp. 123–140. Ed. by D. Pauly and G. I. Murphy. ICLARM Conference Proceedings 9.

Lieth, H. 1975. Historical survey of primary productivity research. In: Primary productivity of the biosphere. pp. 7–16. Ed. by H. Lieth and R. H. Whittaker. Springer-Verlag, New York.

Liew, H. C., and Chan, E. H. 1987. ECOPATH model of a tropical shallow-water community in Malaysia. Manuscript submitted to International Development Research Centre (IDRC)/ Singapore.

Mann, K. 1984. Fish production in open ocean ecosystems. In: Flows of energy and materials in marine ecosystems. pp. 435–458. Ed. by M. J. R. Fasham. Plenum Press, New York.

Marr, J. C. 1976. Fishery and resource management in Southeast Asia. Resources for the Future (RFF)/Program of International Studies of Fishery Arrangements (PISFA) Paper 7. Washington, DC.

Menasveta, D. 1980. Resources and fisheries of the Gulf of Thailand. SEAFDEC Training Dept. Text/ Reference Book Series No. 8, Bangkok.

Menasveta, D., Shindo, S., and Chullarsorn, S. 1973. Pelagic fishery resources of the South China Sea and prospects for their development. South China Sea Fisheries Development and Coordinating Programme. SCS/DEV/73/6. Rome.

Menasveta, P. 1986. Review of the biology and fishery status of cephalopods in the Gulf of Thailand. In: North Pacific workshop on stock assessment and management of invertebrates. pp. 291–297. Ed. by G. S. Jamieson and N. Bourne. Can. Spec. Publ. Fish. Aquat. Sci. 92.

Murdy, E. O., and Ferraris, C. J., Jr. 1980. The contribution of coral reef fisheries to Philippine fisheries production. ICLARM Newsletter 3(1):21–22.

Nguyen, Tac An. 1989. Energy balance of the major tropical marine shelf ecosystem of Vietnam. Biologiya Morya (Sov. J. Mar. Biol.) 15(2):78–83.

Olson, R. J., and Boggs, C. H. 1986. Apex predation by yellowfin tuna (*Thunnus albacares*): Independent estimates from gastric evacuation and stomach contents, bioenergetics and cesium concentrations. Can. J. Fish. Aquat. Sci. 43(9):1760–1775.

Pauly, D. 1979. Theory and management of tropical multispecies stocks: A review, with emphasis on the Southeast Asian demersal fisheries. ICLARM Studies and Reviews No. 1.

Pauly, D. 1980. A new methodology for rapidly acquiring basic information on tropical fish stocks: Growth, mortality, and stock recruitment relationships. In: Stock assessment for tropical small-scale fisheries. pp.154–172. Ed. by S. B. Saila and P. M. Roedel. Proceedings of an international workshop held September 19–21, 1979, at the University of Rhode Island. International Center for Marine Resource Development, University of Rhode Island, Kingston.

Pauly, D., Christensen, V., and Sambilay, V., Jr. 1990. Some features of fish food consumption estimates used by ecosystem modellers. ICES CM 1990/G:17 Sess. Q.

Pauly, D., and Chua, Thia Eng. 1988. The overfishing of marine resources: Socioeconomic background in Southeast Asia. Ambio 17(3):200–206.

Piyakarnchana, T. 1989. Yield dynamics as an index of biomass shifts in the Gulf of Thailand. In: Biomass yields and geography of large marine ecosystems. pp. 95–142. Ed. by K. Sherman and L. M. Alexander. AAAS Symposium 111. Westview Press, Inc., Boulder, CO.

Polovina, J. J. 1984. Model of a coral reef ecosystem. I. The ECOPATH model and its application to French Frigate Shoals. Coral Reefs 3:1–11.

Polovina, J. J., and Ow, M. D. 1983. ECOPATH: A user's manual and program listings. Southwest Fisheries Center, Honolulu, HI. Adm. Rept. H-83-23.

Ricker, W. E. 1968. Food from the sea. In: Resources and man. A study and recommendations. pp. 87–108. Ed. by the Committee on Resources and Man. W. H. Freeman and Company, San Francisco.

Rodin, L. E., Bazilevich, N. I., and Rozov, N. N. 1975. Productivity of the world's main ecosystems. In: Patterns of primary productivity in the biosphere. pp. 294–299. Ed. by H. Lieth. Benchmark papers in Ecology V8. Dowden, Hutchinson and Ross, Stroudsburg, PA.

Rowe, G. T. 1981. The deep-sea ecosystem. In: Analysis of marine ecosystems. pp. 235–267. Ed. by A. R. Longhurst. Academic Press, San Diego.

Rowe, G. T., and Gardner, W. 1979. Sedimentation rates in the slope waters of the northwest Atlantic Ocean measured directly with sediment traps. J. Mar. Res. 37:581–600.

Sambilay, V., Palomares, M. L., Opitz, S., and Pauly, D. 1990. Estimates of relative food consumption by fish and invertebrate populations, required for modeling the Bolinao reef ecosystems, Philippines. Paper presented at the UPMSI/UNDP First National Symposium in Marine Science, 16–18 May, Manila/Bolinao, Philippines.

SEAFDEC (Southeast Asian Fisheries Development Center). 1981. Fishery statistical bulletin for South China Sea area, 1979. Southeast Asian Fisheries Development Center, Bangkok.

Sherman, K. 1990. Large marine ecosystems as global units for management: An ecological perspective. ICES CM 1990/L:24.

Sherman, K., and Alexander, L. M. (Editors). 1986. Variability and management of large marine ecosystems. AAAS Selected Symposium 99. Westview Press, Inc., Boulder, CO.

Sherman, K., and Alexander, L. M. (Editors). 1989. Biomass yields and geography of large marine ecosystems. AAAS Selected Symposium 111. Westview Press, Inc., Boulder, CO.

Sherman, K., and Gold, B. D. 1990. Perspective: Large marine ecosystems. In: Large Marine Ecosystems: Patterns, processes, and yields. p. vii–xi. Ed. by K. Sherman, L. M. Alexander, and B. D. Gold. AAAS Symposium. AAAS, Washington, DC.

Shindo, S. 1973. General review of the trawl fishery and the demersal fish stocks of the South China Sea. FAO Fish Tech. Pap. 120. FAO, Rome.

Silvestre, G., Miclat, E., and Chua, T. E. 1989. Towards sustainable development of the coastal resources of Lingayen Gulf, Philippines. ICLARM Conference Proceedings 17.

Silvestre, G., Selvanathan, S., and Salleh, A. H. M. 1990. Preliminary trophic model of the coastal fishery resources of Brunei Darussalam, South China Sea. ICES CM 1990/L:41 Sess. Q.

Siti, K.D., and Taha, M.S.M. 1986. Stomach contents of selected demersal fish species from South China Sea. In: Ekspedisi Matahari '85: A study of the offshore waters of the Malaysian EEZ. pp. 187–192. Ed. by A.K.M. Mohsin, M.I.H. Mohamed, and M.A. Ambak. Faculty of Fisheries and Marine Science, University Pertanian Malaysia Occasional Publication No. 3, Serdang, Malaysia.

Smith, S. V. 1978. Coral-reef area and the contributions of reefs to processes and resources of the world ocean. Nature 273:225–226.

South China Sea Programme (SCSP). 1978. Report of the workshop on management of resources in the Sunda Shelf, Malacca Strait and related areas. South China Sea Fisheries Development and Coordinating Programme. SCS/GEN/78/18, Manila.

Sparre, P. 1991. Introduction to multispecies virtual population analysis. ICES Mar. Sci. Symp. 193:12–21.

Tandog-Edralin, D. D., Cortes-Zaragoza, E. C., Dalzell, P., and Pauly, D. In press. Some aspects of the biology of skipjack (*Katsuwonus pelamis*) in Philippine waters. Asian Mar. Biol., Hong Kong.

Toft, C. A., and Mangel, M. 1991. Discussion: From individuals to ecosystems; the papers of Skellam, Lindeman and Hutchinson. Bull. Mathematical Biol. 53(1):121–134.

Ulanowicz, R. E. 1986. Growth and development: Ecosystem phenomenology. Springer-Verlag, New York.

Walsh, J. J. 1975. A spatial simulation model of the Peru upwelling ecosystem. Deep Sea Res. 22:1–13.

Walsh, J. J. 1981. A carbon budget for overfishing off Peru. Nature 290:300–302.

White, A. T. 1983. Valuable and vulnerable resources. In: Atlas for marine policy in Southeast Asia seas. pp. 26–39. Ed. by J. R. Morgan and M. J. Valencia. University of California Press, Berkeley.

White, A. T. 1989. Two community-based marine reserves: Lessons for coastal management. In: Coastal area management in Southeast Asia: Policies, management strategies and case studies. pp. 85–96. Ed. by T. E. Chua and D. Pauly. ICLARM Conference Proceedings 19. Ministry of Science, Technology and the Environment, Kuala Lumpur; Johore State Economic Planning Unit, Johore Bahru, Malaysia; and International Center for Living Aquatic Resources Management, Manila, Philippines.

Winberg, G. G. 1956. Rate of metabolism and food requirements of fishes. Trans. Fish. Res. Board Can. 253.

Wyrtki, K. 1961. Physical oceanography of the Southeast Asian waters. Naga Reports 2. Scripps Institute of Oceanography. La Jolla, California.

Yamashita, Y., Piamthipmanus, N., and Mochizuki, K. 1987. Gut contents analysis of fishes sampled from the Gulf of Thailand. In: Studies on the mechanism of marine productivity in the shallow waters around the South China Sea, with special reference to the Gulf of Thailand. pp. 33–55. Ed. by K. Kawagushi. Grant-in-Aid No. 61043019 for Overseas Scientific Survey, Ministry of Education, Science and Culture of Japan, Tokyo.

Yeh, S. A. 1981. Dynamics of the demersal fish resources in the Sunda Shelf area of the South China Sea. Ph.D. diss., University of Washington, Seattle.

Chapter 16

Marine Biogeographic Provinces of the Bering, Chukchi, and Beaufort Seas

G. Carleton Ray and Bruce P. Hayden

Introduction

Our objective in this paper is to encourage the use of biogeography to define large marine ecosystems (LMEs) more explicitly in ecological terms. We present a description of the biogeographic provinces of the contiguous continental shelf region of Beringia. The provinces are derived from a statistical analysis of range data for selected invertebrates, fishes, birds, and mammals. Our data source is the Bering, Chukchi, and Beaufort Seas Coastal and Ocean Zones Strategic Assessment: Data Atlas (NOAA, 1988).

The term "Beringia" was coined by Hopkins (1967) to describe the contiguous continental shelf region of the Bering, Chukchi, East Siberian, and Beaufort seas that was laid almost bare by a fall in sea level of more than 100 m during Pleistocene glaciations (Fig. 1).

Today, this region accounts for about three-quarters of the U.S. Exclusive Economic Zone (EEZ) and a major portion of its fisheries. The region is dominated by the annual formation and deformation of sea ice, which remains possibly its dominant physical force. Some organisms are positively adapted to sea ice and some avoid it, and sea ice dynamics also helps explain why this region is extraordinarily productive.

LMEs have been defined as extensive areas of ocean space, measuring about 200,000 km^2 (60,000 nm^2) or greater, and characterized by distinct hydrographic regimes, submarine topographies, productivity, and trophically dependent populations. Forty-nine LMEs have been identified mostly based on oceanographic criteria and fisheries' distributions. However, as it is the intent to delineate each LME as an ecologically defined unit of ocean space (Sherman, 1991), it seems reasonable to examine how biogeographic patterns may conform to these physical attributes.

The 49 LMEs presently identified encompass 95% of the annual global fishery yield (Sherman, 1991). Thus, the determination of the relevance of biogeographic pattern to LMEs, as currently delineated, is pertinent to a multispecies, ecological approach to fisheries and other marine resources. This presents a dual challenge: first, to enter environmental variables into the equation of determining the status of multiple-species stocks; and second, to become predictive (Morgan, 1987; Sherman, 1988).

The results presented here must be considered preliminary. Determination of the "fit" of biogeographic pattern within a larger regional or oceanic context is no small task. According to Beddington (1986, p. 15), "Given the complexity of most marine systems, it is quite clear that models of such systems are likely to possess an extremely complicated dynamic landscape with a substantial number of alternate states." This reflects the realities that ecosystem dimensions are difficult to establish and that species assemblages are variable. Yet, solutions to fishery management problems will depend to a great extent on how ecosystems are delineated and defined. They will also depend on the "functional diversity" that results from altered species patterns within the LMEs (Steele, 1991).

Methods

Ricklefs (1990) suggested that "The interaction of geological and biological processes of this

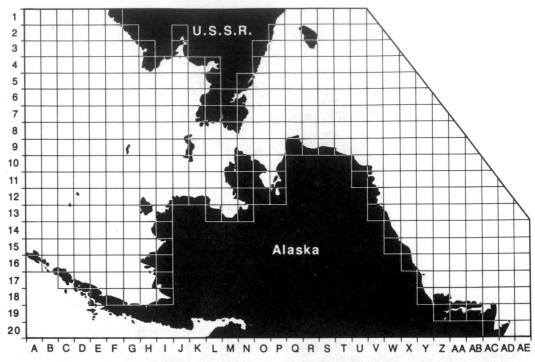

Figure 1. The study area from NOAA (1988). The continental shelf is outlined by the 200 m contour. The projection is Universal Transverse Meractor. The analysis grid is composed of 391, 50 x 50 mi² segments, encompassing 852 x 10³ mi² in total area. Some segments cover land areas. The portions of land included are eliminated from the calculations of total area in Table 2.

extent is evident in the geographical distribution of taxonomic groups of high rank." We interpret this to mean that physical boundaries are often evident in the distributions of groups of commercially important species of invertebrates, fishes, birds, and mammals. The emphasis is on "groups," as the natural history of each species is unique to its environment. Therefore, each species may be an "indicator" of a different set of variables.

A fundamental variable of a species' environment is other species within a higher level of organization called a community. If it is true that individual species respond to discrete environmental variables, then it follows that communities of species respond to aggregates of variables. Thus, a statistical examination of how species are correlated in time and space may yield important insights into ecosystem time-space properties. We call these correlated species "assemblages."

The data

Knowledge of the home range of a species' population is basic to the definition of assemblages. The home range is the overall area

wherein a species' population carries out the sum total of its activities, with the exception of accidental or vagrant individuals. Range is a presence-absence parameter; either members of the population occur within a range or they do not. Range does not describe distribution, that is, the quantifiable abundances of individuals within the range—"major" areas, areas of concentration, and the like. Identification of major areas of concentration would involve a second order of analyses.

Thus, in our analysis we assume that range is the most useful parameter for describing biogeographic patterns. A species' range is often determined by some limiting environmental factor, such as, temperature or salinity. It follows that by aggregating species' ranges into assemblages, we may also help define fundamental ecosystem properties, including ecosystem boundaries. Subsequent analyses would be required to reveal the full suite of spatial and temporal patterns in relation to the subareas or domains of LMEs.

The National Oceanic and Atmospheric Administration (NOAA 1988) catalogues natural history data on Beringia's best-known biota. Four criteria were used to select species for inclusion: (i) representatives from each of the four major groups of concern—invertebrates, fishes, birds, and mammals; (ii) economic value—commercial, recreational, or subsistence; (iii) status as depleted, threatened, or endangered; and (iv) ecological value, for example, as a major food-web component, being especially abundant. These criteria were balanced against adequacy of data for mapping life-history characteristics.

Data for each species were independently derived from both published and unpublished sources. Maps and accompanying descriptions were peer-reviewed and present syntheses of normalized data covering approximately 1 decade. The atlas maps portray life-history data in time and space; accompanying texts provide map interpretation on such subjects as behavior, food and feeding, habitat requirements, community relationships, and growth and reproduction. Some mapped and text data had to be inferred from assumptions about preferred habitats and environmental constraints. Of all data, those for range are most robust.

More than 100 species are included in the Atlas and have been entered in a standard format into NOAA's Computer Mapping and Analysis System, allowing the analysis presented here. Eighty-six of these species were appropriate for the present analysis (Table 1). We believe that these species reasonably represent the species diversity of Beringia.

The statistical method

Levin (1990) suggests that, for understanding the complexity of LMEs, the approach first is statistical analysis of physical and biological variables. Simple techniques such as composites or overlays are not suitable for interpreting complex ecological properties such as biogeographical assemblages. Therefore, we turn to Principal Components Analysis (PCA) for our interpretation. The procedures involved here are diagrammed in Fig. 2.

PCA is a frequently used, multivariate ordination statistical procedure wherein the variance in a complex data set is represented as a few new factors for relatively easy interpretation (Legendre and Legendre, 1983). Multivariate ordination techniques generally fall under the class of statistical methods called factor analysis, as they are based on the extraction of a few eigenvectors (components or factors) from a large association matrix. Ordination techniques are increasingly being applied to the analysis of community structure across environmental gradients, as is our intent here, and by biologists to quantify the underlying patterns of covariation among a set of variables assumed to be interdependent. A common

viewpoint, which we share, stresses the importance of ordination for hypothesis generation, rather than hypothesis testing (Gauch, 1982). Hence, ordination is widely regarded as a valuable exploratory tool.

PCA is an objective technique, since the factor scores (i.e., the original data points projected onto the several principal component axes) are derived solely from the input data matrix (Gauch, 1982). Basically, the analysis transforms a set of intercorrelated variables into a new coordinate system in which the axes are linear combinations of the original variables and are mutually orthogonal. Thus, the original multiplicity of variables is transformed into a new set of mutually independent variables, many fewer in number, but retaining most of the variance in the original data.

For the PCA in this study, the 86 species listed in Table 1 were treated as 86 interdependent variables. For each of these variables, presence or absence was scored as 0 or 1 in each of the 391, 50 x 50 mi^2 cells into which we subdivided the study area (Fig. 1). Thus, the analysis was performed on an 86 x 391 incidence matrix. PCA of incidence matrices is not without problems and may introduce certain biases (Gower, 1966; Swan, 1970; Williamson, 1978). We performed the pair-wise, graphical analysis to reveal bias, as suggested by Williamson (1978), and there were no biases evident in any of the principal components retained for our interpretation.

In this study, we applied varimax-rotated PCA to the problem of species affinities across the study area. Varimax rotation is a rigid orthogonal rotation of the principal component axes, such that the variance explained is optimized for all principal components retained in the analysis. Varimax rotations are applied because we are especially interested in the resolution of transitions between one zone and the next.

Results

The PCA resulted in 15 components, each of which accounted for as much as 17.7% to as little as 1.2% of the total variance. However, only the first six components together explained greater than 5% each, so only these are considered to be statistically significant at the 95% level. Table 2 shows that these six explain 69% of the total variance among the 86 species' ranges. Table 2 also gives the areas covered by each component. Table 1 includes numerical loadings (correlation coefficients) above 0.5 and below °0.5 for each species. These cutoffs are based on the experience that loadings above or below these values are especially

Table 1. Species used for analysis.[*]

	INVERTEBRATES		C1	C2	C3	C4	C5	C6
1	Arrow worm	Sagitta elegans						
2	Euphausiid (1)	Thysanoessa inermis				-0.7		
3	Euphausiid (2)	Thysanoessa raschii				-0.7		
4	False calanus copepod	Pseudocalanus spp.						
5	Feathery calanus copepod	Neocalanus plumchrus				0.7	0.5	
6	Dragonfly amphipod	Themisto libellula						0.5
7	Common two-spined crangon shrimp	Cragon communis	0.6					
8	Ridged crangon shrimp	Crangon dalli						
9	Sand (gray) shrimp	Crangon septemspinosa						
10	Large crangonid shrimp (1)	Argis dentata		0.5				0.6
11	Large crangonid shrimp (2)	Argis lar		0.5				0.5
12	Large crangonid shrimp (3)	Sclerocrangon boreas						
13	Large crangonid shrimp (4)	Sabinea septemcarinata						
14	Northern Pink shrimp	Pandalus borealis	0.5					0.5
15	Side-stripe shrimp	Pandalopsis dispar					0.9	
16	Humpy shrimp	Pandalus goniurus						0.7
17	Yellow-legged pandalid	Pandalus tridens						
18	Opossum shrimp (1)	Mysis litoralis		0.5	0.5			
19	Opossum shrimp (2)	Mysis oculata		0.5	0.5			
20	Opossum shrimp (3)	Mysis polaris				-0.6		-0.6
21	Korean hair crab	Erimacrus isenbeckii	0.9					
22	Red king crab	Paralithodes camtschatica	0.6					
23	Golden king crab	Lithodes aequispina					0.8	
24	Blue king crab	Paralithodes platypus						
25	Opilio tanner crab	Chionoecetes opilio						0.7
26	Bairdi tanner crab	Chionoecetes bairdi	0.8					
27	Red squid	Berryteuthis magister					0.8	
28	Chalky macoma	Macoma calcarea		0.7				
29	Greenland cockle	Serripes groenlandicus		0.7				
30	Iceland cockle	Clinocardium ciliatum		0.8				
31	Neptuneid whelk	Neptunea heros		0.7				
32	Lyre whelk	Neptunea lyrata	0.7					
33	Pribilof whelk	Neptunea pribiloffensis	0.7					
34	Fat whelk	Neptunea ventricosa		0.7				
35	Angular whelk	Buccinum angulossum		0.8				
36	Ladder whelk	Buccinum scalariforme		0.7				
	FISHES							
37	Pacific herring	Clupea harengus pallasi						0.8
38	Pink salmon	Oncorhynchus gorbuscha					0.6	0.5
39	Chum salmon	Oncorhynchus keta					0.6	0.6
40	Coho salmon	Oncorhynchus kisutch					0.6	0.7
41	Sockeye salmon	Oncorhynchus nerka					0.6	0.7
42	Chinook salmon	Oncorhynchus tshawytscha					0.6	0.7
43	Capelin	Mallotus villosus		0.5				0.5
44	Eulachon	Thaleichthys pacificus	0.9					
45	Rainbow smelt	Osmerus mordax			0.8			
46	Saffron cod	Eleginus gracilis						0.6
47	Pacific cod	Gadus macrocephalus	0.6					0.5
48	Walleye pollock	Theragra chalcogramma	0.5			0.6		
49	Yellowfin sole	Limanda aspera	0.5					0.5
50	Alaska plaice	Pleuronectes quadrituberculatus						0.7
51	Starry flounder	Platichthys stellatus						0.6
52	Greenland turbot	Reinhardtius hippoglossoides					0.9	
53	Rock sole	Lepidopsetta bilineata	0.9					
54	Arrowtooth flounder	Atheresthes stomias					0.8	
55	Flathead sole	Hippoglossoides elassodon	0.9					
56	Bering flounder	Hippoglossoides robustus						0.7
57	Pacific halibut	Hippoglossoides stenolepis	0.6					
	BIRDS							
58	Northern fulmar	Fulmarus glacialis					0.6	
59	Shearwaters (Short-tailed and Sooty)	Puffinus tenuirostris, P. griseus					0.6	

60	Oldsquaw	*Clangula hyemalis*			0.9			
61	Common eider	*Somateria mollissima*			0.9			
62	King eider	*Somateria spectabilis*			0.9			
63	Red phalarope	*Phalaropus fulicaria*			0.6	0.7		
64	Glaucous-winged gull	*Larus glaucescens*				0.6		0.7
65	Glaucous gull	*Larus hyperboreus*		0.5		0.5		
66	Ross' gull	*Rhodostethia rossea*		0.7				
67	Black-legged kittiwake	*Rissa tridactyla*		0.8				
68	Red-legged kittiwake	*Rissa brevirostris*	0.5				0.5	
69	Arctic tern	*Sterna paradisaea*		0.5		0.7		
70	Aleutian tern	*Sterna aleutica*				0.6		0.6
71	Murres (Common and Thick-billed)	*Uria aalge, U. lomvia*		0.6		0.6		
72	Black guillemot	*Cepphus grylle*		0.7				
73	Horned puffin	*Fratercula corniculata*		0.8				
74	Parakeet auklet	*Cyclorrhynchus psittacula*						0.7
75	Least auklet	*Aethia pusilla*						0.5
	MAMMALS							
76	Northern fur seal	*Callorhinus ursinus*	0.5			0.6		
77	Northern sea lion	*Eumetopias jubatus*						0.8
78	Pacific walrus	*Odobenus rosmarus divergens*		0.7				
79	Pacific harbor seal	*Phoca vitulina richardsi*	0.5					
80	Spotted seal	*Phoca largha*						0.7
81	Ringed seal	*Phoca hispida*		0.9				
82	Ribbon seal	*Phoca fasciata*		0.7				
83	Bearded seal	*Erignathus barbatus*		0.9				
84	Bowhead whale	*Balaena mysticetus*		0.7				
85	Gray whale	*Eschrichtius robustus*						0.5
86	White whale	*Delphinapterus leucas*		0.9				

*Loadings (correlation coeficients) are listed only when they exceed 0.5 (rounded from three decimal places) or are less than –0.5 (also rounded). Very few species with strong loadings occur in more than one component.

important for interpretation. Figure 3 presents the six components in mapped form. Contour lines on the maps indicate the strength of assemblage gradients and are perhaps also indicative of environmental ecotones.

Interpretation of the derived patterns, which we will now term "provinces," depends on knowledge of the oceanography of these areas and of the life histories and ecological associations of strongly correlated species within each of them. Some general characteristics of each province are summarized below, in order of the amount of variance explained by each.

Beringian shelf province (Component 2)

This area accounts for the greatest amount of variance of any component—17.7% (Table 2). It extends over almost the entire Beringian shelf and is also the largest of the Beringian region's provinces. Seven of its 11 invertebrates with strong loadings are benthic molluscs. One fish, seven birds, and six mammals, most of them benthic-feeding species, are also strongly loaded on this province. A distinct shelf province is strongly indicated, but it is interesting to note that this province does not extend deep into Bristol Bay, but centers on the southwest shelf of the Bering Sea and on the central part of the Chukchi Sea.

Southeast Bering Sea province (Component 1)

This province encompasses the best-known and most heavily utilized Beringian fisheries area. It is defined largely on the basis of a varied assortment of north-temperate species: seven invertebrates, seven fish, one bird, and two mammals. Many other species also occur within the area, but also range widely elsewhere, so are not strongly loaded on this area.

There is an even gradient pattern around this province, indicating that it is transitional to the Beringian shelf northward, the North Pacific southward, and Bristol Bay eastward. That this area does not appear to extend farther westward is probably a consequence of currents that upwell over the western Bering Sea shelf (Coachman, 1986; Incze and Schumacher, 1986; NOAA, 1988).

North Pacific province (Component 4)

The North Pacific Ocean strongly influences the Bering Sea, which is, to a great extent, a sub-

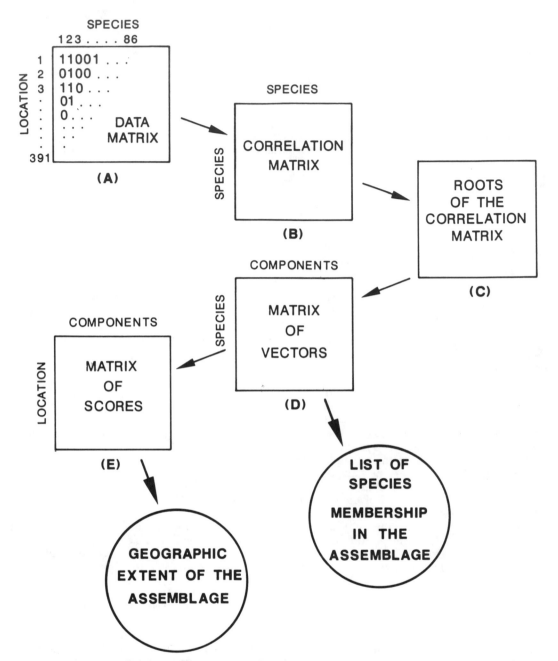

Figure 2. Procedures for Principal Components Analysis (PCA). (A) Presence-absence range data for each species are entered into a location matrix (Fig. 1). (B) The matrix is factored to derive species' correlations. (C) Roots of the matrix are derived. (D) A matrix of vectors (components) yields loadings (correlation coefficients) for each species as members in each component. (E) A matrix of location scores is derived for each assemblage and mapped so that contours can be drawn, defining each assemblage as a biogeographic province.

Table 2. Variances explained and areas of provinces.

	C2	C1	C4[*]	C5	C3	C6[**]		Totals
% of total variance	17.7	13.2	8.9	7.2	7.0	15.0	=	69
Area in mi^2 x 10^3	406	162	113	81	249	[472]	=	1,483
% of total study area (852 x 10^3 mi^2)	47.7	19.0	13.3	9.5	29.2	[55.4]	=	174.1

[*] Size of this province is underrepresented, as the study area (Fig. 1) includes only a small portion of the North Pacific.

[**] This province is subdivided into two parts (Fig. 3): The Bering Sea part is 241 x 10^3 mi^2 and the Chukchi-Beaufort part is 231 x 10^3 mi^2.

arctic, northward extension of that ocean. Therefore, it is to be expected that one Beringian province would be a North Pacific extension.

This assemblage accounts for 8.9% of the total variance. The species that are strongly loaded on this province almost all range widely over the North Pacific (e.g., salmon, cod, flatfishes, and seabirds). The strong negative loadings (Table 1) are for species least likely to be members of this assemblage.

The contours that define this province are tightly packed at about the shelf break, indicating a very strong environmental ecotone there. However, the northward limit is well onto the shelf as a consequence of incursions of fauna during summer.

Bering Sea slope province (Component 5)

This is an outer shelf-to-slope province with very strong gradients at both its northern and southern boundaries; it is the smallest of the provinces (Table 2).

The species in this component are mostly benthic slope species with very heavy loadings of 0.8–0.9 (e.g., side-stripe shrimp, golden king crab, arrowtooth flounder). These species are few, mostly because slope species are underrepresented in our database. However, these species may be indicators of a rich fauna, as continental slopes become increasingly recognized as distinct and comprised of high faunal diversity (Grassle, 1991).

Beringian inner shelf province (Component 3)

This is a nearshore province, with strong loadings for only a few nearshore species, especially opossum shrimp and waterfowl. This paucity of species may result from a poor nearshore-species database, but also may result from the wide variety of offshore species that share a nearshore habitat at some stage in their life history. This province reflects the strong terrestrial-marine interactions of nearshore, inner-shelf environments.

Bering-Chukchi-Beaufort seas shelf province (Component 6)

The assemblage that defines this province is dual, the first part being Beringian (positive contours on Fig. 3) and the second being Chukchi-Beaufort (negative contours on Fig. 3). Thus, this component appears to account for a disproportionate percent of total variance—14.5%.

There are two reasons for this finding. First, PCA has detected seasonal faunal shifts between winter and summer (e.g., salmon and migratory birds and mammals, which also occur in the North Pacific province). Second, there is a strong negative species loading (e.g., *Mysis polaris*) for this component, indicating the strong unlikelihood of its—and other Chukchi-Beaufort species'—occurrence farther south.

Discussion

This paper illustrates the utility of biogeographical analysis for further defining the spatial and structural characteristics of LMEs. The results of our analysis indicate that each of the defined provinces is distinct, both spatially and in its species composition. This approach clearly has the potential to define multispecies groups within an ecological context, as is the intent of the LME concept.

Second, the provinces we describe are overlapping. Each is of the order of magnitude in size proposed for an LME (Table 2), and the total coverage of all provinces together exceeds the total study area by more than half. This reflects the hierarchical structure of ecosystems

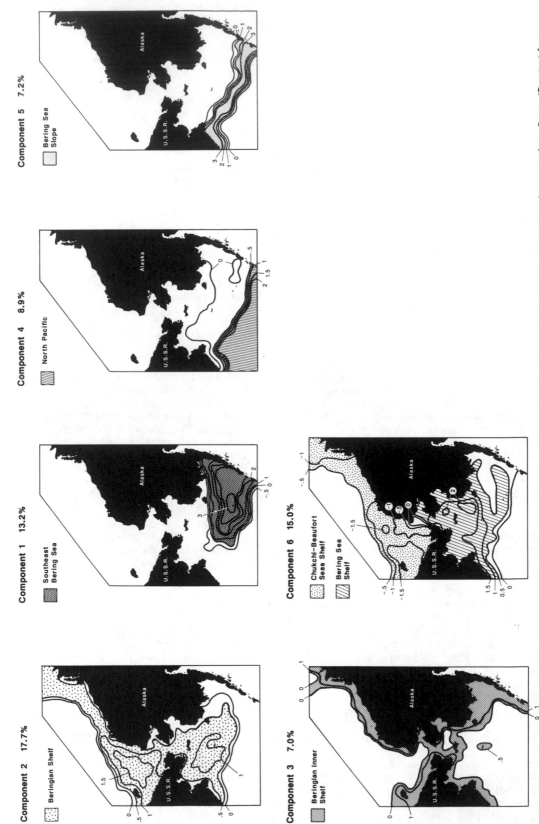

Figure 3 a–f. Provinces of Beringia, as derived by PCA. Contours that delineate each province indicate the strength of possible environmental gradients. (See text for explanation of each province.)

and the overlaps of community structures in nature.

Third, the assemblages that comprise these provinces are distinct. Table 1 shows that few strongly loaded species occur in more than one assemblage: None with loadings of 0.8–0.9 and only 15 with loadings of 0.5–0.7 occur in more than one. Therefore, we suggest that certain species or species groups may be used as "indicators" of the defined provinces.

A recurring problem is the integration of biogeographic pattern with environmental properties. For the Bering Sea, oceanographic fronts have been proposed to separate outer, middle, and inner shelf domains (Coachman, 1986; McRoy *et al.*, 1986). The temptation may be to use these fronts as boundaries for three discrete communities of organisms. But as we see, biogeographic pattern does not appear to reflect this oceanographic pattern. Rather, assemblages are organized into a complex, overlapping fabric that results from complex environmental interactions. For example, Ray and Hufford (1989) performed a PCA for Beringian marine mammals and found that biotic assemblages seemed to be determined, at least in part, by sea ice distribution and type. Nevertheless, sea ice is only one of many physical parameters that may strongly influence marine mammal distributions. We must seek additional attributes for better explanations.

Sherman (1991) depicts two LMEs for the Bering Sea, one eastern and one western, with another for the Gulf of Alaska southward. Useful as these LME distinctions may be for regional management purposes (Morgan, 1987), we believe that some improvements are possible by using species assemblages as indicators of the complexities of ecosystem boundary conditions and to define multispecies groups within provinces.

In all respects, we emphasize the exploratory nature of this work. Because of the incomplete and inconsistent data on species' natural history, our results should be considered as hypotheses for subsequent testing. Are these provinces real in time and space? May they be used for monitoring human and natural perturbations? Do changes in assemblages reflect cyclic or stochastic environmental change? Only the results of further, long-term research can provide answers.

Acknowledgments

We wish to thank the Strategic Assessment Branch of the National Atmospheric and Oceanic Administration for providing support for our work. In particular, we thank our NOAA colleagues, D. J. Basta, who provided the leadership within NOAA to complete this work, and T. F. LaPointe, who spearheaded database development and the computer system used in these analyses. We also are indebted to Allison D. Denton, formerly of the Department of Environmental Sciences of the University of Virginia, for many aspects of the analysis.

References

Beddington, J. R. 1986. Shifts in resource populations in large marine ecosystems. In: Variability and management of large marine ecosystems. pp. 9–18. Ed. by K. Sherman and L. M. Alexander. AAAS Selected Symposium 99. Westview Press, Inc., Boulder, CO.

Coachman, L. K. 1986. Circulation, water masses, and fluxes on the southeastern Bering Sea shelf. Cont. Shelf Res. 5(1/2):23–108.

Gauch, H. G. 1982. Multivariate analysis in community ecology. Cambridge Univ. Press, Cambridge, U. K.

Gower, J. C. 1966. Some distance properties of latent root and vector methods used multivariate analysis. Biometrika 53(3,4): 325–338.

Grassle, J. F. 1991. Deep-sea benthic biodiversity. BioScience 41(7):464–469.

Hopkins, D. M. 1967. The Bering land bridge. Stanford Univ. Press, Stanford, CA.

Incze, L., and Schumacher, D. 1986. Variability of the environment and selected fisheries resources of the eastern Bering Sea ecosystem. In: Variability and management of large marine ecosystems. pp. 109–143. Ed. by K. Sherman and L. M. Alexander. AAAS Selected Symposium 99. Westview Press, Inc., Boulder, CO.

Legendre, L., and Legendre, P. 1983. Numerical ecology. Elsevier Scientific Publishing Co., Amsterdam.

Levin, S. A. 1990. Physical and biological scales and the modelling of predator-prey interactions in large marine ecosystems. In: Large marine ecosystems: Patterns, processes, and yields. pp. 179–187. Ed. by K. Sherman, L.M. Alexander, and B.D. Gold. AAAS Symposium. AAAS, Washington, DC.

McRoy, C. P., Hood, D. W., Coachman, L. K., Walsh, J. J., and Goering, J. J. 1986. Processes and resources of the Bering Sea shelf (PROBES): The development and accomplishments of the project. Cont. Shelf Res. 5(1/2):5–21.

Morgan, J. R. 1987. Large marine ecosystems: An emerging concept of regional management. Environment 29(10):4–9, 29–34.

NOAA (National Oceanic and Atmospheric Administration). 1988. Bering, Beaufort, and Chukchi Seas coastal and ocean zones strategic assessment: Data atlas. U.S. Dept. of Commerce, NOAA, National Ocean Survey, Strategic Assessment Branch, Rockville, MD.

Ray, G. C., and Hufford, G. L. 1989. Relationships among Beringian marine mammals and sea ice. Rapp. P.-v. Reun. Cons. int. Explor. Mer 188:225–242.

Ricklefs, R. E. 1990. Scaling pattern and process in marine ecosystems. In: Large marine ecosystems: Patterns, processes, and yields. pp. 169–178. Ed. by K. Sherman, L.M. Alexander and B.D. Gold. AAAS Symposium. AAAS, Washington, DC.

Sherman, K. 1988. Large marine ecosystems as global units for recruitment experiments. In: Toward a theory on biological-physical interactions in the world ocean. pp. 459–476. Ed. by B. J. Rothschild. Kluwer Acad. Publishers, Dordrecht, The Netherlands.

Sherman, K. 1991. Sustainability of resources in large marine ecosystems. In: Food chains, yields, models, and management of large marine ecosystems. pp. 1–34. Ed. by K. Sherman, L. M. Alexander, and B. D. Gold. AAAS Symposium. Westview Press, Inc., Boulder, CO.

Sherman, K., and Alexander, L. M. (Editors). 1986. Variability and management of large marine ecosystems. AAAS Selected Symposium 99. Westview Press, Inc., Boulder, CO.

Steele, J. H. 1991. Marine functional diversity. BioScience 41(7):470–474.

Swan, J. M. A. 1970. An examination of some ordination problems by use of simulated vegetational data. Ecology 51(1):89–102.

Williamson, M. H. 1978. The ordination of incidence data. J. Ecol. 66:911–920.

Chapter 17

Effects of Climatic Changes on the Biomass Yield of the Barents Sea, Norwegian Sea, and West Greenland Large Marine Ecosystems

Johan Blindheim and Hein Rune Skjoldal

Introduction

This paper focuses on some biological effects of the temperature and salinity anomaly that occurred in the northern North Atlantic during the 1960s and the 1970s. Several authors have described how this event occurred in various regions (Malmberg, 1969; Blindheim, 1974; Dickson and Blindheim, 1984; Dooley et al., 1984; Gammelsrød and Holm, 1984). Speculations on the reason for the anomaly have also been presented, and probably the most plausible was put forward by Dickson et al. (1988), who also summarized the whole event. Therefore, the anomaly will be described only briefly oceanographically.

Also, the major biological effects in different regions associated with the anomaly have been dealt with by various authors. The impact on Norwegian spring-spawning herring was discussed by Devold and Jakobsson, 1968; Jakobsson, 1969; and Dragesund et al., 1980; and it was extensively discussed in the International Council for the Exploration of the Sea (ICES) (Anonymous, 1979). Similarly, the state of the West Greenland cod stock has been described in various reports, one of which recently noted that yields have been related to the environment (Hovgaard and Buch, 1990). In the early 1980s, disturbances occurred in the Barents Sea LME with effects on the stocks of capelin, cod, herring, and mammals (Hamre, 1988, 1990, 1991; Skjoldal and Rey, 1989). The scope of this paper is to relate all these biological effects to the same climatic event—the "great salinity anomaly in the northern North Atlantic."

Similar environmental conditions were observed in the northern North Atlantic during the first decade of the twentieth century. At that time, observations from the western North Atlantic were rather scarce, but we have more detailed information from Norwegian waters.

The Oceanographic Anomaly

The relative magnitude of this oceanographic anomaly was first realized when it was observed in the Faroe-Shetland Channel, where a long time-series (Fig. 1) revealed that it was the largest on record in this century (Dooley et al., 1984). When the anomaly appeared in the Barents Sea, a time series of about the same length in the Russian Kola section showed a similar pattern (Fig. 2).

Dickson et al. (1988) connect the anomaly to events in the Greenland and Iceland seas during the 1960s, when an intense and persistent high-pressure anomaly cell was established over Greenland (Rodewald, 1967). This high pressure gave rise to anomalous northerly winds over the Greenland Sea, increasing the transport in the East Greenland Current of cold and fresh water as well as sea ice from the Arctic Ocean. Malmberg (1972) described how the East Icelandic Current, which previously had been ice-free for many years, attained a polar character in 1965–1971 (Fig. 3) and carried the largest quantities of ice in this century to the north and east coasts of Iceland.

The expanded volume of cold and fresh water in the East Greenland Current proceeded further, partly mixing into Atlantic waters of the Irminger Current, but mainly flowing into the West Greenland and Labrador Currents (Blindheim, 1974; Buch, 1982; Buch and Stein, 1988). Observations during the 1960s off West

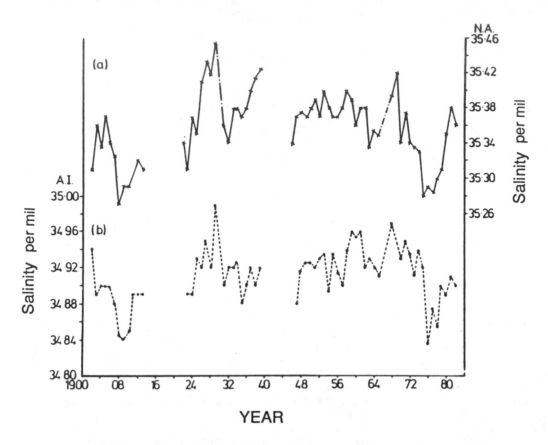

Figure 1. Time series of salinity in (a) North Atlantic (N.A.) water, and (b) Arctic Intermediate (A.I.) water in the Fareo-Shetland Channel from 1902 to 1982 (After Dooley *et al.* 1984).

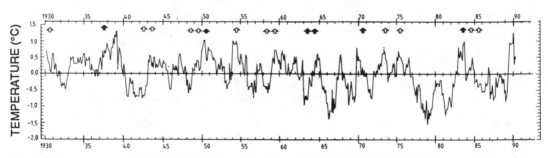

Figure 2. Temperature anomalies in the Kola section (along 33° 30' E), 1930–1988. Filled arrows indicate strong year-classes of cod; open arrows indicate year-classes of medium strength. Year-classes were low in all other years (From Loeng, 1989b).

Greenland indicated a steady decline in temperature and salinity in the upper waters (Fig. 4). This cooling and freshening, which intensified the stratification of the upper layers and restricted the exchange of heat from deeper layers (Fig. 5), brought about a positive feedback to the cooling of the upper layer in this area. (The same was also the case in north Icelandic waters.)

When this pulse reached the Newfoundland area via the Labrador Current, it mixed into the Atlantic waters of the North Atlantic Current and reached the Faroe-Shetland Channel in the mid-1970s (Figs. 1 and 6). From here, it was carried further northward by the Norwegian Atlantic Current and reached the Barents Sea and the Svalbard area by 1979–1980 (Bochkov, 1982; Dickson and Blindheim, 1984; Loeng, 1989a).

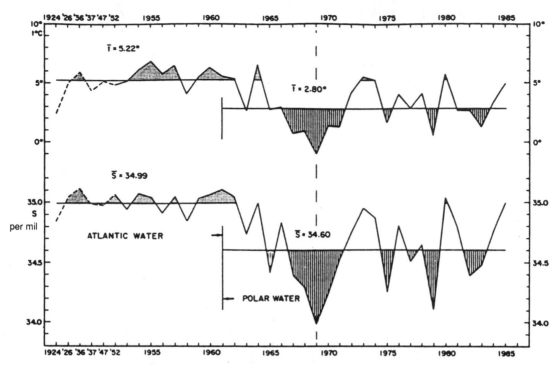

Figure 3. Temperature and salinity at 50 m on a hydrographic station, 66° 32' N, 18° 50' W, in North Icelandic waters in May–June 1924, 1926, 1936, 1937, 1947, and 1952–1985 (From Malmberg, 1986). t̄ and s̄ are mean temperature and mean salinity, respectively.

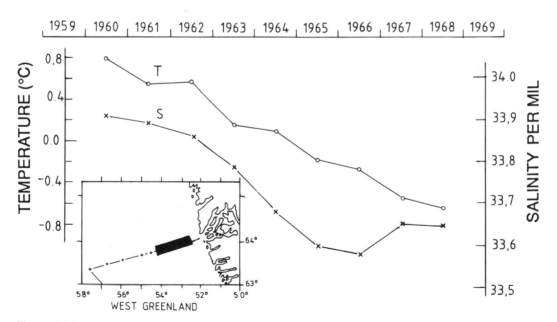

Figure 4. Three-year moving average of mean temperature and salinity in waters of salinity below 34 per mil in the section across the Fylla Bank, West Greenland. Open circles = temperature curve; x's = salinity curve (From Blindheim, 1974).

At their minima, temperatures were 0.5°C to 2°C lower than the average for the preceding 10–20 years, and the salinity anomaly amounted to 0.1–0.15 units. The propagation of the anomalous pulse around the subpolar gyre in the northern North Atlantic is indicated in Fig. 6 (Dickson *et al.*, 1988).

Biological Effects

Norwegian spring-spawning herring

The Norwegian spring-spawning herring (*Clupea harengus*) was the first large stock that

was affected. Prior to the mid-1960s, the adult herring fed in the area northeast of Iceland during summer and autumn. In late autumn, the mature stock congregated in a wintering area east of Iceland, where it remained until about the end of the year when the spawning migration toward the Norwegian coast started. Spawning took place along the Norwegian west coast in early spring (Fig. 7).

When the temperature decline developed in the East Icelandic Current during the 1960s, feeding conditions for the herring probably deteriorated, and part of the stock grazed further to the northeast in the Norwegian Sea. After 1967, the whole stock shifted to a feeding area in the northern Norwegian Sea during the sum-

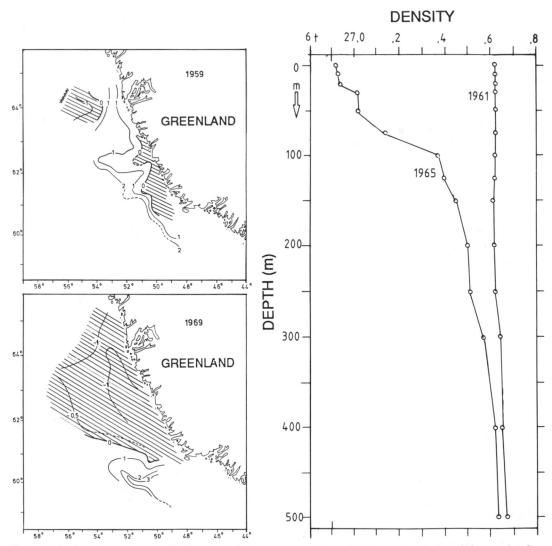

Figure 5. Surface temperatures off West Greenland as observed in April 1959 and April 1969 (left) and density gradients as observed in 1961 and 1965 at a hydrographic station on the slope, 65° N (right).

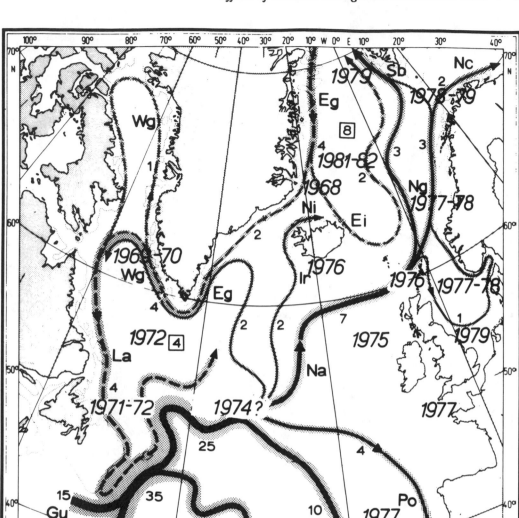

Figure 6. Transport scheme for the 0–1,000 m layer of the northern North Atlantic with the salinity minimum superimposed (From Dickson *et al.*, 1988).

mer. During the entire period, however, the same wintering area east of Iceland was retained. The migration route between the feeding and wintering areas was then more than 1,000 km. Although there is no evidence, it is unlikely that this change in migration and feeding pattern favored the reproduction potential of the stock.

At the same time, the fishing pressure on the stock increased substantially during the 1960s, as purse-seining technology advanced

and the stock was exploited during most of the year. Thus, the stock was exploited to depletion, and since 1970 until quite recently, no herring has been observed in the traditional feeding areas in the Norwegian and Iceland seas.

West Greenland cod

The beginning of the last period with cod

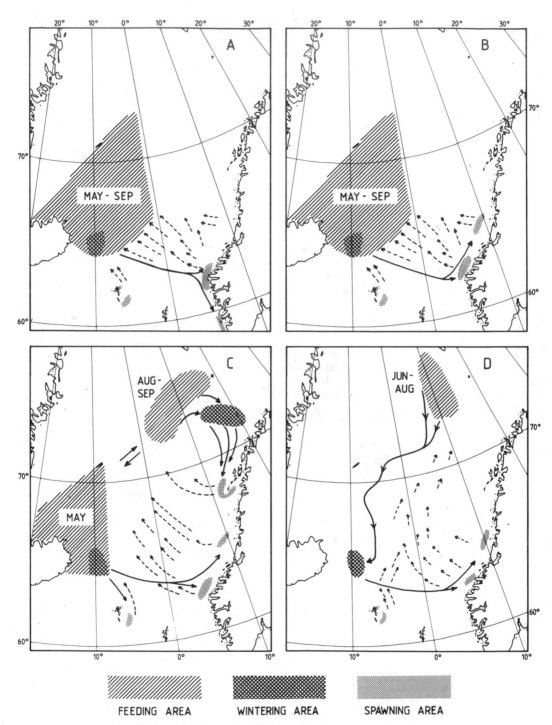

Figure 7. Migration routes of Norwegian spring-spawning herring. (A) 1950–1959; (B) 1960–1962; (C) 1963–1966; (D) 1967–1968. (Modified from Anonymous, 1979).

(*Gadus morhua*) in West Greenland waters dates back to about 1920. The stock gradually built up during the warming period, which started at about the same time. Based on sea-surface temperatures, this large-scale warming, which in West Greenland waters lasted until the mid-1950s, is described by Smed (1974). The buildup of the cod population and various other effects of the warming were described by Jensen (1939) and recently by Hovgaard and Buch (1990). The temperature decrease in the Arctic component water masses of the West Greenland Current on the Fylla Bank (Fig. 4.) had drastic and almost immediate effects on the cod. The stock, which during the 1950s and 1960s had sustained an annual fishing yield of about 300,000 tonnes, declined considerably. From 1968 to 1970, landings were reduced from about 400,000 to 100,000 tonnes. During the following years, this decline continued, and in 1976 landings were down to about 30,000 tonnes.

As in other stocks with heavy fishing pressure, it is difficult to decide to what extent the stock reduction was caused by the deterioration of environmental factors or by fishing mortality. Hovgaard and Buch (1990) have discussed in greater depth how this reduction occurred. Besides fishing mortality, they point out the importance of migration from West to East Greenland waters and to some extent even to Icelandic areas. Reduction in reproduction from local spawning is believed to be of less importance, because they think that a major part of the larvae from this spawning drifted out of the area also in earlier years. The coincidence between the climatic cooling and the abrupt decline in yields of the cod fisheries off West Greenland strongly suggests that the climate played a decisive role.

Northeast Atlantic blue whiting

Some indications of climate-dependent recruitment were also observed in the stock of blue whiting (*Micromesistius poutassou*) in the northeast Atlantic. This stock spawns to the west of the British Isles and at least one northern component feeds into the Norwegian Sea, as indicated in Fig. 8a. The longest series of investigations on this stock has been undertaken by scientists from the former U.S.S.R. During the period 1970–1988, Belikov *et al.* (1990) found no abundant year-classes between 1972 and 1982. Also, the 10-year period with moderate and poor year-classes coincided with the cold period of the anomaly, whereas in 1982 there was an abundant year-class and considerable warming (Fig. 8b).

The Barents Sea LME

Characteristic features of the Barents Sea ecosystem

The Barents Sea is a relatively well-known ecosystem because of extensive investigations of fish stocks and recent ecological research.

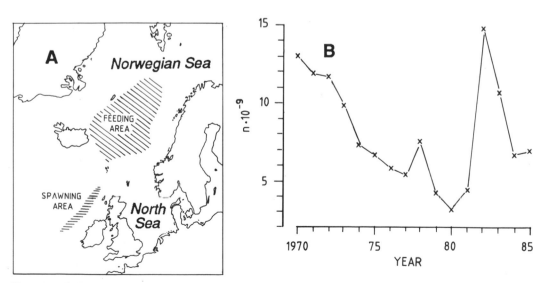

Figure 8. (A) Feeding and spawning area of blue whiting, *Micromesistius poutassou;* (B) Year-class strength by number (n), 1970–1985. (Data from Belikov *et al.*, 1990).

There have been several recent reviews of the Barents Sea ecosystem (Dragesund and Gjøsæter, 1988; Loeng, 1989a; Skjoldal and Rey, 1989; Hamre, 1990; 1991), and only the main characteristics will be recapitulated here.

The Barents Sea is a transition zone where inflowing and relatively warm Atlantic water is being cooled and transformed into Arctic and Polar water. There is great variability in ocean climate, presumably reflecting variable inflow activity (Loeng *et al.*, 1992). There is also much seasonal and interannual variability in ice, which covers from one-third to two-thirds of the area of the Barents Sea at maximum extension during winter. A rapidly developing ice-edge phytoplankton bloom occurs in response to ice melting. This bloom follows the retreating ice edge as a sweeping band of high production across the sea as the ice retreats northward during summer (Sakshaug and Skjoldal, 1989; Skjoldal and Rey, 1989).

Advection plays a major role for the stocks of zooplankton in the southern and central Barents Sea. A key species is the copepod *Calanus finmarchicus*, which is transported with the current of Atlantic water from the Norwegian Sea into the Barents Sea. *C. finmarchicus* performs an extensive seasonal vertical migration, with the major part of the overwintering population residing deeper than

500 m in the Norwegian Sea during autumn and winter (Østvedt, 1955). This seasonal pattern of vertical distribution interacts with the Atlantic inflow activity to strongly influence the amount of *C. finmarchicus* and zooplankton biomass in the southern and central Barents Sea (Skjoldal and Rey, 1989; Skjoldal *et al.*, 1992) (Fig. 9).

The Barents Sea ecosystem contains two of the world's largest fish stocks, the Barents Sea capelin (*Mallotus villosus*) and Northeast Arctic cod (*Gadus morhua*) (Fig. 10). Capelin is a plankton-feeder that performs a large-scale seasonal feeding migration to exploit the plankton production of the northern Barents Sea. This can be viewed as an ecosystem adaptation whereby capelin trails the retreating ice with a time delay (Sakshaug and Skjoldal, 1989; Skjoldal and Rey, 1989). Herring is also a part-time, but important, member of the Barents Sea ecosystem. The southern and central areas of the Barents Sea are important nursery grounds for young and adolescent herring, especially when the Atlantic influence is high and the year-class is strong (Hamre, 1990, 1991).

There is strong empirical evidence that recruitment success of cod, haddock, and herring depends on the climatic oscillation in the Barents Sea, with strong year-classes typically

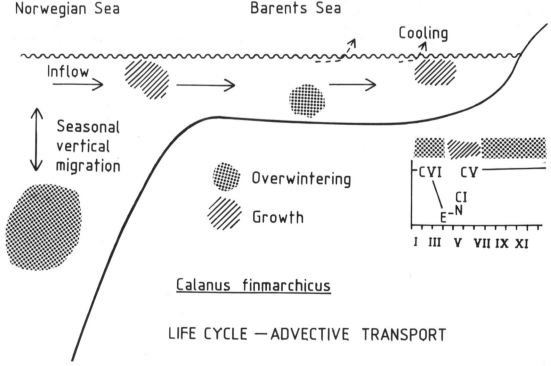

Figure 9. Life cycle and transport of *Calanus finmarchicus* from the Norwegian Sea into the Barents Sea. Seasonal vertical migration and Atlantic inflow interact to control the advective transport of *Calanus*.

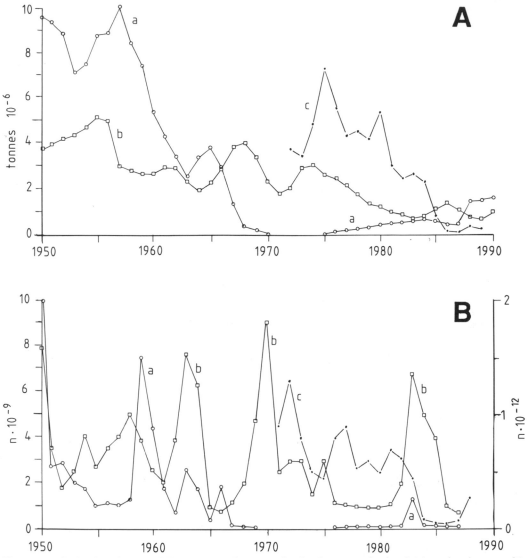

Figure 10. (A) Stock estimates of Norwegian spring-spawning herring, spawning stock (a); cod at the age of 3 years old and older (b); capelin (c). (B) Recruitment estimates by number of herring at 1 year of age (a), n x 10⁻⁹; cod at age of 3 years old and older (b); and capelin (c), n x 10⁻¹² (Modified from Hamre, 1990).

formed upon transition from cold to warm phases of the climatic periods (Fig. 2) (Sætersdal and Loeng, 1987; Loeng, 1989b). There are also strong interactions between the stocks of capelin, herring, and cod, and strong year-classes can have marked influences on other components of the ecosystem. Capelin and herring are major prey for cod, and their dominance in the diet of cod alternates depending on sizes and distribution pattern of the stocks (Hamre, 1988, 1990, 1991; Mehl,

1989, 1992). Capelin spawns along the coasts which border the southern Barents Sea, and the larvae drift and spread with the currents eastward and northward.

Predation by juvenile herring on capelin larvae may lead to recruitment failure of capelin (Mehl, 1989, 1992; Skjoldal and Rey, 1989; Hamre, 1990, 1991). Predation by cod on immature capelin can also have a marked impact on the capelin stock (Mehl, 1989, 1992; Hamre, 1990, 1991).

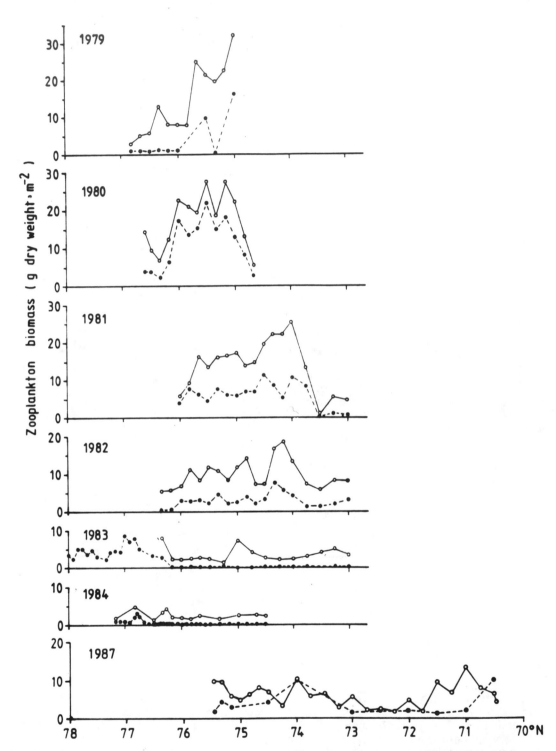

Figure 11. Zooplankton biomass (g dry weight • m⁻²) along a N–S transect in the central Barents Sea during summer for the years 1979–1984 and 1987 (From Skjoldal *et al.*, 1987, with additional data from cruise with R/V *G.O. Sars* in May–June 1987). Values are for the whole water column and the upper 40 m in 1979 and for the upper 200 m and 50 m in 1980–1987 (∘———∘, 0 m-bottom; •- -•, 0-40 m in 1979 and 0–50 m in 1980–1987).

Climatic impact on events during the 1980s

Dramatic ecological events took place in the Barents Sea ecosystem during the 1980s. Details of these events have been described and discussed by Hamre (1988, 1990, 1991), Skjoldal and Rey (1989), Skjoldal *et al.* (1992), and Mehl (1992). These events occurred in response to strong climatic forcing that can be traced back to the salinity-temperature anomaly. When the anomaly reached the Barents Sea in the late 1970s (Fig. 6), it gave rise to an exceptionally long cold period in the area (Fig. 2) (Loeng *et al.*, 1992). This cold period, which lasted from 1977 to 1981, was followed by warmer conditions during 1982–1984. Following this warm period, with highest temperatures in 1982, there was again a colder period, which lasted until 1988. After this, temperatures in the southwestern Barents Sea increased considerably in 1989, and there was a shift to a new period with warm climate.

The transition from cold to warm climate in 1982 set the stage for a sequence of interlinked events. The zooplankton biomass in the central Barents Sea showed a pronounced minimum in 1983–1984 (Fig. 11) (Skjoldal *et al.*, 1987). This has been explained as a result of the interaction between current activity and seasonal vertical migration of *C. finmarchicus* (Fig. 9). Winter water that was poor in this copepod filled the southwestern and central Barents Sea toward the end of the inflow event in late winter 1982–1983 (Skjoldal and Rey, 1989). At this time, *C. finmarchicus* was deeper than the lower range of the Atlantic water, which flowed into the Barents Sea and did not, therefore, follow the inflowing water masses.

Another major consequence of the shift to warmer conditions was the formation of strong year-classes of cod and herring in 1983 (Fig. 2). Through various interactions, these strong year-classes contributed to a collapse of the capelin stock (Fig. 12). Predation from juvenile herring was probably a major cause for recruitment failure of capelin in 1984 and 1985 (Skjoldal and Rey, 1989; Hamre, 1990, 1991; Fossum, 1992). Juvenile cod preyed heavily on capelin and caused a marked increase in the natural mortality of the stock of immature capelin (Mehl, 1989, 1992). These factors, along with the poor individual growth of capelin and too-late relief from fishing pressure (Hamre, 1991), caused the capelin stock to collapse to a stock size of less than 100,000 tonnes in 1986 (Fig. 10).

The decline of the capelin stock had drastic effects on other components of the Barents Sea ecosystem because of the capelin's key role as an intermediate link in the food web. Among the effects from food shortage were poor growth and condition of cod (Mehl, 1992), massive seal invasions along the Norwegian coast, and heavy mortality of seabirds (Fig. 12) (Skjoldal, 1990).

Events similar to those of the 1980s occurred also in the Barents Sea during the first decade of this century (Skjoldal, 1990). A cold period occurred at the beginning of the century, probably with minimum temperature and maximum ice cover in the Barents Sea in 1902 (Helland-Hansen and Nansen, 1909). Indicating that advection also played a role in this case, Helland-Hansen and Nansen (1909) were able to trace the temperature minimum from the southern Norwegian Sea to the Barents Sea with a time delay of about 2 years. Strong ecological disturbances coincided with this cold period. The cod was exceptionally small and in poor condition in 1903 and arrived late at the spawning grounds. The Norwegian coast was invaded by seals and mass mortality of seabirds was observed (Hjort, 1903). This was followed by a shift in ocean climate to a warm phase, and strong year-classes of cod and herring were formed in 1904 (Helland-Hansen and Nansen, 1909; Hjort, 1914; Skjoldal, 1990).

The development of the stocks of cod, capelin, and herring during the recent decades (Fig. 10) provides an example of strong biological interaction associated with strong climatic forcing. It also represents a likely coupling among climatic oscillations on different temporal scales. The historical "herring periods" of approximately 50 years' duration are probably related to large-scale shifts in meteorological conditions (Cushing and Dickson, 1976; Skjoldal, 1990). The collapse of the herring stock during the 1960s (Fig. 10) represented the end of a herring period that started during the first decades of this century. The coincidence between this collapse and "the great salinity anomaly" indicates that the latter may be the signature of a low-frequency (decadal) climatic oscillation. The 1983 year-class, and subsequent strong year-classes of herring during the warm years of 1989–1991 (Anonymous, 1992; Loeng *et al.*, 1992), hopefully represent the start of another herring period.

Superimposed on the long-term changes are important short-term oscillations in the ocean climate of the Barents Sea, with a mean cycle length of about 4 years (Loeng *et al.*, 1992). The inflow events assumed to be associated with the warming phase of these oscillations are related to the formation of strong year-classes of cod, haddock, and herring (Fig. 2) (Sætersdal and Loeng, 1987). This was exemplified in 1983 and later during 1989–1990. Through trophic interactions, the strong year-classes have marked impact on the ecology and fish stocks of the Barents Sea ecosystem.

Discussion

The biological events associated with the large temperature-salinity anomaly in the 1960s and 1970s strongly indicate that ocean climate is an important factor in marine ecosystems dynamics. In high-latitude LMEs, ocean climate variability may be a particularly important driving force. The development of the West Greenland cod population is an example of a pronounced response to change in climatic conditions from a species living in the marginal zones of their distributional range.

The collapse of the Norwegian spring-spawning herring stock during the 1960s represented the end of a herring period that started during the first decades of this century. Undoubtedly, overfishing contributed to the rapid and strong decline of the herring stock (Dragesund *et al.*, 1980; Hamre, 1988, 1990, 1991). There was also, however, an environmental influence. The temperature-salinity anomaly coincided with a change in the migration pattern and recruitment success of herring. The historically known herring periods of approximately 40–60 years' duration are probably related to long-term climatic fluctuations (Cushing and Dickson, 1976; Skjoldal, 1990). From about 1930 onward, there was a long-term temperature decrease (Figs. 1 and 2). In the 1970s, the temperature-salinity anomaly added considerably to this cooling trend. It is possible that such a strong anomaly, occurring in the cold phase of longer climate cycles, has a triggering effect on ecological events. Until a better understanding of the climate system and its impacts on the LMEs is obtained, this remains speculative.

The herring stock of the Norwegian Sea LME has had a large impact on the Barents Sea LME (Hamre, 1988, 1990, 1991). The cod stock of the Barents Sea was elevated around 1950, but has since shown an overall decrease to a very low stock size in 1987 (Fig. 10). There are indications that the growth conditions for the cod stock was poor in the cold climatic period between 1880 and 1920 (Godø and Skjoldal, 1991). During the warmer period from 1920 when the herring stock was high, the conditions for cod were better, as judged from a substantial increase in the individual size of the Barents Sea cod (Rollefsen, 1956). The state of the cod stock, therefore, seems to depend on the state of the herring stock.

The ocean climate has a strong direct influence on the recruitment success of fish species such as cod and herring. The Kola time

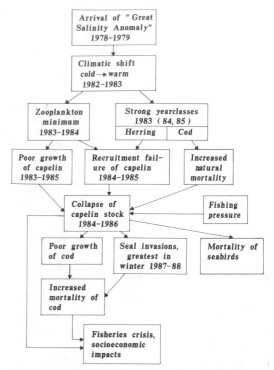

Figure 12. Flowchart of ecological events in the Barents Sea ecosystem from the arrival of the salinity anomaly in the late 1970s to the fisheries crisis in the late 1980s.

series in the Barents Sea demonstrates that strong year-classes coincide with the warming phase of the short-term climate fluctuations on a time scale of approximately 4 years (Fig. 2) (Sætersdal and Loeng, 1987; Loeng *et al.*, 1992). The strong empirical relationship between recruitment and climate fluctuations, and the coincidence of strong year-classes of cod, haddock, and herring, suggest large-scale climatic forcing. The mechanisms behind this are not well understood. It is probable that temperature has a direct effect through its influence on the duration of larval development and resulting surviving recruits (Skjoldal and Melle, 1989). In addition, increasing temperature may be associated with increased current activity, which gives a more-rapid transport and greater geographical spread of the larval population. The availability of zooplankton food organisms (Ellertsen *et al.*, 1989) and short-term climatic events affecting turbulence and larval feeding rate (Sundby and Fossum, 1989) may also be of importance in recruitment success.

The events in the Barents Sea during the 1980s and in the beginning of the twentieth century show that climate also influences ecosystems indirectly. The collapse of the capelin stock in 1985 was not caused by direct climatic effects on the capelin, but mainly by a drastic change in predator-prey relationships, which was caused by earlier climatic effects on the cod and herring stocks. Further, it was to some extent the result of changes in plankton dynamics, which also were related to climatic pulsations. More generally, this demonstrates that when climate gives rise to a sudden change in the species composition of an ecosystem, or otherwise alters the structure of the food web, indirect effects may be drastic and have longer time-scales than the triggering climatic event itself.

References

Anonymous, 1979. The biology, distribution and state of exploitation of fish stocks in the ICES area, Part II. Int. Counc. Explor. Sea Coop. Res. Rep. 86.

Anonymous, 1992. Ressursoversikt for 1992, Havforskningsinstituttet. Fisken Hav. S. No. 1. In Norwegian.

Belikov, S. V., Tereshchenko, E. S., and Isaev, A. 1990. Population fecundity and year-class strength of blue whiting in the Northeast Atlantic. In: Proceedings of a fourth Soviet-Norwegian symposium on biology and fisheries of the Norwegian spring-spawning herring and blue whiting in the Northeast Atlantic, Bergen, 1989. pp. 281–289. Ed. by T. Monstad. Institute for Marine Research, Bergen, Norway.

Blindheim, J. 1974. On the hydrographic fluctuations in the Labrador Sea during the years 1959–1969. Fiskeridir. Skr. Ser. Havunders. 16:194–202.

Bochkov, Yu. A. 1982. Water temperature in the 0–200 m layer on the Kola-Meridian in the Barents Sea, 1900–1981. Sb. Nauchn. Trud. PINRO, Murmansk 46:113–122. In Russian.

Buch, E. 1982. Hydrographic conditions off West Greenland in 1980 and 1981. Int. Counc. Explor. Sea C:5. (Mimeo).

Buch, E., and Stein, M. 1988. Time series of temperature and salinity at the Fylla Bank section, West Greenland. Int. Counc. Explor. Sea C:4. (Mimeo).

Cushing, D. H., and Dickson, R. R. 1976. The biological response in the sea to climatic changes. Adv. Mar. Biol. 14:1–122.

Devold, F., and Jakobsson, J. 1968. The formation and disappearance of a stock of Norwegian herring. Fiskirdir. Skr. Ser. Havunders. 15:1–22.

Dickson, R. R., and Blindheim, J. 1984. On the abnormal hydrographic conditions in the European Arctic during the 1970s. Rapp. P.-v. Reun. Cons. int. Explor. Mer 185:201–213.

Dickson, R. R., Meincke, J., Malmberg, S. Aa., and Lee, A. J. 1988. The "great salinity anomaly" in the northern North Atlantic 1968–1982. Prog. Oceanogr. 20:103–151.

Dooley, H. D., Martin, J. H. A., and Ellett, D. J. 1984. Abnormal hydrographic conditions in the Northeast Atlantic during the 1970s. Rapp. P.-v. Reun. Cons. int. Explor. Mer 185:179–187.

Dragesund, O., and Gjøsæter, J. 1988. The Barents Sea. In: Ecosystems of the world. Continental shelves. pp. 339–361. Ed. by H. Postma and J. J. Zilstra. Elsevier, Amsterdam.

Dragesund, O., Hamre, J., and Ulltang, Ø. 1980. Biology and population dynamics of the Norwegian spring-spawning herring. Rapp. P.-v. Reun. Cons. int. Explor. Mer 177:43–71.

Ellertsen, B., Fossum, P., Solemdal, P., and Sundby, S. 1989. Relation between temperature and survival of eggs and first-feeding larvae of northeast Arctic cod (*Gadus morhua* L.). Rapp. P.-v. Reun. Cons. int. Explor. Mer 191:209–219.

Fossum, P. 1992. The recovery of the Barents Sea capelin *Mallotus villosus* from a larval point of view. ICES J. Mar. Sci. 49(2):237–244.

Gammelsrød, T., and Holm, A. 1984. Variations of temperature and salinity at Station M (66° N, 2° E) since 1948. Rapp. P.-v. Reun. Cons. int. Explor. Mer 185:188–200.

Godø, O. R., and Skjoldal, H. R. 1991. Fluctuations in abundance of the Barents Sea cod related with environmental and ecological changes. Cod Symposium, September 1991. Serial No. N2010. NAFO SCR Doc. 91/117.

Hamre, J. 1988. Some aspects of the interaction between the herring in the Norwegian Sea and the stocks of capelin and cod in the Barents Sea. Int. Counc. Explor. Sea H:42. (Mimeo).

Hamre, J. [1990]. Biodiversity and exploitation of the main fish stocks in the Norwegian-Barents Sea Ecosystem. International Symposium, Our Common Natural Heritage, Bergen 1990. In press.

Hamre, J. 1991. Interrelation between environmental changes and fluctuating fish populations in the Barents Sea. In: Proceedings from an international symposium on long-term variability of pelagic fish populations and their environment, Sendai, Japan, 1989. pp. 259–270. Ed. by T. Kawasaki, S. Tanaka, Y. Toba, and A. Taniguchi. Pergamon Press, Oxford, U.K.

Helland-Hansen, B., and Nansen, F. 1909. *The Norwegian Sea.* Rep. Norw. Fish. Mar. Invest. 2(2).

Hjort, J. 1903. Investigations and observations. In: Annual report regarding the Norwegian fisheries in 1903. pp. 33–34. Norges Fiskeristyrelse, Oslo. In Danish.

Hjort, J. 1914. Fluctuations in the great fisheries of northern Europe viewed in the light of biological research. Rapp. P.-v. Reun. Cons. int. Explor. Mer 20:1–228.

Hovgaard, H., and Buch, E. 1990. Fluctuations in the cod biomass of the West Greenland Sea ecosystem in relation to climate. In: Large Marine Ecosystems: Patterns, processes, and yields. pp. 36–43. Ed. by K. Sherman, L. M. Alexander, and B. D. Gold. AAAS Symposium. AAAS, Washington, DC.

Jakobsson, J. 1969. On herring migrations in relation to changes in sea temperature. Jökul 19:134–145.

Jensen, Ad. S. 1939. Concerning a change of climate during recent decades in the arctic and subarctic regions, from Greenland in the west to Eurasia in the east, and contemporary biological and geographical changes. Biol. Meddr. 14(8):1–76.

Loeng, H. 1989a. Ecological features of the Barents Sea. In: Proceedings of the sixth conference of the comité Arctique international, 1985. pp. 327–365. Ed. by L. Rey and V. Alexander. E. J. Brill, Leiden, Netherlands.

Loeng, H. 1989b. The influence of temperature on some fish population parameters in the Barents Sea. J. Northw. Atl. Fish. Sci. 9:103–113.

Loeng, H., Blindheim, J., Adlandsvik, B., and Ottersen, G. [1992]. Climatic variability in the Norwegian and Barents seas. ICES Mar. Sci. Symp. 195 (In press).

Malmberg, S. A. 1969. Hydrographic changes in the waters between Iceland and Jan Mayen in the last decade. Jökul 19:30–43.

Malmberg, S. A. 1972. Annual and seasonal hydrographic variations in the East Icelandic Current between Iceland and Jan Mayen. In: Sea Ice. Proc. int. conf. Reykjavik. Vol. 4. pp. 42–54. Ed. by T. Karlsson. Nat. Res. Counc. Reykjavik, Iceland.

Malmberg, S. A. 1986. The ecological impact of the East Greenland Current on the north Icelandic waters. In: The role of freshwater outflow in coastal marine ecosystems. NATO ASI Series, Vol. G7. pp. 389–404. Ed. by S. Skreslet. Springer-Verlag, Berlin.

Mehl, S. 1989. The Northeast Arctic cod stock's consumption of commercially exploited prey species in 1984–1986. Rapp. P.-v. Reun. Cons. int. Explor. Mer 188:185–205.

Mehl, S. 1992. The Northeast Arctic cod stock's place in the Barents Sea ecosystem in the 1980s—an overview. In: Proceedings from the Pro Mare symposium on polar marine ecology, Trondheim, 12–16 May 1990.

pp. 525–534. Ed. by E. Sakshaug, C. C. E. Hopkins, and N. A. Øritsland. Polar Research 10. Norwegian Polar Research Institute, Oslo.

Østvedt, O. J. 1955. Zooplankton investigations from weather ship M in the Norwegian Sea, 1948–1949. Skr. Nor. Vidensk. Akad. Hvalråd. 40:1–93.

Rodewald, M. 1967. Recent variations of North Atlantic sea surface temperatures and the "type tendencies" of the atmospheric circulation. Int.

Comm. Northwest Atl. Fish. Redb., 1967 (Part 4). pp. 6–23.

Rollefsen, G. 1956. The stock of "skrei." Ann. Biol. Copenh. 11:95.

Sætersdal, G., and Loeng, H. 1987. Ecological adaptation of reproduction in Northeast Arctic cod. Fish. Res. 5:253–270.

Sakshaug, E., and Skjoldal, H. R. 1989. Life at the ice edge. Ambio 18:60–67.

Skjoldal, H. R. 1990. Management of marine living resources in a changing ocean climate. In: Papers presented on the session "Research on natural resources management" of the Conference "Sustainable development, science and policy," Bergen 8–12 May 1990. Bergen Foundation of Science Rep. No. BFS A90005. pp. 1–17. Bergen, Norway.

Skjoldal, H. R., Gjøsæter, H., and Loeng, H. [1992]. The Barents Sea ecosystem in the 1980s: Ocean climate, plankton, and capelin growth. In: Proceedings from a joint Russian-Norwegian symposium, Murmansk, August 1991. pp. 1–20. Ed. by S. Tjelmeland. ICES Mar. Sci. Symp. 195, (In press).

Skjoldal, H. R., Hassel, A., Rey, F., and Loeng, H. 1987. Spring phytoplankton and zooplankton reproduction in the central Barents Sea in the period 1979–1984. In: Proceedings of the third Soviet-Norwegian symposium, Murmansk, 26–28 May 1986. pp. 59–90. Ed. by Harald Loeng. Institute of Marine Research, Bergen, Norway.

Skjoldal, H. R., and Melle, W. 1989. Nekton and plankton: Some comparative aspects of larval ecology and recruitment processes. Rapp. P.-v. Reun. Cons. int. Explor. Mer 191:330–338.

Skjoldal, H. R., and Rey, F. 1989. Pelagic production and variability of the Barents Sea ecosystem. In: Biomass yields and geography of large marine ecoystems. pp. 241–286. Ed. by K. Sherman and L. M. Alexander. AAAS Selected Symposium 111. Westview Press, Inc., Boulder, Co.

Smed, J. 1974. Monthly anomalies of the sea surface temperature in areas of the northern North Atlantic in 1973. Ann. Biol. 31:11–12.

Sundby, S., and Fossum, P. 1989. Feeding conditions of Northeast Arctic (Arcto-Norwegian) cod larvae compared to the Rothschild-Osborn theory on small-scale turbulence and plankton contact rates. ICES CM 1989/G:19. (Mimeo).

Chapter 18

The California Current, Benguela Current, and Southwestern Atlantic Shelf Ecosystems: A Comparative Approach to Identifying Factors Regulating Biomass Yields

Andrew Bakun

Introduction

One particularly promising use of the large marine ecosystem (LME) concept is as a basis for applications of the comparative method (Bakun, 1986). The experimental method and the comparative method are the "two great methods of science" (Mayr, 1982). The comparative method is the method of choice in situations not amenable to controlled experiments; for example, it has been the basis for "nearly all of the revolutionary advances in evolutionary biology" (*ibid.*). Because marine ecosystems are hardly amenable to experimental controls, the comparative method presents an available alternative for assembling the multiple realizations of a process needed to draw scientific inference.

When the editors of this volume solicited a chapter on this particular set of three LME complexes (Fig. 1), it seemed an attractive opportunity to construct an example of the rather indirect inferential process involved in drawing insights from an interregional comparative study. Moreover, this particular trio of ecosystems provides a convenient framework for presenting some emerging generalizations concerning environmental regulation of the population dynamics of the small, pelagic-spawning, clupeoid fishes that are a vital trophic component of many of the LMEs of the world's oceans.

At first glance, the three regions do not comprise an obvious natural grouping. Two of them, the California and Benguela Current systems, are classic eastern ocean coastal upwelling systems (Wooster and Reid, 1963; Parrish *et al.*, 1983), with narrow to moderate continental shelves and generally broad, rather diffuse, equatorward surface flow (Figs. 2 and 3). In contrast, the area off eastern South America contains a very broad expanse of continental shelf (Fig. 4) and is under the influence of persistent, intense boundary flows typical of western ocean boundaries (Stommel, 1966). Such western boundary currents are known to advect massive amounts of internal baroclinicity and momentum and are characterized by sharp gradients of properties and highly energetic, episodically varying flow interactions (e.g., Olson *et al.*, 1988). However, such differences are not necessarily a disadvantage; in applications of the comparative method, differences may be as useful to the inferential process as similarities.

All three regional systems are extremely rich in fishery resources of a variety of types. Also, important, highly visible seabird and marine mammal populations occupy segments of each. All three feature anchovies (genus *Engraulis*) and sardines (genus *Sardinops* or *Sardinella*) as key links in the local trophic systems. The California Cooperative Oceanic Fisheries Investigations (CalCOFI), having recently celebrated its 40th anniversary (CalCOFI Committee, 1990), and the Benguela Ecology Programme, now in its ninth year (Shannon *et al.*, 1988a), provide a wealth of valuable background material. The international Sardine/Anchovy Recruitment Project (SARP) has now been in existence for about seven years (Anonymous, 1983). SARP was the initial (and consequently most advanced at the present time) focal project of the Programme of Ocean Science in Relation to Living Resources (OSLR), which is jointly sponsored by the Intergovernmental Oceanographic Commission and the Food and Agriculture Organization of the

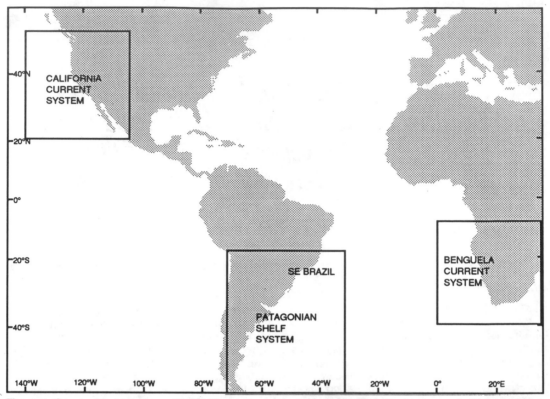

Figure 1. Geographical locations of the LME complexes treated in this chapter. Outlined areas indicate the borders of the three map formats used as bases for the displays presented in Figs. 2, 3, 4, 5, 9, 10, and 11.

United Nations. Although SARP was initially focused in the eastern Pacific, the center of action of SARP activities has recently shifted to the Atlantic, notably the southwestern Atlantic (Anonymous, 1989a). A multinational SARP cruise off Brazil, Uruguay, and Argentina by the research vessel *Meteor* (November–December, 1989) gathered valuable new observational material related to the mechanisms addressed in this chapter.

The California Current System

MacCall (1986) described the history of the pelagic fish populations of the California Current, including the rise of the Pacific sardine (*Sardinops sagax*) fishery, its famous collapse during the 1940s and 1950s, and the subsequent large rise in northern anchovy (*Engraulis mordax*) biomass. Recently, the sardine population off Alta California appears to be rebuilding slowly after many years of nearly total disappearance (Calif. Dept. of Fish and Game, 1990). (Note: In this chapter, the terms "Alta California" and "Baja California" are used to

distinguish between the area of coast occupied by the U.S. state of California and that of the adjacent Baja California Peninsula of Mexico.)

The California Current forms the eastern limb of the North Pacific Subtropical Gyre, flowing generally equatorward off the western United States and northern Mexico and exhibiting a mixture of filaments, mesoscale eddies, and counterflows (Reid *et al.*, 1958). Superimposed on this geostrophic current regime, is a pattern of wind-driven surface Ekman transport directed generally offshore. The offshore movement of the superficial surface layer is replenished near the coast by upwelling of deeper waters. Thus, the neritic ecosystem is dominated by wind-induced coastal upwelling, with the most intense upwelling centered near Cape Mendocino in northern Alta California during the spring and summer (Bakun *et al.*, 1974; note in Fig. 2 the characteristic pattern of closed isotherms, intersecting the coast, surrounding the cool core of upwelled water near the coast). The Baja California coast constitutes a secondary California Current upwelling zone, with a local maximum situated near Punta Baja (Bakun and Nelson, 1977).

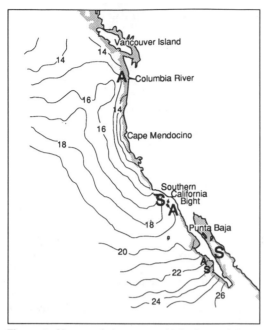

Figure 2. Characteristic summer (July–August) sea-surface temperature (°C) distribution for the California Current region. (Redrawn from Parrish *et al.*, 1983.) The area of the continental shelf is indicated by shading adjacent to the coastline. General locations of principal spawning grounds for anchovies (A) and sardines (S) are indicated.

Upwelling-related enrichment is believed to be a major basis for the high biological productivity of the entire region. Indeed, positive correlation has been noted between indices of upwelling intensity and recruitment success of a variety of commercially important fish stocks of the region (Table 1).

On the other hand, in surveying the spawning habits of a wide variety of California Current fishes, Parrish *et al.* (1981) identified a common pattern of avoidance of releasing pelagic eggs and larvae in geographical and/or seasonal circumstances where they would be rapidly dispersed offshore in the persistent offshore surface Ekman transport that is characteristic of upwelling zones. Thus, although the high nutrient input and associated high primary production may provide the trophic base for the massive fish populations of upwelling ecosystems, the associated wind-driven offshore surface transport appears to present special problems for reproduction.

Larval retention zones

Between the upwelling zones of Alta California and Baja California lies the Southern Califor-

nia Bight, which constitutes the major spawning grounds of the pelagic fishes that dominate the exploitable biomass of the California Current ecosystem. Parrish *et al.* (1981) describe how Pacific sardines and other larger pelagic species (e.g., hake [*Merluccius productus*] and mackerel [*Scomber japonicus*]), which may feed as adults in the upwelling region to the north, migrate long distances to spawn within this coastal bight, where the near-coastal area is sheltered from the large-scale coastwise winds and associated offshore transport. Retention of eggs and larvae within the neritic habitat is aided by a rather enclosed gyral geostrophic circulation pattern within the bight (Fig. 5). Prior to the initial population collapse, larger sardines annually migrated as far as Vancouver Island, where they supported an important Canadian fishery, but returned in large numbers to the Southern California Bight for reproduction. Anchovies, apparently because of their small size and limited migrational capabilities, remain largely resident within (or near) the reproductive area of the bight.

However, anchovies are not totally precluded by their migrational deficiencies from utilizing the rich feeding areas to the north. A northern subpopulation of anchovies spawns successfully in rather special circumstances. During the summer, the outflow (or "plume") of the Columbia River spreads over a wide area of ocean surface in a thin lens of warmed, low-

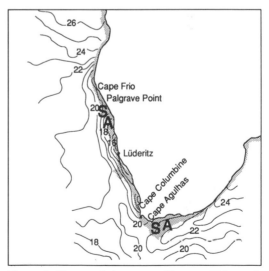

Figure 3. Characteristic summer (January–February) sea-surface temperature (°C) distribution for the Benguela Current region. (Redrawn from Parrish *et al.*, 1983.) The area of the continental shelf is indicated by shading adjacent to the coastline. General locations of principal spawning grounds for anchovies (A) and sardines (S) are indicated.

salinity surface water. In this northern part of the region, the winds diminish sharply during summer and offshore transport is very mild (Bakun *et al.*, 1974). In addition, the lens of lighter water may induce a superficial local anticyclonic circulation around its periphery (Fig. 6), which may aid in maintaining the integrity of the plume and also in retaining larvae and other organisms within the plume area. The interfaces between the surface waters of contrasting densities are zones of convergence where the heavier oceanic surface waters are overridden by the lighter plume waters. The waters mixed at the interfaces between the water types, being more dense than the surface plume waters, slowly slide beneath them in response to gravity. Weakly swimming organisms (such as larval food particles) that are able to resist sinking tend to be concentrated here (Fig. 7).

Except in front of the river mouth, the plume is pushed offshore during periods of active coastal upwelling. During these periods, the offshore surface Ekman flux is sufficient to maintain the slight sea-surface slope implied by the density difference between the upwelled

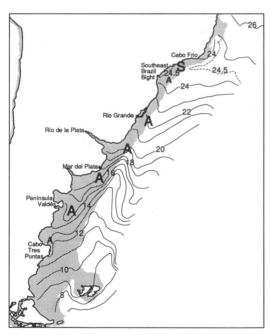

Figure 4. Characteristic summer (December–January) sea-surface temperature (°C) distribution for the region off southeastern South America. (Assembled from distributions presented by Bakun and Parrish, 1990, and by Bakun and Parrish, 1991.) The area of the continental shelf is indicated by shading adjacent to the coastline. General locations of principal spawning grounds for anchovies (A) and sardines (S) are indicated.

water near the coast and the lighter surface-plume waters a short distance offshore. However, during the upwelling season in this region, there are intermittent relaxations or reversals in upwelling-producing wind stress. At these times, the plume tends to collapse coastward (Smith, 1974), overriding the newly upwelled waters and providing a mechanism for returning juvenile fishes and other organisms entrained within the plume waters to the near-coastal zone. This plume environment provides an anchovy spawning habitat sufficient to maintain an endemic northern subpopulation.

Off central Baja California, the California Current swings offshore to join the westward-flowing, southern limb of the Subtropical Gyre. South of this point of deflection, wind-driven offshore surface transport near the coast becomes less intense (Fig. 5), and there is a tendency for the coastal geostrophic current pattern to veer shoreward (Wyllie, 1966). This is the site of an additional successful spawning region for both anchovies and sardines; both species have established substantial southern subpopulations that spawn in this area off southern Baja California (Vrooman and Smith, 1971).

Thus, on the open western coast of the continent, there are two major spawning regions for Pacific sardines. These are displaced from the strong offshore transport conditions of the major upwelling centers to areas where circulation patterns may aid in retaining eggs and larvae. However, they are in each case directly downstream of the upwelling zones. There are three spawning regions for northern anchovy (Fig. 5), two coinciding with those of the sardine, but a third utilizing the special summer conditions of the Columbia River plume. Accordingly, spawning of the northern subpopulation is largely confined to the summer season (Richardson, 1973); in contrast, the bulk of California Current anchovies, spawning in the regions to the south, have a spawning peak in late winter to early spring (Parrish *et al.*, 1983). Thus, with respect to utilization of the adult feeding opportunities in the northern part of the region, what the sardine achieves through superior migrational capabilities, the anchovy appears to at least partially manage via adaptability and flexibility in spawning strategy.

Pacific sardines have also been traditional residents of the interior of the Gulf of California, supporting the growth of a large fishery (more than 230,000 metric tonnes [mt] in 1988) during the period that the California Current population has been in a state of collapse. The situation within the gulf (Hammann *et al.*, 1988) would seem to be basically different from the coastal oceanic LMEs treated in this chapter,

Table 1. California Current: Evidence for Relationships

A. Enrichment
- large biomasses associated with upwelling ecosystems
- positive correlation of recruitment with upwelling index
  * Pacific sardine (Bakun and Parrish, 1980)
  * Pacific mackerel (Parrish and MacCall, 1978)
  * Pacific Bonito (Collins and MacCall, 1977)
  * Dungeness crab (Peterson, 1973)
  * Coho salmon (Gunsolus, 1978)
  * pink shrimp (Rothlisberg and Miller, 1983)

B. Larval retention
- common pattern of avoidance of offshore dispersal
  * avoidance of upwelling centers (Parrish et al., 1981)
  * migration to S. California Bight (Parrish et al., 1981)
- negative correlation of recruitment with upwelling index
  * Pacific hake (Bailey, 1981)

C. Stability
- anomalous food particle concentrations required
  * northern anchovy (Lasker, 1975)
- particle concentrations destroyed by wind mixing
  * northern anchovy (Lasker, 1978)
- negative correlation of recruitment with wind mixing index
  * northern anchovy (Lasker, 1975)
- spawning habits tuned to minimize turbulent mixing of habitat
  * northern anchovy (Husby and Nelson, 1982)
- positive correlation of larval survival rate with calm periods
  * northern anchovy (Peterman and Bradford, 1987)

D. Temperature
- positive correlation with temperature (and El Niño effects)
  * Pacific sardine during period of population decline (Prager and MacCall, 1990)
  * California halibut (Prager and MacCall, 1990)
- negative correlation with temperature
  * Pacific sardine during period of fishery growth (Ware and Thomson, 1991)

probably being more analogous to that of the anchovy and sardine populations that exploit localized upwelling or mixing zones within the Mediterranean (e.g., Regner *et al.*, 1987) and other rather enclosed peripheral seas. Very recently, during a period of anomalous climatic conditions coinciding with heavy fishing pressure on the sardine, an unprecedented establishment of a resident population of northern anchovies within the gulf has occurred (Hammann *et al.*, 1990).

Stability

Another factor that appears to favor spawning success is water-column stability. Lasker (1975, 1978) noted that the average concentration of suitable food particles in the spawning grounds of the Southern California Bight may be too low for newly hatched anchovy larvae to avoid starvation. Apparently, the best place for larvae to survive is within very limited patches where concentrations of food particles are much higher than the larger-scale average. Lasker observed that under stormy conditions, such food particle patches were destroyed by turbulent mixing of the water column. Husby and Nelson (1982) assembled the climatology of an index of rate of addition of turbulent kinetic energy to the ocean by the wind over the entire California Current region. They found that spawning habits of northern anchovy are indeed tuned, seasonally and geographically, in a way that appears to minimize the probability of encountering substantial wind mixing of the upper ocean.

The rate of input of turbulent kinetic energy to the ocean by the wind is roughly proportional to the third power, or cube, of the

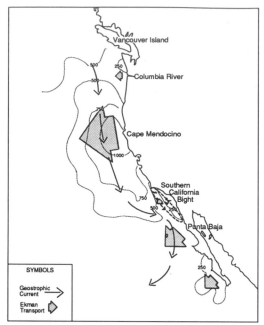

Figure 5. Diagram of characteristic summer flow features and wind-mixing index (m³ • s⁻³, dotted contours) for the California Current region. Broad, shaded arrows indicate magnitude of offshore surface Ekman transport (magnitude proportional to the linear dimensions of arrow symbols). Solid arrows indicate the general trend of underlying geostrophic current flow (no magnitude implied). Gyral circulation within the Southern California Bight is indicated by the closed dashed line. (Ekman transport and wind-mixing index magnitudes are derived from mean July–August distributions presented by Parrish et al., 1983.)

wind speed (Elsberry and Garwood, 1978). According to the summaries of Parrish et al. (1983; see their Appendix charts 9 and 10), a large portion of the Southern California Bight is characterized by values not exceeding 250 $m^3 \cdot s^{-3}$ in either the winter or summer seasons (Fig. 5). To the south, the index increases in the vicinity of the Punta Baja upwelling center, but again drops below 250 $m^3 \cdot s^{-3}$ in both winter and summer seasons in the spawning region off the southern portion of the Baja California Peninsula.

Along the coast to the north of the Southern California Bight, the index increases abruptly, reaching mean values greater than 1,000 $m^3 \cdot s^{-3}$ off Cape Mendocino in both winter and summer. In winter, the values continue to increase to the north, but near the Columbia River during summer, the index drops to a local minimum of much less than 500 $m^3 \cdot s^{-3}$ (Fig. 5). Recall that the northern subpopulation of northern anchovy spawns in

this area during summer only. Also, this is the season when the lens of Columbia River plume water (Fig. 6) imparts particularly strong vertical stability to the near-surface water column, which aids in protecting the larval habitat from turbulent mixing caused by wind action.

Peterman and Bradford (1987) assembled time series of wind data and of larval mortality

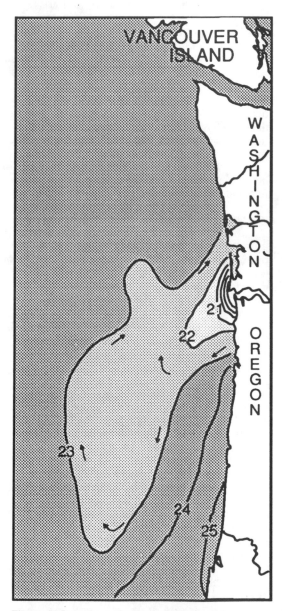

Figure 6. Summer surface-water density (σ,) distribution in the vicinity of the Columbia River Plume. (Redrawn from Landry et al., 1989.) (Small arrows added to indicate direction of inferred superficial geostrophic circulation implied by horizontal density gradients.)

for the Southern California Bight habitat and found that interannual variability in larval anchovy mortality rate was indeed empirically related to frequency of calm periods of sufficient duration for fine-scale strata of food organisms to form.

El Niño

The waters of the Pacific Ocean constitute the greatest mass of heat storage capacity on earth. The Pacific is so large that its atmosphere as a whole is only minimally subject to continental effects, and so is not nearly so strongly forced into seasonal climatic regularity as are the atmospheres of smaller, more continentally influenced oceans. The result is that the coupled ocean-atmosphere system of the Pacific "wanders" from year to year, in response to its own internal dynamics, to a much greater degree than do other ocean regions.

The most dramatic expression of this interyear variability of the Pacific system is the El Niño-Southern Oscillation (ENSO) phenomenon. Because the Pacific is so large, its effect is global. Accordingly, ENSO constitutes the dominant mode of short-term climatic variability throughout the world. Picaut (1985) has provided a useful brief overview, and an entry into the literature, of El Niño dynamics and their effects on eastern ocean regions.

El Niño episodes precipitate by far the strongest interannual fluctuations in conditions affecting marine populations in the California Current region. These deviations from the norm are so radical that they totally dominate any analysis based on standard (i.e., least-squares, etc.) empirical methods. In addition, overshadowing the common El Niño episodes occurring each few years, are infrequent, very extreme El Niño events, such as the ones in 1957–1958 (Sette and Isaacs, 1960) and 1982–1983 (Norton *et al.*, 1985; Glantz *et al.*, 1987). These introduce major transients into the biological community structure that may take many years to dissipate; thus, no one working in the eastern Pacific needs to be convinced that environmental variation regulates marine resource population dynamics.

The El Niño effect is so overwhelming that all variables tend to be intercorrelated, which results in inadequate statistical power to sort out various causative processes and mechanisms. This suggests another potential use of interregional comparative studies. Processes that may be too highly interdependent in one regional system for their effects to be differentiable, may present an informatively different arrangement in another system where large-scale physical constraints (such as land-mass geometry, strength of continental "mon-

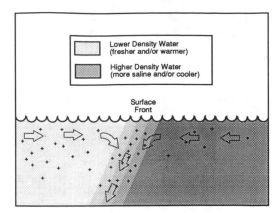

Figure 7. Schematic diagram of a front between waters of differing density. Arrows indicate density-driven flows associated with the front. "Particle" symbols indicate planktonic organisms capable of resisting vertical displacement. (Scales are distorted: vertical scale greatly expanded relative to horizontal; particles greatly magnified, surface waves not to scale, etc.)

soon" effect, degree of Coriolis effect, and degree of influence of the ENSO mode of variability) may differ.

Temperature

Sea temperature is one of the easiest marine environmental parameters to measure. It is routinely reported by ships at sea as part of their weather report transmissions. Temperature has direct physiological effects on organisms and also reflects important ocean processes. For example, upwelling tends to depress sea-surface temperature as it enriches the surface layers with dissolved plant nutrients and enhances plankton productivity.

In the California Current system, the large-scale trend is for temperature to decrease (Fig. 2) and for upper-layer nutrient concentration to increase toward the north (Reid, 1962). Thus, both poleward and with depth, the gradient of temperature is negative, whereas while dissolved nutrient gradients are positive. There is normally also a polewardly increasing gradient of planktonic organisms (Reid, 1962).

In eastern ocean boundary regions, increased coastal upwelling acts to alter the ocean density distribution so as to simultaneously increase equatorward advection (Allen, 1973). Thus, in these systems generally, upwelling and alongshore advection are directly linked such that they vary in the same sense and yield the same qualitative evidence in terms of such parameters as temperature, dissolved nutrients, and plankton productivity. Consequently, local and nonlocal (advective) effects tend to be confounded.

Prager and MacCall (1990) present perhaps the most comprehensive and (allowing for the difficulties alluded to in the previous two paragraphs) credible statistical analysis of environmental influences on California Current fish populations that has been produced to date. One of the clearest results of their analysis, which addresses the period since 1946, is the indication that higher sea temperatures favor sardine recruitment in the Southern California Bight. They point out that interyear temperature fluctuations tend to follow the dominant mode of California Current variability, which is the El Niño mode. This mode includes increased temperature, raised coastal sea level, lowered intensity of upwelling-producing southward wind stress, and increased coastal rainfall, all of which relate in some way to increased poleward advection.

Another recent study (Ware and Thomson, 1991) has assembled wind data for the earlier half of the twentieth century and applied these in an empirical study of the California sardine. This study yields an opposite tendency for increased recruitment to be associated with increased upwelling-producing winds, which they find to be inversely correlated with sea temperature. Thus, they find an unfavorable effect of higher temperature on recruitment, that is, an opposite relationship to that found by Prager and MacCall.

Do these two studies contradict one another? Or can they be reconciled? Certain special aspects of the Benguela Current system bear on these questions.

The Benguela Current System

Crawford et al. (1989) provide an overview of known biological variation and fishery experience in the Benguela Current LME. Sardines (*Sardinops ocellatus*), called pilchards in South Africa, dominated the catches until the late 1960s. The sardine populations collapsed first in the southern Benguela Current and later in the north. Subsequently, the anchovy (*Engraulis capensis*) replaced the sardine in the purse-seine catches in the south. Initially, anchovy catches increased also in the north, but have recently fallen drastically.

The Benguela Current system may possess the strongest sustained locally wind-driven coastal upwelling of any region of the world ocean. Judging from the Ekman transport summaries presented by Parrish et al. (1983), the area near Lüderitz would appear to spew forth, on an annual basis, over two and one-half times the total volume of upwelled water per unit length of coastline as does the Cape Mendocino upwelling core of the California Current system. In addition, there is another upwelling core to the north in the vicinity of Cape Frio (Parrish et al., 1983) that is nearly as intense as that near Lüderitz, and there is a third core to the south near Cape Columbine. (Note in Fig. 3 the presence of packed isotherms indicative of strong coastal upwelling, not only near these "core" features, but essentially along the entire west coast of subtropical southern Africa, although the interaction with the tropical front [Shannon et al., 1987] near Cape Frio complicates the "upwelling signature" in the sea-surface temperature pattern at that location.)

Larval retention zones

In the Benguela system, there is no "Southern California Bight," where coastal morphology can provide a spawning habitat sheltered from the incessant wind-driven turbulent mixing and offshore Ekman transport characteristic of open upwelling coasts. However, the system is so productive and apparently so ideally suited to the life-style of later life stages of small clupeoid fishes that anchovies and sardines do manage to find places to reproduce.

The best available place along the West Coast is apparently the slightly concave stretch of coastline area near Palgrave Point, which lies between the intense upwelling centers of Lüderitz and Cape Frio (Fig. 8). This area supports the major spawning grounds of the northern subpopulations of Benguela Current anchovies and sardines (indicated on Fig. 3). Here, there is a local minimum in the intense wind-driven offshore Ekman transport of the region (Parrish et al., 1983). Also in this location, Nelson and Hutchings (1983) have suggested a weak cyclonic gyre with poleward flow very near the coast. Indeed, this is an area of very strong cyclonic wind stress curl (Bakun and Nelson, 1991), the long-term composite annual mean of which actually has a local curl maximum displaced slightly offshore (i.e., in the ideal position for Ekman pumping to support such a local gyral circulation). Thus, the situation would appear to be as close an analogue to the retention area of the Southern California Bight (compare Figs. 5 and 9) as is available on the Benguela Coast.

The fact that the southward extent of the African continent terminates within the region, allowing free exchange around the southern end of the continent, offers another possibility for escaping the reproductive difficulties of an exposed upwelling coast. The principal spawn-

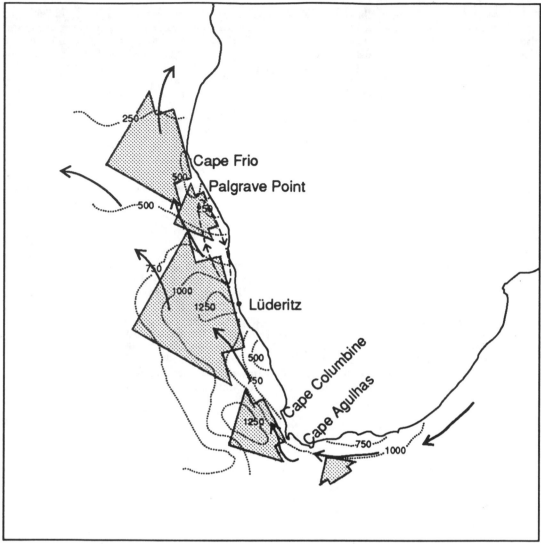

Figure 8. Diagram of characteristic summer flow features and wind-mixing index (m³ • s⁻³, dotted contours) for the Benguela Current region. Broad, shaded arrows indicate surface Ekman transport (magnitude proportional to the linear dimensions of arrow symbols). Solid arrows indicate the general trend of underlying geostrophic current flow (no magnitude implied). Hypothetical gyral circulation south of Palgrave Point is indicated by the closed dashed line. (Ekman transport and wind-mixing index magnitudes are derived from mean January–February distributions presented by Parrish *et al.*, 1983.)

ing habitat of the southern Benguela anchovy and sardine populations is over the Agulhas Bank (the area of very wide continental shelf, indicated in Fig. 3, off the southern extremity of the continent). Here, the wind-driven surface Ekman transport is typically not directed offshore but is actually directed toward the coast (Fig. 8). This serves to retain drifting larvae within the neritic habitat over the Agulhas Bank. In addition, there is evidence of substan-

tial upwelling occurring at the continental shelf break, where the Agulhas Current flow impinges against the shelf edge (W. Peterson, pers. com.).

Various mechanisms are available to cause upwelling to occur, even in the lack of wind-driven offshore surface Ekman flux, where a strong along-shore current interacts with a continental shelf edge (e.g., Hsueh and O'Brien, 1971; Dickson *et al.*, 1980; Paffenhöffer *et al.*,

1984; Mazé et al., 1986; Kinsella et al., 1987). The Ekman transport field would tend to carry the nutrient-laden upwelled water onto the bank habitat to enrich the local trophic system. The geostrophic current pattern over the bank is such that surviving larvae and juveniles are advected westward around the cape peninsula and injected northward into the rich Benguela upwelling system (Shelton and Hutchings, 1982).

Stability

Near the Palgrave Point spawning region, the "wind-cubed" index of wind-mixing intensity falls below 250 $m^3 \cdot s^{-3}$ during the austral summer spawning season (Parrish et al., 1983; again, see their Appendix charts 9 and 10). This value is comparable to the values found for the Southern California Bight and southern Baja California spawning regions of the California Current system. However, in the Agulhas Bank spawning region the values are very much higher, being greater than 600 $m^3 \cdot s^{-3}$ even at its summer minimum. Parrish et al. (1983) attribute the ability of the South African population to successfully reproduce under such energetic wind conditions to a protected stable layer formed at depth by the confluence of warm Indian Ocean surface water from the Agulhas Current with the cooler, more dense Atlantic source water of the Benguela Current (Darbyshire, 1966; see also Shelton and Hutchings, 1990).

The contrast in subsurface stability between the two spawning regions and the upwelling center at Lüderitz is indeed striking (Fig. 9). During all seasons of the year, the water column is rather well mixed in the upwelling center, as it is during the winter in the spawning regions. However, during the summer spawning season, strong stabilization of the water column sets in under the light wind conditions near Palgrave Point. Then, even under the relatively strong summer winds off Cape Agulhas, the advection of Indian Ocean surface water apparently counteracts the substantial local wind-mixing effects (Largier and Swart, 1987) to provide stable conditions and a larval habitat where pronounced subsurface maxima in chlorophyll (Carter et al., 1987) and microplankton (Shelton and Hutchings, 1990) imply accumulation of larval food particles.

In analyzing available time series, Boyd (1979) noted a relationship between increased sea temperature variance and poor anchovy reproduction off Namibia. Such increased vari-

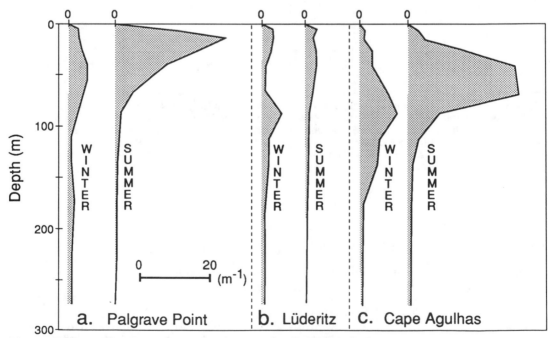

Figure 9. Mean stability [$(\delta\sigma_t)/dz \times 10^3$] versus depth (m=meter) profiles during winter and summer seasons at three locations off southern Africa. (Redrawn from Parrish et al., 1983.) (a) the spawning grounds near Palgrave Point; (b) the upwelling center near Luderitz; (c) the spawning grounds near Cape Agulhas.

ance in temperature might result from increased frequency of mixing events or from rapid alterations of upwelling events and partial restabilizations (i.e., implying lack of continuous periods of stable upper-ocean structure).

Benguela Niño

Interannual warm and cool events of varying intensities do impact the Benguela Current ecosystem (McLain *et al.*, 1985; Walker, 1987). Two such events, occurring in 1963 and in 1984, involved major southward intrusions of warm, highly saline tropical waters into the northern part of the Benguela region. Shannon *et al.* (1986) outlined various similarities of these two events to an El Niño situation, including apparent connections to anomalous situations in the tropical Atlantic. They suggest the term "Benguela Niños" to describe the two events. However, they emphasized that these are the only two events since at least the early 1950s that have this character and that "El Niño–type" episodes therefore seem to be less frequent in the Atlantic than in the Pacific.

Other warm events in the Benguela ecosystem appear to be caused by influxes of warmer Indian Ocean from the poleward end of the system (Fig. 3). Modeling studies have indicated that local wind effects may influence the penetration of Agulhas Current water into the Atlantic (De Ruiter, 1982; De Ruiter and Boudra, 1985).

Temperature

The fact that there are two sources of warm water (Fig. 3), one at each end of the upwelling region, is a very special attribute of the Benguela Current ecosystem (Shelton *et al.*, 1985). Year-class strength of both anchovies and sardines, in both the northern and southern portions of the Benguela system, has been generally positively correlated with temperature (Crawford *et al.*, 1989). However, these authors take pains to state that "a direct link is not suggested" but that the empirical relationships "probably reflect actual changes in the marine environment, such as influxes of the Agulhas Current, of the South Equatorial Current and of the Angola Current" which "may play important roles as regard stratification of spawning areas near the extremities of the Benguela and transport of material from them into the Benguela proper" (*ibid.*, p. 197). For example, the "Benguela Niño" warm events have not been beneficial to epipelagic species (Shannon *et al.*, 1988b).

Thus, anomalous southward advection of less-productive tropical surface water into the system is apparently, as in other systems, nonbeneficial. On the other hand, minor warmings in the north, not associated with one of the infrequent "Niño" episodes, have been associated with good recruitment. These minor warmings may reflect increased stability and less upwelling-related offshore transport in the relatively exposed spawning region between the Cape Frio and Lüderitz upwelling centers. In the case of the southern portion of the Benguela region, warming that reflects increased northward advection of warm Agulhas Current water from the Agulhas Bank spawning grounds would signal increased transport of late larvae and early juveniles to the rich upwelling areas, where they would apparently experience rapid growth and correspondingly enhanced survival. Thus, temperature itself would seem to be an equivocal indicator with respect to environmental regulation of reproductive success, reflecting a mixture of local and nonlocal ecological processes having nonparallel effects.

The Effect of Temperature on the California Sardine

With these results from the Benguela system as background, it is useful to return to the seemingly conflicting results of the two recent empirical studies of factors affecting California sardine recruitment. The study by Prager and MacCall (1990) addresses the period since 1946, during which the population was first in rapid decline, followed by an extended period of collapse and, recently, by the beginning of resurgence (Calif. Dept. of Fish and Game, 1990). Prager and MacCall found higher sea temperatures, less intense upwelling-producing wind stress, and other "El Niño mode" characteristics to be associated with improved recruitment. Ware and Thomson's study (in press), which yields an opposite association of improved recruitment with more intense upwelling-producing wind stress, and correspondingly decreased temperature, encompasses the earlier period of growth and climax of the fishery.

In light of the Benguela experience, it seems that the two studies might be reconciled as follows. During most of the earlier period addressed by Ware and Thomson, the population was very large and was centered off Alta California. It is well known that the El Niño mode corresponds to decidedly lowered plankton concentrations (Bernal, 1981) and other measures of biological productivity (e.g., McGowan, 1985; Pearcy *et al.*, 1985) in the

California Current. The sardine is a plankton-feeding fish that reaches high biomasses in particularly productive regions. Thus, it is intuitively logical that sardines might fare relatively poorly in food-sparse El Niño conditions off California, as they seem to have done in the areas of the Benguela ecosystem affected by the Benguela Niños.

However, the Benguela experience has indicated that warm events may be beneficial to local recruitment when they reflect processes that promote influx into the local area from nonlocal population segments. During the period addressed by Prager and MacCall, the population off Alta California was small, essentially vanishing over some intervals (MacCall, 1986). As we have seen, anomalous poleward advection is a characteristic of California El Niño. Thus, it seems reasonable that the evidence developed by Prager and MacCall of increased recruitment associated with the El Niño reflects an influx (related to the anomalously northward tendencies in flow and related properties) of southern subpopulation fish from their relatively lightly exploited refuge off southern Baja California. Because of the very low resident biomass, even a limited influx of younger fish from the south would show up as a substantial upsurge in local recruitment per unit resident spawning stock.

For example, a relative resurgence of the sardine population off Alta California took place during the intense 1957–1958 El Niño. Various meristic and other types of evidence are indeed indicative of an influx from the southern subpopulation (e.g., Wisner, 1961). In the recent period since about 1976, the eastern Pacific Ocean appears to have fluctuated around a condition of somewhat elevated warm El Niño–type characteristics (e.g., Cole and McLain, 1989).

During this period, the sardine population off Alta California has finally begun to rebound from its long collapse. During the same period, the western North Pacific has been in a generally cool mode (Saiki, 1988), and the Japanese sardine population has undergone an explosive growth many times more rapid than the recent rebound off California. In fact, reports of commercial quantities of young-of-the-year Japanese sardines being found swept one-third of the way across the North Pacific in the rapid flow of the Oyashio-Kuroshio extension (Kenya, 1982) make it conceivable that, rather than witnessing the rebound of the California sardine, we may actually be seeing the arrival of the Japanese sardine.

In any case, the Benguela experience suggests that both the study of Prager and MacCall (1990) and the study of Ware and Thomson (1991) can be correct. That is, rather than being contradictory, their results may reflect very different processes, albeit producing similar local temperature signatures, acting to generate opposite relationships of temperature to local recruitment success during the two periods of different sardine abundance and distribution.

Southern Brazil to Patagonia

The anchovy (*Engraulis anchoita*) of the southwestern Atlantic inhabits an extended stretch of coastal habitat from about Cabo Tres Puntas in south-central Argentina (Ciechomski and Sánchez, 1986) north to Cabo Frio in southern Brazil (Matsuura *et al.*, 1985). The Brazilian sardine (*Sardinella brasiliensis*) is the major sardine-type population of the region. *Sardinella* is a warmer water genus than *Sardinops*, the genus to which the sardines inhabiting the California and Benguela systems belong. The Brazilian sardine population remains largely confined to the coastal bight at the northern end of the region (Saccardo, 1983). This bight, between Cabo Frio and Cabo de Santa Marta Grande in southern Brazil, is identified in this chapter as the Southeastern Brazilian Bight (after Matsuura, 1989).

In contrast to the narrow shelves of the other two regions being discussed, the continental shelf along the western boundary of the South Atlantic is one of the most extensive in the world (Fig. 4). In addition, this coastline is under the influence of narrow, intense western boundary currents, rather than the generally mild eastern ocean boundary flows of the other regions. The southward-flowing Brazil Current dominates the northern part of the region, while the northward-flowing Falkland (Malvinas) Current dominates the southern part. (Note in Fig. 4 the tongue-like isotherm contortions marking these two flows, giving an impression of a "collision" between the two currents off Rio de la Plata.) These oceanic western boundary flows are apparently confined to the shelf break and seaward. The flow over the shelf itself appears to be largely decoupled from the oceanic currents (Olson *et al.*, 1988; Podestá, 1990).

Another contrast is that whereas the California and Benguela systems are characterized by persistent offshore-directed surface wind drift (Ekman transport), most of this region is generally under the influence of onshore wind transport (Bakun and Parrish, 1991). An exception is at the extreme north of the region, where a seasonal, locally wind-driven upwelling regime exists (Bakun and Parrish, 1990). (Note in Fig. 4 the closed isotherm surrounding the upwelling core near Cabo Frio.)

Over this extended neritic habitat, processes of enrichment, stability, and retention combine in several different environmental configurations to yield favorable reproductive habitat for small, pelagic-spawning clupeoids (Bakun and Parrish, 1991). It will be convenient here to segment the discussion in areal sequence.

The Southeastern Brazilian Bight

Bakun and Parrish (1990) find the spawning strategy of the Brazilian sardine to be a nearly exact analogue to that of the population of California sardine spawning within the Southern California Bight. Spawning activity of the Brazilian sardine reaches a peak in austral summer and is essentially confined within the Southeastern Brazilian Bight. Matsuura (1986) describes a massive coastward penetration of cool South Atlantic Central Water (SACW) over the continental shelf floor, occurring during the summer season but absent in winter. The contrast with the overlying warm, tropical surface water results in strong vertical stratification. The SACW water mass also provides a nutrient-rich source for summer seasonal upwelling in the coastal region.

Thus, summer spawning places larvae into the enriched environment downstream of the upwelling center, during a period of continued vigorous upwelling just following the seasonal maximum. Moreover, this is a season when wind-mixing index values of less than 250 $m^3 \cdot s^{-3}$ characterize the entire bight interior (Fig. 10). In addition, the offshore Ekman transport, which is substantial at the upwelling center, drops sharply to a local minimum within the bight. Finally, Bakun and Parrish (1990) infer a closed gyral geostrophic circulation pattern (Fig. 11) that may aid in retaining larvae within the favorable coastal habitat (schematic diagram shown in Fig. 12). In conjunction with the general cyclonic tendency involving the intense Brazil Current flow, cyclonic vortex activity may transfer substantial amounts of rich SACW upward into the surface layers (Matsuura, 1990).

The anchovy also spawns within the bight, but being a temperate rather than a tropical form, the anchovy is spatially segregated by temperature from the Brazilian sardine. The anchovy occur at depth or in upwelling plumes, whereas the sardine are found in warmer surface layers (Matsuura *et al.*, 1985). Anchovy spawning in Brazilian waters occurs from late winter to early summer (Castello, 1989). During winter, the wind-mixing index over most of the bight tends to exceed the 250 $m^3 \cdot s^{-3}$ limit that characterizes most anchovy reproductive habitats, an exception being the Agulhas Bank spawning grounds of the southern Benguela

subpopulations. At this season, spawning in the Southeastern Brazilian Bight is scattered at depth over the outer shelf, where substantial stratification persists (Castro Filho *et al.*, 1987), as in the Agulhas Bank case.

Very recently, the heavily exploited Brazilian sardine population has undergone precipitous decline (Saccardo, 1989) and presently is considered to be in a state of collapse (Anonymous, 1989b). A major recruitment failure of the 1986–1987 year class appears to have contributed to the collapse (Matsuura, pers. com.). The 1986–1987 summer spawning season was particularly notable for an abnormal lack of penetration of temperate, nutrient-laden SACW into the near-coastal habitat of the Southeastern Brazilian Bight. Thus, the area was dominated by relatively infertile tropical surface waters, largely unmitigated by upwelling and mixing of richer waters of the SACW mass. Matsuura attributes the recruitment failure primarily to resulting diminished primary productivity.

The following 1987–1988 spawning season was marked by an even stronger influence of very warm tropical waters near the coast, reflected in an extremely warm sea-surface temperature anomaly. However, in this case it was accompanied by a strong intrusion of cool SACW waters at depth (Rossi-Wongtschowski, 1989a). Accordingly, substantial primary production is inferred (Matsuura, pers. com.). However, evidence of significant successful reproduction in the normal spawning grounds was not found (Rossi-Wongtschowski, 1989a). The seriously depleted state of the adult stock, following the recruitment failure of the previous year class, may in itself explain the apparent lack of spawning success (Matsuura, pers. com.), although it is also possible that either spawning activity or larval survival may have been adversely affected by the anomalous environmental conditions (Rossi-Wongtschowski, 1989a).

By October 1988, the sea-surface temperatures had returned to more normal values, but very few sardines were found in the entire region off southeastern Brazil (Rossi-Wongtschowski, 1989b). Conversely, large quantities of anchovies were noted within the bight (Castello *et al.*, 1991). As indicated earlier, the Brazilian sardine is a tropical *Sardinella* species, whereas the anchovy is a member of the more temperate genus *Engraulis*.

Sea-surface temperature is the most commonly used indicator of temperature effects on marine fish populations. To find the cooler water species doing well and the warmer water species in serious trouble following a period of warm environmental anomalies is not the outcome one would expect if the anomalous sea-surface temperature conditions validly signified

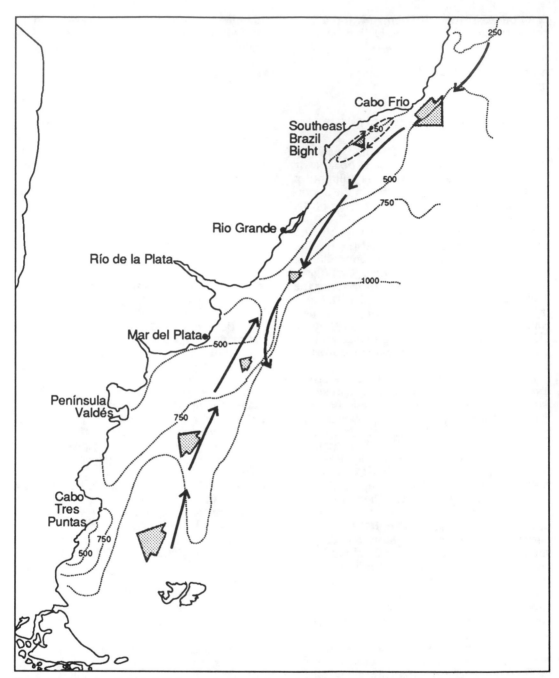

Figure 10. Diagram of characteristic summer flow features and wind-mixing index (m³ · s⁻³, dotted contours) for the region off southeastern South America. Broad, shaded arrows indicate onshore-offshore component of surface Ekman transport (magnitude proportional to the linear dimensions of arrow symbols). Solid arrows indicate the general trend of geostrophic boundary current flow (no magnitude implied). Gyral circulation within the Southeastern Brazilian Bight is indicated by the closed dashed line. (Ekman transport and wind-mixing index magnitudes are derived from mean December–January distributions presented by Bakun and Parrish, 1990, and by Bakun and Parrish, 1991.)

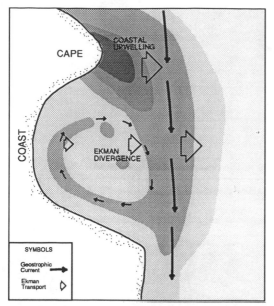

Figure 11. Schematic diagram of spawning habitat within a coastal bight downstream of an upwelling center. Darker shading indicates higher-density (upwelled) surface waters. Upwelled water advects southward and diffuses into the sheltered coastal bight, where wind-induced offshore Ekman transport and turbulent mixing drop sharply. Within the bight, the shoreward gradient in wind stress produces divergent Ekman transport, corresponding upward Ekman pumping (leading to additional local enrichment), and closed gyral circulation (which may be broken into multiple cells) in the bight interior.

ticle concentration, and larval retention. Alternatively, temperature may signal atypical large-scale conditions causing anomalous influxes from nonlocal population segments.

Uruguay and the extreme south of Brazil

Over most of the regions to the south of the Southeastern Brazilian Bight, the Ekman transport tends to be directed onshore (Bakun and Parrish, 1991), rather than offshore. (The Ekman transport symbols shown in Fig. 10 are for midsummer, the time of year when the onshore Ekman transport is least pronounced; at other seasons, the onshore Ekman arrows would be larger.) This onshore surface transport leads to coastal convergence and downwelling, rather than to coastal upwelling. However, as in the Agulhas Bank case, there is

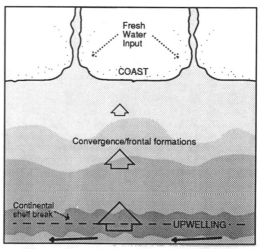

Figure 12. Schematic diagram of coastal pelagic habitat configuration off southern Brazil, Uruguay, and extreme northern coast of Argentina. Darker shading indicates higher-density (upwelled) surface waters. Broad, open arrow symbols indicate surface Ekman transport; symbol sizes connote relative magnitudes. Narrow, heavy arrow symbols at the bottom of the figure indicate the oceanic boundary current flow, which lies seaward of the continental shelf break. Onshore Ekman transport carries water upwelled at the continental shelf break shoreward and helps to retain larvae over the shelf. The wind intensity lessens toward the coast, resulting in Ekman convergence and frontal formations over the shelf between waters of differing densities. Horizontal density differences and vertical stability near the coast are favored by contrasts between upwelled waters and freshwater outflows and by the contrasting water mass characteristics traceable to confluence of strong western boundary flows off the shelf.

direct temperature effects on the two populations. In fact, of course, the situation is quite complex. While the sardine inhabits the surface layer and has its population center and primary reproductive grounds confined to the local Southeastern Brazilian bight area, the anchovy is found in cooler thermocline or subthermocline waters in the bight and also has even larger nonlocal population components and reproductive grounds extending along the coast to the south. In addition, both species may to some degree compensate for direct physiological effects of anomalous temperature conditions by adjusting the depth range occupied.

In any case, this example adds to the impression that a correlation of fish population responses to ocean temperature variation (particularly measured as sea-surface temperature) may not necessarily, or even generally, implicate temperature itself as a direct causative agent. Rather, temperature may act merely as a surrogate variable for dynamic processes affecting trophic enrichment, food par-

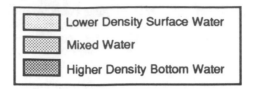

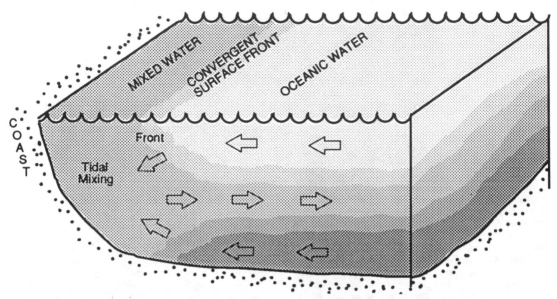

Figure 13. Schematic diagram of a shelf-sea front configuration over the Patagonian shelf (particularly near Península Valdés). Strong tidal mixing results in homogenized water column near the coast. The result is surface water of greater density, and bottom water of lesser density, in the mixed zone relative to the stratified zone offshore. Thus, the lighter offshore surface waters tend to overide the heavier mixed waters, resulting in onshore flow toward a convergent surface front. The heavier offshore bottom water likewise flows under the mixed waters, resulting in onshore flow at the bottom. A balancing offshore flow occurs within the mid-depth thermocline.

abundant evidence (e.g., Servicio de Hidrográfia Naval, 1969; Podestá, 1990) of major upwelling occurring at the shelf break. The onshore-directed Ekman transport in the surface layer would tend to carry the products of the resulting trophic enrichment into the neritic habitat overlying the shelf, where the continent itself serves as a shield from the wind-mixing effects of the large-scale westerly winds.

In this area, there are major freshwater outflows, as well as interleaving of water masses associated with the confluence of the major western boundary flows immediately to seaward, further promoting internal water-column stability. Associated surface fronts (Elgue, 1989) represent convergence zones that serve to concentrate food particle distributions. Here then, in a quite different configuration (schematic diagram shown in Fig. 12) from the more common one represented by the Southeastern Brazilian Bight to the north, the various processes evidently combine to yield similarly favorable conditions for anchovy reproduction (Bakun and Parrish, 1991).

The Patagonian Shelf

Toward the southern end of the reproductive habitat of the anchovy, notably in the vicinity of Península Valdés, strong tidal mixing occurs over the continental shelf. In some shallower locations this mixing acts to homogenize the water column from surface to bottom. The result is that surface water in the mixed areas is of greater density than the surface water of the adjacent stratified areas. Similarly, the bottom water is of lesser density than that of the stratified areas. Thus, the less-dense surface water of the stratified offshore zone tends to override the denser mixed water, leading to onshore flow in the surface layer toward a convergent surface front (Fig. 13). The heavier bottom water of the stratified zone tends to wedge under the lighter mixed water, leading to onshore flow also in the bottom layer. The waters being transported coastward at the surface and bottom layers eventually become entrained in the tidal mixing and become incorporated in the mixed zone. To balance these gravity-driven onshore flows at the surface and bottom, the mixed waters find their equivalent density level within the stratified water column and move offshore in the mid-depth thermocline (Fig. 13).

In this case, the upward mixing of deeper waters in the mixed water column injects dissolved plant nutrients into the illuminated surface layers. Resulting phytoplankton growth may be further concentrated in the convergent frontal zones. Also, the rather stable situation within the thermocline, shielded from surface mixing processes by the lighter waters above and from bottom-generated turbulence by the heavier layer below, may favor patch formation, swarming, and other processes, leading to concentration of small organisms in the trophically enriched waters moving offshore from the mixed zone. Retention of larvae near the frontal zone could be ensured by appropriate vertical migratory behavior to utilize the onshore flow in the surface or bottom layers. Furthermore, on the larger scale, the general onshore wind-driven surface Ekman transport would serve to prevent major loss of larvae from the shelf habitat. (See Bakun and Parrish [1991] for further discussion and documentation.) Association of anchovy larvae with the frontal structures has been demonstrated on the recent SARP cruise of the *Meteor*.

Discussion

This brief examination of three regional LMEs has pointed out a pattern wherein a combination of three general factors is associated with a favorable reproductive habitat for anchovies and sardines. These factors include the following: (i) enrichment of the food web by physical processes (upwelling, mixing, etc.); (ii) opportunity for concentrated patch structure of food particles to accumulate (stability, lack of active turbulent mixing, and/or strong convergence in frontal structures); and (iii) availability of mechanisms promoting retention of larvae within (or transport of larvae to) appropriate habitat.

The processes involved in these three factors are not generally mutually compatible. For example, upwelling is a response to divergent horizontal flow. Upwelling and convergence therefore cannot directly coincide. Likewise, mixing may lead to enrichment, but destroy vital small-scale structure in food particle distributions. Coastal upwelling, being induced by offshore surface transport, is directly linked to loss of larvae from the coastal habitat. The wind that drives the upwelling also mixes the upper ocean layer, destroying its structural stability.

Thus, the three factors can combine favorably only in special environmental configurations. Each factor must be present to a sufficient degree, but not to such an overwhelming degree as to preclude the others. Alternatively, the favorable configurations may involve temporal and spatial lags that allow effects of opposing processes to be mutually supportive. Certainly, a temporal lag is implied between nutrient enrichment and resulting growth of appropriate larval food particles. Also, a temporal lag in a flowing medium implies a spatial lag. If the spatial lag determined by the physical flow configuration is such that food organisms are transported from the divergent, dispersive enrichment zone to a location where convergence and stability promote concentrated food particle structure, then the enrichment and concentration processes no longer oppose each other in their effects on larval feeding conditions. If the transport by the flow field is into a zone where larval retention is favored, then the configuration would seem to be ideal.

Thus, "core" zones of maximum upwelling—such as Cape Mendocino in California, Lüderitz and Cape Frio in Namibia, and Cabo Frio in southeastern Brazil—where enrichment processes are intense but the other two factors are lacking, are avoided in the spawning habits of clupeoid fishes. The most common favorable configuration for eastern ocean coastal regions (Parrish *et al.*, 1983) appears to be that of a coastal bight downstream of one of these upwelling centers (Fig. 11). However, other configurations can apparently provide a similar combination of beneficial effects (Figs. 12 and 13).

Various univariate correlative relationships between recruitment and one or another of these primary factors have been published (see Table 1). It is not surprising that a specific environmental configuration and a specific biological life-style could combine to make variability in a particular single factor stand out from the others in its correspondence to year-class survival; but it is also not surprising that such relationships, which only reflect part of the causal "equation," often break down unexpectedly. For example, an increasing trend in wind speed, by increasing upwelling, might favor good recruitment up to a point where the concentration and the retention factors are unduly disrupted. Continued increases in wind may then become detrimental.

It should be said that, although this chapter draws its examples primarily from three regional LME complexes, the generalizations that emerge are also reflective of findings from other regions. For example, Roy *et al.* (1989) explain the gross seasonal and geographical aspects of the spawning strategies of African clupeoid stocks north of the equator in terms of two limiting factors: planktonic production (related to enrichment processes) and environmental stability (which they associate with both larval retention and lack of turbulent mixing). In fact, Roy (1990) points out that the observed spawning habits of that group of stocks fall into a pattern wherein the spatial locations of spawning grounds appear to particularly optimize larval retention, whereas the seasonal aspects of spawning appear to be tuned in a way that particularly limits exposure to turbulent mixing of the water column. The value of seasonal mean wind that he finds limiting is about 5–6 m · s^{-1}, which, if cubed, falls within the 250 m^3 · s^{-3} contour value (discussed earlier with respect to Figs. 5, 9, and 11).

Cury and Roy (1989) have analyzed recruitment time-series for various clupeoid stocks of West Africa, Peru, and California; their time-series analysis, utilizing new, nonlinear empirical techniques, indicates a consistent "optimal environmental window" (i.e., an intermediate band of wind speeds, neither so high as to cause undue turbulent mixing and offshore transport nor so low as to yield insufficient trophic enrichment via upwelling and mixing) wherein reproductive success is favored. This window is centered near the 5–6 m · s^{-1} mean seasonal wind value (i.e., < 250 m^3 · s^{-3}) that consistently shows up in these studies.

Increased upwelling and mixing are generally reflected in lower sea-surface temperatures, as is increased offshore transport because of its linkage to upwelling. Increased stability, or onshore transport of oceanic surface waters in a coastal upwelling region, tends to be reflected in warmer sea-surface temperature. Thus, temperature is often found to be correlated with year-class success. However, as has been argued in this chapter, the relationship may not be direct, but only a reflection of some dynamic process. Certainly, a pelagic fish would generally be able to compensate for any direct detrimental effects of anomalous temperatures merely by adjusting its depth level. For example, the anchovy inhabits the surface layer off Patagonia, which is toward the poleward end of its range. However, in the Southeastern Brazilian Bight, which is toward the tropical end of its range, it is found within or beneath a shallow thermocline and at the surface only within cool plumes of recently upwelled water (Matsuura *et al.*, 1985). Thus, temperature alone would seem to be an uncertain and somewhat precarious indicator of environmental effects on year-class success.

Tropical outbreaks (e.g., California El Niño, Benguela Niño) and other instances of anomalous dominance of tropical waters and conditions (e.g., lack of normal penetration of productive temperate water masses into the coastal habitat off southeastern Brazil) appear to be generally detrimental to local reproductive success. However, rather than being attributable to direct effects of the associated warm temperatures, the detrimental effects are most often ascribed to suppression of the enrichment factor during these episodes.

Potential effects of global climate change

Recent decades have seen a substantial buildup of CO_2, methane, chlorofluorocarbons, and other greenhouse gases in the earth's atmosphere. The scientific consensus is that substantial global warming is likely (Schneider, 1990). Under these circumstances, one of the major responsibilities of marine scientists must be to provide credible advance insight that can ease adaptation to the resource and ecosystem dislocations to come. If temperature is indeed an equivocal predictor of environmental effects on the population dynamics of important ecosystem components, it will not suffice to merely adjust incrementally upward presently established empirical relationships between temperature and the population responses and distributions of marine organisms. The only situation in which such a procedure would seem to work reliably would be where a species may be at the edge of a range limit that is directly determined by physiological temperature tolerances and where the life-style of the species is such that it is tied to the upper mixed ocean layer or shallow continental shelf regions and thereby prevented from compensating by adjusting its depth level.

One of the global greenhouse effects that we can be most certain about is an increase in temperature contrast between the heated continental land masses and the oceans during the warmer seasons of the year (Anonymous, 1990). This would lead to intensified sea breeze circulations and alongshore coastal winds (Bakun, 1990). These increases in alongshore winds would be in the equatorward direction along eastern ocean boundaries and in the poleward direction along western ocean boundaries (i.e., the directions that would in each case intensify offshore surface transport and coastal up-welling). Increased input of turbulent mixing energy to the upper ocean would also result.

Thus, greenhouse-related intensifications of coastal winds would appear capable of shifting the balance among enrichment, concentration, and larval retention processes in coastal ocean pelagic habitats. The atmosphere, having little heat capacity, will equilibrate very rapidly to greenhouse effects. The ocean, with its very large heat capacity, will respond much more slowly. Consequently, global greenhouse effects might totally reorganize marine reproductive habitat geography long before local ocean temperature increases caused by direct greenhouse heating are evident, particularly because associated intensifications of wind-induced upwelling and vertical mixing would act to cool, rather than warm, the ocean surface layers.

It has been indicated that one major difference between sardine and anchovy life-styles is that the sardine seems to be more adapted for migration. Sardines seem to predominate where reproductive habitats are geographically separated from adult feeding habitats. This seems to be the case in both the California Current and Benguela Current LMEs, where there are limited reproductive grounds and large stretches of highly productive upwelling coasts, which seem to represent inhospitable larval habitat. Both of these LMEs have been dominated by sardines prior to the establishment of fisheries.

In contrast, off eastern South America, the larval anchovy habitat is spread in several different configurations over essentially the entire range of temperature tolerance for *Engraulis* (Bakun and Parrish, 1991). Here, sardines are absent (i.e., anchovies are dominant) over the entire region, except at the northern end where the adult feeding habitat in the upwelling region near Cabo Frio is geographically separated from favorable reproductive habitat within the Southeastern Brazilian Bight. At this northern end of the region, the Brazilian sardine coexists with the anchovy. In the Peruvian system, special features (Bakun *et al.*, 1991) afford extensive favorable larval habitat directly coincident with an extremely rich upwelling system; in the paleosedimentary record of the Peruvian ecosystem (DeVries and Pearcy, 1982), anchovies have apparently usually dominated over sardines. In view of these considerations, one might speculate that if changes in reproductive habitat geography are such as to increase separation between reproductive and adult feeding habitats, then shifts toward sardine dominance might be expected. On the other hand, if such separation is decreased or

if other special opportunities present themselves, one might speculate that the evident adaptability and flexibility of the anchovy life-style could predominate.

Eastern sides of oceans are characterized by much dryer atmospheres than western sides. Because the most important atmospheric greenhouse gas is naturally occurring water vapor, the global accumulation of anthropogenic greenhouse gases will have a larger relative effect in eastern ocean regions than in other parts of the oceans. Indeed, a substantial multidecadal increasing trend in upwelling-producing alongshore wind stress already appears to be occurring in the California, Peruvian, Iberian, and Canary LMEs (Bakun, 1990) and also, although the data are less numerous, in the Chilean and Benguelan LMEs (Bakun, unpublished data).

Therefore, it is likely that biological events observed in our coastal marine ecosystems are already being affected to some degree by global greenhouse effects. Unfortunately, in any particular region, evidence of subtle trends tends to be submerged in large-amplitude natural interyear and interdecadal local variability, and conclusive analysis of individual local events is very difficult. However, greenhouse gases accumulate on a global scale, and it makes sense to look for their effects globally. In this, the LME concept may provide a particularly useful framework. By analyzing events in similarly functioning regional LMEs distributed over the world ocean, global patterns of events related to significant global trends must ultimately appear from within the "noise level" of naturally occurring regional climatic variability. When such patterns are identified and credibly linked to progressive trends, they can constitute a basis for real prognostic capability with which to guide adaptive actions by governments, industries, and individuals.

Acknowledgments

The generous help of Suzana A. Saccardo (IBAMA, Brazil) and of Carmin L.D.B. Rossi-Wongtschowski and Yasunobu Matsuura (University of São Paulo) has contributed greatly to the preparation of this chapter, as have conversations with Jurgen Alheit, Ramiro Sánchez, Guillermo Podestá, and others involved in the international SARP project. Information and comments from National Oceanic and Atmospheric Administration (NOAA) colleagues William T. Peterson, Richard H. Parrish, and Alec D. MacCall have likewise been extremely helpful. C. M. Antoinette Nichols did her usual excellent work in producing the illustrations.

References

Allen, J. S. 1973. Upwelling and coastal jets in a continuously stratified ocean. J. Phys. Oceanogr. 3:245–257.

Anonymous. 1983. Workshop on the IREP component of the IOC programme on ocean science in relation to living resources (OSLR). IOC Workshop Rep. 65. Intergovernmental Oceanographic Commission, UNESCO, Paris.

Anonymous. 1989a. Second IOC workshop on sardine/ anchovy recruitment project (SARP) in the southwest Atlantic. IOC Workshop Rep. 33. Intergovernmental Oceanographic Commission, UNESCO, Paris.

Anonymous. 1989b. Pesquisadores e representatives do setor produtivo discutem situação atual da pesca da sardinha e availiação do estado do estoque. In: CIRM/Inf. 4(2), Jul./Set., 1989. pp. 2–3. Comissão Interministerial para os Recursos do Mar Brasilia.

Anonymous. 1990. The Intergovernmental Panel on Climate Change. U. S. Dept. Commerce. Washington, D.C.

Bailey, K. M. 1981. Larval transport and recruitment of Pacific hake *Merluccius productus*. Mar. Ecol. Prog. Ser. 6:1–9.

Bakun, A. 1986. Definition of environmental variability affecting biological processes in large marine ecosystems. In: Variability and management of large marine ecosystems. pp. 89–108. Ed. by K. Sherman and L. M. Alexander. AAAS Selected Symposium 99. Westview Press, Inc., Boulder, CO.

Bakun, A. 1990. Global climate change and intensification of coastal ocean upwelling. Science 247:198–201.

Bakun, A., and Nelson, C. S. 1977. Climatology of upwelling-related processes off Baja California. CalCOFI Rep. 19:107–127.

Bakun, A., and Nelson, C.S. 1991. Wind stress curl in subtropical eastern boundary current regions. J. Phys. Oceanogr. 21:1815–1834.

Bakun, A., and Parrish, R. H. 1980. Environmental inputs to fishery population models for eastern boundary current regions. In: Workshop on the effects of environmental variation on the survival of larval pelagic fishes. Lima, Peru, 20 April–5 May, 1980. pp. 67–104. Ed. by G. D. Sharp. IOC Workshop Rep. 28. UNESCO.

Bakun, A., and Parrish, R. H. 1990. Comparative studies of coastal pelagic fish reproductive habitats: The Brazilian sardine (*Sardinella aurita*). J. Cons. Int. Explor. Mer 46:269–283.

Bakun, A., and Parrish, R. H. 1991. Comparative studies of coastal pelagic fish reproductive habitats: The anchovy (*Engraulis anchoita*) of the southwestern Atlantic. ICES J. Mar. Sci. 48:343–361.

Bakun, A., Roy, C., and Cury, P. 1991. The comparative approach: Latitude-dependence and effects of wind forcing on reproductive success. Document C.M. 1991/H:45. Theme session on the sardine/anchovy recruitment project (SARP).

79th ICES Statutory Meeting, La Rochelle, France, 26 September–2 October, 1991.

Bernal, P. A. 1981. A review of the low-frequency response of the pelagic system in the California Current. CalCOFI Rep. 12:49–62.

Boyd, A. J. 1979. A relationship between sea-surface temperature variability and anchovy *Engraulis capensis* recruitment off southwest Africa. Fish. Bull. S. Afr. 12:80–84.

CalCOFI Committee. 1990. Fortieth Anniversary Symposium of the CalCOFI Conference. CalCOFI Rep. 31:25–59.

Calif. Dept. of Fish and Game. 1990. Review of some California fisheries for 1989. CalCOFI Rep. 31:9–21.

Carter, R. A., McMurray, H. F., and Largier, J. L. 1987. Thermocline characteristics and phytoplankton dynamics in Agulhas Bank waters. In: Benguela comparative ecosystems. pp. 327–336. Ed. by A. I. L. Payne, J. A. Gulland, and K. H. Brink. Special volume of the South African Journal of Marine Science, Vol. 5. Sea Fisheries Research Institute, Capetown, South Africa.

Castello, J. P. 1989. Synopsis on the reproductive biology and early life history of *Engraulis anchoita*, and related environmental conditions in Brazilian waters. In: Second IOC workshop on sardine/anchovy recruitment project (SARP) in the Southwest Atlantic. IOC Workshop Rep. 65. Annex VII. Intergovernmental Oceanographic Commission, UNESCO, Paris.

Castello, J. P., Habiaga, R. P., Amaral, J. C., and Lima, I. D., Jr. 1991. Prospecção e avaliação de biomassa, por métodos hidroacústicos. In: Distribução e abundância da sardinha e da anchoita na região sudeste do Brasil: Outubro–Novembro de 1988. C. L. D. B. Rossi-Wongtschowski, and S. A. Saccardo (coordenação) Relat. Cruzeiro, N/Oc. Atlântico Sul Inst. Oceanogr. Univ. São Paulo.

Castro Filho, B. M. de, Miranda, L. B. de, and Miyao, S. Y. 1987. Condições hidrográficas na plataforma continental ao largo de Ubatuba: variações sazonais e em média escala. Bolm Inst. Oceanogr., São Paulo, 35(2):135–151.

Ciechomski, J. D. de, and Sánchez, R. P. 1986. Problemática del estudio de huevos y larvas de anchoíta (*Engraulis anchoita*) en relación la evolución de sus efectivos pesqueros. Reseña de veinte años de investigación. Publ. Com. Tec. Mix. Fr. Marl. 1(1):93–109.

Cole, D. A., and McLain, D. R. 1989. Interannual variability of temperature in the upper layer of the North Pacific eastern boundary region. NOAA Tech. Memo. NMFS-SWFC-125. NOAA, Washington, D.C.

Collins, R. A., and MacCall, A. D. 1977. California Pacific bonito resource, its status and management. Calif. Dept. Fish Game Mar. Res. Tech. Rep. 35.

Crawford, R. J. M., Shannon, L. V., and Shelton, P. A. 1989. Characteristics and management of the Benguela as a large marine ecosystem. In: Biomass yields and geography of large marine ecosystems. pp. 67–104. Ed. by K. Sherman and L. M. Alexander. AAAS Selected Symposium 111. Westview Press, Inc., Boulder, CO.

Cury, P., and Roy, C. 1989. Optimal environmental window and pelagic fish recruitment success in upwelling areas. Can. J. Fish. Aquat. Sci. 46:670–680.

Darbyshire, M. 1966. The surface waters near the coasts of southern Africa. Deep-Sea Res. 13:57–81.

De Ruiter, W. 1982. Asymptotic analysis of the Agulhas and Brazil Current systems. J. Phys. Oceanogr. 12:361–373.

De Ruiter, W. P. M., and Boudra, D. B. 1985. The wind-driven circulation in the South Atlantic—Indian Ocean—I. Numerical experiments in a one-layer model. Deep-Sea Res. 32:557–574.

DeVries, T. J., and Pearcy, W. G. 1982. Fish debris in sediments of the upwelling zone off central Peru: A late Quarternary record. Deep-Sea Res. 28:87–109.

Dickson, R. R., Gurbutt, P. A., and Pillai, V. N. 1980. Satellite evidence of enhanced upwelling along the European continental slope. J. Phys. Oceanogr. 10:813–819.

Elgue, J. C. 1989. The physical environment. In: Synopsis on the reproductive biology and early life history of *Engraulis anchoita*, and related environmental conditions in Uruguayan waters. In: Second IOC workshop on sardine/anchovy recruitment project (SARP) in the Southwest Atlantic. IOC Workshop Rep. 65. Annex VI: 1. Intergovernmental Oceanographic Commission, UNESCO, Paris.

Elsberry, R. L., and Garwood, R. W., Jr. 1978. Sea-surface temperature anomaly generation in relation to atmospheric storms. Bull. Am. Meteorol. Soc. 59:786–789.

Glantz, M., Katz, R., and Krenz, M. (Editors). 1987. The societal impacts associated with the 1980–83 worldwide climatic anomalies. United Nations Environmental Program. National Center for Atmospheric Research, Boulder, CO.

Gunsolus, R. T. 1978. The status of Oregon coho and recommendations for managing the production, harvest, and escapement of hatchery-reared stocks. Unpublished report. Oregon Department of Fish and Wildlife, Columbia Region.

Hammann, M. G., Baumgartner, T. R., and Badan-Dangon, A. 1988. Coupling of the Pacific sardine (*Sardinops sagax caeruleus*) life cycle with the Gulf of California pelagic environment. CalCOFI Rep. 24:102–109.

Hammann, M. G., Rosales Casian, J. A., and Segovia Quintero, M. 1990. Distribution and abundance of northern anchovy and Pacific sardine larvae collected with the Kidd-Methot net in the Gulf of California during winter 1990. Poster presented at the Annual CalCOFI Conference, Pacific Grove, California.

Hsueh, Y., and O'Brien, J. J. 1971. Steady coastal upwelling induced by an along-shore current. J. Phys. Oceanogr. 1:180–186.

Husby, D. M., and Nelson, C. S. 1982. Turbulence and vertical stability in the California Current. CalCOFI Rep. 23:113–129.

Kenya, V. S. 1982. New data on the migrations and distribution of Pacific sardines in the Northwest Pacific. Sov. J. Mar. Biol. 8(1):41–48.

Kinsella, E. D., Hay, A. E., and Denner, W. W. 1987. Wind and topographic effects on the Labrador Current at Carson Canyon. J. Geophys. Res. 92:10853–10869.

Landry, M. R., Postel, J. R., Peterson, W. K., and Newman, J. 1989. Broad-scale distributional patterns of hydrographic variables on the Washington/Oregon shelf. In: Coastal oceanography of Washington and Oregon. pp. 1–40. Ed. by M. R. Landry and B. M. Hickey. Elsevier, Amsterdam.

Largier, J. L., and Swart, V. P. 1987. East-west variation in thermocline breakdown on the Agulhas Bank. In: Benguela comparative ecosystems. pp. 263–272. Ed. by A. I. L. Payne, J. A. Gulland, and K. H. Brink. Special volume of the South African Journal of Marine Science, Vol. 5 Sea Fisheries Reserch Institute, Capetown, South Africa.

Lasker, R. 1975. Field criteria for survival of anchovy larvae: The relation between inshore chlorophyll maximum layers and successful first feeding. Fish. Bull. U.S. 73:453–462.

Lasker, R. 1978. The relation between oceanographic conditions and larval anchovy food in the California Current: Identification of the factors leading to recruitment failure. Rapp. P.-v. Reun. Cons. int. Explor. Mer 173:212–230.

MacCall, A. D. 1986. Changes in the biomass of the California Current ecosystem. In: Variability and management of large marine ecosystems. pp. 33–54. Ed. by K. Sherman and L. M. Alexander. AAAS Selected Symposium 99. Westview Press, Inc., Boulder, CO.

Matsuura, Y. 1986. Contribuição ao estudo sobre estrutura oceanográfica da regiáo sudeste entre Cabo Frio (RJ) e Cabo de Santa Marta Grande (SC). Ciencia e Cultura, s. Paulo 38(8):1439–1450.

Matsuura, Y. 1989. Synopsis on the reproductive biology and early life history of the Brazilian sardine, *Sardinella brasiliensis*, and related environmental conditions. In: Second IOC workshop on sardine/anchovy recruitment project (SARP) in the Southwest Atlantic. IOC Workshop Rep. 65. Annex VIII. Intergovernmental Oceanographic Commission, UNESCO, Paris.

Matsuura, Y. 1990. Rational utilization of coastal ecosystem in tropics: Integrated investigation of coastal ecosystem in Ubabtuba region. In: 2nd Simpósio de Ecosystemes da Costa Sul e Sudeste Brasileira, 6–11 April, 1990, Aguas de Lindoa, SP. Vol. 1. pp. 47–52. Academia de Ciências do Estado de São Paulo.

Matsuura, Y., Amaral, J. C., Sato, G., and Tamassia, S. T. J. 1985. Ocorrência de peixes pelágicos e a estrutura oceanográfica da região entre o Cabo de São Tome (RJ) e Cananeia (SP), em Jan–Fev/1979. Ser. Doc. Tec., PDP/SUDEPE, Brasilia, No. 33, pp. 3–70.

Mayr, E. 1982. The growth of biological thought. Harvard University Press, Cambridge, MA.

Mazé, R., Camus, Y., and Le Tareau, J.–Y. 1986. Formation de gradient thermiques à la surface de l'océan, au-dessuss d'un talus, par interaction entre les ondes et le mélange dû au vent. J. Cons. Int. Explor. Mer. 42:221–240.

McGowan, J. A. 1985. El Niño 1983 in the Southern California Bight. In: El Niño north. pp. 166–184. Ed. by W. S. Wooster and D. L. Fluharty. University of Washington Sea Grant Program, Seattle.

McLain, D. R., Brainard, R. E., and Norton, J. G. 1985. Anomalous warm events in eastern boundary current systems. CalCOFI Rep. 26:51–64.

Nelson, G., and Hutchings, L. 1983. The Benguela Current. Prog. Oceanogr. 12:333–356.

Norton, J., McLain, D., Brainard, R., and Husby, D. 1985. The 1982–83 El Niño event off Baja and Alta California and its ocean climate context. In: El Niño north—Niño effects on the Eastern Subarctic Pacific Ocean. pp. 44–72. Ed. by W. S. Wooster and D. L. Fluharty. Washington Sea Grant Program, University of Washington, Seattle.

Olson, D. B., Podestá, G. P., Evans, R. H., and Brown, O. 1988. Temporal variations in the separation of Brazil and Malvinas Currents. Deep-Sea Res. 35:1971–1990.

Paffenhöffer, G.–A., Wester, B. T., and Nicholas, W. D. 1984. Zooplankton abundance in relation to state and type of intrusions onto the southeastern United States shelf during summer. J. Mar. Res. 42:995–1017.

Parrish, R. H., Bakun, A., Husby, D. M., and Nelson, C. S. 1983. Comparative climatology of selected environmental processes in relation to eastern boundary current pelagic fish reproduction. In: Proceedings of the expert consultation to examine changes in abundance and species composition of neritic fish resources. pp. 731–778. Ed. by G. D. Sharp and J. Csirke. FAO Fish. Rep. 291. FAO.

Parrish, R. H., and MacCall, A. D. 1978. Climatic variation and exploitation in the Pacific mackerel fishery. Calif. Dept. Fish Game Fish. Bull. 167.

Parrish, R. H., Nelson, C. S., and Bakun, A. 1981. Transport mechanisms and reproductive success of fishes in the California Current. Biol. Oceanogr. 1:175–203.

Pearcy, W., Fisher, J., Brodeur, R., and Johnson, S. 1985. Effects of the 1983 El Niño on coastal nekton off Washington and Oregon. In: El Niño north. pp. 188–204. Ed. by W. S. Wooster and D. L. Fluharty. University of Washington Sea Grant Program, Seattle.

Peterman, R. M., and Bradford, M. J. 1987. Wind speed and mortality rate of a marine fish, the northern anchovy (Engraulis mordax). Science 235:354–356.

Peterson, W. T. 1973. Upwelling indices and annual catches of dungeness crab, Cancer magister, along the west coast of the United States. Fish. Bull. U.S. 71:902–910.

Picaut, J. 1985. Major dynamics affecting the eastern tropical Atlantic and Pacific oceans. CalCOFI Rep. 26:41–50.

Podestá, G. P. 1990. Migratory pattern of Argentine hake Merluccius hubbsi and oceanic processes in the southwestern Atlantic Ocean. Fish. Bull. U.S. 88:167–177.

Prager, M. H., and MacCall, A. D. 1990. Biostatistical models of contaminant and climate influences on fish populations of the Southern California Bight. Old Dominion University Oceanography Tech. Rep. 90–04.

Regner, S., Regner, D., Marasovic, I., Krsinic, F. 1987. Spawning of sardine, Sardina pilchardus (Walbaum, 1792), in the Adriatic under upwelling conditions. Acta. Adriat. 28:161–198.

Reid, J. L., Jr. 1962. On circulation, phosphate phosphorus content, and zooplankton volumes in the upper part of the Pacific Ocean. Limnol. Oceanogr. 7:287–306.

Reid, J. L., Jr., Roden, G. I., and Wyllie, J. G. 1958. Studies of the California Current system. CalCOFI Rep. 6:28–56.

Richardson, S. L. 1973. Abundance and distribution of larval fishes in waters off Oregon, May–October 1969, with special emphasis on the northern anchovy, Engraulis mordax. Fish. Bull. U.S. 71(3):697–711.

Rossi-Wongtschowski, C. L. D. B. 1989a. Availiação instantanes do tamanho do estoque desovante da sardinnha-verdadeira (Sardinella brasiliensis) na região sudeste do Brasil. In: IOUSP e SUDEPE dessenvolvem projeto integrado sobre "sardinha" no Costa Sudeste do Brasil. CIRM/Inf. 4(1), Jan./Mar., 1989. p. 4. Comissão Interministerial para os Recursos do Mar, Brasilia.

Rossi-Wongtschowski, C. L. D. B. 1989b. Prospecção o availiação dabiomassa de sardina na Costa Sudeste, por metodos hidroacústicos. In: IOUSP e Sudepe dessenvolvem projeto integrado sobre "sardinha" no Costa Sudeste do Brasil. CIRM/Inf. 4(1), Jan./Mar., 1989. pp. 5–6. Comissão Interministerial para os Recursos do Mar, Brasilia.

Rothlisberg, P. C., and Miller, C. B. 1983. Factors affecting the distribution, abundance and survival of Pandalus jordani (Decapoda, Pandalidae) larvae off the Oregon coast. Fish. Bull. U.S. 81:455–472.

Roy, C. 1990. Reponses des stocks de poissons pelagiques à la dynamique des upwellings en Afrique de L'Ouest: Analyse et modelisation. These de Doctorat. Université de Bretagne Occidentale, Brest, France.

Roy, C., Cury, P., Fontana, A., and Belvèze, H. 1989. Stratégies spatio-temporelles de la reproduction des clupéidés des zones d'upwelling d'Afrique de l'Ouest. Aquat. Living Resour. 2:21–29.

Saccardo, S. A. 1983. Biología y disponibilidad de sardina (Sardinella brasiliensis Steindachneer, 1879) en la costa sudeste del Brasil. In: Proceedings of the expert consultation to examine changes in abundance and species composition of neritic fish resources. pp. 449–464. Ed. by G. D.

Sharp and J. Csirke. FAO Fish. Rep. 291. FAO.

Saccardo, S. A. 1989. Programa integrado de estudos biológicos sobre sardinha. In: IOUSP e SUDEPE dessenvolvem projecto integrado sobre "sardinha" no Costa Sudeste do Brasil. CIRM/Inf. 4(1), Jan./Mar., 1989. pp. 4–5. Comissão Interministerial para os Recursos do Mar, Brasilia.

Saiki, M. 1988. Time dependent variability with a few decades. In: Studies on fishery oceanography—Proceedings of the 25th anniversary symposium "Fisheries and fishery oceanography in the coming century," Tokyo, November 10–13, 1986. pp. 358–366. Japanese Society of Fisheries Oceanography.

Schneider, S. H. 1990. The global warming debate heats up: An analysis and perspective. Bull. Amer. Meteor. Soc. 71:1292–1304.

Servicio de Hidrografía Naval. 1969. Datos y resultados de las campanãs "pesquería." Pesquería V. Publicaciones del Proyecto de Desarrollo Pesquero No. 10/V. Mar del Plata, Argentina.

Sette, O. E., and Isaacs, J. D. (Editors). 1960. Symposium on "The changing Pacific Ocean in 1957 and 1958." CalCOFI Rep. 7:13–217.

Shannon, L. V., Agenbag, J. J., and Buys, M. E. L. 1987. Large- and mesoscale features of the Angola-Benguela Front. In: Benguela comparative ecosystems. pp. 11–34. Ed. by A. I. L. Payne, J. A. Gulland, and K. H. Brink. Special volume of the South African Journal of Marine Science, Vol. 5. Sea Fisheries Research Institute, Capetown, South Africa.

Shannon, L. V., Boyd, A. J., Brundrit, G. B., and Taunton-Clark, J. 1986. On the existence of an El Niño phenomenon in the Benguela system. J. Mar. Res. 44:495–520.

Shannon, L. V., Crawford, R. J. M., Brundrit, G. B., and Underhill, L. G. 1988b. Responses of fish populations in the Benguela ecosystem to environmental change. J. Cons. Int. Explor. Mer 45:5–12.

Shannon, L. V., Shackleton, L. Y., and Siegfried, W. R. 1988a. The Benguela ecology programme: The first five years. S. Afr. J. Mar. Sci. 84:472–475.

Shelton, P. A., Boyd, A. J., and Armstrong, M. J. 1985. The influence of large-scale environmental processes on neritic fish populations in the Benguela Current system. CalCOFI Rep. 25:72–92.

Shelton, P. A., and Hutchings, L. 1982. Transport of anchovy, *Engraulis capensis* Gilchrist, eggs and early larvae by a frontal jet. J. Cons. Int. Explor. Mer 40:185–198.

Shelton, P. A., and Hutchings, L. 1990. Ocean stability and anchovy spawning in the southern Benguela Current region. Fish. Bull. U.S. 88:323–338.

Smith, R. L. 1974. A description of current, wind, and sea level variations during coastal upwelling off the Oregon Coast, July–August 1972. J. Geophys. Res. 79:435–443.

Stommel, H. 1966. The Gulf Stream. 2d ed. Univ. California Press, Berkeley.

Vrooman, A. M., and Smith, P. E. 1971. Biomass of the subpopulations of northern anchovy, *Engraulis mordax* Girard. CalCOFI Rep. 15:49–51.

Walker, N. D. 1987. Interannual sea surface temperature variability and associated atmospheric forcing within the Benguela system. In: Benguela comparative ecosystems. pp. 121–132. Ed. by A. I. L. Payne, J. A. Gulland, and K. H. Brink. Special volume of the South African Journal of Marine Science, Vol. 5. Sea Fisheries Research Institute, Capetown, South Africa.

Ware, D. M., and Thomson, R. E. 1991. Link between long-term variability in upwelling and fish production in the northeast Pacific Ocean. Can. J. Fish. Aquat. Sci. 48:2296–2306.

Wisner, R. L. 1961. Evidence of a northward movement of stocks of the Pacific sardine based on the number of vertebrae. CalCOFI Rep. 8:75–82.

Wooster, W. S., and Reid, J. L. 1963. Eastern boundary currents. In: The sea. Vol. 2. pp. 253–280. Ed. by M. N. Hill. Interscience Pub., New York.

Wyllie, J. G. 1966. Geostrophic flow of the California Current at the surface and at 200 m. CalCOFI Atlas No. 4. Scripps Institution of Oceanography, La Jolla, CA.

Part Three:
Sustainability and Management of Large Marine Ecosystems

Chapter 19

Regional Approach to Large Marine Ecosystems

Lewis M. Alexander

Large Marine Ecosystems (LMEs) are regional phenomena in two respects. First, they conform to the geographical concept of regions, that is, areas of the earth's surface, possessed of certain distinguishing characteristics that differentiate them from neighboring areas. The distinguishing characteristics, in this case, are unique bathymetry, oceanography, and productivity, within which marine populations have adapted reproductive growth and feeding strategies. It is the task of marine scientists to identify such areas of unique characteristics.

Second, LMEs are regional phenomena in that they exist within areas of the world ocean that are characterized by certain geographical processes. Some systems are produced by upwelling currents along the west coast of continents, others occur in extensive continental shelf areas. Still others are found in semi-enclosed seas. There are similarities, as well as differences, among these various ecosystems.

From a regional perspective, three aspects of LMEs are significant. First are the definitional issues: Which marine fisheries areas, including their coastal systems, constitute distinct ecosystems? This process of identification on a global basis is still ongoing. Some current-induced systems are now well understood, as for example, the Canary Current, Benguela Current, Humboldt Current, and California Current systems. There is also clear evidence of a U. S. Northeast Continental Shelf ecosystem, and a Patagonian Shelf, system among others.[1] Within semi-enclosed seas, marine

scientists have identified Black Sea, Yellow Sea, Baltic, and North Sea ecosystems.

The identification of a single system becomes more difficult along the west coast of Africa. How extensive is the Gulf of Guinea system? Does a single ecosystem extend northward to Senegal and Mauritania? Is the Mediterranean Sea, stretching for over 2,000 miles from west to east, composed of one ecosystem or several ecosystems? What is the situation in the Arabian Sea? Already, more than two dozen LMEs have been firmly established, but the problem of identification is still an evolving one.

In some situations, after the ecosystem has been identified, there remains the problem of defining its geographical boundaries. The seaward limits of the Northern California Current ecosystem, for example, remain undetermined. In the case of the Antarctic ecosystem, the northern limit is taken to be the Antarctic Convergence Front, about which little precise geographical data yet exist. How far north does the North Sea ecosystem extend?

A third definitional problem relates to subsystems. The Northeast Continental Shelf ecosystem is now believed to be composed of four subsystems: Gulf of Maine, Georges Bank, Southern New England, and the Mid-Atlantic Bight. Is the Adriatic ecosystem an LME in its own right, or a part of a larger Mediterranean system or systems? Are there two ecosystems between the southern United States and northern South America—a Gulf of Mexico and a Caribbean—or are the two part of a larger system? Further data collection and analysis will be necessary in order to answer such questions.

A second regional issue concerns jurisdictional patterns. Relatively few LMEs are

---

1. For a description and map of 49 LMEs, see Sherman Chapter 1, this volume.

within the exclusive economic or fisheries zones of a single state. Ecosystems off southwestern Alaska and eastern Australia are among the exceptions. Where two or more countries are involved in potential joint research and management programs, the success of such programs will depend in part both on the state of relations existing among the neighboring countries, and on the relative levels of marine scientific knowledge and concern within the states. In the North and Baltic seas, for example, cooperative marine scientific investigations and management programs are already well advanced. In a "worst case" situation, that in the Gulf of Guinea area, 16 independent countries exist between Mauritania to the north and Angola to the south, with different languages and colonial histories, and with little regional interactions among all or some of the countries.

Superimposed on the general nature of national differences may be territorial controversies, such as regarding island groups in the South China Sea, or maritime boundary disputes, such as between the People's Republic of China and the Republic of Korea or Japan. Under such conditions, joint management programs may be difficult to arrange, which does not negate the general ecosystem management concept.

A third regional issue involves multiple-use conflicts within an LME. Along with fisheries, there may be offshore oil developments, the use of the ocean for dumpsites, tanker shipping lanes, or coastal pollution from land-based sources. One potential problem of such multiple uses is pollution of fisheries habitats, particularly of spawning and nursery grounds, with consequent impacts on the available fisheries stocks. The nature and intensity of pollution problems also vary from one region to another.

It is frequently noted in LME literature that it is important to obtain baseline and time-series data in order to distinguish between natural and man-induced changes in the composition of an ecosystem. It will also be necessary over time to consider the impacts of global change on specific ecosystems. Such impacts will probably not be uniform, but will vary from region to region, particularly from tropical to mid-latitude to polar areas, and for each climatic group, with respect to current-induced, continental shelf, or semi-enclosed seas ecosystems. What is needed now is a great deal more scientific data in order to move forward, within the conceptual framework of LMEs, in further identifying, describing, and comparing marine ecosystems throughout the world ocean.

References

Sherman, K. Chapter 1, this volume.

Chapter 20

Legal Regimes for Management of Large Marine Ecosystems and Their Component Resources

Martin H. Belsky

Introduction

In 1984, at a symposium on Large Marine Ecosystems (LMEs), I first suggested that international law had been evolving to a mandate for a comprehensive ecosystem approach to ocean management (Belsky, 1985). More recently, I have indicated that this evolution is complete (Belsky, 1989a) and includes not only regulation, but also basic marine research, assessment, and monitoring (Belsky, 1990). Nation-states are now bound by international law to apply this holistic model in their domestic law and practice and in their foreign policy (Belsky, 1989b).

This chapter will give a short review of the evolution of this ecosystem approach into a binding rule of international law, followed by a discussion of how application of this model can and should lead to sensible decisionmaking.

The Mandate for an Ecosystem Model

Until the 1960s, nation-states and international law considered most of the ocean to be under no state or international control (Knight, 1975; MacRae, 1983). Each nation had the right to establish rules for its coasts and a small band of territorial sea adjacent to its coasts (S.S. *Lotus* case, 1927). Activities beyond this were regulated, if at all, on a voluntary basis, either through cooperative agreements or by controls by each nation over nationals or vessels (United States v. Flores, 1933; Churchill and Lowe, 1983).

International law was premised on national sovereignty and high-seas freedoms

(Shaw, 1986; Janis, 1988). Marine policy was focused on freedom of navigation and maximum exploitation of resources (Simon, 1984). Environmental risks and conservation of resources were not major concerns (Hoban and Brooks, 1987). Multinational action, or even cooperation, was viewed as a potential threat to national sovereign rights (Wenk, 1981).

A new environmental awareness emerged in the late 1960s (Belsky, 1984). This led many nations to begin to manage resources and control pollution in their own coastal areas and territorial seas (Belsky, 1985). It also led to a few multistate arrangements to share resources and ocean space (Friedheim, 1975).

When nation-states acted, however, either individually or collectively, they adopted an ad hoc approach—responding to a particular perceived pollution or marine species conservation problem. Ecosystems, as such, remained largely unregulated (Belsky, 1985).

During the 1970s, increased environmental sensitivity and a desire for more nation-state control over adjacent ocean space resulted in exponential growth in domestic legislation (Belsky, 1984). States enacted laws claiming resources over a 200-mile Exclusive Economic Zone (EEZ), establishing policies to regulate the commercial and endangered species in these zones, requiring assessment and monitoring of activities, providing more controls over pollution of coastal and ocean waters, and mandating reconciliation of conflicting uses of coastal and ocean space (Lutz, 1976; Copes, 1981).

Government leaders soon recognized that increased jurisdictional zones had the potential of increased interstate conflict and began to face the impact of individual resource, development, and pollution policies on the oceans. As a result, the world community started to address the inherent problems of a case-by-case, nation-by-nation approach to the oceans ecosystems (Belsky, 1989c).

By the late 1970s, the international rhetoric had changed from unlimited nation-state power to "responsible stewardship" (Goldie, 1975). A comprehensive multistate approach was accepted as sound national and international policy and soon became accepted as binding on nation-states as law.

The "Ecosystem Model" as Law

The "rules" of international law develop informally. They begin with explicit or implicit acknowledgment by the community of nations as shown by state practice or custom, proceed to acceptance as general principles of law, and finally as international conventions (Reisman, 1986; Restatement, 1987).

Since the early 1980s, the past decade, nation-states have been moving toward a recognition of the need to coordinate separate ocean policies to assure both maximum protection of their oceans and coasts and future continued exploitation of the oceans resources (Lutz, 1976; Boczek, 1986; Roe, 1987; Levy, 1988). This individual state practice of applying a comprehensive approach has been paralleled by multistate actions.

Nation-states accept and now apply an obligation to prevent harm to their own environment and resources, as well as to the environment of other nation-states and to adjacent and shared resources (Trail Smelter case, 1941; Charter of Economic Rights and Duties of States, 1975). They also accept as binding customary law an obligation of rational and equitable utilization of their resources (Bilder, 1972; Handl, 1978; Restatement, 1987). Prevention of harm and "rational and equitable use," of course, mean that resources and uses must be studied and managed in a comprehensive manner, focusing on the LMEs in which resources exist.

In 1972, this multinational responsibility was accepted by the members of the United Nations in endorsing the Stockholm Declaration on the Human Environment (Stockholm Declaration, 1972). The declaration placed an international obligation on all nation-states to adopt, individually and collectively, an integrated approach in the development of resources and control of adverse environmental effects (Smith, 1984). In the later Draft World Charter for Nature, the member states of the United Nations reaffirmed this mandate by demanding that states and citizens act in such

a way so as not to threaten the "integrity of the ecosystems and organisms with which they coexist" (Draft World Charter for Nature, 1980).

Evidence of the emergence of this new mandate for a comprehensive ecosystem approach can also be found in recent multilateral agreements such as the 1980 Convention for the Conservation of Antarctic Marine Living Resources (CCAMLR, 1980; Convention for the Protection of the Natural Resources and Environment of the South Pacific Region (Convention for the Conservation of Antarctic Living Marine Resources, 1980; Convention for the Protection of the Natural Resources and Environment of the South Pacific Region, 1987).

Legal scholars, looking at this individual and collective state practice, now argue that a comprehensive ecosystem approach to the oceans is the evolved rule of international law. They point to the 1982 United Nations Convention on the Law of the Sea (UNCLOS) as proof that the evolution is now complete (Oxman, 1986; Blischenko et al., 1988; Levy, 1988).

In earlier papers, I indicated how UNCLOS requires each nation-state to take all appropriate actions to preserve and protect the marine environment and to manage its resources based on the interdependence of species. I also indicated how the convention requires nation-states to integrate its ocean policies and laws and to act collectively, as well as individually. All the provisions of the convention are intended to be read as a whole. When read in such a manner, the provisions provide for the "ecosystem approach" (Belsky, 1989a; Belsky, 1989c).

For those nations that have ratified UNCLOS, a comprehensive approach is mandated (Vienna Convention on the Law of Treaties, 1969). The ecosystem approach, however, is also binding on nonsignatories—as a statement of customary international law. The basic objectives and obligations of the convention for comprehensive ecosystem management are accepted by all nation-states and are being followed in nation-state legislative and treaty practice (MacRae, 1983; Sohn, 1985; Restatement, 1987; Kwaitkowski, 1988).

Even when nation-states have nonconforming practice, they attempt to justify their conduct in terms of the provisions of the convention (Blischenko et al., 1988; Kwaitkowski, 1988). Finally, the scholarly consensus and the most recent decisions of international courts accept the relevant provisions of the treaty as representing the new "state practice" (Case concerning delimitation of the maritime boundary of the Gulf of Maine, 1984; Sohn, 1984; Restatement, 1987).

Science and the Ecosystem Model Mandate

The "ecosystem model" is nothing more than a shorthand for holistic or comprehensive ocean management. The mandate for use of this model seeks to force government leaders to apply scientific principles to domestic and international law and policymaking (Belsky, 1986).

Recent symposia have confirmed that the only way to deal with the resources in and the activities affecting the oceans is to view the ocean space as a series of LMEs (Sherman and Alexander, 1986; Sherman and Alexander, 1989; Sherman *et al.*, 1990). Actions that affect any part of a marine ecosystem should only be undertaken upon recognition that such actions will necessarily affect the whole ecosystem (Belsky, 1989c).

Management must consider all the species in an ecosystem as well as the impact on that ecosystem of marine-related activities and the pollution caused by those activities on that ecosystem (Gordon, 1981; Kindt, 1986). The term "ecosystem" itself is defined as "the pattern of relationships between all biotic (living) and abiotic (non-living) entities within a defined boundary of space and time" (Hoban and Brooks, 1987, p. 5). An ecosystem management model, therefore, is simply a set of research and regulatory decisions that recognizes this "pattern of relationships" (Ray, 1970).

The model is not limited to just regulatory decisions. It also requires a coordinated effort in basic research, assessment, and monitoring as a basis for comprehensive ecosystem regulation. Adequate regulatory controls on activities and limitations on pollutants must first be based on assessment of the present status of the marine environment. Changes to that environment must then be monitored. A continual evaluation of current information must be tied into periodic revision of restrictions (Belsky, 1989d; Belsky, 1990).

Living marine resources are constrained by ecological laws. Biological and nonbiological factors interact and affect each other within the same ecosystem. Environmental studies must therefore not only analyze the impact of activities on such resources and their habitats, but also on the productivity levels of organisms in particular food chains and the effects of various allowable levels of exploitation on this productivity (Yuru, 1985).

As living resource management must be based on an ecosystem model, basic research and then assessment and monitoring must also focus on the whole ecological mosaic in a region—the impact of pollution and pollution controls and the impact of exploitation of one species on another (Lie, 1985; Yuru, 1985; Roe, 1987).

Applying the Ecosystem Model

The ecosystem comprehensive model must be incorporated into each nation-state's own legal system, as part of any new bilateral or multilateral agreement and in the establishment of any formal or informal marine regulatory or management program. Government leaders, scientists, developers, and fishery managers must study the ecosystem of each state's coastal areas and expanded EEZ, and then base their actions on the study findings. They also must apply the model to regional seas or multinational arrangements and policies (Belsky, 1985).

As a practical matter, such applications may be difficult but not impossible. In most jurisdictions, there are numerous statutes, policies, rules, plans, or treaty obligations that deal separately with marine pollution problems and that deal individually with particular marine species. However, few, if any, of these statutes, plans, rules, or policies are directly inconsistent with the comprehensive ecosystem model. Enough discretion is given to each country's regulators to maximize environmental protection and balance resource management issues so that these requirements can be interpreted in a way that is consistent with the ecosystem model (Belsky, 1989c).

At the international level, nation-states are continually considering new arrangements for dealing with resources and the environment. Under UNCLOS, they have accepted a mutual obligation to "take into account" the ecosystem impact of their policies. The nations of the world, both individually and collectively, should now, within their individual and joint capabilities, make all efforts to reach the scientific ideal of total ecosystem research and management (Belsky, 1989d).

Applying the Ecosystem Model in International Policy

How can the ecosystem model be applied in policy and decisionmaking? As evidenced by the Law of the Sea Convention and other more specific conventions for geographical regions, nation-states must adopt a comprehensive approach to research, pollution, and resource management decisions.

As described above, UNCLOS mandates an ecosystem approach by nation-states in their domestic law and in their multinational arrangements. Nation-states are to join together to consider the issues of pollution and management in specific geographical regions. When doing so, UNCLOS specifically provides that nation-states are to act cooperatively to minimize and control pollution in shared jurisdictional zones and the high seas (UNCLOS, 1982; Miovski, 1989).

Nation-states must secure the "best scientific evidence" and adequately provide for assessment of risks and monitoring of risks and effects (UNCLOS, 1982). They must manage on an ecosystem model in order to avoid overexploitation, to adequately consider the environmental impacts on habitats, and to sufficiently consider the interrelationships of species (Boczek, 1983; Burke, 1985).

This UNCLOS mandate is practical. It is premised on a regional approach, focusing on the LMEs that may be under the control of several governments and must be studied and managed by those governments in a coordinated manner, preferably under a regional organization or structure (Miovski, 1989; Qadar, 1989).

The regional approach is already being applied. Under CCAMLR, for example, cooperative research, assessment, and management is to be based on a total ecosystem conservation standard. In addition, the convention requires signatory states to conduct their affairs so as to minimize risks to the Antarctic marine ecosystem (CCAMLR, 1980).

Antarctica is, of course, unique, but this regional approach is not. In recent years, nation-states have been joining together to control pollution for particular geographic regions where jurisdictions overlap, such as in the Caribbean, the Baltic, the Mediterranean, the North, and the Yellow seas (Johnston and Enomoto, 1981; Keckes, 1981; Boxer, 1983; Convention for the Protection and Development of the Marine Environment of the Wider Caribbean Region, 1984; Valencia, 1988). Now, they appear more willing to address the interaction of species management and pollution in the regional context (Blischenko *et al.*, 1988; Miovski, 1989).

In a 1986 South Pacific convention, for example, the nations of the South Pacific region "recognized" the special ecological nature of the region and the potential threat to the region's "ecological equilibrium." They agreed in their domestic laws and in international arrangements to "take all appropriate measures," including research, assessment, monitoring, and regulation to "control pollution . . . and to ensure sound environmental management and development of natural resources" (Convention for the Protection of the Natural Resources and Environment of the South Pacific Region, 1987, Arts. 4[1], 5[1]). This language, though not using the exact words, is the ecosystem model.

The ecosystem model and these precedents of nation-state cooperation could serve as a call for a broad application of the regional approach and a commitment to LME research and management (Johnston, 1978). It could be applied in such regions as those governed by the European Community (Koers, 1978; Haagsma, 1989); in North America (Wilkinson and Connor, 1983); in the Yellow Sea (Valencia, 1988); in the Bering Sea (Miovski, 1989); and in the North Sea (Peet, 1987).

Marine Research

As indicated above, the ecosystem model—upon which adequate science is premised—requires that ocean management actions be based on adequate research, assessment, and monitoring. Scientists and governments must prepare an agenda for research in each nation's EEZs and in those areas of common jurisdiction (Draft World Charter for Nature, 1980; Smith, 1984). Nations and the international community must provide funding for this "ocean information policy" (UNCLOS, 1982; Creech, 1986).

The mechanism for implementing this mandate should be domestic laws—which, in many cases, already exist—that provide for environmental impact statements about proposed activities (NEPA, 1986) and regional agreements that provide for assessment and monitoring (Carroll and Mack, 1982; Convention for the Protection and Development of the Marine Environment of the Wider Caribbean Region, 1984).

Application of the new ecosystem research requirements should also be read into the various domestic laws and international agreements that provide for the conduct of research by nation-states (MFCMA, 1976; OCSLAA, 1981) or that provide for cooperative multistate action (Stockholm Declaration, 1972; UNEP Working Group, 1987). The problem, of course, will be adequate funding. Still, when existing research dollars are allocated, the ecosystem model mandate should be used as a vehicle for ecosystem-based research, and the model should also be used to make demands for adequate funding (Byrne, 1986).

Applying the Ecosystem Model to Fisheries Regulation

Obviously, a key aspect of a comprehensive ecosystem approach to oceans management is the establishment "of an ecosystem-wide management program for fish and other living marine resources which permits multi-species management of those resources" (Roe, 1987, p. 33). In each nation, and in bilateral and multilateral arrangements, the model must be integrated into the numerous statutes that provide for the study, protection, and management of living marine resources.

The means to achieve this goal can be found in UNCLOS itself. The provisions for management of the living resources of the sea adopt the "maximum sustainable yield" standard, but say that this standard has to be qualified by "other relevant environmental and economic factors" and to take into account, the "interdependence of stocks" (UNCLOS, 1982, Art. 61[3]). In addition, nation-states, in managing specific resources, must "take into consideration the effects of species associated with or dependent upon harvested species with a view to maintaining or restoring populations of such associated or dependent species above levels at which their reproduction may become seriously threatened" (UNCLOS, 1982, Art. 194[5]). Finally, the convention requires nation-states to establish all the pollution measures necessary "to protect and preserve rare or fragile ecosystems as well as the habitat of depleted, threatened or endangered species and other forms of marine life" (UNCLOS, 1982, Art. 61[1], 119[1][a], 194[1]).

In almost all nations, specific laws establish procedures for the protection of endangered species, marine mammals, and the habitats of all fisheries; programs for research and development of resources; and standards for management of commercial fisheries (Lutz, 1976; Office of the Special Representative of the Secretary-General for the Law of the Sea, 1986; Blischenko *et al.*, 1988).

Historically, the study, protection, and management of these resources has been on a species by species basis—without regard to the interaction of species or the impact of other ocean activities on their habitat or life cycle (Batchelor, 1988). This ad hoc approach must give way to the new mandate for a comprehensive approach, and new laws must be promulgated or old statutes interpreted to include an ecosystem approach (Batchelor, 1988; Belsky, 1989c).

The application of the new international law mandate to living marine resource management is relatively straightforward. In many countries, one agency is responsible for most marine fisheries management and can adopt the ecosystem mandate and apply it (MFCMA, 1976); but even in those nations where there is a division of responsibility among agencies or among local and national governments, the international law mandates reconciliation of competing and overlapping programs on a national basis (Belsky, 1989c).

Conflicting Uses and Pollution

As described earlier, fisheries management did not consider the impact of other ocean uses and pollution on their plans and regulations. In many nations, the various ocean uses, and the pollution that can result from them and from onshore activities, are governed by multiple laws and policies. The resulting pollution and resource management problems can be reconciled on an ecosystem model. In fact, laws that call for a "balancing" of interests must be interpreted to require that the balance incorporate the ecosystem model. This can be applied, for example, to policies for offshore mineral exploitation, ocean dumping, and vessel pollution (Center for Ocean Management Studies, 1984; Johnston, 1984).

One unique application of the ecosystem model can be for select areas needing special attention. For example, the United States has laws that allow the creation of "marine sanctuaries" to preserve or restore special ocean areas for their conservation, recreational, ecological, or aesthetic values (Brooks, 1984). The United States and other nations can use the ecosystem model as a means to support a unified, comprehensive approach to such select areas.

Coastal Activities and Uses

The ecosystem approach, of course, is not limited to activities and pollution in far offshore areas (Ray, 1989). UNCLOS specifically requires each nation-state to deal with its coastal areas—to perform adequate assessment and monitoring, to control activities so as to avoid pollution, and to regulate resource management (UNCLOS, 1982).

In this zone, the conflicts between development and conservation are often acute (Simon, 1978). Because of the access to the ocean, the adjacent coastal space is often highly developed, leading to more intense exploitation of resources and disposal of wastes. Unfortunately, the coastal ecosystems, especially estuarine systems, are also often the most fragile (Ross, 1982).

Moreover, activities that occur in or affect this coastal area, even though often within the jurisdiction of one nation-state, are frequently subject to conflicting policies by multiple jurisdictions (Myers, 1981; Gibson and Halliday, 1987). Yet, only recently have there been any substantial attempts to coordinate coastal research and management (Gibson and Halliday, 1987).

The ecosystem model gives a legal and policy framework for action. Even if there are multiple jurisdictions within a nation-state that have responsibility for coastal use control, international law prescribes that *the nation-state* is responsible for implementing its international mandate (Borchard, 1940; Restatement, 1987). To avoid an international delict, all inconsistent local laws must give way (Henkin, 1984) and national laws must be interpreted in a manner consistent with international law (Shaw, 1986). Thus, the ecosystem model is a mandate for nation-states to resolve the conflicting uses of the coastal area, reconcile the various laws affecting coastal activities, and provide for protection of fragile coastal areas (Agoes, 1987).

As indicated above, the ecosystem model provides a framework for multistate regional arrangements and actions. Similarly, it serves as a basis for each nation-state to require a national response to coastal problems—a response that recognizes the risks to coastal ecosystems from ad hoc decisionmaking (Gibson and Halliday, 1987; Ray, 1989). In these nation-states that have explicit coastal zone management programs (Amarasinghbe and Alwis, 1980; Hershman, 1981), those programs are to provide for ecosystem research and management (Belsky, 1989c). However, even in these nations that have no such explicit programs, coastal ecosystems must, under international law, be managed as ecosystems. The ecosystem model should be the vehicle either for creation of explicit coastal zone programs, or, as a minimum, establishing arrangements that provide for comprehensive study and management of coastal zones (Campbell and Yoskin, 1980; Carlberg, 1981; Waite, 1981).

Estuarine research and management is a prime example of the ecosystem model. These coastal water bodies, considered to be the richest and most vulnerable sector of the coastal waters (Clark, 1977), face threats from development, physiological changes, and pollution (Myers, 1981). Still, management of such coastal space provides opportunities: to resolve conflicts; to propose comprehensive management; to establish sanctuaries, if necessary; and to test generally an ecosystem model (Evans, 1980).

Conclusion

In this chapter, and in previous papers, I have intentionally advocated the use of the ecosystem approach as both a preferable rule and a mandatory legal doctrine binding on all nation-states. The model has been criticized, however, as being too costly and beyond the capacity of scientists and government leaders to implement (Burke, 1985).

I disagree. I am not satisfied with continuation of a species-by-species, pollutant-by-pollutant research and management scenario. I am, in fact, concerned that continued advocacy of an ad hoc approach, even for practical reasons, may be detrimental to the future of fisheries research and marine environmental and resource protection.

Separate rules for each species, and for each pollutant, may, in fact, be worse than no management at all. Establishing standards that may be based on incomplete, or more likely, inaccurate information may result in long-term harm to a species, and certainly to the ecosystem that supports it. Implementing rules for individual species, without adequate consideration of the impact, within an ecosystem, of activities and resource development, of coastal and ocean pollution, and of the interaction of species, perpetuates the myth, consistently refuted by science, that we can consider problems piecemeal. It reinforces, though in more sophisticated terms, the "tragedy of the commons" (Hardin, 1968).

The ecosystem approach described in this paper does not establish onerous burdens. Defining what is an ecosystem, what species are in that system, how the species that are part of that ecosystem interrelate, and how ocean and coastal activities affect the ecosystem and the living resources in that system, are complex issues, and the state of scientific investigation of marine ecosystems is still at an early stage (Alexander, 1986).

The "ecosystem model," however, does not call for precise determinations of ecosystems, nor precise determinations of all interactions. Rather, it mandates that environmental and resource planners and regulators take a comprehensive look at how their policies and rules interact.

The language used in UNCLOS illustrates the process. All the elements of an ecosystem must "be taken into account" in decisionmaking. Obligations are to be implemented by nation-states through "the best practicable means at [their] disposal and in accordance with their capabilities" (UNCLOS, 1982, Arts. 61[1]; 119[1][a]; 194[1]). The ecosystem model, therefore, is not unduly burdensome on nation-states. Rather, it is a call on such nation-states to make, within their capabilities, all efforts to reach the scientific ideal of total ecosystem management.

An ecosystem management approach, recognizing the present scientific and fiscal limitations in providing absolute answers, is clearly the preferred international policy. Nation-states should apply the international law mandate of comprehensive ecosystem approach to marine management. Scientists and

other interested individuals should press nation-states and the international community to
accept the model and use it as a basis for research, planning, and regulations.

References

Articles and books

Agoes, E. 1987. Management of the seas and coastal
zones: A brief outlook on recent developments in
Indonesia. In: The UN convention on the Law of
the Sea: Impact and implementation. pp. 452–
463. Ed. by E. Brown and R. Churchill. Law of the
Sea Institute, Honolulu.

Alexander, L.M. 1986. Large marine ecosystems as
regional phenomena. In: Variability and
management of large marine ecosystems. pp.
239–240. Ed. by K. Sherman and L.M. Alexander.
AAAS Selected Symposium 99. Westview Press,
Inc., Boulder, CO.

Amarasinghbe, S., and Alwis, R. 1980. Coastal zone
management in Sri Lanka. In: Coastal zone '80.
pp. 2828–2842. American Society of Civil
Engineers, New York.

Batchelor, A. 1988. The preservation of wildlife
habitat in ecosystems: Towards a new direction
under international law to prevent species
extinction. Florida Intl. L. J. 3:307.

Belsky, M. H. 1984. Environmental policy law in the
1980's: Shifting back the burden of proof.
Ecology Law Quarterly 12:1.

Belsky, M. H. 1985. Management of large marine
ecosystems: Developing a new rule of customary
international law. San Diego Law Review 22:733.

Belsky, M. H. 1986. Legal constraints and options for
total ecosystem management of large marine
ecosystems. In: Variability and management of
large marine ecosystems. pp. 241–261. Ed. by K.
Sherman and L.M. Alexander. AAAS Selected
Symposium 99. Westview Press, Inc., Boulder,
CO.

Belsky, M. H. 1989a. Developing an ecosystem
management regime for large marine ecosystems. In: Biomass yields and geography of large
marine ecosystems. pp. 443–459. Ed. by K.
Sherman and L.M. Alexander. AAAS Selected
Symposium 111. Westview Press, Inc., Boulder,
CO.

Belsky, M. H. 1989b. Marine ecosystem model: Law
of the Sea's mandate for comprehensive
management. In: New developments in marine
science and technology: Economic, legal and
political aspects of change. pp. 115–134. Ed. by
L.M. Alexander, S. Allen, and L. Hanson. Law of
the Sea Institute, Honolulu.

Belsky, M. H. 1989c. The ecosystem model—
mandate for a comprehensive United States
ocean policy and Law of the Sea. San Diego Law
Review 26:417.

Belsky, M. H. 1989d. The marine ecosystem
management model and the Law of the Sea:

Requirements for assessment and monitoring.
In: The international implications of extended
maritime jurisdiction in the Pacific. pp. 236–262.
Ed. by J. Craven, J. Schneider, and C. Stimson.
Law of the Sea Institute, Honolulu.

Belsky, M. H. 1990. Interrelationships of science and
law in the management of large marine ecosystems. In: Large Marine Ecosystems: Patterns,
processes, and yields. pp. 224–234. Ed. by K.
Sherman, L. M. Alexander, and B. D. Gold. AAAS
Symposium. AAAS, Washington, D.C.

Bilder, R. B. 1972. International law and natural
resources policies. Natural Resources Journal
20:451.

Blischenko, I. P., Gureyev, S. A., and Andreyev, E. P.
(Editors). 1988. The international Law of the Sea.
Progress Publishers, Moscow, U.S.S.R.

Boczek, B. 1983. The protection of the Antarctic
ecosystem: A study in international environmental law. Ocean Development and International
Law 13:347.

Boczek, B. 1986. The concept of regime and the
protection of the marine environment. In: Ocean
yearbook 6. pp. 271–297. Ed. by E. Borgese, N.
Ginsberg, J. Baylson, N. Dunning, and D. Dzurek.
University of Chicago Press, Chicago, IL.

Borchard, E. 1940. The relationship between
international law and municipal law. Virginia
Law Review 27:137.

Boxer, B. 1983. The Mediterranean Sea: Preparing
and implementing a regional action plan. In:
Environmental protection. pp. 267–309. Ed. by D.
Kay and H. Jacobson. Allanheld, Osmun & Co.
Publishers, Inc., Totowa, NJ.

Brooks, D. L. 1984. America looks to the sea. Jones
and Bartlett Publishers, Inc., Boston, MA.

Burke, W. 1985. The Law of the Sea convention and
fishing practices of non-signatories, with special
reference to the United States. In: Consensus and
confrontation: The United States and the Law of
the Sea convention. Ed. by J. VanDyke. Law of the
Sea Institute, Honolulu.

Byrne, J. 1986. Large marine ecosystems and the
future of ocean studies: A perspective. In:
Variability and management of large marine
ecosystems. pp. 299–308. Ed. by K. Sherman and
L. M. Alexander. AAAS Selected Symposium 99.
Westview Press, Inc., Boulder, CO.

Campbell, A., and Yoskin, Y. 1980. Coastal zone
management in Japan and France. In: Coastal
zone '80. pp. 2814–2827. American Society of
Civil Engineers, New York.

Carlberg, E. 1981. The coastal policy issue in
Sweden.
In: Center for Ocean Management Studies, com-
parative marine policy. pp. 47–55. Ed. by V. K.
Tippie. Praeger, Brooklyn, NY.

Carroll, J. E., and Mack, N. B. 1982. On living
together in North America: Canada, the United
States and international environmental relations.
Denver Journal of International Law and Policy
12:1.

Center for Ocean Management Studies (COMS).
1984. Resource use and use conflicts in the

exclusive economic zone. Times Press, Wakefield, RI.

Churchill, R. R., and Lowe, A. V. 1983. The Law of the Sea. Manchester University Press, Dover, NH.

Clark, J. 1977. Coastal ecosystem management. In: Ecology. pp. 1–49. Conservation Foundation, Washington, DC.

Copes, P. 1981. Marine fisheries management in Canada: Policy objectives and development constraints. In: Center for Ocean Management Studies, comparative marine policy. pp. 135–136. Ed. by V. K. Tippie. Praeger, Brooklyn, NY.

Creech, H. 1986. In search of an ocean information policy. In: Ocean yearbook 6. pp. 15–28. Ed. by E. Borgese, N. Ginsberg, J. Baylson, N. Dunning, and D. Dzurek. University of Chicago Press, Chicago, IL.

Evans, N. 1980. Conflict resolution: Lessons from Grays Harbor. In: Coastal zone '80. pp. 776–790. American Society of Civil Engineers, New York.

Friedheim, R. L. 1975. Ocean ecology and the world political system. In: Who protects the ocean? pp. 151–190. Ed. by J. Hargrove. West Publishing Co., St. Paul, MN.

Gibson, J., and Halliday, J. 1987. In pursuit of an enigma: The coastal zone in management and law. In: The UN convention on the Law of the Sea: Impact and implementation. pp. 441–451. Ed. by E. Brown and R. Churchill. Law of the Sea Institute, Honolulu.

Goldie, L. F. E. 1975. International maritime environmental law today—an appraisal. In: Who protects the oceans? pp. 63–121. Ed. by J. Hargrove. West Publishing Co., St. Paul, MN.

Gordon, W. 1981. Management of living marine resources: Challenge of the future. In: Center for Ocean Management Studies, comparative marine policy. pp. 145–167. Ed. by V. K. Tippie. Praeger, Brooklyn, NY.

Haagsma, A. 1989. The European community's environmental policy: A case study in federalism. Fordham Intl. L. J. 12:311.

Handl, G. 1978. The principle of "equitable use" as applied to internationally shared natural resources: Its role in resolving potential international disputes over transboundary pollution. Revue Belge De Droit International 14:40.

Hardin, G. 1968. Tragedy of the commons. Science 162:1243.

Henkin, L. 1984. International law as law in the United States. Michigan Law Review 82:1555.

Hershman, M. 1981. Coastal zone management in the United States: A characterization. In: Center for Ocean Management Studies, comparative marine policy. pp. 57–63. Ed. by V. K. Tippie. Praeger, Brooklyn, NY.

Hoban, T., and Brooks, R. 1987. Green justice: The environment and the courts. Westview Press, Inc., Boulder, CO.

Janis, M. W. 1988. An introduction to international law. Little, Brown & Co., Boston, MA.

Johnston, D. (Editor). 1978. Regionalization of the Law of the Sea. Ballinger Publishing Co., Cambridge, MA.

Johnston, D. 1984. Conservation and management of the marine environment: Responsibilities and required initiatives in accordance with the 1982 U.N. convention on the Law of the Sea. In: Law of the Sea Institute, The developing order of the oceans. pp. 133–179. Ed. by R. Krueger and S. Riesenfeld. Law of the Sea Institute, Honolulu.

Johnston, D., and Enomoto, L. 1981. Regional approaches to the protection of the marine environment. In: The environmental Law of the Sea. pp. 285–385. Ed. by D. Johnston. International Union for the Conservation of Nature and Natural Resources, Gland, Switzerland.

Keckes, S. 1981. Regional seas: An emerging marine policy approach. In: Center for Ocean Management Studies, comparative marine policy. pp. 17–20. Ed. by V. K. Tippie. Praeger, Brooklyn, NY.

Kindt, J. 1986. Marine pollution and the Law of the Sea. William S. Hein & Co., Inc., Buffalo, NY.

Knight, H. 1975. International fisheries management: A background paper. In: The future of international fisheries management. pp. 1–49. Ed. by H. Knight. West Publishing Co., St. Paul, MN.

Koers, A. 1978. Internal aspects of the common fisheries policy of the European community. In: Regionalization of the Law of the Sea. pp. 81–91. Ed. by D. Johnston. Ballinger Publishing Co., Cambridge, MA.

Kwaitkowski, B. 1988. Conservation and optimum utilization of living resources. In: The Law of the Sea: What lies ahead? pp. 245–275. Ed. by T. Clingan. Law of the Sea Institute, Honolulu.

Levy, J. P. 1988. Towards an integrated marine policy in developing countries. Marine Policy 12:326.

Lie, U. 1985. Marine ecosystems: Research and management. In: Managing the oceans: Resources, research, law. pp. 311–328. Ed. by J. Richardson. Lomond Publications, Inc., Mt. Airy, MD.

Lutz, R. 1976. The laws of environmental management: A comparative study. American Journal of Comparative Law 24:447.

MacRae, L. M. 1983. Customary international law and the United Nations Law of the Sea Treaty. California Western International Law Journal 13:181.

Miovski, L. 1989. Solutions in the convention on the Law of the Sea to the problem of overfishing in the central Bering Sea: Analysis of the convention, highlighting the provisions concerning fisheries and enclosed and semi-enclosed seas. San Diego L. Rev. 26:525.

Myers, J. 1981. America's coasts in the 80's: Policies and issues. Coast Alliance, Washington, DC.

Oxman, B. 1986. Antarctica and the new Law of the Sea. Cornell International Law Journal 19:211.

Peet, G. 1987. Sea use management for the North Sea. In: The UN convention on the Law of the Sea: Impact and implementation. pp. 430–440. Ed. by E. Brown and R. Churchill. Law of the Sea Institute, Honolulu.

Qadar, M. A. 1989. Extended maritime jurisdiction—a case for regional cooperation for the management of fisheries resources in the Bay of Bengal. In: The international implications of extended maritime jurisdiction in the Pacific. pp. 292–305. Ed. by J. Craven, J. Schneider, and C. Stimson. Law of the Sea Institute, Honolulu.

Ray, C. 1970. Ecology, law and the "marine revolution." In: Biological conservation. pp. 7–17. Elsevier Publishing Co., Ltd., London.

Ray, C. 1989. Sustainable use of the global ocean. In: Changing the global environment. pp. 71–87. Ed. by D. Botkin, M. Caswell, J. Estes, and A. Orio. Academic Press, Inc., Boston, MA.

Reisman, M. 1986. The teaching of international law. International Lawyer 20:987.

Restatement, foreign relations law of the United States (Third). 1987. 2 vols. American Law Institute Publishers, St. Paul, MN.

Roe, R. 1987. Some thoughts on the management of interjurisdictional fisheries. In: Coastal states are ocean states. pp. 33–40. Ed. by Gerard J. Mansone. University of Delaware, Center for Marine Studies, Newark.

Ross, D. 1982. Introduction to oceanography. 3d ed. Prentice Hall, Inc., Englewood, NJ.

Shaw, M. N. 1986. International law. 2d ed. Grotius Publications Ltd., Cambridge, England.

Sherman, K., and Alexander, L. M. (Editors). 1986. Variability and management of large marine ecosystems. AAAS Selected Symposium 99. Westview Press, Inc., Boulder, CO.

Sherman, K., and Alexander, L. M. (Editors). 1989. Biomass yields and geography of large marine ecosystems. AAAS Selected Symposium 111. Westview Press, Inc., Boulder, CO.

Sherman, K., Alexander, L. M., and Gold, B. D. (Editors). 1990. Large marine ecosystems: Patterns, processes, and yields. AAAS Symposium. AAAS, Washington, DC.

Simon, A. 1978. The thin edge—coast and man in crisis. Harper & Row, New York.

Simon, A. W. 1984. Neptune's revenge. Franklin Watts, New York.

Smith, G. 1984. The United Nations and the environment: Sometimes a great notion? Texas Intl. L. Rev. 19:335.

Sohn, L. B. 1984. The Law of the Sea crisis. St. John's Law Review 58:237.

Sohn, L. B. 1985. The Law of the Sea: Customary international law developments. American University Law Review 34:271.

Valencia, M. 1988. The Yellow Sea: Transnational marine resource management issues. Marine Policy 12:382.

Waite, C. 1981. Coastal management in England and Wales. In: Center for Ocean Management Studies, comparative marine policy. pp. 65–74. Ed. by V. K. Tippie. Praeger, Brooklyn, NY.

Wenk, E. 1981. Global principles for national marine policies: A challenge for the future. In: Center for Ocean Management Studies, Comparative Marine Policy. pp. 3–16. Ed. by V. K. Tippie. Praeger, Brooklyn, NY.

Wilkinson, C., and Connor, D. 1983. The law of the Pacific salmon fishery: Conservation and

allocation of a transboundary common property resource. Kansas Law Review 32:17.

Yuru, L. 1985. Amassing scientific knowledge to preserve the marine environment. In: Managing the ocean: Resources, research, law. pp. 125–129. Ed. by J. Richardson. Lomond Publications, Inc., Mt. Airy, MD.

Cases

Case concerning delimitation of the maritime boundary of the Gulf of Maine (Gulf of Maine case). 1984. I.C.J. Rep. 246 (para. 94). Reprinted in International Legal Materials 23:1197.

S. S. *Lotus* case (France v. Turkey). 1927. P.C.I.J., Ser. A, No. 10.

Trail Smelter case (U.S. v. Canada). 1941. 3 U.N. Rep. Arb. Awards 1911.

United States v. Flores. 1933. 289 U.S. 137.

Other materials

Charter of Economic Rights and Duties of States. 1975. G.A. Res. No. 3281. Reprinted in International Legal Materials 14:252.

Convention for the Conservation of Antarctic Marine Living Resources (CCAMLR). Done May 7, 1980. T.I.A.S. No. 8826. Reprinted in International Legal Materials 19:841 (1980).

Convention for the Protection and Development of the Marine Environment of the Wider Caribbean Region. 1984. Treaty Doc. No. 98–13. 98th Cong. Second Session. International Legal Materials 22:227.

Convention for the Protection of the Natural Resources and Environment of the South Pacific Region (November 25, 1986). 1987. International Legal Materials 26:38.

Draft World Charter for Nature. 1980. G. A. Res. 35/7. 35 U.N. GAOR Supp. (No.48). U.N. Doc. a/35/48. Reprinted in International Legal Materials 20:462.

MFCMA. Fishery Conservation and Management Act of 1976 (later retitled Magnuson Fishery and Conservation Management Act). 1976. Public Law No. 94–265. 90 Stat, 331 (codified in 16 U.S.C. Sections 1801–1882) (1976, 1982, and Supp. III 1985).

NEPA (National Environmental Policy Act) of 1969. 1986. Public Law No. 91–90. 83 Stat. 852 (codified at 42 U.S.C. Sections 4321–4370) (1986).

OCSLAA (Outer Continental Shelf Lands Acts Amendments) of 1978. 1981. Public Law No. 95–372. 92 Stat. 629. 16 U.S.C. Sections 1456–56a, 1464; 43 U.S.C. 1331–56, 1801–1866 (Supp. 1981).

Office of the Special Representative of the Secretary-General for the Law of the Sea. 1986. The Law of the Sea: National legislation on the exclusive economic zone, the economic zone, and the exclusive fishery zone. U.N. Sales No. E.85.V.10.

Stockholm Declaration. Report of the U.N. conference on the human environment. 1972. U.N.

Doc. A/Conf. 48/14. Reprinted in International Legal Materials 11:1416.

UNCLOS (United Nations Convention on the Law of the Sea). 1982. U.N. Doc. A/Conf. 62/121. Reprinted in International Legal Materials 21:1245.

UNEP (United Nations Environmental Program) Working Group. 1987. Report of the working group of experts on environmental law on its second session on environmental impact assessment. UNEP.WG.152/4.

Vienna Convention on the Law of Treaties. 1969. U.N. Doc. A/Conf. 39/27. Reprinted in International Legal Materials 8:679.

Chapter 21

Ocean Management and the Large Marine Ecosystem Concept: Taking the Next Step

Robert W. Knecht and Biliana Cicin-Sain

Introduction

The goal of this chapter is to begin to bridge the gap between two evolving concepts—large marine ecosystems (LMEs) and ocean management. Application of the LME concept to the management of ocean resources and ocean space, potentially a highly desirable notion, must overcome a number of obstacles related to existing political and jurisdictional realities. Creative measures are needed to ensure that these political and jurisdictional realities do not impede the timely application of the LME concept to ocean management in individual nations, as well as to multinational efforts at ocean management.

First, the chapter explores the concept of ocean management, focusing on some of the differences among traditional approaches to management of individual marine resources and the more integrated ocean management approach now gaining favor. The goals and functions of ocean management are examined in some detail and early attempts at integrated management are discussed. The regional dimension of ocean issues is then examined as a prelude to a discussion of the application of the LME approach to ocean management.

The Concept of Ocean Management

Inherent difficulties with the present system of ocean resources management

The system for managing ocean resources in the United States has evolved over the last 2 to 3 decades. Although the system is functional in the sense that management decisions are be-ing made (except, perhaps, in the case of oil and gas leasing in frontier areas), it is beset by a number of problems. First, the ocean management system in this country is fragmented both geographically and functionally. Different management schemes exist for the same resource within state and federal waters, and within both state and federal waters, management regimes tend to be single-purpose or single-use oriented. The pieces of legislation, sets of regulations, and institutional approaches that are found in the schemes to manage oil and gas in state waters are different from those of the systems used to manage the same resource in federal waters. The same applies to fisheries, navigation, marine protected areas, and the like.

The fragmented, single-purpose orientation of the existing management system poses a number of difficulties and shortcomings. First, and perhaps most important, it is difficult to examine the effects of management decisions in one resource sector on other resource sectors and uses. It is also difficult to approach the question of multiple uses in a given ocean area. Presently, there is no neutral forum or framework within which the "highest and best use" or uses of a given area can be hammered out. Similarly, it is difficult to make trade-off decisions among uses, given that each management decision regarding a use or resource is embedded within a particular (and different) management framework (Cicin-Sain and Knecht, 1985).

Another substantial shortcoming of the present system of ocean resource management is the weak scientific basis upon which such management decisions often rest. Although our understanding of the behavior of the ocean and its processes has advanced significantly in recent years, it is still woefully inadequate for predicting the effects and consequences of particular ocean activities. Clearly, one of the big attractions of the LME concept is the possibility that a management system based on such an approach would be one in which the management decisions would be solidly grounded.

Many challenges exist along the way to the development of a more rational approach to ocean management. We believe that the first challenge is to move away from the single-purpose ocean management framework toward a multiple-use approach. It is clear that the coastal ocean adjacent to the United States will see increased use in the coming years, with subsequent increases in conflicts and a need for multiple-use management approaches.

One of the major problems in devising an equitable multiple-use management approach relates to the specialized knowledge and expertise currently associated with single-purpose management frameworks. How can this valuable expertise be retained within the management system while, at the same time, the system is broadened to include a multiple-use orientation? Also, how can a multiple-use management system properly reflect the public and private values that are at stake (Cicin-Sain, 1992)? Finally, politically, it will be difficult to overcome the entrenched interests who favor single-use/resource approaches (Juda and Burroughs, 1990). One can easily imagine, for example, the intensity with which commercial fishing interests would fight to maintain a free-standing management regime devoted solely to fisheries.

Characteristics of ocean management

Before going further, it would seem wise to define "ocean management" in the broader multiple-use context. Borrowing from Juda and Burroughs (1990, p. 27):

> Ocean management seeks, in accordance with some system of politically determined values which is either explicit or implicit, to increase the benefits which may be derived from the resource and non-resource uses of the ocean. At the same time, ocean management attempts to ameliorate conflict of use situations. In general, ocean management tries to provide for a directed balance between and among the various uses of ocean space, as well as to protect the ocean environment from damage to its viability.

Cicin-Sain *et al.* (1990) differentiate between ocean management and marine resources management. Ocean management, as they see it, has a multiple-use focus, an area orientation, and involves policy integration. Marine resources management, on the other hand, focuses on the control and allocation of single resources and or uses. Ocean management, of course, involves the management of human uses and government actions and not the resources directly.

It is useful to consider the governmental functions that an ocean management framework should perform. We see four specific functions in this regard:

- To exercise stewardship responsibilities over the resources found in the ocean and the ocean environment itself on behalf of current and future generations, past generations with a special claim on ocean resources, various publics in the nation-state, private lease holders, and the international community.
- To maintain public safety and order.
- To control and regulate interactions among multiple uses (and users) with equity as a guideline.
- To ensure an adequate return for the public from the exploitation of publicly owned ocean resources.

With regard to the stewardship function listed above, as we have discussed elsewhere (Cicin-Sain and Knecht, 1985), when nation-states proclaim "sovereign rights" over the resources of their 200-mile zones (as they do when they proclaim exclusive economic zones [EEZs]), they accept certain obligations and duties as well. Those obligations include managing renewable resources so that they are as available to future generations as they are to the present one.

Finally, with regard to the development of conceptual models for ocean management, relevant examples can be found in several areas. For example, multiple-use management schemes exist in a number of marine protected areas such as the Great Barrier Reef in Australia (Kenchington, 1991). In such areas, zoning has been used to control and manage multiple uses within fixed areas. Relevant experience in multiple-use management is also being gained in the United States within the National Estuary Program as it seeks to put water-body management programs in place in a number of the nation's important estuaries (EPA, 1990). At the international level, some of the regional seas programs are beginning to attempt mul-

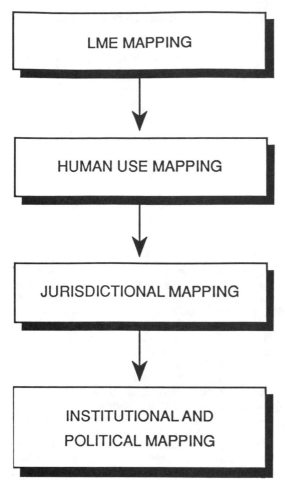

Figure 1. Steps in the application of the LME approach to ocean management.

tiple-use ocean management on a region-wide basis. A good example is in the South Pacific, where the nations of the South Pacific Forum have recently constructed regimes involving access to fisheries, ocean dumping, and other environmental issues, and freedom from nuclear activity within a common ocean space (Cicin-Sain and Knecht, 1989).

Regional Approaches and Ocean Management

Because LMEs are defined regionally, we now turn to the question of regional approaches to ocean management. Recently, several U.S. states and territories conducted studies of regional issues confronting them. These studies were conducted in the Pacific among American flag territories and the state of Hawaii (Pacific Basin Development Council, 1989)

and along the West Coast and Alaska (Cicin-Sain *et al.*, 1990). These studies suggest that ocean-related problems in a region tend to be of two types—common problems and shared problems. Common problems are those similar problems faced by a number of states (such as coastal erosion, nonpoint source pollution problems, and inappropriate coastal development). Shared problems, on the other hand, are problems where the states or territories are physically or economically linked by the resources, environments, or activities concerned (for example, sea-level rise, oil transportation planning, and management of shared resources such as migratory species).

Indeed, even in the absence of the emergence of the LME approach, institutional actors both at the national and subnational levels are discovering that actions taken on a regional basis to address ocean issues frequently make the most sense. The conservation and management of living marine resources and their habitats is often best done on a regional basis, given the transboundary nature of many resources. Risks associated with offshore developments adjacent to one political entity (such as oil spills) can affect sensitive areas in other entities. Economic development of EEZ resources off one state or territory can have consequences (both positive and negative) for the economies of adjacent states (Cicin-Sain *et al.*, 1990).

Hence, beginning in the mid-1970s (in the United States with the 1976 enactment of the Magnuson Fishery Conservation and Management Act), there has been a trend toward the use of regional approaches to the management of certain ocean resources. Internationally, the evolving Regional Seas Program has set the stage for the application of regional approaches in many other parts of the world.

The LME Approach and Ocean Management

We now turn to the question, "What can the LME approach provide to an ocean management framework?" Based on our understanding of the LME approach and what it offers, we see five potential benefits.

- Almost by definition, a better scientific understanding exists within an LME than in any other management area. Indeed, to the extent that this is true, it will provide an improved scientific foundation for all aspects of resource and environmental management.
- Orienting a management framework to an LME should permit more accu-

rate determination of appropriate boundaries for the management area. To the extent that management areas can be based on scientific understanding and not political subdivisions, improved management should take place.

- An LME approach provides a framework for understanding the physical and biological properties of a marine area. Gaps in understanding should be more easily discernable within the LME framework and, hence, research should be more readily guided to those questions.
- The LME brings greater scientific understanding to the ocean management process and with it the potential for improved approaches to conflict resolution.
- Finally, the improved understanding of the important processes (physical and biological) occurring within a management area should allow possible increases in overall resource benefits (for example, safely taking larger numbers of a particular fish species).

In Fig. 1, we illustrate a number of steps that will be helpful in attempting to link the LME concept to ocean management. We can visualize three additional kinds of mapping in addition to the LME itself. These involve (i) human-use mapping, (ii) jurisdictional mapping, and (iii) institutional and political mapping. These steps, in effect, provide the linkage between the physical and biological ocean system, on one hand, and the legal, institutional, and political systems associated with various uses within that ocean system on the other.

With regard to human-use mapping, the following variables need to be examined:

- What uses are occurring where (within the LME)?
- From where are these uses emanating?
- How are they being regulated/managed and by whom?
- With what other uses are they interacting and how?
- How are the other uses being regulated and by whom?

With regard to jurisdictional mapping, the following questions need to be examined:

- What (legal) jurisdictions are involved in the LME? What regulatory activity exists within each jurisdiction? What cross-jurisdictional efforts exist, if any?

- What areas are involved in each jurisdiction?

Concerning the political and institutional mapping, the following questions need to be examined:

- What agencies/institutions have responsibility for jurisdictions within the LME? What are their goals and values for the ocean area involved? What incentives can be devised to entice different political jurisdictions to cooperate with one another?
- How can political cooperation be fostered within an LME?

In our judgment, it is difficult to overestimate the importance of political and institutional factors when considering the potential for improved ocean management. We fear that the great promise LMEs could bring to improved management could come to naught unless equal attention is paid to this side of the equation. The 1980 Canberra Convention (the Convention on the Conservation of Antarctic Marine Living Resources) provides a case in point (Scully, this volume). While the convention was very forward looking in its embrace of the ecosystem approach, its handling of institutional and political issues was less than optimum. The fact that a consensus is needed among nation-state parties in order to take management action has resulted in a very slow start under the convention (Friedheim and Akaha, 1988).

Some Suggestions for Next Steps

In our judgment, the next step in encouraging the application of the LME concept in ocean management would be to test the approach in various advantageous settings. Three such settings are described below.

- *LMEs within a nation-state.* A number of already delimited LMEs exist entirely within (or almost entirely within) national EEZs. For example, in the United States, LMEs exist along the east and west coasts, in the Gulf of Mexico, and in Alaskan waters. It should be possible to convince regional groupings of U.S. coastal states and appropriate federal agencies to experiment with the LME approach to improve management in their jurisdictions. The current Gulf of Maine initiative, though not exclusively domestic in nature, could provide an example of a successful "marriage" of regional management and the LME approach.

- *Enclosed or semi-enclosed seas.* Here, promising regional seas programs would be identified and technical assistance for potential application of the LME approach provided. Conceivably, the Humboldt Current system along the west coast of South America might offer a suitable context for such an effort.

- *Special focus on situations where an LME coincides with regional political groupings.* It stands to reason that one of the best places to test the usefulness of the LME in ocean management would be in a place where the physical dimensions of the LME coincide roughly with the dimensions of a political grouping (such as in the Gulf of Mexico, the Gulf of Maine, or perhaps, the Caribbean Sea). Where such congruence exists, special efforts should be mounted to encourage existing regional management entities to consider the LME concept. At the very least, such efforts could lead to more focused research and improved understanding of the important physical and biological systems present.

To move the LME approach beyond the present level (principally involving scientific conferences and publications) probably will require that an existing international organization embrace the concept, because the steps outlined above will require communication, organization, and international initiatives. Of the various possible organizations, the World Conservation Union (formerly the International Union for the Conservation of Nature [IUCN]) may be most appropriate for this purpose, because it has an established scientific reputation and a tradition of forward-looking initiatives of this kind. Furthermore, the World Conservation Union, being a hybrid organization consisting of both national governments and inter- and nongovernmental organizations, is in a good position to disseminate the LME concept along a number of pathways.

Conclusion

Humankind and its institutions, values, and uses of the ocean are as much a part of LMEs as are ocean currents or biological trophic levels. We must appreciate that our social inventions are very much a part of the challenge—as are our economic and political systems and their operation, our regulatory organizations (often under the influence of the regulated), and our national and international management institutions with their healthy respect for bureaucratic turf.

We must be concerned with "funding chains" as well as "food chains," with "patterns of power and influence," as well as "patterns of predator and prey," with "value systems" as well as "current systems." We must strive to understand the dimensions of the human and social aspects of LMEs with the same enthusiasm and zeal that we map the parameters that describe the physical characteristics.

We are not calling for less science, but rather for more work on the human and social side of the problem—economics, law, policy, and institutions—where, after all, much of the problem lies.

References

Cicin-Sain, B. 1992. Introduction to "Values and the American ocean." Special issue of Ocean and Shoreline Management 17(3–4):193.

Cicin-Sain, B., Hershman, M., Hildreth, R., and Isaccs, J. 1990. Improving ocean management capacity in the Pacific coast region: State and regional perspectives. National Coastal Resources Research and Development Institute Report, July 1990. Newport, OR.

Cicin-Sain, B., and Knecht, R. W. 1985. The problems of governance of the U.S. ocean resources and the new exclusive economic zone. Ocean Dev. Int. Law 15(3–4):289–320.

Cicin-Sain, B., and Knecht, R. W. 1989. The emergence of a regional ocean management in the South Pacific. Ecol. Law Q. 16(1):171–215.

EPA (Environmental Protection Agency). 1990. Progress in the national estuary program: Report to Congress. EPA 503/9–90–005, February 1990. Washington, DC.

Freidheim, R., and Akaha, T. 1988. Antarctic resources and international law: Japan, the U.S., and the future of Antarctica. Ecol. Law Q. 16(1):119–154.

Juda, L., and Burroughs, R. H. 1990. The prospects for comprehensive ocean management. Mar. Policy 14(1):23–35.

Kenchington, R. 1991. Tourism development in the Great Barrier Reef Marine Park. Ocean and Shoreline Management 15(1):57.

Pacific Basin Development Council. 1989. Pacific basin management of the 200-mile exclusive economic zone: Implications for the EEZ for the American flag Pacific Islands. Honolulu.

Scully, R.T. 1993. Convention on the Conservation of Antarctic Marine Living Resources. This volume, chapter 21.

Chapter 22

Convention on the Conservation of Antarctic Marine Living Resources

R. Tucker Scully

Introduction

The Convention on the Conservation of Antarctic Marine Living Resources (CCAMLR) represents an international effort to apply an ecosystem-wide approach to the conservation of living resources. It is the first international agreement that defines its area of application and objectives by specific reference to a marine ecosystem. CCAMLR applies to the populations of all species of living organisms found south of the Antarctic Convergence or Polar Front. Its conservation objective includes maintenance of ecological relationships between harvested, dependent, and related populations; restoration of the health of depleted populations; and prevention of changes in the marine ecosystem that are not potentially reversible over 2 or 3 decades.

The convention, which entered into force in 1982, establishes both a commission and a scientific committee for CCAMLR, as well as a secretariat to serve both. To date, the commission and scientific committee have held 10 annual meetings, with the most recent scheduled from October 21 to November 1, 1991, at its headquarters in Hobart, Tasmania, Australia.

This chapter examines the ongoing effort to implement CCAMLR with reference to the lessons it may offer for the methodology of managing large marine ecosystems (LMEs).

The Origins of the Convention

The origins of CCAMLR are to be found in scientific research carried out in the Southern Ocean. The Antarctic scientific community played an important role in identifying the need for the conservation of living resources found in Antarctic waters. Research coordinated by the Scientific Committee on Antarctic Research (SCAR), during and following the International Geophysical Year (IGY) of 1957–1958, began to sketch out a picture of the structure and components of the Antarctic marine ecosystem. SCAR sponsored major symposia on Antarctic oceanography in 1966 and 1970.

The late 1960s and early 1970s also saw the emergence of commercial interest in the fish stocks found in Antarctic waters. These waters, which earlier in the century had been the area of intense harvesting and overharvesting of marine mammal populations (seals and whales), began to attract distant-water fishing fleets, particularly from the former U.S.S.R. Perceptions of the fisheries' potential—in the short run for finfish, in the long run for krill—stemmed in part from research activities, as did realization of the possible effects of uncontrolled harvesting of fisheries resources, particularly in light of the emerging understanding of the vulnerability of the Antarctic marine ecosystem to harvesting, with the heavy dependence of its predators on a single species, Antarctic krill (*Euphausia superba*).

As a consequence, SCAR initiated steps to orient research in Antarctic waters toward living resources and an understanding of the Antarctic marine ecosystem. In August 1972, SCAR's working group on biology established a subcommittee on the living resources of the Southern Ocean. This group received the official cosponsorship of the Scientific Committee on Oceanographic Research (SCOR) in 1975 and was upgraded by SCAR to the "group of specialists on the living resources of the Southern Ocean" in 1976. The group identified the wise management of the Antarctic marine ecosystem as a primary criterion in determining research priorities.

The imperative for building the empirical basis for conservation of Antarctic marine living resources, first elaborated within the scientific community, was acknowledged on the political level by the Antarctic Treaty Consultative Parties (ATCPs) at the Eighth Antarctic Treaty Consultative Meeting (ATCM VIII) in 1975. Recommendation VIII–10 (ATCM, 1975) of that meeting called for encouragement of studies which could lead to the development of effective measures for the conservation of Antarctic marine living resources and urged SCAR to continue its scientific work on these matters.

More specifically, ATCM VIII endorsed the convening of a meeting by SCAR to address programs for the study and conservation of Antarctic marine living resources. The resulting First International Symposium on Living Resources of the Southern Ocean, took place in Woods Hole, Massachusetts, in August 1976, immediately followed by a meeting of the SCAR/SCOR group of specialists. From these meetings emerged the Biological Investigations of Marine Antarctic Systems and Stocks (BIOMASS) program. BIOMASS was designed as a 10-year cooperative international and interdisciplinary research program with the principal objective of gaining a deeper understanding of the structure and dynamic functioning of the Antarctic marine ecosystem as a basis for future management of potential living resources.

In response, the ATCPs initiated discussion at the political level of mechanisms to meet the conservation requirements of the Antarctic marine ecosystem. The possibility of negotiating an agreement to deal with Antarctic marine living resources was raised at meetings—in Paris in 1976 and in London in July 1977—to prepare for the Ninth Antarctic Treaty Consultative Meeting (ATCM IX). By the time of ATCM IX (September 1977, also in London), the ATCPs were prepared to commit themselves to conclusion of a definitive regime for the conservation of Antarctic marine living resources (ATCM, 1977, Recommendation IX–2). Recommendation IX–2 recognized both the need to establish a good scientific foundation for appropriate conservation measures and the urgency of ensuring that these resources are protected by the establishment of sound conservation measures that will prevent overfishing and protect the integrity of the Antarctic ecosystems.

Recommendation IX–2 also provided for the establishment of a special negotiating process (a special consultative meeting) to con-

clude the definitive regime and *inter alia* directed that the regime provide for the effective conservation of the marine living resources of the Antarctic ecosystem as a whole. The emphasis upon providing for the effective conservation of the Antarctic ecosystem as a whole was also reflected in the description of the area to be covered by the definitive regime. Paragraphs III 3(d) and (e) of Recommendation IX–2 (ATCM, 1977) provide:

(d) The regime should cover the area of specific competence of the Antarctic Treaty.

(e) The regime should, however, extend north of 60° south latitude where that is necessary for the effective conservation of species of the Antarctic ecosystem, without prejudice to coastal state jurisdiction in that area.

Negotiation of the Convention

The negotiations outlined in Recommendation IX–2 were initiated 5 months after its adoption in Canberra, Australia (February–March 1978). The negotiations resulted in the adoption of CCAMLR, also in Canberra, on May 20, 1980. Members of the Antarctic scientific community, as members of ATCP delegations, played an important part in the negotiations—particularly in the elaboration of the objectives of CCAMLR, the definition of the area of the convention, and the functions of the CCAMLR scientific committee.

CCAMLR's provisions reflected understanding of the nature of the Antarctic marine ecosystem, the importance of regional approaches to conservation, and the necessity of establishing a sound scientific basis for research management decisions. From one perspective, these facets mirrored the development of international obligations to conserve living resources, articulated in the emerging provisions of the then draft United Nations Convention on the Law of the Sea (UNCLOS).

Three aspects of CCAMLR are significant in this regard: (i) the definition of the area to which it applies; (ii) the objective of the Convention (Art. II, which seeks to articulate the ecosystem management approach); and (iii) the data requirements (both in reporting obligations and in the functions of the institutions established by CCAMLR).

Area

Article I of CCAMLR defines the area and the resources to which the convention applies. It is believed to be the first international agreement that identifies its area of application by reference to an ecosystem and seeks to describe the components and spatial extent of that ecosystem. As provided for in Recommendation IX–2, CCAMLR applies to Antarctic marine living resources south of 60° south latitude (the area of the Antarctic Treaty) and to such resources of "the area between that latitude and the Antarctic Convergence which form(s) part of the Antarctic marine ecosystem" (CCAMLR, 1980, Art. I, para. 1).

The northern limit of CCAMLR's area of application therefore is established as the Antarctic Convergence. The Convergence or Polar Front, as it is often called, is a transition zone within which colder Antarctic waters from the south mix with and sink below warmer sub-Antarctic waters from the north. It represents a significant environmental barrier that many species do not cross and has been viewed as the northern boundary of purely Antarctic populations. Although the Convergence is an oceanographic phenomenon, a mixing zone that varies in time and space in response to physical conditions, CCAMLR sets forth geographic coordinates to approximate its location for its regulatory purposes (CCAMLR, 1980, Art. I, para. 4). Antarctic marine living resources are, in turn, defined as the populations of all species of living organisms found south of the Convergence (Art. I, para. 2), and the Antarctic marine ecosystem as "the complex of relationships of Antarctic marine living resources with each other and with their physical environment" (Art. I, para. 3).

The manner in which the area of application of CCAMLR is defined reflects growing emphasis upon regional approaches to management of living resources. The articles of UNCLOS relating to the conservation of living resources, including Art. 61 on the conservation of the living resources of the Exclusive Economic Zone (EEZ); Art. 63 on "straddling stocks"; Art. 64 on highly migratory species; and Arts. 118 and 119 on the conservation of the living resources of the high seas, all refer to appropriate regional and subregional organizations as vehicles for managing living resources. CCAMLR takes the concept a step further in defining a region by reference to an ecosystem. CCAMLR is the first international agreement to attempt to delineate an LME for resource management purposes.

Objective

An ecosystem approach to management is articulated in more operational terms in Art. II of CCAMLR (1980), which sets forth the convention's objective. The objective of CCAMLR is defined as the conservation of Antarctic marine living resources, where conservation is understood to include the "rational use" of such resources. Article II, paragraph 3, prescribes three principles of conservation in accordance with which any harvesting and associated activities are to be conducted.

The first of these principles sets forth a standard with respect to populations that are the targets of harvesting:

(a) Prevention of decrease in the size of any harvested population to levels below those which ensure its stable recruitment. For this purpose, its size should not be allowed to fall below a level close to that which ensures the greatest net annual increment.

The second principle establishes a standard for populations dependent upon or related to harvested populations and for depleted populations:

(b) Maintenance of the ecological relationships between harvested, dependent, and related populations of Antarctic marine living resources and restoration of depleted populations to the levels defined in subparagraph (a), above.

The third principle elaborates a standard applicable to the marine ecosystem as a whole, introducing the need to avoid irreversible changes in that ecosystem.

(c) Prevention of changes or minimization of the risk of changes in the marine ecosystem which are not potentially reversible over two or three decades, taking into account the state of available knowledge of the direct and indirect impact of harvesting, the effect of the introduction of alien species, the effects of associated activities on the marine ecosystem, and of the effects of environmental changes, with the aim of making possible the sustained conservation of Antarctic marine living resources.

CCAMLR's ecosystem approach to management, reflected in particular in the definition of its area of application and its objective, although justifiably characterized as innovative, also reflects general trends toward multispecies management driven by evolution in scientific knowledge and capability. The provisions of UNCLOS illustrate these trends. In fulfillment of their obligations to conserve living resources both in EEZs and on the high seas (United Nations Convention on the Law of the Sea, 1982), nation-states are called upon to take measures, based on the best scientific evidence available, to

> Maintain or restore populations of harvested species at levels which can produce the maximum sustainable yield as qualified by relevant environmental and economic factors ... and taking into account fishing patterns, the interdependence of stocks and any generally recommended international minimum standards, whether subregional, regional, or global; and (to) take into consideration the effect on species associated with or dependent upon harvested species with a view to maintaining or restoring populations of such associated or dependent species above levels at which their reproduction may become seriously threatened.

Data

As illustrated by the provisions of CCAMLR, prosecution of an ecosystem management approach is data-dependent. The convention's data collection and reporting obligations and the functions ascribed to its institutions emphasize the necessity of adequate databases for making management decisions. Members of the commission, established by CCAMLR, are required "To the greatest extent possible, (to) provide annually to the commission and to the scientific committee such statistical, biological and other data and information as the commission and scientific committee may require in the exercise of their functions" (CCAMLR, 1980, Art. XX, para. 1). More specifically, commission members are obligated to provide "in the manner and at such intervals as may be prescribed, information about their harvesting activities, including fishing areas and vessels, so as to enable reliable catch and effort statistics to be compiled" (Art. XX, para. 2). Finally, commission members agree that in their harvesting activities "advantage ... be taken of opportunities to collect data needed to assess the impact of harvesting" (Art. XX, para. 4).

CCAMLR's provisions here again parallel the more general evolution of international legal obligations to base management decisions on scientific data. Articles 61 (conservation of the living resources [of the EEZ]) and 119 (conservation of the living resources of the high seas) of UNCLOS (1982) both include the following provision:

Available scientific information, catch and fishing effort statistics, and other data relevant to the conservation of fish stocks shall be contributed and exchanged on a regular basis through competent international organizations whether subregional, regional, or global, where appropriate and with participation by all states concerned.

On this point, CCAMLR's obligations are more elaborate, extending to the collection, as well as to the reporting, of necessary data and information.

Recognition of the data-intensive nature of the ecosystem management approach is also reflected in the functions of CCAMLR's institutions. CCAMLR provides for a commission (composed of the original signatories plus acceding parties during such time as they are engaged in research on or harvesting of Antarctic marine living resources—now numbering 20) and a scientific committee (in which all commission members are entitled to participate), as well as a secretariat to serve both.

From one perspective, the institutions provided for by CCAMLR resemble those of traditional multilateral fisheries agreements: a commission consisting of the parties that decides upon management measures (in this case, by consensus) and a scientific committee in which the players are again the parties rather than independent scientists. What is unique in CCAMLR is the emphasis upon data and information requirements in the functions of the institutions.

The scientific committee is called upon to carry out such activities as may be directed by the commission. However, it is also accorded independent functions to develop the basis for implementing CCAMLR's ecosystem approach. It is to provide a forum for consultation and cooperation concerning the collection, study, and exchange of information with respect to Antarctic marine living resources and to encourage and promote cooperation in the field of scientific research in order to extend knowledge of the marine living resources to the Antarctic marine ecosystem.

The specific functions of the scientific committee (CCAMLR, 1980, Art. XV, Para. 2) are as follows:

(a) Establish criteria and methods to be used for determinations concerning the conservation measures referred to in Article IX of this Convention.

(b) Regularly assess the status and trends of the populations of Antarctic marine living resources.

(c) Analyze data concerning the direct and indirect effects of harvesting on the populations of Antarctic marine living resources.

(d) Assess the effects of proposed changes in the methods of levels of harvesting and proposed conservation measures.

(e) Transmit assessments, analyses, reports, and recommendations to the Commission as requested or on its own initiative regarding measures and research to implement the objective of this Convention.

(f) Formulate proposals for the conduct of international and national programs of research into Antarctic marine living resources.

To some extent, it is the specificity and independence in scientific committee functions, rather than the references to scientific data and information, per se, that are significant. The importance of sound scientific data and information to the achievement of CCAMLR purposes is more striking in the catalogue of the functions of the commission. The overall function of the commission—as the political, policy-making institution—is to effect the objective and principles set out in Art. II (see above). Specifically, the commission is called upon to be responsible for the following tasks (CCAMLR, 1980, Art. IX, Para. 1):

(a) Facilitate research into and comprehensive studies of Antarctic marine living resources and of the Antarctic marine ecosystem.

(b) Compile data on the status of and changes in populations of Antarctic marine living resources and on factors affecting the distribution, abundance, and productivity of harvested species and dependent or related species or populations.

(c) Ensure the acquisition of catch and effort statistics on harvested populations.

(d) Analyze, disseminate, and publish the information referred to in subparagraphs (b) and (c) above and the reports of the scientific committee.

(e) Identify conservation needs and analyze the effectiveness of conservation measures.

(f) Formulate, adopt, and revise conservation measures on the basis of the best scientific evidence available, subject to the provision of paragraph 5 of this Article.

(g) Implement the system of observation and inspection established under Article XXIV of this convention.

(h) Carry out such other activities as are necessary to fulfill the objective of this convention.

The first four of these tasks are specifically directed toward establishing the scientific data and information bases and analytical capability necessary to pursue CCAMLR's ecosystem approach. Efforts to implement these functions by the commission and scientific committee are essential aspects of the operation of CCAMLR.

The Operation of CCAMLR

Much of the effort within the institutions of CCAMLR (particularly the scientific committee) has been directed at developing the capability to carry out the assessments required to give effect to CCAMLR's objective and principles. This effort has proceeded on a number of fronts: identification of data needs and formats, addressing of methodological problems, examination of research priorities, and creation of necessary institutional structure. At the past two annual meetings, the commission and scientific committee have also sought to come to grips with the crucial link between the application of assessment capability to management objectives—the need to address management policies or conservation strategies.

At its second annual meeting, in 1983, the CCAMLR scientific committee established an ad hoc working group on data collection and handling. The ad hoc working group developed formats for the presentation of inventories of past data from commercial fisheries and on scientific data from research activities in Antarctica, as well as a suggested format for data on fishing operations in the convention area—with separate provisions for fish and for krill. The commission endorsed these formats and called for members to provide data in accordance with them.

This initial work recognized the need for the acquisition of time series of detailed catch and effort data, and scientific data and information, as a basis for population assessments—hence the emphasis upon provision of historic data and agreed criteria for collection of future data. There also emerged a perception that the different characteristics of fish and krill necessitated different data protocols for each.

The ad hoc working group on data collection and handling was disbanded at the third annual CCAMLR meetings (1984), with three successor groups established, to pursue the lines of endeavor identified at the previous session. These new groups were (i) the ad hoc group on krill research priorities, (ii) the ad hoc group on fish stock assessment, and (iii) the ad hoc working group on ecosystem monitoring.

Krill

The ad hoc group on krill operated during the third meeting of the scientific committee. As a result of its deliberations, it was recognized that the unique characteristics of krill, as well as uncertainties regarding its life history, required the development of new methodologies of population assessment and, thus, would generate new ways of collecting and reporting data and information. As an initial step toward addressing this question, the scientific committee convened a workshop on krill catch per unit effort (CPUE) prior to its fourth meeting in 1985 and authorized a krill CPUE simulation study to examine development of models for krill populations.

At its sixth meeting (1987), the scientific committee proposed, and the commission approved, upgrading the working group to the ad hoc working group on krill, with a mandate, *inter alia*, to recommend actions with respect to krill stock assessment and ecosystem monitoring. A year later (1988), the commission decided to establish it as a permanent working group on krill, with the following terms of reference (CCAMLR Report of the 7th Meeting of the Scientific Committee, 1988, Para. 55, p. 14; CCAMLR, Report of the 7th Meeting of the Scientific Committee, Para. 2.26, p. 10):

(1) Review and evaluate methods and techniques for estimating krill abundance, taking note of the effects of patchiness and the influences of the physical environment.
(2) Review and evaluate information concerning the size, distribution, and composition of commercial krill catches, including likely future trends in these catches.
(3) Liaise with the working group for the CCAMLR Ecosystem Monitoring Program for assessing any impact of changes in krill abundance and distribution on dependent and related species.
(4) Evaluate the impact on krill stocks and krill fisheries of current and possible future patterns of harvesting, including changes brought about through management action, in order that the committee may formulate appropriate scientific advice on krill to the commission.
(5) Report to the scientific committee on information and data required from commercial krill fisheries.

The krill working group, at an intersessional meeting held in 1989 (between the seventh and eighth annual meetings), con-

sidered the results of the krill simulation study. Resulting recommendations made to the scientific committee led to commission endorsement of provisions on reporting of fine-scale catch data for krill in the primary statistical subareas where such fishing takes place.

Fish

The ad hoc group on fish stock assessment, established in 1984, was converted into a formal standing working group at the 1987 meetings. Its terms of reference (CCAMLR Report of the 6th Meeting of the Commission, 1987, Para. 49, p. 14; CCAMLR Report of the 6th Meeting of the Scientific Committee, 1987, Para. 5.71; p. 34–35) are to:

(a) Apply and develop methodologies for fish stock assessment, including
  (i) procedures for monitoring fish stock abundance and population structure;
  (ii) protocols for the collection and analysis of fishery-related data, including the relevant operations of the CCAMLR data base; and
  (iii) analytical procedures for the estimation and projection of fish stock trajectories.
(b) Review and conduct assessments of the status and potential yield of fish stocks in the convention area.
(c) Evaluate the actual and potential impact on fish stocks and fisheries of past, present, and possible future management actions.

As a result of its work, that of the other groups, and of the predecessor group on data collection and handling, the commission has taken a series of decisions, pursuant to Art. XX of CCAMLR, to elaborate the legal obligations of parties to collect and report data. These include the following:

- Detailed specifications of finfish data to be collected and archived and of finfish data to be submitted annually to the commission (fourth annual meeting, 1985).
- The initiation of routine annual reporting of fine-scale catch and effort data on finfish (fifth annual meeting, 1986).
- Refinements in reporting fine-scale catch and effort data on finfish (sixth annual meeting, 1987).
- Specification of fine-scale catch and fishing effort data on krill for the inte-

grated study areas designated for ecosystem monitoring (sixth annual meeting, 1987).

At its most recent meetings, the scientific committee has concentrated on identifying data needs for specific fisheries, with an emphasis on new and developing fisheries (e.g., the long-line fishery for *Dissostichus eleginoides* and the fishery for the myctophid, *Electrona carlsbergi*). Although these steps lay the groundwork for future collection and reporting, difficulties remain in obtaining historic data in usable form.

Ecosystem monitoring

The ad hoc working group on ecosystem monitoring, established in 1984, was converted into a formal standing working group the following year, with the following terms of reference (CCAMLR Report of the 4th Meeting of the Scientific Committee, 1985, Para. 7.14, p. 38):

(1) To plan, recommend, coordinate, and ensure the continuity of a multination CCAMLR ecosystem monitoring program within the convention area.

(2) To identify and recommend research, including theoretical investigations to facilitate design and evaluation of the recommended ecosystem monitoring program.

(3) To develop and recommend methods for the collection and storage and analysis of data, including data formats for submission to CCAMLR.

(4) To facilitate the analysis of data and their interpretation, and to identify the management implications.

(5) To report progress to each meeting of the scientific committee with recommendations for further work.

As a result of its work, the scientific committee has identified a number of potential indicator species (prey and predator) and identified integrated study areas for monitoring predator-prey interactions. As with the work on assessment of krill populations, the CCAMLR efforts in ecosystem monitoring address basic methodological questions posed by multispecies management, including how to detect and distinguish between environmentally driven and harvesting-induced changes in ecological relationships.

Consideration of ecosystem monitoring issues within CCAMLR also has raised important issues as to scientific research priorities and coordination of such research relevant to CCAMLR. (The provisions of CCAMLR—its objectives and principles—in effect establish a far-reaching agenda for scientific research activities.) One clear conclusion is the necessity of integrating monitoring activities (again to obtain time series of data to detect ecological changes) into Antarctic research programs.

A second conclusion has been recognition of the need to pool efforts and divide areas for research emphasis. In response to this need, the CCAMLR scientific committee initiated consideration of its long-term program of work (at its fourth annual meeting in 1985). This program is considered and updated at each meeting and is designed to provide an informal means of identifying CCAMLR research priorities and for coordinating the conduct of such research.

Science and Policy

The record of the implementation of CCAMLR, viewed from the perspective of addressing and articulating the scientific and technical basis for an ecosystem management approach is impressive. There has been generated—both in the institutional and substantive sense—a major coordinated effort to develop the information base and analytical tools to carry out the convention's objective. Particular difficulties remain in obtaining historical data (largely pre-1982) on fisheries, but generally very significant progress has been achieved.

Equally important will be the manner in which the parties to CCAMLR respond to the challenge of integrating the tools that are being developed into the political process of management decisions. The effort to address this issue is of more recent origin. The issue, perhaps somewhat misleadingly, has become associated with the questions of elaborating a conservation strategy for CCAMLR. At its fifth annual meeting (1986), the commission recognized the importance of developing a process for defining a strategy for the progressive achievement of the objective of the convention (Art. II) and established a working group to examine this issue. It is significant that the working group was formed by the commission rather than the scientific committee—a recognition that management policy is more a political than a scientific issue.

The working group on conservation strategy developed the following terms of reference (CCAMLR Report of the 6th Meeting of the Commission, 1987, Para. 107, p. 27):

(1) To develop a common understanding as to the management implica-

tions of Article II of the convention.

(2) To develop possible conservation approaches for achieving the objectives of Article II by means contained in Article IX.

(3) To select and apply performance criteria for assessing each approach.

(4) To identify, for preferred approaches, specific short- and long-term goals consistent with the objectives of the convention.

(5) To formulate the framework of a strategy for managing activities in order to achieve these goals.

(6) To report to the commission recommending appropriate action.

At the seventh annual meeting (1988), it was agreed to change the name of the group to the working group for the development of approaches to conservation of Antarctic marine resources, in order to clarify the nature of its work. The group initiated examination of these items at the sixth annual meeting (1987) with an emphasis upon developing a common understanding of the term "rational use" in relation to "conservation," as those terms are used in CCAMLR. At the eighth and ninth annual meetings (1989 and 1990), the working group emphasized the need to articulate methodologies for dealing with new and developing fisheries and to devise criteria for setting bycatches for depleted populations at levels that would promote their restoration.

On a separate track, the scientific committee (also in 1987) indicated that, with respect to specific issues relating to fish populations, it had difficulty in providing advice to the commission because of lack of guidance on management policy. In essence, the committee served notice to the commission that the time had come to provide such guidance. In response, the commission noted the scientific committee's points relating to the need for management strategies and, *inter alia*, requested advice from the committee on a number of specific matters, taking into account the multispecies characteristic of ongoing fisheries. It also noted the relationship of these matters to the work of the group on conservation strategy.

Since that time, an increasingly detailed dialogue in elaborating management policy has evolved between the commission and the scientific committee. This dialogue not only has given direction for the elaboration of specific conservation measures, it has also identified emerging priorities for management action. At the ninth annual meeting (1990), for example, the commission, on the basis of scientific committee responses, laid the groundwork for management policies for krill and for new and developing fisheries.

With respect to krill, although consensus has continued to elude the commission on precautionary catch quotas for krill, it noted, as the basis for initiating a management policy, four general concepts suggested by the scientific committee (CCAMLR Report of the 9th Meeting of the Commission, 1990, Para. 4.17, p. 9):

(1) Aim to keep the krill biomass at a higher level than might be the case if only single-species harvesting considerations were of concern.

(2) Given that krill dynamics have a stochastic component, focus on the lowest biomass that might occur over a future period, rather than the mean biomass at the end of that period, as might be the case in a single-species context.

(3) Ensure that any reduction in food to predators that may arise because of krill harvesting is not such that land-breeding predators with restricted foraging ranges are disproportionately affected in comparison with predators in pelagic habitats.

(4) Examine what level of krill escapement would be sufficient to meet the reasonable requirements of krill predators.

With respect to new and developing fisheries, the commission adopted a conservation measure requiring reporting of biological and effort data on the recently initiated long-line fishery for *Dissostichus eleginoides*. They also endorsed a scientific committee recommendation on modification of practices in the fishery, pending availability of these data, as well as pending provision of data on the best means of avoiding incidental mortality of seabirds.

In addition, the commission called for members to conform with the basic idea of advance notification of any new fishery. In taking this step, it agreed with the principle that the development of a new fishery should be directly linked with the process of elaborating scientific advice and management with respect to the fishery. It further recognized the need to assess the potential yield of a new fishery before it begins to develop in order to satisfy the subobjectives of CCAMLR's conservation objective (CCAMLR, 1980, Art. II).

The significance of this dialogue resides not only in its initial results, but also in the fact that the parties to CCAMLR have committed themselves to identify management policies on the basis of clear delineation of the relationship between science and policy, in the conceptual as well as the institutional sense. Although the consensus decision-making procedures of the commission are sometimes slow and frustrating, the parties to CCAMLR have wisely resisted efforts to transfer decision-making prerogatives to the scientific committee. As a result, members of the scientific committee have been increasingly able to identify the implications of management measures in objective and nonpolemical fashion. The importance of avoiding politicization of the scientific advisory function is another important lesson to be learned from the experience of CCAMLR to date.

What may augur well for the future implementation of CCAMLR is that the interaction between science and policy is taking place on several fronts. Elaboration of a definitive interpretation of the three principles incorporated in Art. II, which constitute CCAMLR's conservation objective, is not likely in itself to be a productive endeavor. What is necessary is the recognition that the principles of Art. II require interpretation and refinement on a continuing basis to develop short- and medium-term, as well as long-term, goals. Further, this continuing interpretation and refinement constitute, and should be viewed as, the process that integrates the scientific and political requirements of CCAMLR. In this context lies the significance of the dialogue between the commission and the scientific committee.

Addendum—Conservation Measures

Beginning at its third annual meeting, CCAMLR adopted a number of conservation measures, primarily aimed at conserving fish stocks in the convention area. There have been expressions of dissatisfaction with the substance and timeliness of these measures, both within and outside the commission. Frequently, also, the CCAMLR consensus decision-making system is cited as the source of what are described as the convention's shortcomings.

I have not specifically discussed the actual conservation measures adopted under CCAMLR, because in my view, the identification of possible precedents for the management of LMEs offered by CCAMLR lies in the process through which it originated, was elaborated, and has operated. The ongoing efforts to establish and institutionalize the relationship between the scientific and technical requirements for management and the political process for making management decisions seem more relevant to this chapter.

I would not wish to conclude, however, without a word on the specific measures that have emerged from the operation of CCAMLR to date. First, as a participant in the deliberations of the commission, I share to some extent the view that the elaboration of conservation measures has been slow and insufficiently comprehensive and anticipatory; at the same time, I believe that CCAMLR's record is a remarkably good one. Instant gratification ordinarily does not result from the process of multilateral lawmaking. For the Antarctic marine ecosystem, as with many large systems, this process offers the only means of preserving conservation options.

There are 14 conservation measures in force under CCAMLR (1990). These include

(1) minimum mesh sizes for pelagic and bottom trawl fisheries for species of finfish, as well as regulations on mean-size measurement;

(2) total allowable catches for the three directed fisheries around the island of South Georgia (statistical subarea 48.3);

(3) prohibition on directed fisheries for other finfish populations in statistical area 38.3, as well as limits on the bycatch of such species resulting from permitted directed fisheries;

(4) imposition of a closed season (1 April– 4 November) on the primary directed fishery (for *Champsocephalus gunnari*) in statistical subarea 48.3, as well as prohibition on the use of bottom trawls in that fishery;

(5) operation of a catch- and effort-reporting system applicable in statistical subarea 48.3;

(6) prohibition on all directed fishing for finfish in the areas of the Scotia Sea other than that around South Georgia (statistical subareas 48.1 and 48.2);

(7) total allowable catches for the directed fisheries on Ob and Lena Banks (statistical subarea 58.4); and

(8) procedures for according protection for CCAMLR ecosystem monitoring program sites.

In addition, the commission has elaborated and placed into operation a detailed system of observation and inspection and has agreed that there should not be expansion of pelagic driftnet fishing in the area of CCAMLR.

These achievements have all been agreed upon by consensus—or I should say negotiated, because consensus is far more a process of negotiation than a formal method of making decisions. I would note that no international institution, whatever its formal decision-making rules, can operate successfully without relying primarily upon consensus. Given the basic legal and political differences over territorial sovereignty and maritime jurisdiction in Antarctica among the parties to CCAMLR, there was and is no alternative to a consensus system in which measures cannot be adopted over the serious objection of an individual party. The issue facing the parties to the Antarctic Treaty and to the instruments it has spawned, such as CCAMLR, is to sustain the will to make the consensus system work.

References

ATCM (Antarctic Treaty—Report of the Eighth Consultative Meeting). 1975. 9–20 June 1975. Oslo, Norway. Norweigan Ministry of Foreign Affairs.

ATCM (Antarctic Treaty—Report of the Ninth Consultative Meeting). 1977. September 1977. London, England.

Convention on the conservation of Antarctic marine living resources (CCAMLR). 1980. Arts. I–XXXIII. Complete copy of treaty may be obtained by writing: CCAMLR, 25 Olde Wharf, Hobart, Tasmania 7000, Australia.

CCAMLR. 1985. Report of the 4th meeting of the scientific committee.

CCAMLR. 1987. Report of the 6th meeting of the commission.

CCAMLR. 1987. Report of the 6th meeting of the scientific committee.

CCAMLR. 1988. Report of the 7th meeting of the scientific committee.

CCAMLR. 1988. Report of the 7th meeting of the scientific committee.

CCAMLR. 1990. Report of the 9th meeting of the commission.

United Nations Convention on the Law of the Sea (UNCLOS). 1982. Arts. 61 and 119.

Chapter 23

Simulation Study of Effects of Closed Areas to All Fishing, with Particular Reference to the North Sea Ecosystem

Niels Daan

Introduction

Fisheries management has traditionally concentrated on implementing harvesting strategies that have the objective of limiting competition among fishermen and enhancing total fish yields. Thus, all measures imposed so far on fisheries in the North Sea, including total allowable catch (TAC), mesh-size regulations, minimum landing sizes, and closed areas and seasons are aimed at controlling fishing mortality on (components of) all commercially important stocks. If these measures have conservation effects on the fish species involved, these are secondary and subordinate to the expected effects on the long-term yields. So far, nature conservation has not been an objective of fisheries management by itself.

In recent years, multispecies considerations have been taken into account when providing management advice, but these have been typically related to direct interactions between exploited fish species through predation (Anonymous, 1989). Because of the complexity of the marine ecosystem and the uncertainties about the quantitative interactions that may exist through non-commercially important components of the food web, a general ecosystem approach to fisheries management is severely hampered (Daan, 1986). Still, increasing concern is being expressed about possible adverse side effects of fishing (Anonymous, 1988).

Trawls, in particular, are basically unselective fishing gear, which potentially catch everything that is available in the path of the trawl. Indeed, trawling gear generally results in large bycatches of organisms—fish, as well as invertebrate bottom fauna—that have no eco-

nomic value and are discarded at sea. Conservative estimates indicate that on average each square meter of the bottom in the southern North Sea is touched at least once per year by a beam trawl, and some areas may on average be hit seven times or more (Welleman, 1989). These beam trawls are used to catch flatfish and sole, in particular, and carry a large number of heavy chains to chase the fish out of the sand. In practice, the actual number of times the bottom is plowed will vary considerably in space because skippers may favor particular stretches within an area. Although trawling may obviously have profound effects on the benthic system, particularly regarding long-living or tender species, it is difficult to study such effects (Anonymous, 1988) because there are no areas that have been guaranteed void of fishing and that can serve as controls.

Other fishing gear pose different problems. Gill nets employed in the cod fishery on wrecks have been reported to catch significant numbers of cetaceans, and observations by divers indicate that ruined nets are sometimes left in position, which continue to catch fish. With the growing awareness of environmental issues, there is an increasing demand for a more integrated approach to marine ecosystem management (Daan, 1989), which is exemplified in the declaration of the Third International Conference on the Protection of the North Sea, held in March 1990 in The Hague. This appears to be even more urgent because, until now, fisheries management in the North Sea cannot be said to have achieved its aim of controlling effort. The general trend observed in North Sea fish stocks is that the TAC regulations have not been able to reduce or stop overfishing. On the contrary, the situation in many stocks has become worse (Daan, 1989). As a consequence, fisheries management should be open to new suggestions regarding possible measures aimed at improving the balance between fisheries and the maintenance of a stable ecosystem.

The creation of protected areas that are closed to all forms of fishing might be a first step in this direction (Daan, 1989). Clearly, such areas may protect particular components within the system that are highly sensitive to fishing, although the ultimate effect will depend upon the spatial exchange of individual species between areas and the spatial dependence of various stages during their life history. For instance, the conservation effect on a sessile benthic organism will be different from the effect on marine mammal species, which roam over vast areas. Whatever the objectives of creating reserves may be, it would be appropriate to evaluate the side effects on fisheries in order to make these consequences acceptable to the industry. In this respect, it was also relevant that within the European Parliament a discussion has been initiated about the possibility that closed areas may present an alternative strategy to fisheries management. This idea is based on the assumption that protected areas are bound to reduce the fishing mortality on the commercial fish stocks and that, if the areas created are only large enough, TAC regulations may no longer be required.

This chapter addresses the question of whether the introduction of protected areas would have significant effects on commercial fish stocks, based on a general simulation model of typical dispersion rates observed among commercial fish species. Of course, the actual effects will depend strongly on the location of these areas in relation to the distribution of each species. Without a definite proposal, the number of options for evaluating the effects of closures on each species would obviously be infinitely large. Therefore, the characteristics of the starting point in this study are (i) that the protected areas are chosen more or less at random with respect to an arbitrary species; (ii) that the total fishing effort is not affected by the closure, and thus fishing is more intense in the unprotected area after the closure; and (iii) that the dispersion rate determines the flow of fish between the protected and the unprotected areas.

Methods

The effect of closing an area to all fishing on the yield of one species depends on the relative size of the reserve, the relative abundance in the reserve, the status quo distribution of fishing effort, and the rate of dispersion of the fish within the system. Realistic data on all these aspects are not readily available; thus, we are not in a position to make a proper quantitative evaluation for any particular North Sea fish species. However, to investigate the potential effects on a species, we may start from a homogeneous distribution of a fish stock over the entire area and homogeneous effort distribution. For this purpose, the North Sea was split into 186 squares corresponding to International Council for the Exploration of the Seas (ICES) statistical rectangles of approximately 30 x 30 miles, each containing an equal and arbitrary number of fish that are exploited at equal rates.

The initial abundance of fish in each rectangle was put at 1,000; then, with time-steps of 1 week, the numbers surviving in each rectangle were calculated for various rates of dispersion and for different area closures. To simulate dispersion, it was assumed that 25%, 50%, or 75% of the fish present in a square would stay there in the next time-step and that the remainder were split equally among the eight surrounding squares. These values correspond roughly to dispersion coefficients (Jones 1959; Anonymous, 1971) of 196, 83, and 41 miles^2 · day^{-1}, respectively. In the case of squares bordering the coastline, the dispersion proportions were adjusted so that rates of exchange between adjoining squares would be in equilibrium. The original annual fishing mortality coefficient on the total North Sea population was set at 0.5 or 1.0, which means that with a homogeneous effort distribution, the fish within each rectangle are exploited at exactly these mortality rates. Depending on the relative size of the protected area (where fishing mortality was reduced to zero), the fishing mortality experienced by the populations in fished squares was raised by the total number of rectangles divided by the number of unprotected rectangles. This mimics the assumption that the fishing effort originally present in the protected area would be transferred and redistributed over the unprotected area. In catchability terms, it is thus assumed that the catchability is constant within squares, but measured over the entire population the catchability will change as a result of the area closure.

The simulations were run over 52 weeks, and the effective fishing mortality coefficient (F), as influenced by the creation of a protected area, was estimated from the number of fish surviving over the entire North Sea after 1 year. The model was designed as a spreadsheet. Figure 1 shows an example of the numbers of survivors in each rectangle after 52 time-steps for a 10% contiguous closure of the entire area, a fishing mortality of 1.0, and a dispersion of

Iteration: **52**

Scattered rectangles - 25% protected

Iteration	E6	E7	E8	E9	F0	F1	F2	F3	F4	F5	F6	F7	F8
52	345	311	303	304	323	304	301	307	339				
51	311	311	320	304	304	305	318	305	304				
50	303	320	306	304	303	317	305	303	300				
49	305	305	304	318	304	302	302	317	303				
48	324	305	304	305	317	302	302	304	324				
47	305	307	319	304	302	303	315	300	298				
46	300	322	308	304	303	315	301	298	294				
45		314	308	319	304	301	299	311	296	292			293
44	370	317	308	306	316	301	299	299	310	295	293	296	320
43			322	303	302	302	314	298	297	298	310	297	295
42			299	301	301	315	301	299	298	312	300	297	294
41		289	297	315	303	301	300	313	300	298	298	316	
40				306	316	301	300	301	313	298	297	295	
39			338	305	302	302	314	299	299	300	311	293	287
38				303	305	314	301	301	298	312	299	297	294
37					303	299	299	313	299	298	299	317	305
36					319	296	299	299	313	295	297	306	344
35					296	301	312	298	294				
34						315	298	296	290				
33						293	295	307	289				
32						293	295	295					
31						292	318						

Norway

Britain

Netherlands

Parameters:

Nr of squares	186
Nr protected	46
F	1.0
M	0.2
Dispersion	83

Total surviving:

25% protected	56748
No protection	56022

Figure 1. Schematic representation of North Sea statistical rectangles with the survivors of an original 1,000 fish-per-rectangle after 52 iterations, when a contiguous area representing 10% of the total area is closed to all fishing (shaded area with numbers). Parameter values used in the simulation are given, as well as the total number of survivors compared with the number surviving when there is no closure. F represents the fishing mortality coefficient before the closure, but the true Fs in each rectangle are adjusted assuming that the effort in the closed area is redistributed over the remaining area. M represents the natural mortality coefficient.

50%. Simulations were also run with a contiguous area of 25% of the North Sea closed to fishing (Fig. 2) and with a 25% closure of individual rectangles scattered over the entire North Sea (Fig. 3). The actual locations of the closed areas were completely arbitrary.

Results

Table 1 summarizes the estimated annual fishing mortality coefficients for the different closed-area options.

These simulations indicate that a contiguous closed area representing 10% of the total North Sea leads, at most, to a 5% reduction in F for the lowest dispersion rate. For higher dispersion rates, the reduction is lower because it takes less time before the fish emerge from the protected area and become available to the fishery. Although there are more fish left in the protected area (Fig. 1), the protection of the stock is largely counteracted by the higher mortality in the remaining area. For a 25% contiguous protected area, the reduction in fishing mortality ranges between 10% and 14% because it takes a longer time before the average fish spreads into the unprotected area. However, if one out of four squares scattered over the entire North Sea is closed to all fishing, the reduction in fishing mortality would be lower, on the order of 0.5, and the likely reduction in F would then become marginal.

Discussion

The results of these simulations cannot be interpreted directly in terms of the effect of protected areas on, for instance, the cod stock, because we have assumed that the stock is distributed homogeneously, and thus so is the effort, whereas in reality, these assumptions are undoubtedly not met for any species. Moreover, after the introduction of the protected area, the stock is no longer homogeneously distributed, whereas the effort is kept constant. It would be more likely that effort would be concentrated in the rectangles bordering the closed area because there would be more fish there. Despite these imperfections, some general observations can be made.

From tagging experiments with southern and central North Sea cod, the observed dispersion rate of about 60 miles$^2 \cdot$ day^{-1} during the first 3 months of release (Anonymous, 1971) lies within the range of values used here. Thus, the simulations suggest that, given an average fishing mortality of 0.8 for cod, one might need to close about one-quarter of the main distribution area of cod in order to get a 10% reduction in fishing mortality on the stock. Plaice appear to be more stay-at-home, with dispersion less than 30 miles$^2 \cdot$day^{-1} (DeVeen, 1970), but because this species is less intensively exploited, the effect of closed areas would be correspondingly smaller. For other species, such as herring, which migrate extensively throughout the North Sea, area closures would have even less effect on the exploitation rate.

Generally, it seems fair to conclude that closing even a relatively large proportion of the North Sea, accomplished using small, separated areas scattered throughout, would not affect the exploitation rate significantly. Still, from a management point of view, any effects that reduce the exploitation rate must be considered to be positive rather than negative, at least for overexploited fish species.

The results presented here can be interpreted in two ways. First, area closures to all fishing cannot be considered an effective alter-

Table 1. Summary of estimated annual fishing mortality coefficients for various closed-area options.

Dispersion Coefficient	Fishing Mortality Coefficients*			
	No Protection	10% Protection*	25 Protection[†]	25%* Protection[‡]
186	1.0	0.97 (−3%)	0.90 (−10%)	0.99 (−0.8%)
93	1.0	0.96 (−4%)	0.88 (−12%)	0.99 (−1.4%)
41	1.0	0.95 (−5%)	0.86 (−14%)	0.97 (−3%)
93	0.5	0.49 (−2%)	0.48 (−4%)	0.50 (0%)

*The percent change in protected versus unprotected situations for each coefficient is given in parentheses.

[†]Contiguous area closed

[‡]Scattered rectangles closed

Iteration:
52

Contiguous area - 25% protected

Parameters:
Nr of squares	186
Nr protected	46
F	1.0
M	0.2
Dispersion	83

Total surviving:
25% protected	62638
No protection	56022

	E6	E7	E8	E9	F0	F1	F2	F3	F4	F5	F6	F7	F8
52	217	217	217	217	217	217	217	217	217				
51	217	217	217	217	217	217	217	217	217				
50	217	217	217	218	218	218	218	218	218				
49	217	217	218	218	218	219	219	219	219				
48	218	218	218	219	220	220	220	221	221				
47	218	219	220	221	222	223	225	226	226				
46	218	220	221	223	226	228	231	234	235				
45		222	224	228	232	236	242	247	254	261			265
44	221	225	229	234	240	248	257	267	275	283	288	286	280
43			235	242	253	266	282	297	309	315	316	310	303
42			245	252	268	291	318	344	363	373	370	353	334
41		244	252	263	284	319	377	418	447	465	470	434	
40			258	273	301	346	418	472	512	540	557	584	679
39			262	285	315	365	446	509	558	595	625	656	709
38				296	329	378	461	529	585	630	666	693	731
37					340	380	462	532	594	646	686	716	745
36					347	376	432	521	582	641	698	728	
35					341	372	419	499	561				
34						367	405	476	512				
33						359	388	449	488				
32						345	367	425					
31						331	349						

Norway · Britain · Netherlands

Figure 2. As for Fig. 1, but for a 25% contiguous closed area.

Iteration: 52

Contiguous area - 10% protected

	E6	E7	E8	E9	F0	F1	F2	F3	F4	F5	F6	F7	F8
52	269	269	269	269	269	269	269	269	269				
51	269	269	269	269	269	269	269	269	269				
50	269	269	269	269	269	269	269	269	269				
49	269	269	269	269	269	269	269	269	269				
48	269	269	269	269	269	269	269	269	269				
47	269	269	269	270	270	270	270	270	270				
46	269	269	269	270	270	271	270	271	271				
45		269	270	271	271	273	274	276	279	282			285
44	269	270	270	273	273	275	278	281	285	289	292	293	291
43			271	274	275	278	283	289	295	300	303	302	300
42			273	276	277	282	289	299	311	321	325	319	312
41		272	274	277	280	287	297	313	334	356	372	348	
40			274	279	282	291	305	327	361	414	448	485	
39			275	281	285	295	312	340	384	453	507	553	585
38					287	298	316	348	399	478	542	589	615
37					289	298	316	347	401	488	560	612	639
36					290	297	312	339	383	475	571	627	657
35					288	294	304	322	357				
34						289	295	305	309				
33						284	287	291	293				
32						279	281	284					
31						277	278						

Britain

Norway

Netherlands

Parameters:

Nr of squares	186
Nr protected	19
F	1.0
M	0.2
Dispersion	83

Total surviving:

10% protected	58320
No protection	56022

Figure 3. As for Fig. 1, but for a 25% closure of scattered rectangles throughout the North Sea.

native to the present fisheries management regime of using catch quotas. The likely effects are too small, unless specific areas of high densities could be selected, and even then additional management measures would probably be required to control fishing effort in the remaining area. Second, it would appear that closing areas in order to protect a particular habitat against the effects of fishing, either for reasons of nature conservation or for scientific purposes, would hardly affect the fisheries because the fish will be caught at some later time in the unprotected area.

Of course, closed areas can be much more effective if they are specially designed to protect juvenile fish, but such measures have already been evaluated (e.g., Anonymous, 1987) and introduced. The use of area closures has proven to be an effective management measure, and, therefore, such analyses are beyond the scope of this chapter. Evaluating the extent to which closed areas that are designed to protect ecosystem components other than fish would affect fish stocks and fisheries is an essentially different matter.

Obviously, the present evaluation could be greatly improved if reliable data were available about the distribution of particular species, the distribution of effort, and the rate of dispersion and migration of fish between different areas. The kinds of data that are presently collected by the Working Group on Technical Measures of the Scientific and Technical Committee for Fisheries of the European Community could prove extremely useful for quantitative evaluation of effects from area closures, particularly if these data are supported by reliable tagging data.

References

Anonymous. 1971. Report of the North Sea roundfish working group on North Sea cod. ICES C.M. 1971/F:5.

Anonymous. 1987. Report of the ad hoc meeting of the North Sea flatfish working group, IJmuiden, 2–5 February 1987. ICES C.M. 1987/Asses:14.

Anonymous. 1988. Report of the study group on the effects of bottom trawling. ICES C.M. 1988/B:56.

Anonymous. 1989. Report of the multispecies assessment working group, Copenhagen, 7–16 June 1989. ICES C.M. 1989/Assess:20.

Daan, N. 1986. Results of recent time-series observations for monitoring trends in large marine ecosystems with a focus on the North Sea. In: Variability and management of large marine ecosystems. pp. 145–173. Ed. by K. Sherman and L. M. Alexander. AAAS Selected Symposium 99. Westview Press, Inc., Boulder, CO.

Daan, N. 1989. The ecological setting of North Sea fisheries. Dana 8:17–31.

DeVeen, J. F. 1970. On the orientation of the plaice (Pleuronectes platessa L.) I. Evidence for orientating factors derived from the ICES transplantation experiments in the years 1904–1909. J. Cons. Int. Explor. Mer 33(2):192–227.

Jones, R. 1959. A method of analysis of some tagged haddock returns. J. Cons. Int. Explor. Mer 25:58–72.

Welleman, H. 1989. Literatuurstudie naar de effecten van de bodemvisserij op de bodem en het bodemleven. Netherlands Institute for Fishery Investigations, MO 89–201 (Internal Report).

Chapter 24

Research and Management in the Northern California Current Ecosystem

Daniel L. Bottom, Kim K. Jones, Jeffrey D. Rodgers, Robin F. Brown

Introduction

Strategies for ocean management

The expanding influence of human disturbance in large coastal and marine regions has increased awareness of the need to manage entire ocean ecosystems as units (Sherman and Alexander, 1986; Sherman *et al.*, 1990). Management of coastal and ocean resources in the United States is divided between state and federal jurisdictions and among numerous public agencies with narrowly defined and often conflicting responsibilities. These responsibilities have evolved through separate acts of legislation intended to manage particular economic resources or to address specific environmental concerns. Continued overexploitation of fisheries, widespread coastal pollution, and human-induced climate change are indicators that numerous technological and regulatory solutions applied case-by-case to environmental problems in the 1970s are inadequate to resolve the regional and global concerns of the 1990s.

Management at the ecosystem level is complicated by the inability to accurately forecast and measure ecological responses to human disturbance. Despite progress in environmental assessment during the past decades, prediction of ecological effects in complex marine ecosystems remains an elusive goal. Furthermore, because ecological data are costly to collect and difficult to interpret, and because management interests are highly specialized, environmental research is often directed toward single issues and regulatory functions—for example, to establish quantitative discharge standards for specific pollutants or to estimate maximum sustained yield for regulating harvest of individual species. The adequacy of reductionist controls to sustain ecosystem health is highly uncertain, however, because ecological data are often ignored (Lewis, 1980), and the response of complex ecosystems to human disturbance is unpredictable.

This circularity underscores a critical need to devise strategies for ecosystem-level management that are not dependent upon precise prediction of whole-system effects. Holling (1973, p. 21) proposed a qualitative, "resilience" approach to environmental management that "would emphasize the need to keep options open, the need to view events in a regional rather than a local context, and the need to emphasize heterogeneity." This philosophy is expressed in "adaptive management," which treats environmental actions as experiments that can be modified as results provide a better understanding of ecological responses (Holling, 1978; Lee, 1989). Bella and Overton (1972) recommended a strategy for localizing the distribution and intensity of human activity in order to minimize the likelihood of large-scale environmental change. Such management strategies accept uncertainty as a fundamental characteristic of complex ecosystems and emphasize prevention of serious environmental effects that may be impossible to correct after the fact. From this perspective, the goal of environmental assessment shifts from the quantitative proof or prediction of change to the prudent management of human activities to avoid unanticipated and irreversible consequences (Bella and Overton, 1972).

Recent proposals to develop a large marine ecosystem (LME) along the northwest coast of the United States have focused attention on the need for an integrated program of ocean management. We recently prepared a research plan to identify the kinds of information that are needed to support prudent management in the region (Bottom et al., 1989). This chapter describes our general approach to environmental assessment in a highly variable and unpredictable marine ecosystem.

Need for a regional research plan

The continental margin of the United States off the coasts of northern California, Oregon, and Washington is a highly productive coastal region that has long supported commercial harvest of a diversity of living marine resources. Increasing demand for oil, gas, and strategic minerals has stimulated recent proposals to explore and develop nonliving resources of the Exclusive Economic Zone in the area. The entire region was included in the U.S. Department of Interior's proposed 1987–1992 oil and gas leasing program (MMS, 1987). Chromite-bearing black sands off the southern Oregon coast and titanium-rich sands off the northern Oregon and southern Washington coasts also are under consideration for future placer mining (U.S. Bureau of Mines, 1987). Dredging of sand and gravel deposits off the Washington and Oregon coasts also may become economically feasible in the future, because local onshore supplies in some coastal areas are dwindling.

Development proposals such as these raise a large number of questions about environmental consequences that previously have not been of concern in this region. Most oceanographic and fisheries data have not been collected consistently or systematically throughout the area, and thus, information is inadequate to evaluate the risks of offshore development to many marine populations, habitats, and communities. Several workshops have been conducted to review environmental studies related to oil and gas development and have generated long "wish lists" of activities organized by scientific discipline (U.S. Bureau of Land Management, 1977; MMS, 1988). This laundry-list approach provides no theoretical or applied context to explain why individual recommendations were chosen or how they may be integrated into a cohesive program for ecosystem management. To improve regional environmental assessment, we prepared a research plan for the U.S. continental margin off Washington and Oregon (Bottom et al., 1989). Most of our research recommendations apply to the region south to Cape Mendocino, California (40° N latitude), where ecological characteristics are similar to the Washington-Oregon region.

The California Current Ecotone

Large regions of the Pacific Ocean contain unique assemblages of pelagic species and have been described as distinct ecosystems (Fager and McGowan, 1963; McGowan, 1971; Wahl et al., 1989). The boundaries of these ecosystems correspond to large water masses whose physical and chemical properties are conserved by recirculation in semi-enclosed gyres (Favorite et al., 1976). Along the boundaries between water masses are transition zones where faunal groups overlap and experience dramatic fluctuations in the physical environment (Fig. 1).

The California Current region off the west coast of the United States is a transition environment between subarctic and subtropic water masses and the freshwater systems that enter the ocean along its landward boundary (Fig. 1). Unlike the large oceanic gyres that circulate in the Central and North Pacific, the California Current is an "open" system. Physical and biological properties vary with seasonal and interannual fluctuations in the currents that transport water from adjacent water masses. In the North Pacific region, a branch of the Subarctic Current turns southward to become the California Current, which, in the spring and summer, occurs at the surface along the west coast of the United States. When upwelling subsides in the fall, the northward-flowing California Undercurrent (Davidson Current) appears at the surface and carries warm equatorial water inshore (Favorite et al., 1976).

Regions of eastern boundary currents such as the California Current region have been described as nonequilibrium systems in which physical forces may be more important determinants of species composition and abundance than biological interactions such as competition and predation (Bernal, 1981). Biological systems in these regions are more spatially heterogeneous and temporally unpredictable than in other marine ecosystems (Chelton et al., 1982). The region of the California Current is an ecotone composed of a small number of endemic coastal species and a larger mixture of subarctic and subtropic species, many near the periphery of their distributional range (Johnson and Brinton, 1962; McGowan, 1971). Marine populations within this ecotone experience dynamic fluctuations (MacCall, 1986) that may reflect complex interactions of local, regional, and global scales of physical processes. Annual fluctuations in the physical environment influence recruitment of commercial species throughout the California Current region (Peterson, 1973; Lasker, 1978; Mysak, 1986; Nickelson, 1986). Marked variations in fish stock abundance that long preceded commercial harvest in the region have been docu-

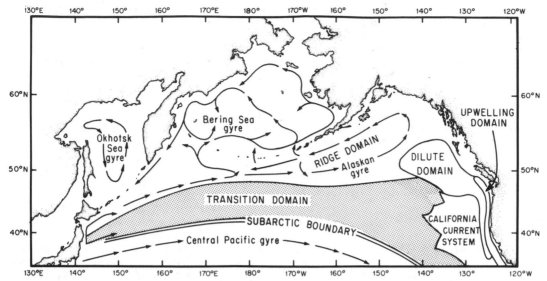

Figure 1. Diagram of domains and current systems in the Subarctic Pacific region (adapted from Favorite *et al.*, 1976).

mented from fish scales preserved in sediments off California (Soutar and Issacs, 1974).

The causes of natural variability throughout the California Current are poorly understood, in part because of differences in the timing and scales of oceanographic research conducted in different locales. Along the central Oregon coast, for example, detailed studies were conducted in the 1970s to describe coastal upwelling processes caused by local winds. In this region, spatial patterns of primary productivity and biomass (Small and Menzies, 1981), zonation of zooplankton (Peterson *et al.*, 1979), and year-to-year variations in zooplankton abundance (Peterson and Miller, 1975) were strongly influenced by upwelling processes. Off central and southern California, broad-scale, systematic surveys have described interannual variations in zooplankton biomass that are not explained solely by coastal upwelling. Large-scale advection of nutrient-rich water from the subarctic region (Wickett, 1967; Chelton, 1981; Chelton *et al.*, 1982) and strong offshore transport of high phytoplankton biomass, particularly near coastal headlands (Abbott and Zion, 1987), have been suggested as alternative mechanisms to explain these low-frequency, large-scale patterns.

Landry *et al.* (1989) conclude that interannual variations off Washington and Oregon are likely the result of both regional and global-scale processes. Annual strength of flows from the north may influence the location of the subarctic boundary and biological produc-

tion processes off Washington and Oregon. Shifts in boundary location may explain annual variations in zooplankton composition and biomass (Fulton and LeBrasseur, 1985) or production of subarctic salmonids near the southern edge of their distributional range (Bottom *et al.*, 1986).

The influence of global-scale processes off the Washington and Oregon coasts is most evident during unusually strong poleward flows associated with El Niño-Southern Oscillation (ENSO) events in the eastern tropical Pacific. During these periods, biologists have reported range extensions of fish, birds, and plankton (McClain and Thomas, 1983; Pearcy *et al.*, 1985; Mysak, 1986); reduced reproductive success of Oregon seabirds (Graybill and Hodder, 1985); changes in migration routes of adult sockeye salmon (*Oncorhynchus nerka*) to the Fraser River, British Columbia (Wickett, 1967; McClain and Thomas, 1983); and reduced size, fecundity, and survival of adult coho salmon (*O. kisutch*) off Oregon (Johnson, 1984).

For purposes of management, it is appropriate to divide the California Current ecosystem into two major subregions, north and south of Cape Mendocino, California (MacCall, 1986). Wind stress south of the cape is equatorward throughout the year. To the north, winds are poleward and unfavorable for upwelling during the winter. Local winds account for most variability in alongshore currents in the north off Oregon, but may be less important off southern California (Huyer, 1983). The percentage of subarctic water and the abundance of subarctic species in the California Current region also decline toward the south. These trends are reflected in a latitudinal gradient in the composition of marine fishes that inhabit estuaries throughout the region (Bottom and Jones, 1990). The region between Cape Mendocino and Puget Sound is described as a zoogeographic transition between Californian and Aleutian biological provinces to the south and north, respectively (Ekman, 1967). For these reasons, we define a northern California Current ecosystem (NCCE)—the continental margin from Cape Mendocino, California, to Vancouver Island, British Columbia (48° N latitude)—as an appropriate ecological unit for regional management.

Implications for resource management

In an ecosystem such as the NCCE, where remote physical forces cause profound and unpredictable fluctuations, resource management strategies that depend upon environmental prediction and control may be inherently risky. Such strategies are fundamental to fisheries management, for example, which seeks to maintain a stable, maximum yield of commercial species. However, simple stock recruitment relationships, traditionally used to establish harvest quotas, ignore the role of environmental variation on fish production, as well as the potential effects of intensive fish management on ecosystem functions. The well-documented collapse of the Pacific sardine fishery off California may have been the effect of harvest combined with environmental conditions that control the reproductive success of stocks throughout the region (Smith, 1978). Severe reductions in population abundance related to overexploitation may increase the risk of irreversible change in the structure of marine fish communities (May, 1977; Beddington, 1986). Debate continues regarding whether the slow recovery of Pacific sardine in the California Current is evidence for irreversible replacement of these stocks by other pelagic species (MacCall, 1986).

Resilience of the biological system in the California Current may depend upon a diversity of individual populations and species that respond independently to a fluctuating environment. The hierarchical organization of ecosystems dampens oscillations that otherwise might be destructive if all stocks and species responded identically to environmental variation (O'Neill *et al.*, 1986; Johnson, 1988). Individual stocks of pelagic fishes off California have fluctuated widely, but independently, in response to environmental change over the last several hundred years so that their combined biomass has remained relatively stable (Smith, 1978).

Paradoxically, attempts to stabilize individual components of an LME may cause greater biological instability. For example, efforts to boost the yield of adult salmon in the Pacific Northwest through the release of large numbers of fish from hatcheries may have contributed to a substantial decline in the diversity and abundance of locally adapted stocks (Goodman, 1990). Reduced heterogeneity of coho salmon populations in Oregon waters, in turn, may explain a decrease in annual survival of juveniles and increased variability of fisheries harvest since the mid-1970s (Bottom *et al.*, 1986; Nickelson, 1986). Numerous native stocks of other salmonid species within the NCCE also have been lost or reduced to very low levels through a variety of human causes (Nehlsen *et al.*, 1991). Reduced stock diversity throughout the region may diminish the capacity of salmonids to dampen effects of a highly variable physical environment or could cause depressed populations to be replaced by other species.

A fundamental risk of increased human disturbance in marine ecosystems is that entirely new patterns of temporal and spatial variability may be imposed for which local populations and communities have no prior evolutionary experience. Biological organization in the NCCE reflects persistent patterns of environmental change. Life-history strategies for marine fishes, birds, and mammals, for example, may be linked to strong seasonal signals in water temperature, upwelling, coastal currents, or river flows. Understanding patterns of physical and biological variability is necessary if human disturbance patterns are to be directed in ways that are compatible with the established organization of an LME. A comprehensive conservation strategy for the NCCE must address multiple scales of environmental variability that affect multiple levels of biological organization.

A Research Plan for the NCCE

Concern about environmental uncertainty in the NCCE is reflected in two fundamental goals of ecosystem management discussed by Lewis (1980): to maintain ecological diversity and to protect the organization and health of the system as a whole. To satisfy the first goal, selected subsystems, habitats, or species may be protected to prevent piecemeal loss of diversity from physical destruction, overexploitation, or misuse. Designation of protected areas cannot prevent exposure to pollutants transported in water or global changes that may alter entire systems. However, judicious placement of management zones, buffer areas, genetic preserves, and the like can help to minimize risks. Classification of the diversity of habitats and communities in an LME, therefore, should be a fundamental objective of environmental assessment.

The second goal—to protect the organization and health of the ecosystem—requires a more detailed understanding of ecological interrelationships, patterns, and processes. Biological organization of the NCCE is very dynamic, and the state of the ecosystem at a particular time may affect its sensitivity to disturbance. Static measures of distribution and abundance of marine communities, therefore, may be inadequate to design effective protective strategies or to later distinguish effects caused by human influence. The larger the scale and intensity of human intervention proposed in the NCCE (from single to cumulative sources), the more important it will be to understand the causes of natural variation that may determine the alternative environmental consequences.

Because research cannot provide certainty about the outcome of management decisions in the NCCE, the purpose of environmental assessment should be to improve understanding of the possible risks so that measures can be taken to maintain ecological diversity and to protect the organization and health of the ecosystem. Our research plan was developed specifically to evaluate the risks of oil, gas, and placer mineral development in the NCCE (Bottom *et al.*, 1989). However, our approach should apply generally to resource management issues in this and other LMEs.

We organized the plan by the categories of information needed to understand and avoid environmental risks. Quantitative risk assessment procedures are inappropriate in LMEs because the probabilities of various environmental outcomes will be unknown. We define risk in qualitative terms according to the following three general categories of information needed to evaluate the possibility of adverse environmental effects.

(i) What is the likelihood that a population, community, or ecosystem will be exposed to a disturbance or disturbances that could cause harm?

(ii) How sensitive is a particular organism, population, or community to exposure?

(iii) What are the possible alternative outcomes in the natural environment if organisms, populations, or communities are sensitive and they are exposed?

In the case of oil pollution, for example, the chance of exposure is the possibility that a spill will occur and that certain populations or communities might encounter the oil; the sensitivity of an organism, population, or commu-

nity to exposure is an evaluation of the sublethal or toxic effects of hydrocarbons; and finally, the outcome is all alternative responses of the marine ecosystem that might result, given, for example, the sensitivities of each population or community that is exposed, the persistence of oil that is released into the marine environment, the long-term changes in the biological community or ecosystem that could result, and so on.

Environmental restrictions that regulate offshore development are intended to decrease the chances of an undesirable outcome by reducing one or more of these components of risk (Fig. 2). For example, the surest method to minimize risk is to limit the chance of exposure. In the most extreme case, this can be accomplished by a complete prohibition of the activity. A less-restrictive approach is to limit the specific location of the activity to minimize the likelihood that critical or sensitive areas and organisms will be exposed. When the decision has been made to allow an activity in an area, then risk must be regulated by controlling the extent or magnitude of exposure. Stipulations to a leasing agreement, for example, may regulate the discharge of contaminants or the timing of activity to protect organisms in the area of disturbance. When an activity is permitted and stipulations have been established, little can be done to affect the final component of risk—the ecological outcome of exposure to a disturbance. At this stage of the decision process, research may be designed to monitor certain indicators of the results, and in some cases, changes in the management program may be possible to prevent continuation of an undesirable consequence.

Figure 3 lists the steps we followed to identify research needs for the NCCE. First, we reviewed the types of offshore development that are most likely to occur, their potential locations, and what is currently known about the environ-

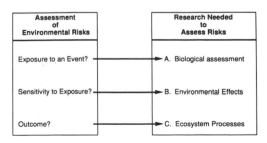

Figure 2. Three principal factors in an assessment of environmental risk. Program elements A, B, C of the research plan (Fig. 4) correspond generally to the three categories of information needed to address each of these factors.

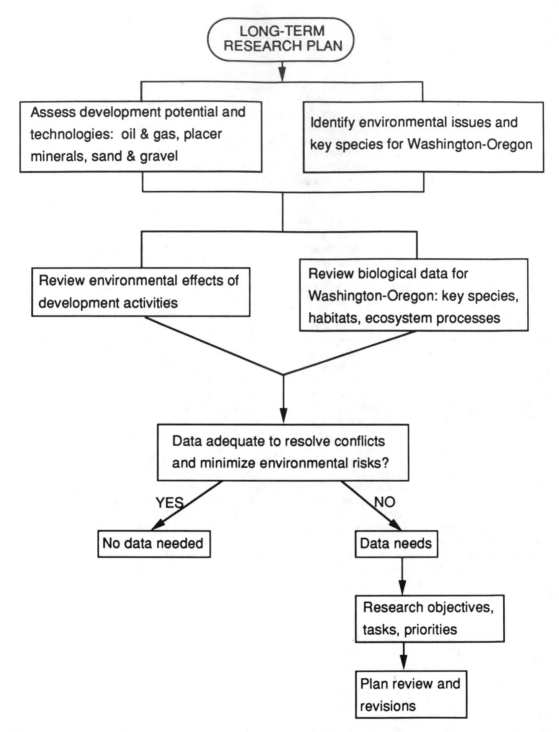

Figure 3. Methodology for developing an environmental research plan for the NCCE.

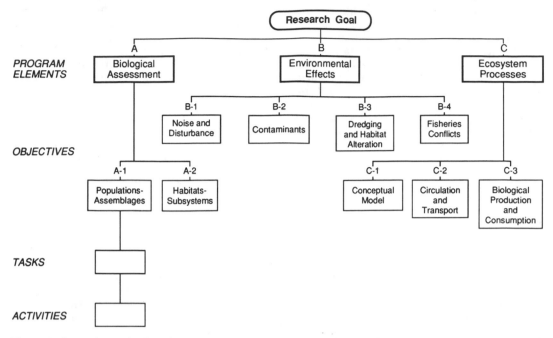

Figure 4. General organization of program elements and objectives in an environmental research plan proposed for the NCCE. Specific objectives, tasks, and activities in each program element are described by Bottom *et al.* (1989).

mental effects of each development activity. Second, we reviewed whether existing information is adequate to evaluate environmental risks in the NCCE. We organized research needs in the plan into three major program elements that generally correspond to the three components of risk described above (Fig. 4).

Program element A: Biological assessment

The risk of exposure to development activity can be minimized and options for the future can be preserved by protecting critical habitats and production areas and maintaining undisturbed examples of the diversity of communities represented in the NCCE. Research in this program element is needed to classify the diversity of habitats and communities throughout the region and to describe the distribution of important or susceptible populations. We developed a list of key species and assemblages for the NCCE as a guide in order to review whether existing inventory data are adequate to define candidate areas for protection. The plan lists research needs for many key species or groups of species for which little life-history or distributional information is available.

Program element B: Environmental effects

To establish environmental standards for offshore operations, research is needed to understand the sensitivity of organisms, populations, or communities to development activities. Program element B includes experimental studies in the laboratory or field to test the effects of proposed offshore disturbances. Baseline and monitoring studies are also listed in the plan for those activities that have not been widely studied elsewhere or for results that cannot be readily transferred to this region. These studies may be needed to evaluate whether lease stipulations provide adequate long-term protection. Four major categories of potential effects were identified in the plan that may require research to manage placer mining or oil and gas operations in the NCCE: noise and disturbance (from aircraft, vessels, and geophysical surveys); contaminants; dredging and habitat alteration; and conflicts with sport and commercial fisheries.

Program element C: Ecosystem processes

The outcome of disturbance at the ecosystem level is the most difficult component of risk to evaluate. Pollutants are transported beyond a local site by ocean currents so that the deferral of critical habitats from a lease sale cannot guarantee that they will be free from exposure to contaminants. Biological effects may be transferred via food chains or magnified by shifts in community structure. The cumulative effects of many small activities thus may be greater than the sum of their parts. Information is needed about the natural processes that control the transfer of materials, community dynamics, and production and flow of energy through the ecosystem. This research is essential in order to interpret the results of experimental studies and apply them to natural systems. It is also cost-effective, because the information is needed to address a broad range of environmental issues, whether the specific concern is oil spills, dredge spoils, mining discharges, or ocean outfalls. An environmental studies program must strike a reasonable balance between issue-oriented studies needed to describe sensitivities to individual environmental disturbances and process-oriented studies needed to interpret the possible outcomes of these effects.

Research Priorities

The National Research Council (NRC, 1985) concluded that the lack of understanding about natural variation and the inability to apply laboratory studies to the field are among the factors that most seriously undermine assessment of pollution effects in marine environments. These difficulties are accentuated in the NCCE, where extreme fluctuations in the physical environment and the lack of systematic inventory data limit understanding of the causes of natural variability and the risks of human disturbance. Because oceanographic information for the NCCE has not been collected consistently in time and space, regional scales of variability are poorly understood. Lack of basic inventory data for many habitats and species limits understanding of the specific resources that could be at risk from offshore development.

Although the research plan also includes experimental studies to evaluate the effects of individual development activities on selected populations and communities (Fig. 4), we conclude that understanding of natural variability is the most critical limitation to environmental assessment in the NCCE. More consistent monitoring of representative populations and communities is needed. Emphasis should be placed on expanding the number of habitats and communities for which there are data on population fluctuations. Concurrently, research on physical processes at a variety of scales is needed to interpret biological fluctuations. The interactive effects of local upwelling events and large-scale climatic change on biological production are particularly important. Satellite imagery may be useful to provide a synoptic view of regional oceanographic conditions through time.

Future research should also emphasize variability of entire aggregations of species. Much of the biological information for Washington and Oregon is analyzed and reported for single species of economic importance and has limited application to many broader ecological questions. In addition, environmental decisions will frequently focus on the development of specific locations where all species that co-occur may be directly affected. An important research task in the NCCE will be to summarize information in a geographic and multiple-species format that will facilitate site-specific management decisions. Assemblage-habitat distributions in the region can be evaluated through multivariate analysis of previous biological surveys, as well as new surveys designed at the multiple-species level.

Computerized mapping systems may provide a useful method to describe geographic relationships among species, environmental factors, and proposed offshore development activities. The Oregon Department of Fish and Wildlife is currently analyzing data from logbooks maintained by commercial fishermen to describe the distribution of harvest and fishing efforts and thereby identify areas critical to fisheries (Starr and Saelens, 1987). A geographic information system is being developed to allow these results to be compared with the locations of proposed oil and gas lease tracts and other development proposals in the NCCE.

A Conceptual Model

Understanding variability among the diverse components of an LME is a difficult undertaking. A conceptual model of the NCCE is needed, not as a means of prediction, but to classify the diversity and organization of the system and to define the critical questions that research should address. We recommend a hierarchical model to describe the processes that are believed to control biological variability for a variety of temporal and spatial scales.

Figure 5 is an example of a regional classification of the NCCE that could provide the general organization for a conceptual model. In this example, we define a generalized coastal domain as the continental margin to the edge of the shelf break (200 m contour). The coastal domain is further divided into three subregions that reflect north-south gradients in geological

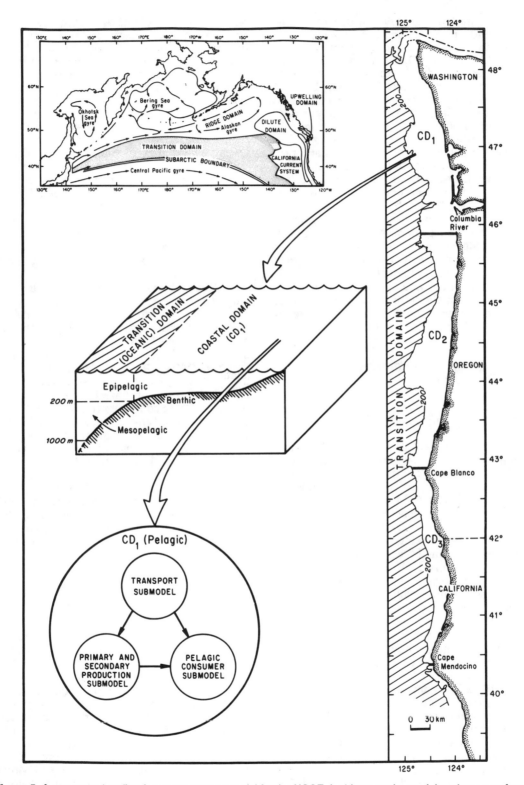

Figure 5. A conceptual outline for an ecosystem model for the NCCE. In this example, spatial scales range from the entire subarctic Pacific to benthic and pelagic habitats within subsystems of the NCCE. The division between coastal and transition (oceanic) domains is approximately the slope break of the continental shelf (200 m contour). The coastal domain is divided into three major subsystems (CD_1, CD_2, and CD_3) (Bottom *et al.*, 1989).

features; patterns of variation in winds, currents, and upwelling; and presumably, differences in production and consumption processes (Bottom *et al.*, 1989). A conceptual model should help to identify the components and processes of the ecosystem that may be most sensitive to disturbance and the deficiencies in existing knowledge that most severely restrict understanding of population fluctuations.

A critical interpretive step must follow completion of environmental studies and precede planning for offshore development. That is, the results of many unrelated studies (often conducted for reasons other than environmental assessment) must be compiled, analyzed collectively, and translated into a program of ecosystem management. A conceptual model can provide the framework needed to integrate and interpret information at an ecosystem level. Synthesis of environmental information, revision of the model, and update of the management program should be an ongoing process to reflect the evolution of scientific understanding that will result from new research.

Integrating Management and Research

Setting standards for environmental decisions

Management proposals and decisions determine the technical issues that research should address and when the results will be needed. Timeliness of research is critical if results are to keep pace with development proposals. All too often, however, environmental studies are a response to development decisions that already have been made. As a result, research becomes a means of legitimizing rather than making decisions. Standards for information collection must be built into a decision process to assure that results are timely and beneficial to ocean management.

The prelease phase of oil and gas development is a critical period when risks to an entire planning region are evaluated. Prelease decisions determine which areas within an entire planning region will be made available for leasing. Despite a detailed series of administrative steps that precede a federal lease sale, there are no specific standards to define the quality or quantity of environmental information necessary to make prelease decisions. Minimum standards should be established to ensure that critical leasing decisions do not precede collection and evaluation of the environmental information needed to support them.

The experimental approach of "adaptive management" (Holling, 1978; Lee, 1989) offers one method for building evaluation into a management program. Large-scale ecological experiments, however, may be difficult or impossible to evaluate and may increase the risk of whole-system effects. Wherever possible, resource "experiments" should be replicated and scaled appropriately to allow separation of multiple environmental and management effects (Walters and Holling, 1990). As a general rule, management decisions that can be broken into small, incremental stages with careful evaluation at each stage may help to minimize environmental risks. Similarly, decisions that occur over short temporal and spatial scales are usually less likely to have large-scale impacts. In this respect, the federal leasing program, which opens entire regions to oil and gas development for a period that could last several decades, is inconsistent with prudent management of environmental risks.

Uncertainty and the potential for irreversible change in LMEs suggest that conservative criteria should be applied to environmental decisions. Statistical detection of ecological changes frequently cannot be achieved; however, the costs of failure to detect them are often quite high. Peterman (1990a; 1990b), therefore, recommends the burden of proof be changed to require that resource users demonstrate that their actions will *not* have adverse consequences. Bella (1979) proposes the use of "environmental safety factors" to compensate for continued uncertainty about the outcome of environmental decisions in complex ecosystems. The greater the ignorance about biological resources and their response to a proposed disturbance, the greater the safety factor that may be necessary (Holt and Talbot, 1978; Bella, 1979). As research improves the understanding of environmental risks, safety factors applied to environmental decisions should be adjusted accordingly.

Strategic planning

A major weakness of environmental assessment programs is that system-wide effects are frequently dismissed after an initial, superficial review. Emphasis quickly shifts to the development and protection of selected areas or the control of pollutants at a particular development site. The ability to avoid large-scale effects decreases, however, as the management program becomes increasingly focused on local operational standards. Progress has been made in developing ecological approaches to resource development (e.g., Clark and Munn, 1986). Nonetheless, most existing management programs lack both the conceptual framework or model needed to guide ecosystem-level planning and the specific administrative machinery to require that local decisions are evaluated in a regional context.

A regional management plan is needed to guide resource management activities in the NCCE. In order to achieve a desired regional effect, the plan should describe the activities that will be allowed and the locations where they may occur (Regier and Baskerville, 1986). Because the consequences of decisions are often unpredictable, management strategies may be based on general ecological principles and experience (rather than on "hard data") to minimize the possibility of large-scale effects. Examples of such strategies might include (i) maintaining ecological diversity region-wide by limiting the total area of any one habitat, community, or population that will be affected by development; (ii) separating environmental disturbances in time and space; (iii) limiting or prohibiting certain categories of use to protect highly diverse or productive habitats and communities; (iv) minimizing or prohibiting disturbances in shallow and poorly flushed areas; and (v) protecting keystone species that tend to regulate community structure and stability. A combination of broad policies, offshore management zones, and performance standards may be used to direct local actions consistent with regional strategies.

Fragmentation of marine resource and regulatory responsibilities in the United States remains a fundamental impediment to management of the NCCE. Separate agencies are responsible for managing marine mammals, marine birds, and sport and commercial fisheries; pollutant discharges; ocean dredging and disposal; and oil, gas, and mineral development. Management is further segregated geographically between adjacent states and along the 3-mile boundary that divides state and federal jurisdictions. The concept of total ecosystem management becomes even more complicated in areas where international boundaries are involved.

Research activities in the NCCE reflect this lack of coordination among management interests and funding sources in the region. High costs of research and narrow institutional interests diffuse research activities into an array of individual, short-term, and local issues. Ironically, the lack of basic ecological understanding that has resulted from research on a series of unrelated issues is universally limiting to the management capabilities of the many resource interests in the region. A cooperative program of research institutions and state and federal resource agencies is needed to plan, review, and coordinate environmental studies throughout the NCCE.

The recent development of a draft Oregon Ocean Resources Management Plan (Oregon Ocean Resources Management Task Force,

1988) is one hopeful step toward integrated management in the Pacific Northwest. Although the plan does not address management issues for the entire NCCE, it does develop general policies for the use of both living and nonliving resources in federal and state waters of the Oregon continental margin.

The most severe environmental threats in LMEs may be the result of unconscious decisions that, alone or in combination with others, produce system-wide responses. Uncontrolled, large-scale "experiments" are being conducted in the NCCE without whole-system management strategies in place to preserve future options. Large-scale experiments in progress with unknown ecological effects include intensive harvest of multiple species in commercial fisheries; system-wide releases of large numbers of artificially produced salmon; low-level, chronic pollution from multiple sources; and global climate change. Ecosystem planning cannot guarantee that irreversible environmental consequences of human activities will be prevented. At the very least, however, strategic planning is needed to ensure that future disturbances are the result of conscious decisions that include deliberate safeguards to mitigate the effects of human ignorance.

Decisions to direct the intensity and patterns of human use in marine ecosystems are not purely technical in nature. They will require difficult ethical choices about the quality of life for present and future generations. Science has an obligation to provide understanding of the alternative risks and to recommend safeguards, but it cannot make the choices.

References

Abbott, M. R., and Zion, P. M. 1987. Spatial and temporal variability of phytoplankton pigment off northern California during Coastal Ocean Dynamics Experiment I. J. Geophys. Res. 92:1745–1755.

Beddington, J. R. 1986. Shifts in resource populations in large marine ecosystems. In: Variability and management of large marine ecosystems. pp. 9–18. Ed. by K. Sherman and L. M. Alexander. AAAS Selected Symposium 99. Westview Press, Inc., Boulder, CO.

Bella, D. A. 1979. An inquiry into the rationale and use of environmental safety factors. Oregon State University, Department of Civil Engineering, Corvallis.

Bella, D. A., and Overton, W. S. 1972. Environmental planning and ecological possibilities. Journal of the Sanitary Engineering Division, Proceedings of the American Society of Civil Engineers 98:579–592.

Bernal, P. A. 1981. A review of low-frequency response of the pelagic ecosystem in the

California Current. CalCOFI Report XXII:49–62.

Bottom, D. L., and Jones, K. K. 1990. Species composition, distribution, and invertebrate prey of fish assemblages in the Columbia River Estuary. Prog. Oceanogr. 25:243–270.

Bottom, D. L., Jones, K. K., Rodgers, J. D., and Brown, R. F. 1989. Management of living marine resources, a research plan for the Washington and Oregon continental margin. Publication No. NCRI-T-89-004. National Coastal Resources Research and Development Institute, Newport, OR.

Bottom, D. L., Nickelson, T. E., and Johnson, S. L. 1986. Research and development of Oregon's coastal salmon stocks. Oregon Department of Fish and Wildlife Progress Reports (Fish), Project number AFC-127, 30 September 1985 to 29 September 1986, Portland.

Chelton, D. B. 1981. Interannual variability of the California Current—Physical factors. CalCOFI Report XXII:34–48.

Chelton, D. B., Bernal, P. A., and McGowan, J. A. 1982. Large-scale interannual physical and biological interaction in the California Current. J. Mar. Res. 40:1095–1125.

Clark, W. C., and Munn, R. E. (Editors). 1986. Sustainable development of the biosphere. Cambridge University Press, Cambridge, England.

Ekman, S. 1967. Zoogeography of the sea. Sidgwick and Jackson, London.

Fager, E. W., and McGowan, J. A. 1963. Zooplankton species groups in the North Pacific. Science 140:453–460.

Favorite, F., Dodimead, A. J., and Nasu, K. 1976. Oceanography of the subarctic Pacific region, 1960–71. International North Pacific Fisheries Commission Bulletin Number 33. Vancouver, Canada.

Fulton, J. D., and LeBrasseur, R. J. 1985. Interannual shifting of the subarctic boundary and some of the biotic effects on juvenile salmonids. In: El Niño North: Niño effects in the eastern subarctic Pacific Ocean. pp. 237–247. Ed. by W. S. Wooster and D. L. Fluharty. Washington Sea Grant Program. University of Washington, Seattle.

Goodman, M. L. 1990. Preserving genetic diversity of salmonid stocks: A call for federal regulation of hatchery programs. Environ. Law 20:111–166.

Graybill, M., and Hodder, J. 1985. Effects of the 1982–83 El Niño on reproduction of six species of seabirds in Oregon. In: El Niño north: Niño effects in the eastern subarctic Pacific Ocean. pp. 205–210. Ed. by W. S. Wooster and D. L. Fluharty. Washington Sea Grant Program. University of Washington, Seattle.

Holling, C. S. 1973. Resilience and stability of ecological systems. Annual Review of Ecology and Systematics 4:1–23.

Holling, C. S. (Editor). 1978. Adaptive environmental assessment and management. John Wiley and Sons, New York.

Holt, S. J., and L. M. Talbot. 1978. New principles for the conservation of wild living resources. Wildlife Monographs No. 59. The Wildlife Society, Inc., Louisville, KY.

Huyer, A. 1983. Coastal upwelling in the California Current System. Prog. Oceanogr. 12:259–284.

Johnson, L. 1988. The thermodynamic origin of ecosystems: A tale of broken symmetry. In: Entropy, information, and evolution: New perspectives on physical and biological evolution. pp. 75–105. Ed. by B. H. Weber, D. J. Depew, and J. D. Smith. MIT Press, Cambridge, MA.

Johnson, M. W., and Brinton, E. 1962. Biological species, water masses, and currents. In: The sea. Volume 2. pp. 381–414. Ed. by M. N. Hill. Interscience Publishers, New York.

Johnson, S. L. 1984. The effects of the 1983 El Niño on Oregon's coho and chinook salmon. Oregon Department of Fish and Wildlife Information Reports (Fish) 84–8, Portland.

Landry, M. R., Postel, J. R., Peterson, W. K., and Newman, J. 1989. Broad-scale distributional patterns of hydrographic variables on the Washington/Oregon shelf. In: Coastal oceanography of Washington and Oregon. pp. 1–40. Ed. by M. R. Landry and B. M. Hickey. Elsevier Publishing Company, New York.

Lasker, R. 1978. Ocean variability and its biological effects—Regional review—Northeast Pacific. Rapp. P.-v. Reun. Cons. int. Explor. Mer 173:168–181.

Lee, K. N. 1989. The Columbia River Basin: Experimenting with sustainability. Environment 31:6–33.

Lewis, J. R. 1980. Options and problems in environmental management and evaluation. Helgo. Meeresunters. 33:452–466.

MacCall, A. D. 1986. Changes in the biomass of the California Current ecosystem. In: Variability and management of large marine ecosystems. pp. 33–54. Ed. by K. Sherman and L. M. Alexander. AAAS Selected Symposium 99. Westview Press, Inc., Boulder, CO.

May, R. M. 1977. Thresholds and breakpoints in ecosystems with a multiplicity of stable states. Nature 269:471–477.

McClain, D. R., and Thomas, D. H. 1983. Year-to-year fluctuations of the California countercurrrent and effects on marine organisms. CalCOFI Report XXIV:165–181.

McGowan, J. A. 1971. Oceanic biogeography of the Pacific. In: The micropaleontology of oceans. pp. 3–74. Ed. by B. M. Funnell and W. R. Riedel. Cambridge University Press, Cambridge, England.

MMS (Minerals Management Service). 1987. Interior Department releases 5-year offshore oil and gas leasing program. In: POCS Current Events, May 1987. pp. 1–2. Minerals Management Service, Pacific OCS Region, Los Angeles.

MMS. 1988. Proceedings: Conference/workshop on recommendations for studies in Washington and Oregon relative to offshore oil and gas development, May 1988. Minerals Management Service, OCS Study MMS 88–0090, Pacific OCS Region, Los Angeles.

Mysak, L. A. 1986. El Niño, interannual variability and fisheries in the Northeast Pacific Ocean. Can. J. Fish. Aquat. Sci. 43:464–497.

Nehlsen, W., Williams, J. E., and Lichatowich, J. L. 1991. Pacific salmon at the crossroads: Stocks at risk from California, Oregon, Idaho, and Washington. Fisheries, A Bulletin of the American Fisheries Society 16:4–21.

Nickelson, T. E. 1986. Influences of upwelling, ocean temperature, and smolt abundance on marine survival of coho salmon (*Oncorhynchus kisutch*) in the Oregon production area. Can. J. Fish. Aquat. Sci. 43:527–535.

NRC (National Research Council). 1985. Oil in the sea: Inputs, fates, and effects. National Academy Press, Washington, DC.

O'Neill, R. V., DeAngelis, D. L., Waide, J. B., and Allen, T. F. H. 1986. A hierarchical concept of ecosystems. Monographs in Population Biology 23. Princeton University Press, Princeton, NJ.

Oregon Ocean Resources Management Task Force. 1988. Managing Oregon's ocean resources. Interim report to the Joint Legislative Committee on Land Use, July 1, 1988. Oregon Department of Land Conservation and Development, Portland.

Pearcy, W., Fisher, J., Brodeur, R., and Johnson, S. 1985. Effects of the 1983 El Niño on coastal nekton of Oregon and Washington. In: El Niño north: Niño effects in the eastern subarctic Pacific Ocean. pp. 188–204. Ed. by W. Wooster and D. Fluharty. Washington Sea Grant Program. University of Washington, Seattle.

Peterman, R. M. 1990a. Statistical power analysis can improve fisheries research and management. Can. J. Fish. Aquat. Sci. 47:2–15.

Peterman, R. M. 1990b. The importance of reporting statistical power: The forest decline and acidic deposition example. Ecology 71:2024–2027.

Peterson, W. T. 1973. Upwelling indices and annual catches of Dungeness crab, *Cancer magister*, along the west coast of the United States. Fish. Bull. U.S. 71:902–910.

Peterson, W. T., and Miller, C. B. 1975. Year-to-year variations in the planktonology of the Oregon upwelling zone. Fish. Bull. U.S. 73:642–653.

Peterson, W. T., Miller, C. B., and Hutchison, A. 1979. Zonation and maintenance of copepod populations in the Oregon upwelling zone. Deep-Sea Res. 26A:467–494.

Regier, H. A., and Baskerville, G. L. 1986. Sustainable redevelopment of regional ecosystems degraded by exploitive development. In: Sustainable development of the biosphere. pp. 75–101. Ed. by W. C. Clark and R. E. Munn. Cambridge University Press, Cambridge, England.

Sherman, K., and Alexander, L. M. (Editors). 1986. Variability and management of large marine ecosystems. AAAS Selected Symposium 99. Westview Press, Inc., Boulder, CO.

Sherman, K., Alexander, L. M., and Gold, B. D. (Editors). 1990. Large Marine Ecosystems: Patterns, processes, and yields. AAAS Symposium. AAAS, Washington, DC.

Small, L. F., and Menzies, D. W. 1981. Patterns of primary productivity and biomass in coastal upwelling regions. Deep-Sea Res. 28A:123–149.

Smith, P. E. 1978. Biological effects of ocean variability: Time and space scales of biological response. Rapp. P.-v. Reun. Cons. int. Explor. Mer 173:117–127.

Soutar, A., and Issacs, J. D. 1974. Abundance of pelagic fish during the 19th and 20th centuries as recorded in anaerobic sediment off California. Fish. Bull. U.S. 72:257–273.

Starr, R. M., and Saelens, M. R. 1987. Identification of important marine habitat. Oregon Department of Fish and Wildlife, Project Completion Report, Portland.

U.S. Bureau of Land Management. 1977. Recommendations for baseline research in Washington and Oregon relative to offshore resource development. Research Triangle Institute, NC.

U.S. Bureau of Mines. 1987. An economic reconnaissance of selected heavy mineral placer deposits in the U.S. Exclusive Economic Zone. Open File Report 4–87, Washington, DC.

Wahl, T. R., Ainley, D. G., Benedict, A. H., and Degange, A. R. 1989. Associations between seabirds and water masses in the northern Pacific Ocean in summer. Mar. Biol. 103:1–11.

Walters, C. J. and Holling, C. S. 1990. Large-scale management experiments and learning by doing. Ecology 71:2060–2068.

Wickett, W. P. 1967. Ekman transport and zooplankton concentration in the North Pacific Ocean. J. Fish. Res. Board Can. 24:581–594.

Chapter 25

Sustainable Development of the Great Barrier Reef as a Large Marine Ecosystem

Graeme Kelleher

> The environment does not exist as a sphere separate from human actions, ambitions and needs.
> —Gro Harlem Brundtland

Introduction

The concept that development should be sustainable is not new. It has existed in virtually every group of humans who have lived and depended on the earth's natural bounty. One of the most important factors in eroding the commitment to sustainability in theory and practice in the twentieth century has been the application of modern economic analysis—incorporating the methods of benefit-cost analysis, net present worth, and discount rates. Taken together, these methods tend to lead to decisions that state tacitly or explicitly that anything that happens more than 20 years hence is irrelevant.

Recently, there has been a dramatic realization by world leaders that we must change our approach to development. Margaret Thatcher, Mikhail Gorbachev, George Bush, and Australia's former Prime Minister, Bob Hawke, have recognized the absolute dependence of development on protection of the natural environment (Kelleher, 1989).

The Great Barrier Reef Marine Park Act was one of the first pieces of legislation in the world to apply the concept of sustainable development to the management of a large natural area. So far, the approach has been successful—overexploitation of the Great Barrier Reef has largely been prevented.

We are entering a new and more difficult phase: Direct use of the marine park is increas-

ing; government expenditure as a proportion of gross domestic product is decreasing; there are proportionately fewer resources for management; management agencies are being forced to recover costs from users who are reluctant to pay; and there is evidence that nutrient levels in the waters of some parts of the marine park are at times above those at which some corals can thrive.

The next 10 years will determine whether there is sufficient commitment in the minds and hearts of Australians to ensure that the Great Barrier Reef is protected from insidious degradation.

The Great Barrier Reef

The Great Barrier Reef (the reef) is the largest system of corals and associated life forms anywhere in the world. It is encompassed in a marine park within the Great Barrier Reef Region (Fig. 1) covering an area of about 350,000 km^2 on the Australian continental shelf—larger than the United Kingdom. The reef stretches for almost 2,000 km along the northeastern coast of Queensland in a complex maze of approximately 2,900 individual reefs, ranging in area from less than 1 hectare to more than 100 km^2. In the north, the reef is narrow and its eastern edge is marked by a series of narrow "ribbon" reefs, but in southern areas it broadens out and presents a vast wilderness of "patch" reefs.

The reef is diverse not only in the form and size of its individual reefs and islands, but in its inhabitants. Six species of turtle occur in the region, and it is believed that there are more than 1,500 species of fishes (GBRMPA, 1981).

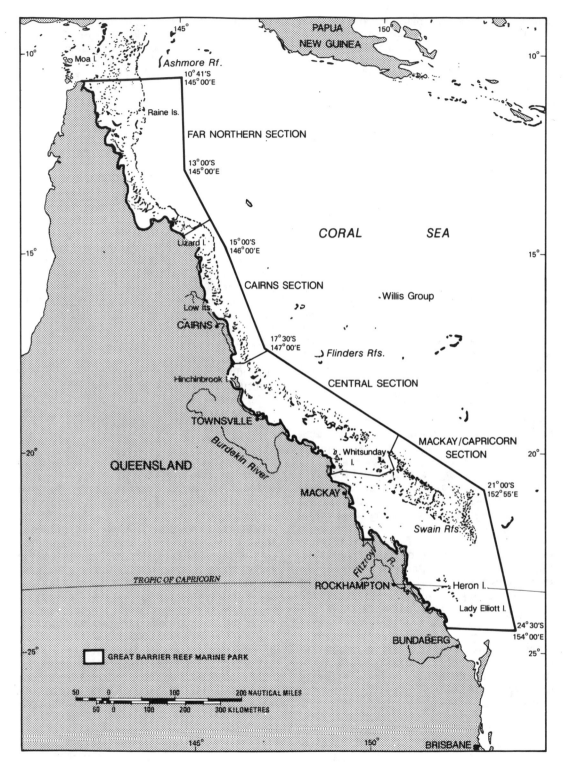

Figure 1. Great Barrier Reef Region of Australia.

The reef may be the last place on earth in which the dugong (*Dugong dugong*), an endangered species, is still common and is not in jeopardy (Bertram, 1979). About 350 species of reef-building coral have been identified on the reef (C. Wallace, pers. com.), and the islands are inhabited or visited by more than 240 species of birds (Kikkawa, 1976).

Human Use of the Great Barrier Reef

It has been estimated that the value of reef-related activities (on the reef and on the adjacent mainland) approximates $1 thousand million (Australia) per annum.

Commercial fishing and tourism, recreational pursuits (including fishing, diving, and camping), traditional fishing, scientific research, and shipping all occur within the Great Barrier Reef Region. The only activity which is prohibited throughout the region is oil drilling.

Resort tourism is the largest commercial activity in economic terms. In 1986–1987, there were an estimated 162,000 visitor trips resulting in 1,018,000 visitor nights spent at the 24 island-resorts, and the visitors spent around $175.6 million (Australia) at the resorts (Queensland Tourist and Travel Corporation, 1988). Resort guests make extensive use of reefs and waters for recreational activities including fishing, diving and snorkeling, water sports, sightseeing, reef-walking, and collecting.

The popularity of the reef and the adjacent coastal region as a destination for tourists increased 40-fold over the period from the 1940s to 1980 and is continuing to increase. Recently, accommodation establishments have been built directly on reefs. The last 10 years has seen the introduction of many large, stable, high-speed catamarans, providing day trips to islands and outer reefs. The latest development is wave-piercing catamarans, which provide faster, more comfortable transport.

There are conflicts among the various users of the reef and between some users and those who wish to see the reef maintained in its pristine state forever. Some uses of parts of the reef have already reached levels that may fully exploit the productive capacity of the system—bottom-trawling for prawns is an example. Run-off from islands and the mainland contains suspended solids, herbicides, pesticides, nutrients, and other materials. The magnitude of their effects on the reef is not yet known, but investigations are under way.

This description applies also to other reef systems throughout the world's tropical seas. The need for and the difficulties of managing uses so that they are sustainable forever are also common. Perhaps the system of management that has been developed on the Great Barrier Reef could be applied elsewhere, although the acceptability of any management system is likely to be diminished where there are very high levels of usage and economic dependence on reef areas, for instance in many parts of Asia.

The Goal and Aims of the Authority

The Great Barrier Reef Marine Park Authority has derived a primary goal and a set of aims from the provisions of the act and recognition of the political, legal, economic, sociological, and ecological environment in which it operates.

The authority believes that any use of the reef or associated areas should not threaten the reef's essential ecological characteristics and processes. Activities depending on the reef's renewable resources should generally be held at or below maximum sustainable intensities indefinitely. This belief has led the authority to adopt the following primary goal:

> To provide for the protection, wise use, understanding and enjoyment of the Great Barrier Reef in perpetuity through the development and care of the Great Barrier Reef Marine Park. (GBRMPA, 1991, p. 7)

However, not only the physical aspects of the reef need to survive. If the reef is to be protected, administrative arrangements also must be durable. Failure of the authority would not necessarily or even probably be followed by the creation of new, more effective arrangements. In Australia, the major determinant of administrative survivability of organizations such as the authority is public support. In the long run, government support flows from it. Recognizing that the authority and the marine park concept already have a degree of public support, the authority must act in ways that sustain or increase that support. What are those ways? It seems clear that the groundwork has been well established in the act through the formal requirements for public participation, the provisions for a consultative committee, the composition of the authority itself and its functions, as well as the ability to perform those functions in association with Queensland or its agencies.

Generally speaking, the public is likely to continue to support the marine park and the authority if the primary goal is perceived as being achieved efficiently. In order for this to occur, the public will have to (i) be aware of what the authority and its day-to-day management agencies are doing and the way they are doing it; (ii) be aware of the effectiveness and costs of their programs and the reasons for them; and (iii) to the extent practicable, be involved in the establishment and management of the marine park. A set of aims has been derived from these criteria and related observations. They are subordinate to the primary goal and must be read in conjunction with it and with each other.

The authority's aims

The nine aims of the authority are described below. (GBRMPA, 1991, pp. 7–8)

Aim 1: To achieve its goal and other aims by employing people of high calibre, assisting them to reach their full potential, providing a rewarding, useful and caring work environment and encouraging them to pursue relevant training and development opportunities.

Aim 2: To involve the community meaningfully in the development and care of the marine park.

Aim 3: To achieve management of the marine park primarily through the community's commitment to the protection of the Great Barrier Reef and its understanding and acceptance of the provisions of zoning, regulations and management practices.

A further aim has been derived from the requirement for the marine park and the authority to be seen as serving the public interest in the long term. The rate of change of the social and physical environments are increasing. There is a great need for adaptability and flexibility in all that governments do.

Aim 4: To adapt actively the marine park and the operations of the Authority to changing circumstances.

The public purse is by no means infinite. Public support will continue to be dependent on the careful use of public resources.

Aim 5: To minimize the costs of caring for and developing the marine park consistent with meeting the goal and other aims of the Authority.

The social contract's success depends on the benefits exceeding the costs. Any form of regulation is, in the mind of a free agent, a fetter and an imposition.

Aim 6: To minimize regulation of, and interference in, human activities consistent with meeting the goal and other aims of the authority.

The authority is the agency responsible for reconciling the many conflicting demands on the reef. It must be viewed as competent and fair, in reality as well as in perception.

Aim 7: To achieve competence and fairness in the care and development of the marine park through the conduct of research, and the deliberate acquisition, use and dissemination of relevant information from research and other sources.

The act makes it clear that the authority must provide for reasonable use, consistent with conservation. The following two aims are essential facets of this responsibility.

Aim 8: To protect the natural qualities of the Great Barrier Reef, while providing for reasonable use of the reef region.

Aim 9: To provide for economic development consistent with meeting the goal and other aims of the authority.

Taken together, this goal and these aims provide a policy framework for evaluation of proposed programs and actions. We believe that much of the success of the marine park is the result of applying these policies. We believe further that they can (and should) be applied in natural resource management anywhere, not only in Australia. Much of the thrust of policy is communication with and involvement of the public in the development and care of the marine park. Educational and informational programs are, therefore, critically important elements of the authority's work.

The Marine Park and the Zoning System

The Great Barrier Reef Marine Park is not a national park. It is a multiple-use protected area, fitting the definition of Category VIII of the classification system used by the IUCN/the World Conservation Union (1982). It also meets the criteria for selection and management as a Biosphere Reserve (Category IX), although it has not been formally proposed or established as one. The reef was inscribed on the World Heritage List in 1981 as a natural site (Category X).

Through the use of zoning, conflicting activities are separated, areas are provided that are suitable for particular activities, and some areas are protected from use. Levels of protection of zones within the park vary from almost complete absence of restriction on activity to nearly complete protection, where almost no human activities are permitted. The only activities that are prohibited throughout the park are oil exploration, mining (other than for approved research purposes), littering, spearfishing with scuba gear, and the taking of large specimens of certain species of fish.

In the zoning plans that have been developed so far, there are three major categories of zones:

(1) Preservation and Scientific Research—Equivalent to IUCN Category I, Scientific Research/Strict Nature Reserve. The only human activity

permitted is strictly controlled scientific research.

(2) Marine National Park—Equivalent to IUCN Category II, National Park. The major uses permitted are scientific, educational, and recreational.

(3) General Use—Equivalent to IUCN Categories IV, Managed Nature Reserve, and VI, Resources Reserves. Uses are held at levels that do not jeopardize the ecosystem or its major elements. Commercial and recreational fishing are generally permitted, although bottom-trawling is prohibited in one of these two zones.

The zones are fixed during the life of a zoning plan (generally 5 to 7 years). They are complemented by generally smaller areas that give special protection from time to time to animal breeding or nesting sites, to sites in general use and to other zones that must be protected to allow appreciation of nature—free from fishing or collecting, and to sites suitable for scientific research.

Because there has been a dramatic increase in the use of the marine park by tourists, the existing zoning system, which focuses on fishing, is proving inadequate. There is increasing competition for tourism use of particular sites. Usually, these sites are near major areas of coastal development (e.g., Cairns or Townsville) or have particular attributes that make them suitable for tourism—the Whitsunday Islands, for example.

At a special conference arranged by the authority in late 1988, participants agreed that there was a need to incorporate a tourism strategy into the zoning system. This strategy would identify those areas that are particularly suited to tourism development and those that should be retained in their natural state, undisturbed by such development. The strategy is being implemented through the zoning system, initially in the rezoning of the Cairns Section of the marine park. It will be extended to the other three sections as they are rezoned during the next 5 years. Initial zoning of the entire marine park was completed on schedule in 1988—Australia's bicentenary.

A major constraint in zoning for tourism has been that many tourists like to observe the natural qualities of the Great Barrier Reef undisturbed by fishing. Modern technology allows them to do this from semisubmersible vessels and from underwater observations. If carried out with care, these activities can have very little, if any, effect on the reef's ecosystem. They are, in other words, compatible with the authority's goal. The need to provide for such activities in zones that are protected from fishing has led the authority to consider adopting a slightly different zoning system.

New names have been designed to describe the qualities of the zones more accurately. The major change proposed to the zoning system is to divide each of the zones, other than the Preservation Zone, into two categories:

Category 1—No structures (for example, floating hotels, pontoons, or mariculture) will be permitted. Mooring buoys may be permitted.

Category 2—Such structures are permitted provided they meet environmental guidelines. A permit is required.

The adoption of the new zoning scheme will allow the authority, in association with interested members of the public and with other agencies, to develop and apply tourism strategy for the whole of the Great Barrier Reef. The aim will be to ensure that the whole reef will not become dotted with tourist and other structures, while at the same time providing for careful development on reefs that are suitable for that purpose. The strategy should allow the authority to protect the Great Barrier Reef while providing for sustainable development.

Linkage with the land

It has always been the aim of the authority to ensure that the Great Barrier Reef is managed as a single ecosystem, including all the waters of the Great Barrier Reef Region and the 900 islands within its outer boundaries.

This has been achieved largely through the following mechanisms:

- Coordination of policy and action between the federal and state (Queensland) governments by a four-person ministerial council
- The state government being represented by one member on the three-person authority (the chairman represents the federal government and a third member is independent)
- Day-to-day management of the waters and islands by a single management group, with costs shared by the two governments
- The policies of the authority (described previously)
- A strong commitment of the authority and Queensland agencies to work together

Deleterious effects on the Great Barrier Reef caused by human activities on the mainland are more intractable.

The major issue probably is run-off of nutrient-enriched water from farmland. However, even in this case, high levels of cooperation in research have been achieved among the authority, farmers' organizations, and state government agencies responsible for primary industry. Our experience has been that if cooperation is achieved in carrying out research into a problem, then that cooperation is likely to extend into defining and applying ways of solving the problem.

Future Challenges

Apart from tourism, there are three major areas of challenge in our task of ensuring that the Great Barrier Reef is sustained into the twenty-first century:
(1) Achieving sustainable fishing and collecting
(2) Understanding the "crown-of-thorns" phenomenon
(3) Protecting the Great Barrier Reef from pollution

Sustainable fishing

The Great Barrier Reef prawn fishery is overcapitalized, as is nearly every major world fishery. It is going through the now familiar process of declining catch-effort ratios. This problem is being addressed in two ways.

Through the zoning system, some areas of the marine park are closed to trawling. These amount to about 30% of the total area. They act as reference areas, so that the long-term effects of trawling can be determined. They also ensure that the species that occupy them are fully protected.

The second approach is being taken by the fishing industry itself. It consists of traditional fisheries practices, including seasonal closures, protection of nursery areas, and attempts at limiting the total effort expended in the fishery.

We are fairly confident that these two complementary approaches together will ensure that any damage to the Great Barrier Reef system from trawling can be repaired. Additionally, the authority is working with the commercial fishing industry in a major study of the effects of all kinds of fishing on the reef ecosystem.

Crown-of-thorns starfish

The federal government and its agencies have already spent more than $5 million on research into the crown-of-thorns starfish. Not only are we convinced that this money was well spent, but we are equally sure that a high level of research effort must continue to be applied to this issue, at least until we have answered un-

equivocally the question of whether human activity is affecting the intensity and frequency of population outbreaks.

If it is demonstrated scientifically that human activity is a major factor, then the authority would react in two ways. First, it would reconsider its policy of limiting control of population numbers to areas of particular scientific or tourist value. Second, it would have to move strongly to ensure that the human activities that were shown to be causative factors were modified.

If, on the other hand, the scientific evidence that continues to accumulate indicates that outbreaks have not changed in intensity or frequency since Europeans arrived in Australia, the authority would maintain its existing policies. Those policies are to limit control of outbreaks to areas of particular value to tourism or to science, and to continue to study scientifically the whole phenomenon and its role in the reef system.

The sooner we can resolve this issue, the better. Although it is true that crown-of-thorns research answers many scientific questions of a general nature, scarce research funds could be spent more effectively on answering these general questions if crown-of-thorns starfish were not an issue. Further, it is conceivable that postponement of resolution of this issue might lead to progressive degradation of the Great Barrier Reef, requiring hundreds of years for recovery.

Pollution control

Monitoring has shown that the levels of toxic chemicals in virtually the whole Great Barrier Reef Region are very low—close to the limit of detectability. Such chemicals are not likely to be a problem in the near future, but monitoring will be continued.

There is an ever-present risk of an oil spill in the Great Barrier Reef. The precautions that have already been taken to minimize this risk include (i) the recommendation by the International Maritime Organization that all ships using the waters of the Great Barrier Reef Region be piloted and (ii) the development of an oil spill contingency plan entitled: "Reef Plan." This plan is a component of the National Plan to Combat Pollution of the Sea by Oil. In the event of a spill, an on-scene coordinator is appointed to take direct charge of operations to combat the pollution incident. He is assisted by an operations controller who is responsible for all administrative actions in support of the on-scene coordinator. The on-scene coordinator is advised by a committee of experts on actions to be taken and on materials and methods that are best suited for combatting a specific oil pollution situation. The GBRMPA is a member of this committee. We have developed a unique program for Apple Macintosh computers that incorporates bathymetric and environmental information covering the whole region and a predictive capability for the movement of spilled oil under various weather conditions.

In light of experiences in other parts of the world, it would be naive to expect even the best oil-spill contingency plan to prevent damage to the marine park by an oil spill. The authority will continue to work with other agencies in reducing the risk of oil spills and in improving the capacity for a response to such events.

A more insidious, but potentially no less serious, concern for the future is the increase in nitrogen and phosphorus in waters of the Great Barrier Reef. Work on the Great Barrier Reef, in the Great Barrier Reef Aquarium, and in other parts of the world has shown that reef-building corals are very vulnerable to increases in the levels of these nutrients. Monitoring has shown that the porosity of some nearshore reefs of the Great Barrier Reef has been increasing over the past few decades because of an increase in phosphorus levels. At times, the levels of nitrogen in some parts of the Great Barrier Reef significantly exceed levels that have been shown to cause death in branching corals permanently subjected to them. Already the authority is expending almost half of its annual research funds in defining the seriousness of this problem. We are seeking financial support from private enterprise to establish a more comprehensive reef-wide monitoring program and to develop a complete understanding of the origins of these nutrients so that corrective action can be taken if necessary. I believe that protection of the Great Barrier Reef from increasing nutrient levels may be the greatest challenge facing the authority in the next 2 decades.

Conclusions

The Great Barrier Reef Marine Park is an example of the practical application of the principles defined in the World Conservation Strategy (IUCN/UNEP/WWF, 1980). It can be seen as a model for development of the kind described in the report of the World Commission on Environment and Development—"Our Common Future" (Brundtland, 1987).

Our experience has shown how difficult the attainment will be of sustainable development throughout the world. In Australia, we are blessed with a comparatively small population, great natural resources, and a strong emotional and intellectual commitment by the people to the protection of the Great Barrier Reef. Even with these advantages, the authority is continually struggling to prevent insidious degradation of the Great Barrier Reef ecosystem. The pressures for overexploitation are not confined to the direct uses of the Great Barrier Reef. They come also from the effects of activities that release pollutants into the reef's waters.

More than ever before, the authority needs the support of the Australian public if their primary goal—protecting the Great Barrier Reef forever—is to be achieved. The need for public and governmental support for environmental protection is evident everywhere. Fortunately, there is unprecedented recognition of this need throughout the world. This conference is an indication of that recognition. Increasingly, we must expect and demand that developmental industries such as tourism voluntarily take action to ensure that development is sustainable.

At national and international levels, the adverse effects of land-based human activities on marine ecosystems must be recognized and actions must be taken to reduce them. This will require higher degrees of cooperation than have been attained historically among governments, levels of government, and agencies within government. Integrated planning and management processes will have to be developed. The success of such approaches will be assisted if planning and management are applied at the scale of large marine ecosystems.

References

Bertram, G. C. L. 1979. Dugong numbers in retrospect and prospect. In: The Dugong, proceedings of a seminar workshop held at James Cook University, 8–13 May. Ed. by H. Harsh. James Cook University, Townsville, Australia.

Brundtland, Gro Harlem. 1987. Our common future. In: Report of the World Commission on Environment and Development. Foreword. p. xi. Oxford University Press, Oxford, UK.

GBRMPA (Great Barrier Reef Marine Park Authority). 1981. Nomination of the Great Barrier Reef by the Commonwealth of Australia for inclusion in the World Heritage List. GBRMPA, Townsville, Australia

GBRMPA. 1991. Annual report, 1990–1991. December 1991. p. 7. GBRMPA, Townsville, Australia.

IUCN-The World Conservation Union (formerly the International Union for the Conservation of Nature and Natural Resources). Commission on National Parks and Protected Areas. 1982. Categories, Objectives, and Criteria for Protected Areas. IUCN, Gland, Switzerland.

IUCN/UNEP (United Nations Environment Program) (UNEP)/World Wildlife Fund (WWF). 1980. World conservation strategy: Living resource conservation for sustainable development. Gland, Switzerland.

Kelleher, G. 1989. Graduation ceremony address. Griffith University. Brisbane, Australia.

Kikkawa, J. 1976. The birds of the Great Barrier Reef. In: Biology and geology of coral reefs. Vol. 3. Biology 2. pp. 282–291. Ed. by O. A. Jones and R. Endean. Academic Press, New York.

Queensland Tourist and Travel Corporation. 1988. Major survey research programme. Brisbane, Australia.

Chapter 26

Role of National Political Factors in the Management of LMEs: Evidence from West Africa*

Victor Prescott

Introduction

The concept of LMEs is undoubtedly sound from a biological standpoint, but it may under certain extreme conditions prove difficult to apply the idea of an LME to problems of fisheries management (Morgan, 1989, p. 378).

Political factors will provide one set of difficulties to the application of the LME concept to problems of fisheries management. Alexander (1989) has drawn attention to the political difficulties encountered when the LME includes areas subject to different international oceanic regimes or the authority of different coastal states; but political division of the LME is not the sole political factor. Difficulties may also arise from the nature of a country's political, economic, and administrative capacities and its relations with neighboring countries.

This chapter attempts to describe the range of political factors that might make coordinated management of an LME difficult, despite the obvious advantages of scientific cooperation. It was decided to focus on the LMEs along the west coast of Africa—from the Strait of Gibraltar to the Cape of Good Hope—largely because of a greater familiarity on the author's part with the politics of this Atlantic littoral than with any other of the Atlantic's continental fringes.

An earlier paper identified three potential LMEs along the west coast of Africa (Prescott, 1989). They were located along the coasts of southwest Africa, the Gulf of Guinea, and northwest Africa, and were associated, respectively, with the Benguela, Guinea, and Canary Currents. Crawford *et al.* (1989) and Garcia (1982) have produced useful studies of the Benguela and Canary Current fisheries, respectively.

The Development of LMEs as Functional Management Units

There appear to be at least six stages in the process of establishing a management system for any LME: (1) the extent and hydrographic and biological dynamics of the LME must be established by careful scientific study, assisted by the collection of accurate statistics about the

---

* Editors' Note—In emphasizing the divisive political forces associated with LME management, Dr. Prescott has picked a "worst-case" scenario, the Guinea Current ecosystem. No other LME is bordered by so many coastal countries, with such a variety of languages, cultures, and colonial histories, and, for most of the states, with so few traditions of maritime use.

From a scientific perspective, the prognosis is good in a cooperative program aimed at monitoring changes in the productivity, yield levels, and overall "health" of the Gulf of Guinea LME. The World Bank supported a mission in April-May 1992 that visited the Ivory Coast, Ghana, Nigeria, Benin, and Cameroon in preparation for providing financial support for a multiyear Gulf of Guinea ecosystem monitoring project. The project will obtain new information on the ecosystem and make recommendations for mitigation of any stressors identified in the course of the study. UNEP and FAO, along with NOAA (U.S.), Natural Environmental Research Council (NERC) (U.K.), ORSTOM (France), and IUCN, will maintain close collaboration with the principal investigators from each of the participating countries to aid in the accomplishment of long-term program objectives.

types and quantities of fish caught; (2) it then is necessary to establish which countries can make national claims to all or part of the LME. If part of the LME lies outside the 200-nautical mile (nm) Exclusive Economic Zone (EEZ) that coastal states can claim, and therefore forms part of the high seas, any country with a distant fishing fleet would have the right to fish in that part of the LME; (3) there must be agreement among the countries with legitimate claims to fish in the LME to prepare management plans. It seems likely that depending on the assumptions agreed upon by the countries concerned, there might be more than one possible management plan; (4) this stage would involve scientists, economists, and administrators drafting plans that could then be submitted for the review and approval of the various governments; (5) the approved plan must then be implemented; and (6) the status of the resources should be monitored so that if unexpected results occur, the plan can be modified.

Political Problems and LMEs

Political problems could disrupt any stage of this process, and there is no reason to believe that there would be different political problems at different stages. For example, serious political friction between two neighbors could pervade each stage, and the administrative inability to provide reliable statistics could adversely affect stages (1), (4), and (6). For this reason, we will first identify the most common political problems in a broad sense, and then explore the extent to which they exist in the three LMEs off the west coast of Africa.

Although this proposition cannot be stated as a rule, it will generally be the case that as the number of claimant states in an LME increases, the difficulty in securing agreement on the need for a functional LME management plan will also increase. There are obvious reasons for this likelihood. First, a number of countries will have differing interests in offshore resources and differing priorities for developing them. Second, as the number of bureaucracies involved increases, there will be more chance of delay from those who give the proposals a low priority or who are grossly inefficient. Third, as the number of countries increases, the risk of any two having quarrels over other issues will rise, and such friction might retard the development of fisheries management schemes.

If involvement is limited to those states in the vicinity of the LME, we will at the very least know the maximum number of players engaged in the effort to secure a functional agreement. If part of the LME lies in the high

seas and can be fished by distant fleets from other countries, it will be harder to identify those who have a legitimate interest in that LME. Further, the interests of the local and distant countries might not always coincide.

It is necessary to consider the impact of political friction between legitimate claimants of any LME. For example, agreements involving Israel and Lebanon, China and South Korea, or Britain and Argentina appear to be unlikely; but, we do not have to focus on these extreme examples. From time to time, friction connected with a whole range of issues might cause a cooling in relations between neighbors and hinder negotiations regarding fisheries programs.

Morgan (1989) observed that jurisdictional disputes might generate a high level of friction that can hinder cooperation and planning activities. This generalization will apply whether the dispute is about land or about sections of the maritime domain.

The ability of any claimant state to play a constructive role in the development of management plans will be influenced by its political, economic, and administrative capacities.

We tend to think of countries as stable political entities that can maintain a whole series of different negotiations with different partners at the same time, despite domestic difficulties that any particular government might experience. In short, there tends to be a continuity of the political process regardless of changes caused by governmental elections. That is not the invariable situation. In some countries, domestic crises can produce a paralysis of government that would place international maritime planning procedures at risk. West Africa Liberia, during the first 7 months of 1990, provides a depressing example of this situation; but it is unnecessary to use the extreme example of a civil war or military coup. Some governments that are facing daily struggles to remain in power and are coping with serious economic or sectional problems have little time and few resources to devote to issues of international planning that are not perceived to be important to the main issues.

The economic status of any country might also influence its ability to participate in the formulation of management plans for LMEs. Apart from the primary need to provide funds for research, collection of fishery statistics, consultants, bureaucrats, and conferences, there is also the question of the costs that might be associated with implementation of any management plan. Such plans for LMEs must contain provisions for management in the long term. It will not be surprising if some of the plans recommend a reduction of fishing effort for some areas or for various species. This could create problems for a government, whether the reduction concerns the domestic industry or foreign fleets who pay commercial fees for fishing in national waters. It is obvious that countries that have a significant fishing industry will have strong reasons for developing resource management plans for the LME and cooperating in their implementation; but the very significance of that fishing industry to present economic activity sometimes might make it difficult for governments to impose restrictions on production without some form of international subsidy.

However, countries for which fishing is a marginal activity might be reluctant to spend funds and impose restrictions that might provide benefits mainly to their neighbors. This raises the interesting question of whether a management plan for an LME must be an all-or-nothing proposition—in other words, will there be circumstances when the development of a management plan can be useful even if all countries do not become completely involved? There seem to be two limiting cases. If the biological dynamics of the LME are uniform throughout the region, without any significant migration of important species, it might be possible to develop a management plan that was forced to exclude some areas. That might be the situation if the areas were small and marginal to the LME. The other case would involve significant migration of important stocks on a random or seasonal basis, where there could be a strong possibility that the plan would fail if some of the migrations were taking place through waters that were not subject to management.

The administrative capacity of countries would have a significant impact on the extent to which the country could collect data, participate in a plan construction, or enforce regulations required by the completed plan. Lawson and Robinson (1983) demonstrated that in the early 1980s there were some artisanal fisheries in West Africa where governments appeared to exert very little control on the total number of boats, techniques used, and levels of harvesting.

Although the logistics of monitoring the activities of foreign fishing vessels operating legally in national waters are simpler than for artisanal fishermen, Kaczynski (1989) has shown that in some West African cases, governments have failed to derive the expected financial and other benefits from agreements with foreign fleets.

Political problems in the West African LMEs

The distribution of national claims and LMEs is shown in Fig. 1. The Canary Current LME of northwest Africa consists mainly of six national claims with a small area of high seas. Because the exact boundaries of the LME are not known,

it is possible that the entire LME is occupied by national claims of the mainland states, including Morocco, Mauritania, and Senegal, with the lion's share subject to Moroccan jurisdiction, with Portugal and Spain claiming sections of the LME from their offshore islands of Madeira and the Canary Islands, respectively. The sixth state is Cape Verde, which appears to have access to only a tiny section of the LME.

Some of the main characteristics of the fisheries of four African states are given in Table 1.

Although six states appear to have claims to sections of this LME, it seems likely that Morocco, Mauritania, and Senegal are the most important for three reasons: (i) It is along their shores that the comparatively narrow continental shelf is located, (ii) Cape Verde only has claim to a very small portion of this LME, located in comparatively deep water, (iii) Portugal and Spain, before and after their entry to the European Economic Community (EEC), had negotiated fishing access to the national waters of Morocco and Senegal. It would probably not be too bold to suggest that if these three African, moslem, continental states can agree on proposals to consider a management plan, the other offshore territories will not stand in the way.

The most serious friction between the continental states concerns Mauritania and Senegal. On April 9, 1989, two Senegalese farmers were killed by Mauritanian herders on an island in the Senegal River. The incident sparked riots in Nouakchott, capital of Mauritania, in which Senegalese citizens were the main victims, and then in Dakar, Senegal's capital, where Mauritanians suffered. The net result was opposite flows of refugees amounting to approximately 100,000 in each direction.

The problem arose because of the inadequate definition of the boundary between the two countries. It is based on a French decree of 1933 that defined the boundary as the right or Mauritanian bank of the main arm of the Senegal River. Any bank is usually a poor location for an international boundary and especially a bank on a river with such a gentle gradient as the Senegal, which moves laterally over significant distances each year. In May 1990, a new black opposition group called the United Front for Armed Resistance in Mauritania was formed and killed 51 people near Kaedi, close to the Senegal River. This group, as well as two similar organizations, draws support from Senegalese refugees and black Mauritanians who are at the bottom of Mauritanian society. Slavery was not abolished in that country until 1980.

At last report, efforts at reconciliation involving the Organization of African Unity had not succeeded. As of February 1990, it was estimated that 240,000 Mauritanian refugees from Senegal needed resettlement. Disputes of this

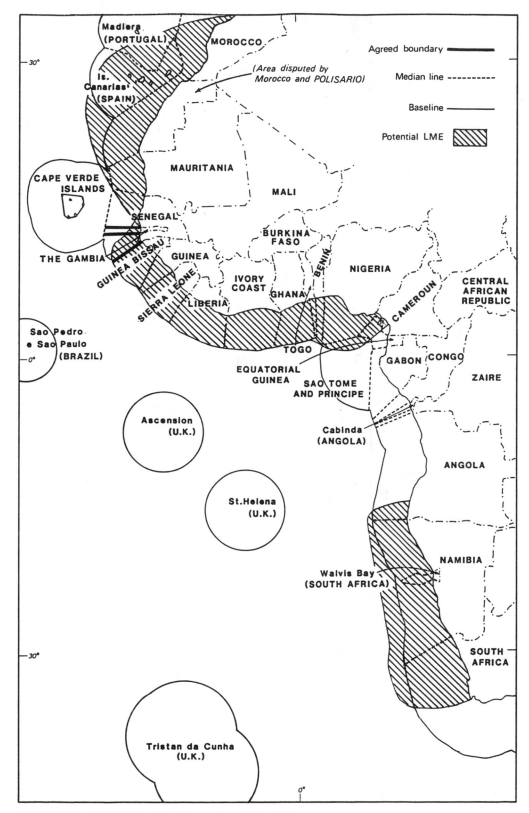

Figure 1. Potential LMEs and maritime claims in West Africa.

Table 1. Some characteristics of the fishing industries of African states that can claim parts of the LME off northwest Africa.

	Total Catch 1987 (metric tonnes)	Main species	Balance of fish 1987 Imports (−) & exports (+)	Foreign debt 1988 ($US million)	Balance of trade 1988 ($US million)
Cape Verde	6925	Yellowfin Tuna (39%) Skipjack Tuna (30%)	+2.6	141	−75
Mauritania	93000	Octopus (55%) Grunts and Sweetlips (6%)	+169	2200	+99
Morocco	455505	European Pilchard (56%) Chub Mackerel (6%)	+351	19800	−900
Senegal	284000	Sardinella (40%) Porgies (5%)	+256	3700	−234

Sources: FAO. 1989. Yearbook: Fisheries statistics, catches, and landings. FAO. Rome. African Research Bulletin. 1964. Political, social, and cultural series; economic, financial, and technical series. Africa Research Bulletin. Exeter.

nature would be serious impediments to proposals for international cooperation on fisheries management plans.

When we examine the political capacity of these three countries, there is no evidence that any of the governments face a serious present domestic threat. It could be argued that Morocco's occupation of Western Sahara, which used to be a Spanish colony, raises problems for its jurisdiction over offshore waters. When Spain withdrew from the colony in February 1976, Western Sahara was partitioned by Morocco and Mauritania. A war of liberation was started by POLISARIO, which represents the indigenous population of Western Sahara and has bases in western Algeria. This organization compelled Mauritania to withdraw from the southern section in 1979, but Morocco simply replaced Mauritanian forces and now holds the entire former Spanish colony. Although the war has continued, there are reasons to expect that Morocco's position will not be seriously challenged. First, Morocco has completed the construction of a defensive wall embellished with detecting equipment which makes POLISARIO penetration difficult, and the wall now runs for about 1,100 kilometers. Second, in 1988, Algeria, POLISARIO's main ally, reestablished friendly relations with Morocco. Third, Portuguese and Spanish fishing fleets recognize Morocco's authority over the waters off the coast of the former Spanish colony, despite an attack on a Portuguese vessel by POLISARIO in 1987.

When the economic capacity of these countries is considered, two points are outstanding. First, the fishing industry is an important sector of the economy of each country because the industry shows a profit in terms of exports and the sale of fishing rights to foreign fleets. Any profitable activity is encouraged by those countries with high levels of debt and population growth rates of about 3% per annum. Second, each of these countries is trying to expand the industry and its profitability. For example, Mauritania has been cooperating with other Arab governments. In 1986, a joint fishing company was established with Tunisia, and three Tunisian trawlers were moved from Bizerte on the Tunisian coast to Nouakchott. A year later, an agreement was signed with Algeria for a joint fishing effort in Mauritanian waters, which involves Algerian investment in training, boat construction and repairs, and fish processing and marketing.

Fishing in Senegal has become more significant in terms of the national economy as the former mainstays of groundnut and phosphate industries have faltered. Groundnuts have suffered from drought and competition with other vegetable oils, while phosphates have suffered a general decline in world prices. The fishing industry employs about 100,000 people and consists of small-scale artisanal fishing and large-scale trawler fishing. Lawson and Robinson (1983) estimated there were 4,100 canoes in operation in 1980, but the number has certainly increased since then. Unfortunately, the Senegalese trawler fleet of 135 vessels had been allowed to deteriorate, and by 1988 all were second- or third-hand and at least 20 years old. Attempts are being made to replace trawlers through aid from European countries, and, in addition, efforts are proceeding to encourage artisanal fishermen to use waters further away from the main centers that are overfished. In 1987, Senegal signed a new fisheries agreement with the EEC. The agreement provided for payment of $11 million (U.S.) in grants for research and training scholarships for Senegalese fishermen. A proportion of the tuna catch also had to be landed in Dakar for local processing or consumption. The very large per capita fish consumption of the Senegalese means that much of the increased production will be consumed locally rather than earning export revenue.

Morocco, with the longest coastline in this LME, not only has a vigorous fishing industry, but has entered into cooperation with Saudi Arabia and the United Arab Emirates and has sold fishing rights to the EEC. The agreement with the Emirates provided for six trawlers to be delivered by the end of 1990 and be based at Agadir. The first two trawlers built in Spanish yards were delivered in September 1989. As part of the arrangements with the EEC, which became necessary with the entrance of Spain and Portugal, Morocco secured fees of $84 million (U.S.) plus license fees for each boat and duty-free access for up to 17,500 tonnes of Moroccan sardines. The agreement covers 700 Spanish boats and 30 Portuguese vessels.

If agreement could be reached to prepare management plans for this LME, it is likely that these countries would insist that one of the premises be the maintenance of present levels of production. It is inconceivable that they would consider a reduction in fishing effort by their own or external fleets. This would only become more likely if markets for phosphates, iron ore, and groundnuts rose significantly, thereby reducing the relative importance of fishing to the national economy. If this happened and a plan was devised and implemented, there is no reason to suppose that there are any failings in the administrative capacities of these states to make the plan work.

Gulf of Guinea LME

The LME associated with the Guinea currents extends from Gambia in the west to Cameroon and Equatorial Guinea in the east. It appears that this LME is entirely contained within the claims to an EEZ 200 nm wide that is claimed by 12 of the 13 coastal states along this section of the coast. Cameroon is the exception. It claims a territorial sea only 50 nm wide, presumably because it is zone-locked on a concave coast by Nigeria and especially by an island called Bioko (Fernando Poo), which belongs to Equatorial Guinea.

If analysis of the physical and biological characteristics of this LME reveal that it must be managed as a single entity, then there will be difficulties in securing the active support of all 13 countries. Although there have been serious international squabbles among pairs of these countries at various times in the past, the level of international friction was not particularly high in 1990. In other years, borders have been closed to traffic, and immigrant workers have been sent back to their homelands from places such as Nigeria. This has created serious problems of resettlement in places such as Benin and Ghana. There have also been periodic allegations that the territory of a neighbor was being used as a base for dissidents seeking to topple Sekou Toure of neighboring Guinea, and the present civil war in Liberia was started by an invasion of rebels from the adjacent Ivory Coast. At this stage, it cannot be predicted what effect the multinational peace-keeping force in Liberia will have on international relations in West Africa.

However, even if international friction does not prove to be a significant obstacle to the development of management plans for this LME, it is likely that difficulties will be presented by the political capacities of some of these countries. Although Côte d'Ivoire still has the same ruler who led the country to independence in 1960, Houphouet Boigny, the remaining 12 countries have experienced a total of 23 successful coups d'état. In the past 30 years, no government in this region has changed via election. It nearly happened in 1967 in Sierra Leone, but the opposition was prevented from taking power by a military coup.

However, the frenetic pace of change has slowed significantly during the 1980s, and apart from Liberia, where it is not clear whether General Doe's rule will be continued, the other 12 countries have had the same government for at least 5 years. Apart from Côte d'Ivoire, there has been the same government for 23 years in Togo and 18 years in Benin. Unfortunately, none of these countries enjoys a level of economic development that allows governments to take long-term views, and in several of the countries there are sudden changes of fundamental policies to meet short-term pressures. Those pressures might come from external aid-donors or from sections of the domestic community, such as farmers or teachers.

Selected characteristics of the fishing industries (Table 2) of these countries reveal the following: (1) Ghana is the largest producer of fish, with Nigeria, Côte d'Ivoire, and Cameroon as secondary producers; (2) although only Ghana and Sierra Leone have a surplus of fish exports over fish imports, the industry is significant for all these countries because the catch supplements food stocks and reduces the level of imports required; and (3) in six of the cases, more than half of the catch is composed of two or three kinds of fish. Those species that have been identified most often are sardinella or various kinds of Bonga shad. Maps of the distribution of fish (FAO Fisheries Department, 1981) show that sardinella are found mainly off the coasts of Guinea-Bissau, Guinea, Sierra Leone, Liberia, Côte d'Ivoire, and Ghana. Bonga shad is an inshore fish found along the coasts of Gambia, Guinea-Bissau, Guinea, Sierra Leone, Togo, Benin, Nigeria, Cameroon, and Equatorial Guinea.

The information about levels of debt and trade balances in Table 2 suggests that none of these countries is in a position to reduce its level of fishing activity in order to allow recovery of fish stocks if that was considered desirable. It is also possible that most of the countries would have to seek external funds to support the level of research and surveillance necessary to produce and implement any management plan for this LME.

Because there are large disparities in the levels of fishing activity, it might be hard to persuade countries with small areas of sea and low catches that a management plan would be for the benefit of all and not just for the major producers. The example of Cameroon might encourage them. Although Cameroon is a geographically disadvantaged coastal state in terms of making maritime claims, its fishing industry is remarkably strong. In 1987, Cameroon entered into an agreement with Confederation of Italian Syndicates Abroad (CIS Estero) Estero of Italy for a project costing $175.3 million (U.S.). This investment was spread over 3 years and involved the formation of a joint company, Society Thonniere du Cameroun, to operate a fleet of four 300-tonne vessels, ten 160-tonne vessels, and 100 artisanal boats of 10-meter length. The aim is to catch tuna, mackerel, and sardines that will be processed at new installations in the Cameroon port of Limbe for the domestic market.

At present, a number of countries earn revenue from allowing foreign vessels to fish in their waters, which they would generally be reluctant to forego.

Table 2. Some characteristics of the fishing industries of African states that can claim parts of the LME in the Gulf of Guinea.

	Total Catch 1987 (metric tonnes)	Main species	Balance of fish 1987 Imports (−) & exports (+)	Foreign debt 1988 ($US million)	Balance of trade 1988 ($US million)
Benin	9916	Bonga Shad (41%), Maderia Sardinella (25%)	−1	800	−165
Cameroon	62529	Bonga Shad (29%), Sardinella (29%), Palaemonid Shrimps (19%)	−34	3360	+500
Côte d'Ivoire	74250	Sardinellas (43%), Bonga Shad (16%)	−10	13000	+700
Equatorial Guinea	3600	Crustaceans (13%), Sharks (11%), Tuna-like fish (11%)	−2.8	189	−2
Gambia	11676	Bonga Shad (27%), Snappers (12%)	−1	354	−37
Ghana	317817	European anchovy (29%), Sardinella (27%)	+13.7	3300	+80
Guinea	28000	Sardinella (71%)	−3	1600	+11
Guinea-Bissau	3500	Natantion Decapods (25%), Big-eye Grunt (20%)	−0.3	340	−32
Liberia	14731	Dentex (13%), West African Croaker (13%)	−6.9	1500	+199
Nigeria	145755	Bonga Shad (26%), Threadfin (9%)	−161	38900	+900
Sao Tome and Principe	2500	Half beaks (10%), Flying fish (9%)	−0.6	72	+7
Sierra Leone	37000	Bonga Shad (58%), Sardinella (18%)	+7	895	−8
Togo	14462	European anchovy (61%)	−10	1200	−62

Sources: FAO. 1989. Yearbook: Fisheries statistics, catches, and landings. FAO. Rome; African Research Bulletin. 1964. Political, social, and cultural series; economic, financial, and technical series. Africa Research Bulletin. Exeter.

Any problems associated with the administrative capacities of any of these states to supervise a management plan would probably relate to the artisanal fisheries. Lawson and Robinson (1983, p. 289) sounded a warning in the following terms:

...canoe fisheries present special difficulties of enforcement because of their wide geographical dispersal, the large number of units and their extreme mobility. In many countries fishery administrations are weak and not well motivated and, except in the most flagrant cases of over-exploitation, do not generally appreciate the need for management. In some countries attempts are being made to operate through traditional fishermen organizations, but while these may work with the distribution of inputs, it does not seem to us that these organizations will be powerful enough to implement and enforce tough management measures unless the fishermen are themselves convinced of the need for action.

Benguela Current LME

Crawford *et al.* (1989) define this LME as lying between 16° S and the Agulhas Reflection area off the southern tip of the continent. As the transition zone between the Gulf of Guinea and Benguela Current LMEs has yet to be delineated, the northern limit of 16°S would mean that Angola could claim a section of the LME along its southern coast measuring 140 km. Spawning and nursery areas for the more abundant species (Crawford *et al.*, 1989) all lie south of the Cunene River, which marks the boundary between Namibia and Angola. For this reason, it is assumed in this discussion that only Namibia and South Africa have claims into that portion of the LME that lies within 200 nm of the continent. It is not clear whether the LME extends more than 200 nm from the continent. If it does, then this would be an added complication, because countries with distant fishing fleets would have rights to fish in these outer sections. Article 63(2) of the 1982 Convention of the Law of the Sea (UNCLOS, 1982) requires coastal states and states fishing the high seas to seek agreement on conservation measures for fish stocks that lie partly in an EEZ and partly in the adjoining seas.

It might be assumed that the advantage of only two countries being involved would be more than offset by the former colonial relationship between them and the long struggle that led to Namibia's independence in 1990. Such an assumption could well be wrong for two main reasons. First, even before the 1989 elections, the leaders of the Southwest Africa Peoples' Organization had abandoned much of the rhetoric about nationalization of the means of production in favor of a mixed economy. In the few months since independence, the government, led by President Nujoma, showed a high degree of pragmatism in the appointments to the cabinet and the policies adopted. The specific policies for fishing will be discussed later.

Second, it is my view that the changes that have occurred in South Africa's political system recently are now irreversible and that sooner or later there will be a racially integrated government. In the meantime, there is no insuperable obstacle to improving relations between Namibia and South Africa that will have mutual advantages. This fact will not be lost on the Namibian authorities, who recognize that African problems are less significant to Western and developed powers now compared to problems and opportunities in eastern Europe. The present South African government has recognized for some time the advantages of achieving the best possible relations with neighbors to the north.

Although South Africa's ownership of Walvis Bay, in Namibia, and the Guano Islands will remain an irritant to relations between the two countries, it is not an insurmountable problem, providing that South Africa allows reasonable access to Walvis Bay and does not insist on making strict maritime claims from the Guano Islands (Prescott, 1989). These two countries have the largest catches of states on the west coast of Africa, thus the fishing industry is important to both countries (Table 3).

In terms of the political capacity, it is plain that the Namibian authorities are determined to address the problems of reestablishing the country's fishing industry.

During the period before Namibia's independence, South Africa controlled only the inshore waters, while the International Commission for South East Atlantic Fisheries governed the catches in more distant waters. This division arose because countries with distant fishing fleets did not recognize South Africa's right to administer Namibia. The effect of this arrangement was to deplete almost all fish stocks to the point that, at present, the Namibian government is considering allowing quotas for horse mackerel only.

The Minister for Agriculture, Fisheries, and Water Development recognizes that the stocks were exploited to a level where the prime objective must be to restore them (Minister G. Hanekom, pers. com.). To that end, member countries of the Commission for South East Atlantic Fisheries were requested to cease fishing in Namibia's waters by March 31, 1990, and that appears to have happened. Namibia has proclaimed an EEZ of 200 nm and is establishing a

Table 3. Some characteristics of the fishing industries of coastal states that might be able to claim parts of the Benguela LME.

	Total Catch 1987 (metric tonnes)	Main species	Balance of fish 1987 Imports (−) & exports (+)	Foreign debt 1988 ($US million)	Balance of trade 1988 ($US million)
Angola	73157	Cunene Horse Mackerel (28%) Sardinella (11%)	−33.6	1800	+400
Namibia	519368	Southern African anchovy (73%) Southern African pilchard (13%)	n.a.	n.a.	+47
South Africa	898799	Southern African anchovy (66%) Cape hake (15%)	+60*	1100	+3000

*This figure relates to entire South African fishery. Sources: FAO. 1989. Yearbook: Fisheries statistics, catches, and landings. FAO. Rome.

surveillance force of three patrol vessels and two aircraft. The services of a Norwegian research vessel have been secured to assess stocks, and trends and quotas will eventually be allocated on the basis of that information. Foreign companies will have to pay for quotas when they are granted, whereas bona fide Namibian firms will be granted quotas without any fee. The minister outlined a proposal to encourage the growth of industry as stocks recover:

> To encourage the growth of our own industry our system of quota levies allows further rebate on fish caught and landed and processed in Walvis Bay, and we hope that through this policy we may induce certain companies to erect factories onshore for the processing of fish and to minimize transhipment from our waters. Transhipment will, however, in any case, only be allowed in harbour so that we can keep a proper control thereof. (Minister G. Hanekom, pers. com.)

Although these are times of change in South Africa, there is no evidence of any paralysis of the day-to-day operations in those government departments dealing with fisheries, agriculture, transport, or foreign affairs. These departments should be quite capable of conducting negotiations regarding any joint fishery arrangements and Namibia's continued access to Walvis Bay. Unlike Namibia and the other major continental fishing nations off northwest Africa, South Africa has major fisheries in the Indian Ocean, thus its fishing effort is not confined to a single LME for which neighbors might want to develop a comprehensive management plan. This situation might make it easier for South Africa to consider involvement in the development of a unified plan for the Benguela LME. Indeed, Crawford *et al.* (1989) refer to attempts to develop a complex ecosystem model for the Benguela LME. It is not known how successful these attempts have been.

There are two characteristics of this LME that suggest that the administrative capacities of these two states are sufficient to cope successfully with the implementation of any management plan. First, there are very few ports along this section of coast; along the 2,000 km of coastline, there are only two major ports, Walvis Bay and Cape Town, and two minor ports, Luderitz and Port Nolloth. Second, the population densities along this coast are extremely low, and there is no significant artisanal fishing activity.

Conclusions

This analysis has pursued the idea raised by Alexander (1989) and Morgan (1989) that po-

litical problems might make it difficult to implement management plans for LMEs. These types of problems will probably present a major obstacle to management plans for the Gulf of Guinea LME, which, because of the great diversity among the coastal states, is something of a special case among LMEs. There are 13 states involved, and there are wide disparities between the extent of development of their fishing industries and the importance of those industries to the national economies. There has also been a high level of political instability in West Africa up to the mid-1980s, and recent events in Liberia show that instability might return or might develop in a stable country, such as Côte d'Ivoire when Houphouet Boigny retires or dies. There are also many large and small ports in this section of West Africa, and the high densities of coastal populations give rise to a significant artisanal sector. This sector will make it hard to collect reliable data and to monitor quotas, if such quotas are set. If management plans were to call for a reduction in fishing activity to allow stock levels to rise, this would create economic burden for most of the countries concerned.

There are fewer political problems associated with the Canary Current LME. There are only three mainland states involved, and they all have important fishing industries. There is a higher level of political stability in these three countries than in West Africa generally, and all the states have Moslem majorities. There are fewer ports along this northwest coast, and, although artisanal fishing increases in importance in the southern section along the Senegalese coast, it is fairly restricted in geographical extent at present.

There is still an unresolved territorial dispute between Mauritania and Senegal along the Senegal River, and there have been significant counterflows of refugees across that line. There is still the possibility of further fighting between Moroccan troops and POLISARIO fighters, and the latter have sometimes made foreign fishing vessels their target.

Contacts between these countries and the EEC seem to be satisfactory at present, and it would be of benefit if the management plan involved Portugal and Spain operating from the Azores and Canary Islands, respectively. However, if the management plan called for reduced fishing effort, this would run counter to the short-term economic interests of the African states.

The most serious political difficulty facing Namibia and South Africa, if they decide to explore the possibility of a management scheme for the Benguela Current LME, is the territorial dispute over South Africa's Walvis Bay and the Guano Islands. In the period since

Namibia's independence, this problem has not surfaced in any significant way. The prospects for developing a management plan are aided by the fact that only two countries are involved, and Namibia has already acknowledged that there is a need to let the fish stocks recover following the period before independence when they were overexploited. The almost total absence of artisanal fishing in this LME and the scarcity of ports along this section of coast should make the monitoring required by any plan simpler than in other parts of West Africa. We can expect a period of political stability in Namibia, and, although changes are occurring in South Africa, they do not seem to be interfering at present with the administration of fisheries and foreign affairs departments. If improved race relations in South Africa allow improvements in the economy, South Africa, with its Indian Ocean fisheries and its mining sector, could consider the possibility of reduced fishing activity in the Southeast Atlantic with more equanimity.

There is an obvious disparity between the apparent precision of the biological and hydrographic data collected by those who define the extent and dynamics of LMEs and the political data presented in this analysis. That disparity cannot be avoided. The specific biological data are necessary before a management plan can be developed, and the collective political will of the coastal states is necessary if the plan is to be implemented. The connection between the plan and its acceptance is the clear demonstration of the economic benefits. Real progress will probably only be possible after those benefits have been proved in models and one group of countries has been persuaded to produce a management plan for a single LME.

References

Alexander, L. M. 1989. Large marine ecosystems as global management units. In: Biomass yields and geography of large marine ecosystems. pp. 339–344. Ed. by K. Sherman and L. M. Alexander. AAAS Selected Symposium 111. Westview Press, Inc., Boulder, CO.

African Research Bulletin. 1964. Political, social, and cultural series; economic, financial, and technical series. Africa Research Bulletin. Exeter.

Crawford, R. J. M., Shannon, L. V., and Shelton, P. A. 1989. Characteristics and management of the Benguela as a large marine ecosystem. In: Biomass yields and geography of large marine ecosystems. pp. 169–220. Ed. by K. Sherman and L. M. Alexander. AAAS Selected Symposium 111. Westview Press, Inc., Boulder, CO.

FAO (Food and Agricultural Organization) Fisheries Department. 1981. Atlas of the living resources of the sea. FAO, Rome.

FAO. 1989. Yearbook: Fishery statistics, catches, and landings. FAO. Rome.

Garcia, S. 1982. Distribution, migration and spawning of the main fish resources in the northern CECAF area. CECAF/ECAF Series 82/25(En):1–9.

Kaczynski, V. M. 1989. Foreign fishing fleets in the sub-Saharan West African EEZ. Mar. Policy 13:2–15.

Lawson, R., and Robinson, M. 1983. Artisanal fisheries in West Africa: Problems of management implementation. Mar. Policy 7:279–290.

Morgan, J. 1989. Large marine ecosystems in the Pacific Ocean. In: Biomass yields and geography of large marine ecosystems. pp. 377–394. Ed. by K. Sherman and L. M. Alexander. AAAS Selected Symposium 111. Westview Press, Inc., Boulder, CO.

Prescott, J. R. V. 1989. The political division of large marine ecosystems in the Atlantic Ocean and some associated areas. In: Biomass yields and geography of large marine ecosystems. pp. 395–442. Ed. by K. Sherman and L. M. Alexander. AAAS Selected Symposium 111. Westview Press, Inc., Boulder, CO.

United Nations Convention on the Law of the Sea (UNCLOS). 1982. Art. 63(2).

Chapter 27

Large Marine Ecosystems of the Pacific Rim

Joseph R. Morgan

Introduction

The concept of large marine ecosystems (LMEs), defined as "regions with unique hydrographic regimes, submarine topography, and trophically linked populations" (Sherman and Alexander, 1986, Preface, p. xxv), is useful for a number of reasons. It is multidisciplinary, has great potential for effective management of important living resources, and provides a sound basis for examining an important topic from both an academic and a practical standpoint.

The fact that LMEs are regions interests geographers; that they are delineated by unique hydrographic regimes and submarine topography brings oceanographers "into the act"; and fisheries biologists are well qualified to examine the trophic linkages. In addition, LMEs are defined and affected by nonliving environmental factors in the oceans, such as currents, sea-surface temperatures, salinities, tides, and nutrient concentrations. This ecosystem approach is the chief concern of scholars in a number of academic disciplines who call themselves "ecologists."

Moreover, although the definition of the term does not imply any specific management methodologies or classifications, LMEs have become the focus of national and international fisheries management, under the provisions of the 1982 Convention on the Law of the Sea. Although the convention's provisions are technically not yet in force, because not enough nations have ratified, they are rapidly becoming customary or de facto international law. An ecosystem approach effectively furthers both the management of fishery resources under the convention's provisions on Exclusive Economic Zones (EEZs) and the admonitions in the convention that countries bordering semi-enclosed seas should cooperate with each other in preventing pollution and exploiting living resources for the maximum benefit of all.

There are a number of ways to classify LMEs, depending on the interests of the classifier. An ecologist might classify systems as predation driven, environmentally driven, or pollution driven (Sherman, 1992). The biomass and yields of predation-driven systems are influenced greatly by predator-prey relationships. When man is the predator, as he is in the case of large-scale commercial fishing activities, management of a predation-driven system can be carried out by setting quotas, establishing seasons, controlling mesh sizes, and so forth.

An environmentally driven LME is under the primary control of environmental factors, such as the occurrence of El Niño, which drastically alters ocean temperatures and thereby controls the distribution of certain species of fish. Less dramatic environmental events can increase or decrease the extent of upwelling, raise or lower salinity, and change both direction and speed of ocean currents.

Pollution-driven ecosystems are directly under human control, because, by definition, pollution is caused by humans. Control of pollution, and the associated process of waste disposal into marine regions where important LMEs occur, is a management problem of national and international proportions.

There are some well-recognized ocean current LMEs, for example the Humboldt, Kuroshio, and Oyashio in the Pacific Ocean, and the Gulf Stream in the Atlantic. The specific oceanographic characteristic of the current is the principal controlling factor, and changes in the force of the current or its temperature or salinity characteristics can have great effects on the yield of the ecosystem.

Other LMEs might be termed enclosed sea systems, because an enclosed or semi-enclosed sea, such as the Mediterranean or the Sea of Japan, is the dominant factor in defining the unique hydrographic regime, which, along with submarine topography and trophically related populations, delineates the ecosystem.

Finally, there are some LMEs that might best be termed "open-sea" systems. In the Pacific Ocean, two examples are the Gulf of Alaska and the Coral Sea (Morgan, 1987).

The categorizations of LMEs by ecological and geographic characteristics are useful, but political factors are even more important. Hence, we might classify marine ecosystems as being completely under a single country's jurisdiction; in a semi-enclosed sea, where regional cooperation is needed for effective management; or international in scope, such as the high-seas ecosystems for tunas. Some combination of classifications by ecology, geography, and politics is needed for optimum LME management.

The Pacific Rim

The Pacific Ocean stretches 15,500 km from the Bering Strait to Cape Adare, Antarctica, and 17,200 km from Panama to Mindanao (Fairbridge, 1966). For the purposes of this chapter, however, the Pacific is considered to extend as far south as the Antarctic Convergence, which forms the northern boundary of the Southern (or Antarctic) Ocean. The large east-west extent of the Pacific affects its hydrographic characteristics to a considerable extent. In general, ocean currents are narrower and swifter in the western Pacific than in the eastern part. In the North Pacific, the Kuroshio is an example of a typical western ocean strong current, whereas in the eastern Pacific, the California Current is broader, more diffuse, and less swift flowing.

The Pacific Rim can be generally classified as seismically active, with numerous volcanoes, earthquakes, and other manifestations of ocean-bottom movements. Deep trenches are found in both the eastern, northern, and western Pacific; the Peru-Chile, Aleutian, Japan, Mariana, and Mindanao Trenches are prime examples. The term "Ring of Fire" has been applied to the well-known volcanic nature of the Pacific's coastal areas.

From the standpoint of managing LMEs, there are considerable differences between the eastern and western Pacific regions. Stretching from the southern tip of South America to the western extremity of the Aleutian Island chain, the coasts are generally smooth, with few marginal seas; the only exception of note is the Gulf of California. On the other hand, the western Pacific is noted for its marginal seas, of which the Yellow Sea, East China Sea, South China Sea, and the Sea of Japan are probably the most well known. In these semi-enclosed seas, there are LME management problems that are mostly of a political nature, with the necessity to work out agreements among a number of countries to control overfishing (human predation), prevent pollution, and manage the resources in accordance with generally accepted international law. The eastern Pacific ecosystems are generally ocean current-dominated and, more importantly from a management standpoint, under the control of single nations.

Two Representative LMEs

The Humboldt Current LME and the Sea of Japan LME are considered representative of both the eastern and western Pacific Rim and of the geographic and management nature of ecosystems in the oceans in general. The Humboldt Current LME in the eastern Pacific is largely environmentally driven, representative of ocean current systems, and effectively under the management control of a single country. The Sea of Japan LME is predator driven, is a semi-enclosed sea system, and is under the management control of four countries that have borders on the sea and claim various marine jurisdictions in the confined waters.

The Humboldt Current LME

The Humboldt (Peru) Current is a broad, generally sluggish, stream of cool water that flows from the Antarctic region along the west coast of Chile and Peru. Average speeds are on the order of one-half knot or less, with an average transport of $15–20 \times 10^6$ m$^3 \cdot$ s^{-1} (Fairbridge, 1966). Some oceanographers distinguish between a coastal current, which they refer to as the Peru Current, and an offshore stream, the Humboldt Current. Others consider the Humboldt and Peru simply alternative names for the same current. It is part of the general South Pacific Ocean circulation and is one of the water movements linking the Southern Ocean circulation with that of the South Pacific, because some of the Humboldt Current water is derived from the Antarctic Circumpolar Flow as it passes through the Drake Passage between Antarctica and South America.

The current hugs the coast as far north as about 4°S latitude, after which it turns west to merge with the South Equatorial Current. The waters of the Humboldt Current are relatively saline, with a salinity of 34.5 parts per thousand or greater, while the waters to the north are warmer and relatively fresh. The zone of transition is to the south of the equator near the Peruvian coast, crossing the equator near the Galapagos Islands (Fairbridge, 1966).

The surface waters along the coast of Peru and Chile are markedly cooler than might be expected from consideration of air temperatures alone, because of the current and the presence of distinct areas of upwelling along the coast. The upwelled water results in lower ocean temperatures near the coast, coupled with low salinities and high nutrient concentrations (Pickard and Emery, 1982). These oceanographic conditions are responsible for the generally high biological productivity of the Humboldt Current LME, which is based primarily on very high phytoplankton concentrations in the cold, low-salinity, nutrient-rich upwelled water. The great concentrations of fish (chiefly the Peruvian anchoveta) both support a productive fishery and provide food for numerous seabirds. Thus, in a sense, the LME can be considered to extend from the ocean into the atmosphere.

The Humboldt Current ecosystem has been known for a long time. For many years, the guano produced by birds feeding on the surface-dwelling fish of the ecosystem was considered to be the most important economic product. Later, a productive commercial fishery was started, and for many years the profits from both guano and fish meal were considerable, contributing greatly to the Peruvian economy. Changes in the direction and strength of the Humboldt Current affect the productivity of the LME, but the largest environmental effect results from the occurrence of El Niño events.

In 1947, a well-known oceanographer, R. Coker (1947, p. 132), had this to say about El Niño and its effects:

Although the Peru Current, or coastal part of the Humboldt Current, is generally constant in rate and direction of flow over most of its course, there sometimes occur significant aberrations in its northernmost reaches. In northern summer a branch of the current crosses the equator to continue northward alongside the Equatorial Countercurrent, which also turns north. In northern winter, the summer of the southern hemisphere, the easterly flowing countercurrent turns southward, when it is known as El Niño (referring to Christmas), to flow a few degrees southward from the equator and close to the coast of Ecuador. Occasionally, but seemingly quite irregularly as to intervals of years, the southward-flowing warm current continues on along the coast of Peru. This unwanted substitution of warm for cool waters has disastrous effects upon the fish and other organisms of the coastal regions. There may ensue such a mortality of organisms and so much decomposition as to create decidedly foul conditions of water and air. The hydrogen sulphide liberated as a product of decomposition may blacken the paint of ships; this unpleasant and expensive phenomenon has long been known to seamen as "The Callao Painter." The valuable guano-producing birds, which feed upon anchovetas and other small fish and crustacea in the Peru Current, may also suffer heavy mortality or they may abandon the islands and the region with consequent great commercial loss.

According to more modern theories, the increase in temperature brought about by El Niño does not result in the death of surface-swimming fish, as was once thought to be the case. Rather, recent studies have shown that the fish merely descend below the abnormally warm surface water. The absence of fish near the surface decreases the food supply of may seabirds and results in their death; at the same time, commercial fisherman suffer from reduced catches (Pickard and Emery, 1982). Early ideas about the causes of El Niño focused on local events and mechanisms, but modern research suggests that more distant phenomena may play an important role.

There are now three separate hypotheses that may explain the El Niño phenomenon. The local-mechanism idea is that reduction of equatorward coastal winds reduces the rate of upwelling and causes warmer-than-normal surface water temperatures. A related hypothesis is that reduction of trade winds in the eastern tropical Pacific will allow the surface waters to warm, because the cooling effect of the southeast trades is reduced. A third hypothesis attributes the occurrence of an El Niño event to increased zonal stress to the west, resulting from stronger-than-normal southeast trade winds. This causes a pile-up of water in the western Pacific, which, after the winds subside, flows eastward, eventually causing an increase in water temperatures in the eastern Pacific (Pickard and Emery, 1982).

As research into the causes of El Niño continues, it may be possible to forecast an event with some degree of accuracy. The movements of warm water along the equator also change the air circulation over the entire Pacific. The conditions of reversal in the atmosphere, called the Southern Oscillation, bring with it changes in sea levels in the Pacific Islands, droughts in Australia, heavy rains and floods in otherwise dry regions, and perturbations in weather patterns that can be felt in India, New Zealand, the United States, and Canada. Using satellite observations and sea-level measurements on Central Pacific islands, predictions of an El Niño are now possible 6 to 9 months before it affects the coast of South America. An accurate forecast of El Niño might be used to predict the intensity of fishing effort that can be tolerated by the LME.

The history of the Peruvian anchoveta fishery provides convincing evidence of the combined effects of an El Niño event and presumed overfishing. After several years of small catches based on moderate fishing intensity (because the catch was used for local consumption only), the government of Peru began an export-oriented industry based on the production of fish meal. Figure 1 indicates the catch of Peruvian anchoveta (*Engraulis ringens* T.) for the years 1964–1983. (The fish meal exports began in 1957.) The peak year of 1970 resulted in a catch of slightly more than 12 million metric tonnes, putting Peru in the position of the world's number one fishing country in terms of catch tonnage. The relatively small catches of the Peruvian anchoveta by Chile, in more southerly waters, fluctuated minimally during El Niño periods, while catches of

Sardina espanola rose as anchoveta catches declined. Although the value of the anchoveta fishery in monetary terms was not the world's highest because almost all of the catch was rendered into low-priced fish meal, nevertheless, Peru's economy thrived on the basis of its commercial fishery operations.

The 1972–1973 El Niño had a disastrous effect on the anchoveta catch in Peru, and the fishery never recovered its previous strength (Holt, 1978). The effects of the 1976 and 1982–1983 El Niños are also evident. Whether the very high catches in the years just prior to the precipitous drop in 1972–1973 are to be considered the norm, or whether they are anomalous, is a matter of conjecture; but it is obvious that El Niño events have greatly influenced the economic health of the anchoveta fishery. It is possible that an accurate forecast of a forthcoming El Niño might be useful in managing the fishery. Because the fishery is effectively under the jurisdiction of Peru, management procedures based on scientific research should be both possible and beneficial. There are no international politics involved, and the structure of the LME seems to be quite clear-cut and understandable.

Although the Humboldt Current LME is primarily environmentally driven—controlled to a great extent by the occurrences of El Niño events, possibly under some influence of the strength or weakness of the coastal upwelling—predation might also play a part. If, as some scholars suppose, too large a catch is taken during years when the stock is absent from surface waters to a significant degree, subsequent annual crops might be affected detrimentally. In the case of an LME such as the Humboldt Current, predation is under the control of man and can be controlled by a combination of government regulations based on scientific research.

The Sea of Japan LME

The relatively simple picture presented by the environmentally driven Humboldt Current LME is not matched by the much more complicated pattern in the Sea of Japan. The LME is more complex, both in terms of numbers of species in still-unknown trophic relationships and in the political associations among the four countries with interests in the commercial fisheries: Japan, North Korea, South Korea, and the former U.S.S.R.

The 1982 Convention on the Law of the Sea urges states bordering enclosed or semi-enclosed seas to cooperate with each other in the exercise of their individual rights under the convention. This is necessary, because most

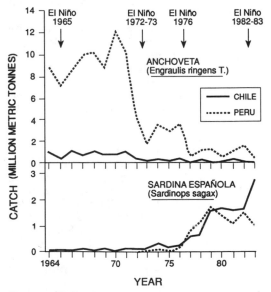

Figure 1. Fluctuations in the catch of anchovies and sardines in the Humboldt Current area off Chile and Peru, 1964 to 1983. (From Canon, 1986.)

semi-enclosed seas contain areas of overlapping maritime jurisdictional claims, owing to the rights of coastal nations to claim EEZs extending to 200 nm from their coastal baselines. The Sea of Japan is no exception, and many of the marine policy decisions of the four countries have been made with a view to reducing possible conflict over areas subject to multiple claims. Specifically, Sea of Japan countries have tended to deliberately claim less area than entitled to under the convention, or to forego EEZ claims in favor of recognizing equidistant lines as de facto maritime boundaries.

Because the convention specifically requires countries to coordinate the management, conservation, and exploitation of living resources of semi-enclosed seas, and because the trophic relationships among the numerous species in the Sea of Japan are complicated, this sea provides an ideal "laboratory" for establishing a system of rational management in accordance with international law. Progress has been made to do just that, but a number of factors militate against a complete solution to the problem. Some of these problems are political, others scientific.

The political problems complicate the management problems, because, although the four regional countries share historical and cultural backgrounds, they differ in their internal political systems. Moreover, there is a lack of complete diplomatic relations among the four: North and South Korea have obvious differences, and Japan and the former U.S.S.R. have a controversy over four small islands at the southern end of the Kuril chain, which makes agreements over fisheries and other marine policy matters more difficult.

There are a number of fishery agreements, however, which is a tribute to the determination of the governments to establish sensible management procedures in spite of a lack of formal relations. Japan and South Korea have made much progress in negotiating de facto maritime boundaries in the region. When the World War II peace treaty was signed in 1952, South Korea established the Peace Line (Rhee Line) in order to protect Korean fisheries in both the Sea of Japan and the Yellow Sea. In 1965, the Japan–Republic of Korea Fishery Agreement Boundary replaced the Peace Line. The new agreement provides more area to the Japanese. Since the signing of the Law of the Sea Convention in 1982, the two countries have become even more cooperative. Both have declined to claim the full extent of territorial seas and EEZs to which they are entitled, because overlapping claims would result. For instance, both South Korea and Japan claim only 3 nm of territorial sea in the Korea Strait, leaving a strip of high seas in the strait. The two

countries have a continental shelf boundary agreement in the southern part of the Sea of Japan, and, in the near future, it may be possible for the two to establish an agreement on EEZs based primarily on equidistance.

There is a fishery agreement between Japan and the former U.S.S.R., despite the fact that the two countries have not yet signed a World War II peace treaty and despite the ongoing dispute over the ownership of islands. The August 1977 Joint Japan-Soviet Interim Fishery Agreement works reasonably well, but further agreements between the two countries will depend on resolution of the Kuril Islands dispute.

Because the Sea of Japan contains both demersal and pelagic resources, both underwater topography and oceanographic characteristics of the water column are important for understanding the nature of the LME. The Sea of Japan is easy to delineate, because it is almost a classical semi-enclosed sea. Only the Mediterranean is more typical. Five straits separate the sea from the Pacific Ocean to the east, the Sea of Okhotsk to the north and northeast, and the East China Sea to the south. The narrowest of the straits is the Shimonoseki, which separates Honshu from Kyushu and provides a narrow passage for shipping between the Sea of Japan and the Inland Sea. The exchange of water through the strait is negligible, and from an oceanographic standpoint it can be ignored. The Tatar Strait, between the Russian mainland and Sakhalin Island is also narrow and shallow; the exchange of water between the seas of Okhotsk and Japan is likewise not great. The other straits are of more consequence, but both the Tsugaru and La Perouse straits provide far less of a water exchange than does the much more important Korea Strait. Considerable inflow and outflow of East China Sea water occurs through the eastern and western channels of the strait.

The Sea of Japan floor consists of the remnants of continental crust, volcanic peaks and lava flows, and various faulted and folded features. Considerable sedimentation, however, has masked the morphology of the peaks and tended to fill in the basins. The Sea of Japan is naturally divided into a northern part, north of 40° 30' N, which contains the large Sea of Japan Basin and the Abyssal Plain, and the ridge and trough topography that predominates in the southern part. The Yamato Rise, Korea Continental Borderland, Tsushima Basin, Yamato Basin, and the ridges and troughs along the continental slope of Honshu are prominent features south of 40° 30' N.

The dominant current in the western Pacific is the Kuroshio, which is a component of the North Pacific Current gyre. One branch of the Kuroshio enters the Sea of Japan as the Tsushima Current through the Korea Strait. It then travels northward along the west coast of Honshu, eventually becoming too weak and diffuse to retain its name and characteristics. Outflow from the Sea of Japan occurs from the Tsugaru Warm Current and the Soya Warm current, which transfer Sea of Japan water into the North Pacific and the Sea of Okhotsk via the Tsugaru and La Perouse (Soya) straits. Other currents of importance are the East Korea Warm Current, the Maritime Province Cold Current, and the North Korea Cold Current. In the winter season, the Liman Current is present in the Tatar Strait. This wind-driven current is absent during the summer months (DMA, 1984).

The relatively great depth of the Sea of Japan facilitates the development of distinct water masses. In general, there is warmer water on the Japan side and colder water on the Korean-Siberian side of the sea. Three distinct water masses form: the surface water, middle water, and deep water. The surface water extends only to a depth of about 25 m; in the summer it is separated from the middle water by a distinct thermocline. In the colder sector, the surface water is formed by the melting of snow and ice, which occurs in the summer and early fall. In the warm sector, the surface water is a mixture of low-salinity, high-temperature water (originating in the East China Sea and entering the Sea of Japan through the Korea Strait) and coastal-water runoff from the Japanese islands (Fairbridge, 1966).

The middle water has high temperatures and high salinities. It originates in the intermediate layers of the Kuroshio off the Kyushu coast, entering the Sea of Japan during the winter and spring seasons. The vertical decrease of temperature with depth is very pronounced; the temperature at 20 m is 17° C, falling to 2° C at 200 m (Fairbridge, 1966).

The deep water is very uniform in character, with temperatures between 0° and 0.5° C and a salinity of 34.0 to 34.1 parts per thousand. There is some evidence of adiabatic warming of the water at depths below 1,500 m, which some oceanographers refer to as bottom water (Fairbridge, 1966).

There are a great many species of commercially important fish in the Sea of Japan, including anchovies, chub mackerels, conger pike eels, flounders, hairtails, jack mackerels, lizard fish, Pacific cod, Pacific herring, Pacific salmon, Pacific saury, red sea bream, round scad, sardines, Sohachi flounder, squid, tanner crab, walleye pollock, and yellowtail. There may be more than one LME; in any case, the intricate trophic relationships have not been completely worked out. Of equal importance is the fact that distributions are decidedly uneven, with some species widespread throughout the sea and others concentrated in small areas.

Japan, North Korea, South Korea, and the former U.S.S.R. have active commercial fisheries, and they catch some species, such as sardines, Pacific salmon, squid, and Sohachi flounder, throughout the Sea of Japan (Morgan, 1987). For that reason, optimal management of the fishery should involve all four countries. In the case of other species, such as jack mackerel and tanner crab, the concentration is primarily on the Japanese side of the sea, and the problems of management for optimum yield are minimized. Among the demersal species, flounder are taken along the coast of the entire Korean peninsula, as well as in coastal waters off Hokkaido and Sakhalin. Hence, management by all four countries is desirable. For species in the Korea Strait (chub mackerel, for example), Japan and South Korea must cooperate.

Although there are fishery agreements, and the four Sea of Japan coastal states are sophisticated in their marine scientific research, particularly in fisheries, there is no evidence that an ecosystem approach to fisheries management is in operation or being seriously considered. Management of the Sea of Japan fisheries as an LME—or perhaps as two or more LMEs—would require that individual fisheries regions that are based on current bilateral agreements give way to a renouncement of separate national claims to fisheries jurisdictions and consideration of the sea as a unit. Following this, a four-nation research organization could be formed with a view toward establishing management procedures for an LME. Catch quotas could then be based on the optimum yield of the ecosystem as well as on agreed-upon needs of the individual countries.

Conclusion

LMEs are regions, with boundaries based on certain critical features of oceanography and frequently complicated ecological relationships. After delineation, they can be managed for optimum economic yields. Such management, however, is frequently difficult, because maritime jurisdictional issues, international political relationships, and the ecology of ecosystems themselves must be considered. Some LMEs are more amenable to management than others; there are examples of "easy" and "difficult" management problems.

The Humboldt Current LME presents a relatively easy management problem. It is completely within the jurisdiction of a single country; hence, political problems and prickly international relations are not factors. It is primarily environmentally driven, and variations in the annual yield of the system are generally predictable, if the principal environmental event, the El Niño, can be forecast. Continued research with the aim of better understanding and forecasting the El Niño is an obvious long-range management tool. Moreover, the trophic relationships involved are *relatively* uncomplicated, with a single species, the Peruvian anchoveta, the principal end product. However, a better understanding of the actual effect of El Niño is needed. Does the El Niño, with its unpleasantly warm water, kill fish, or does it simply drive them away by encouraging them to occupy other surface areas or perhaps stay at depth? Does the El Niño kill plankton or merely reduce upwelling and thereby diminish the supply of nutrients essential for phytoplankton productivity? How is the effect of El Niño coupled with the effects of human predation? Which is more important? Answers to these questions would result in more effective management of the important anchoveta fishery.

The Sea of Japan LME, on the other hand, presents a much more complicated management problem. There are numerous interrelated species, and the oceanographic factors on which the LME depends are complex. Most important, however, is the problem of international relations among the coastal states. Because this LME is primarily predation driven, the most critical management factor is determination of the permissible annual catch, which must be established by the research efforts of four countries with interests in the fisheries. After it is calculated, the catch must be apportioned among the countries according to some agreed-upon principles of equity. In view of the nature of the political situation in the Sea of Japan, it is difficult to be optimistic about prospects for management of this LME in the near future.

References

Canon, J. R. 1986. Variabilidad ambiental en relacion con la pesqueria neritica pelagica de la zona Norte de Chile. In: La pesca en Chile. pp. 195–205. Ed. by P. Arana. Escuela de Ciencias del Mar, Facultad de Recursos Naturales, Universidad Catolica de Valparaiso.

Coker, R. E. 1947. This great and wide sea. The University of North Carolina Press, Chapel Hill.

DMA (Defense Mapping Agency). 1984. Sailing directions (planning guide) for the North Pacific Ocean. Defense Mapping Agency, Hydrographic/Topographic Center, Washington, DC.

Fairbridge, R. W. 1966. The encyclopedia of oceanography. Reinhold Publishing Corporation, New York.

Holt, S. 1978. Marine fisheries. In: Ocean yearbook I. p. 41. Ed. by E. M. Borgese and N. Ginsburg. University of Chicago Press, Chicago, IL.

Pickard, G. L. and Emery, W. J. 1982. Descriptive physical oceanography: An introduction. 4th ed. Pergamon Press, Oxford.

Sherman, K. 1992. Large marine ecosystems. In: Encyclopedia of earth system science. Vol. 2. pp. 653–673. W.A. Nierenberg. Academic Press, San Diego.

Sherman, K., and Alexander, L. M. (Editors). 1986. Variability and management of large marine ecosystems. AAAS Selected Symposium 99. Westview Press, Inc., Boulder, CO.

Part Four:
Technology Applications to the Monitoring Process in Large Marine Ecosystems

Chapter 28

Applications of Advanced Acoustic Technology in Large Marine Ecosystem Studies

D. V. Holliday

Introduction

Many questions one would like to address in order to achieve a better understanding of the dynamics of large marine ecosystems (LMEs) arise from attempts to understand marine ecology on smaller scales. These include the response of phytoplankton to ocean physics and chemistry, the response of zooplankton to phytoplankton dynamics and ocean physics, and the response of higher trophic levels to both food availability and predation. In attempting to understand how small-scale and process studies can be integrated to predict the dynamics of LMEs, we are challenged to design an adequate and affordable long-term sampling strategy for hundreds of thousands of square kilometers and multiple species in several very different physical environments. To achieve adequate sampling, our strategy must consider both the longest and the shortest time scales present in the ecosystem, as well as both the largest and the smallest spatial scales.

There are few long time-series in oceanography and even fewer that cover an area comparable to that encompassed by most LMEs. Fiscal arguments alone, such as the cost of collecting and processing large numbers of samples of organisms, lead us to constantly search for new technologies to apply to the study of the ocean. Our constant hope is that technology will eventually bring affordable ways to sample synoptically with high resolution over large areas. Satellite technology has brought us a capability to obtain quasi-synoptic estimates of some parameters at or near the ocean surface, but measurement of deeper distributions of phytoplankton, secondary production, and fisheries resources have not yet proven susceptible to measurement by satellite-borne sensors. The development of methods for obtaining near-synoptic views of these ocean parameters is likely to be the subject of considerable research and development during the 1990s.

The search for means to remotely detect and measure the distributions, biomass, and dynamics of zooplankton, micronekton, and fish is currently focused on the field of underwater acoustics. This chapter presents a short summary of some of the key technological approaches that use acoustics, followed by some thoughts regarding where research may lead us in the 1990s.

Overview of Key Acoustic Technologies

A number of advances in our understanding of marine ecosystems have resulted from progress in methodology and technology. The development of underwater acoustics for studying the sea's living resources is an example of such progress.

Observation of pelagic nekton with underwater sound dates to the 1930s (Sund, 1935; Balls, 1948). Scientific observations of mesopelagic organisms with underwater sound began after World War II (Duvall and Christiansen, 1946; Eyring et al., 1948; Johnson, 1948). These observations were largely focused on the deep scattering layer, which proved to be caused by mesopelagic fishes, such as myctophids (Hersey and Backus, 1954).

With the development of echo counting (Craig and Forbes, 1969) and echo integration techniques (Ehrenberg and Lytle, 1972), it became practical to use single-frequency echo sounders to make quantitative estimates of biomass in diffuse layers or sparsely packed schools of fish. With the addition of dual-beam (Ehrenberg, 1974; Traynor and Ehrenberg, 1979) and split-beam (Ehrenberg, 1983; Foote et al., 1986) technology, it became possible to estimate the target strengths of the organisms in situ. With only a few exceptions (e.g., Pieper, 1979; Richter, 1985a and 1985b; Simard et al., 1986; Greene et al., 1989; Greene and Wiebe, 1990), applications of these techniques have largely been in fisheries surveys and resource assessments.

Another family of single-frequency techniques for estimating the target strength of individuals involves measurement of the probability density function (PDF) of the envelope (average or peak) of echo returns (Ehrenberg, 1972; Peterson et al., 1976; Ehrenberg et al., 1981; Stanton, 1985a, 1985b). These methods work best when the individuals are spatially resolved and have a relatively narrow distribution of sizes, but can also provide some information when echoes overlap. They also have the potential to provide size-abundance estimates in some situations (Stanton and Clay, 1986). The dual-beam and split-beam methods for estimating target strength require that organism densities be such that individuals are resolved by the basic acoustic system in use, that is, the echoes do not overlap.

In addition to these techniques, which usually rely on a vertical propagation path from the source to the individual fish, school, or layer of organisms under examination, single-frequency techniques have been developed that utilize the quantitatively more difficult, potentially refracted, horizontal path from an acoustic sensor to the target organisms. Fish school sizes and size distributions have been measured with single-beam, side-looking acoustic beam sonars (Hewitt et al., 1976), and fish behavior has been studied with a unique sector-scanning sonar (Harden-Jones and McCartney, 1962). The horizontal acoustic propagation path has not been widely employed in fisheries acoustics because it is difficult to make accurate estimates of range-dependent losses in sound propagation over long distances, even with good measurements of local, vertical-sound speed profiles. Thus, quantitative target-scattering characteristics are more difficult to determine when using this path geometry. One technique is a partial exception to this general rule. It uses a single frequency and must examine the target from a position in the plane in which the animal is swimming. The technique uses the Doppler effect to measure the tail-beat modulation and swimming speed of a fish in order to estimate the fish's size. Because no measurement of target-scattering strength is required, the procedure is relatively insensitive to variability along the horizontal propagation path (Holliday, 1974).

Sensors that operate at different frequencies are often used to examine the same distribution of fish (for example, see Arnone et al., 1990), but the quantitative combination of scattering data at different frequencies to extract biophysical characteristics of the organisms (e.g., size) has lagged significantly behind the practice of collecting the data. One exception was the detection of swimbladder presence in schooling or aggregated fish and the use of the position and shape of the resonance in the frequency spectrum of the scattering to determine fish size distributions (Holliday, 1972).

The acoustic assessment of small zooplankton and micronekton is a relatively recent development. The earliest work dates from the 1960s, but most was done in the 1980s. By using single-frequency acoustic systems, usually echo sounders, it was possible to obtain useful insight into the spatial extent, spatial heterogeneity, and biomass of populations (Northcote, 1964; Bary, 1966a and 1966b; McNaught, 1968; Baraclough et al., 1969; Sameoto, 1976, 1980; Pieper, 1979). The concept of using multiple-frequency echo sounding to obtain size discrimination for zooplankton originated with McNaught (1968). In the mid-1970s, a formal mathematical structure for extracting size from measurements of volume backscattering at multiple frequencies was developed (Holliday, 1977). This formulation has since been used in various forms to acoustically determine the size-abundance distribution for pelagic fish (Holliday, 1985), for mesopelagic fish (Johnson, 1977a and 1977b; Kalish et al., 1986), for micronekton, such as, euphausiids (Greenlaw, 1979), and for assemblages of small zooplankton (Holliday and Pieper, 1980; Costello et al., 1989; Holliday et al., 1989; Pieper et al., 1990). Greenlaw (1979) has also suggested a two-frequency method for estimating the size and abundance for organisms of a single, unknown size.

Each of these techniques depends on either an empirical or an experimentally validated theoretical mathematical model to relate acoustic-scattering cross sections (target strengths) to the size and biophysical characteristics of the target organisms. In general, empirical relations are used for nekton (Love, 1969; Nakken and Olsen, 1977; Foote, 1987). A truncated variant of the fluid-sphere model (Anderson, 1950) has been used successfully in the case of small zooplankton, such as, copepods (Pieper and Holliday, 1984), but does not lead to satisfying results for elongate scatterers such as euphausiids (Greenlaw, 1979). This has inspired recent work that has led to a more elaborate mathematical formulation for a bent cylinder of finite length (Stanton, 1988, 1989a, 1989b). Models are also under development for rough surface cylindrical shapes as well as for various shapes with elastic shells (Stanton, 1990).

Strategies for Sampling LMEs

During the last 2 decades, methods have been advanced that offer scientists and managers an ability to quantify fishery resources by carrying out acoustic surveys over large areas (Midttun and Nakken, 1971; Forbes and Nakken, 1972; MacLennan and Forbes, 1984). Although the effort involved in conducting such surveys is not trivial, the methodology is sufficiently advanced that several fish stocks are managed, either wholly or in part, on the basis of the results of these survey techniques. The instrumentation needed to employ the most advanced of these methods, namely, dual-beam and split-beam echo sounding, is commercially available. These surveys are routinely done at moderate speeds (e.g., 8 knots) on vessels not specifically designed to provide a quiet acoustic measurement platform and at higher speeds on ships with low-noise characteristics. These technologies are directly applicable to and are available for current use in the assessment and monitoring of nekton in LMEs.

In part, at least, because micronekton and zooplankton are further removed from direct exploitation, there has been less interest in the commercial development of off-the-shelf acoustic instrumentation for assessment of these organisms. An exception is for the large euphausiids (*Euphausia superba*) found in Antarctica, where echo integration instrumentation has been adapted from that originally designed for fisheries assessment.

One of the principal differences between acoustic instrumentation used to estimate the abundance of fish and that for examining micronekton and zooplankton is the acoustic frequency required to obtain sufficiently high levels of scattering from the target organisms. For fisheries, although exceptions readily can be found that reflect special circumstances, most acoustic instrumentation designed for studying nekton uses frequencies between 20 kHz and 200 kHz. For large plankton and micronekton, optimal frequencies lie between 50 kHz and 500 kHz. Instruments designed for the study of smaller zooplankton, such as, adult copepods and their developmental stages, demand frequencies that range from those appropriate for micronekton to above 10 MHz.

The use of hull-mounted or towed acoustic sensors requires dedication of a ship for the period of the survey or experiment. The use of such sensors on conventional surveys that involve collection of direct samples with nets, trawls, or pumps can become a valuable supplement to the conventional sampling program. If the acoustic data collection is quasi-continuous, it may strengthen arguments regarding interpolation between spatially discrete conventional samples. In many cases, acoustic data can be used to direct conventional sampling efforts to depths and places of interest. Fronts are good examples of ecologically significant oceanographic features that have an acoustic signature that can be used in defining patterns from which one can develop rational biological sampling strategies (see Arnone *et al.*, 1990, for an example of such patterns). In some circumstances, especially when organism size and mobility can be acoustically determined, an intelligent choice of sampling gear can be made based on the acoustic data.

On the other hand, fiscal constraints and the finite availability of research ship time argue for the development and more extensive use of unattended sensor technology. Moorings have long been a mainstay of sampling programs in physical oceanography, yet they are not widely used in biological oceanography or fisheries research.

There have been two primary barriers to placing acoustic sensors on long-term moorings for sensing fish and plankton. These have been the substantial electrical power requirements for such devices and the problem of how to store and retrieve the large quantities of data that these sensors can generate. A combination of advances in battery technology and lower-power electronic components has now removed many of the barriers to development of moored, active acoustic sensors. Improvements in solar-cell technology now offer a feasible supplement to extend battery life. Low electrical power microcomputers and memory chips now allow reduction of data quantity by making on-board processing practical. Low-power data-storage devices are also now available at moderate costs.

Secondary barriers to the use of active acoustic devices on moorings have included concerns about the reliability of the electronics and the potential for fouling of the acoustic transducer by marine growth. With adequate (electrical and electronic) stress testing, reliability can be reasonably well assured, though the price of high reliability electronic components and the time it takes to adequately test a system can be a significant factor in increasing development costs. Inexpensive access to data from remote systems via satellites (e.g., ARGOS) is still measured in bytes; however, this capability can allow enough information to be transferred to monitor the "health" of the system. This can give the investigator the option of retrieving and fixing a system that has failed a few weeks into a planned year-long deployment, rather than waiting for the entire year to discover the failure. In many environments, marine growth can be controlled sufficiently with a rational choice of materials and antifouling agents. In other areas, often coastal, the proximity to land makes routine maintenance plausible.

Many of the same technological advances that have made moored acoustic systems feasible also lead us to consider other less-conventional modes of deploying acoustic sensors to study marine life. These are principally adaptations of concepts previously used as platforms for other kinds of sensors. Examples include arrays of moorings with tethered, vertically profiling sensors, surface drifters, programmed subsurface drifters (e.g., Rossby *et al.*, 1985; Rossby, *et al.*, 1986; Stommel, 1989), and expendable acoustic probes.

An Example of Instrumentation for the 1990s

We have previously considered a variety of acoustic instrumentation with an emphasis on potential use in fisheries research (Holliday, 1980, 1985). In the remainder of this chapter, we will focus on a different research thrust for the 1990s—an attempt to quantify the distributions and dynamics of micronekton and zooplankton by use of underwater acoustics.

As an example of potential applications of acoustic technology that should become practical in the 1990s, we offer a conceptual design of a sampling system for a hypothetical experiment. The scenario chosen to provide a framework for this discussion is a multiyear program to study the dynamics of zooplankton and the larvae of fish and crustacea in the central Irish Sea.

Acoustic system design considerations

The success one is likely to achieve in designing and employing an acoustic system to study marine life depends on one's knowledge of the physical characteristics and behavior of the target species or assemblage and its environment. If one has such information, it can be used to select specific characteristics of acoustic instrumentation to optimize its performance. Likewise, selection of the mode of employment of the acoustic gear will often depend on these parameters. Thus, it is useful to consider a specific place and problem when attempting to illustrate the kinds of applications of acoustics that are likely to become practical in the 1990s.

The problem we wish to address in this hypothetical experiment is measuring changes in the biomass distribution of zooplankton, by size, over several years' time, in a specific ecosystem. Although the ecosystem to be addressed, the central Irish Sea, is small compared to some of the ecosystems we could have chosen, it has two important features. First, it is sufficiently large to allow us to illustrate several acoustic techniques that, with appropriate modifications (principally in the communications mode), could have applications in virtually any LME and in many limnological ecosystems as well. Second, we can illustrate those techniques with data, even though they were collected by means other than those we suggest here.

In what follows, we ignore several factors that would have to be addressed if one were to actually pursue the implementation of the system. These include (i) conceiving the design based on observations of the environment and animals collected during one cruise, during one season of one year; (ii) a possibly high gear mortality caused by fishing activity and submarine operations in the area; and (iii) the practical aspects of setting up a long-term, cooperative program between the institutions and nations with an interest in this particular ecosystem. We also ignore the practical economic forces that relate to the importance or lack of importance of specific fisheries in the area.

The Irish Sea (Figure 1) is a semi-enclosed, shallow sea that lies between England and Ireland. It is open to the Atlantic on the north through a relatively constricted channel. In the south the opening is much wider. The salinity of the surface waters near the coasts are seasonally influenced by runoff from land. Although there is substantial tidal action throughout the area, the eastern (Welsh) side of the central Irish Sea appears to be more strongly tidally mixed than is the western (Irish) side. In the late spring and early summer, the Irish side of the central Irish Sea often exhibits a weak stratification, with a lens of relatively warm, fresher water forming in the area of weaker mixing. A more detailed description of the area can be found in White *et al.* (1988) and in the references in that paper.

The size range of zooplankton in the Irish Sea is comparable to that found in the open ocean. Much of the plankton biomass lies at sizes between 0.050 mm ESR and 4 mm ESR, where the term ESR refers to the Equivalent Spherical Radius, or the radius of a sphere with the same volume as the actual plankter. The

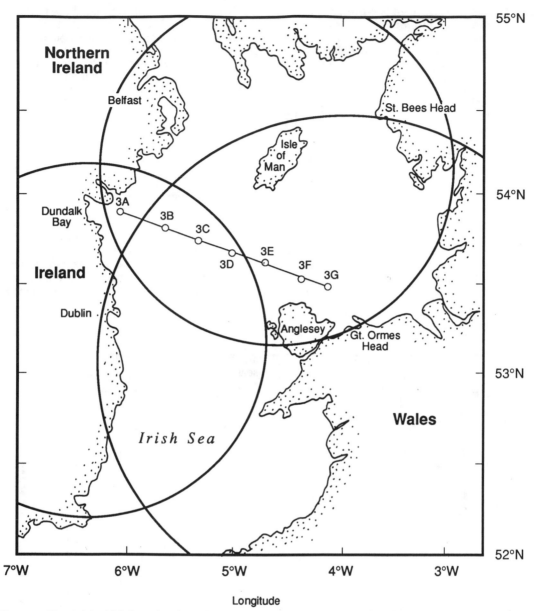

Figure 1. Chart of the Irish Sea, the site of a hypothetical monitoring program for micronekton and plankton. Stations 3A through 3G were occupied in the summer of 1989 by the Multifrequency Acoustic Profiling System (MAPS). The circles indicate the "line-of-sight" coverage for a hypothetical VHF telemetry system. The station in Wales, on Cornedd Llywelyn, is at a height of 1,058 m. The station in Ireland is at 540 m on Great Sugar Loaf Mountain. A single station on Snaefell, on the Isle of Man, at an elevation of 617 m, would cover most of the north central Irish Sea. The communications range calculation assumes a 3-m-high antenna on the buoy.

use of the ESR measure follows from its importance and use in a number of the more widely accepted mathematical models for acoustic scattering. For a zooplankter with a shape similar to a calanoid copepod, multiplying the ESR by between 4 and 5 will give an approximation to the animal's length.

The size spectrum and species distribution is displayed in Figure 2 for a station in the central Irish Sea. The data represent summed biovolumes for eight samples from an oblique Longhurst-Hardy Plankton Recorder (LHPR) tow (mesh size of 0.280 mm) from the surface to a depth of about 50 m near Station 3C (see Figure 1 for station location). The biovolumes were computed by measuring the lengths of the animals and then applying species-specific regressions to obtain an estimate of the actual volume of each animal. The size resolution (bin width) used in preparing this curve was 0.050 mm ESR. Because the mesh size was about 0.070 mm in ESR units, it is probable that the

LHPR sample has underestimated the biomass at smaller sizes. It is also possible that larger organisms have been undersampled because of avoidance. The species identified in Figure 2 accounted for 95% of the total biovolume in the samples. Organisms smaller than 1.1 mm ESR represent 99% of the biovolume collected at this station.

Applicable acoustic technologies

We will consider two different acoustic techniques. The first of these methods, to be used for studying micronekton, fish larvae, and small fish, involves measurement of the target strength, which is a measure of the effectiveness of an organism to scatter sound back toward the source of the sound (Clay and Medwin, 1977). The best available methods for making this measurement are the dual-beam and split-beam methods (Ehrenberg, 1989),

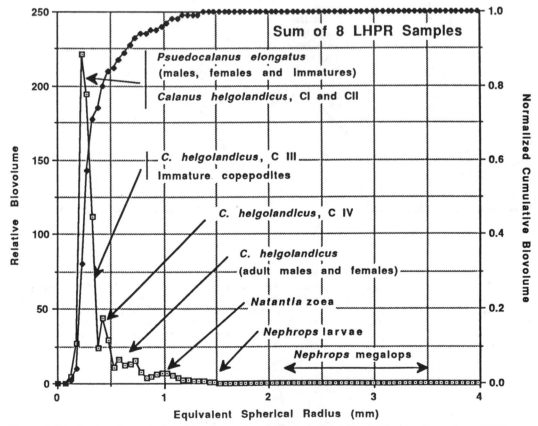

Figure 2. Distribution of zooplankton biovolumes from eight Longhurst-Hardy Plankton Recorder (LHPR) samples at Station 3C in the Irish Sea. The cumulative biovolume is also shown.

both of which demand that the combination of instrument beam width, pulse length, and organism abundance be such that echoes from an individual undergoing examination are distinct from echoes from nearby individuals. This limits the use of these techniques to organisms whose maximum abundance is on the order of the inverse of the acoustic sampling volume (preferably less).

In order to convert the target strengths measured with the dual-beam system to animal size, it would be necessary to make enough measurements on individuals to form a regression relation between target strength and size at the operating frequency of the dual-beam system. This is the traditional approach, used by Love (1969) for fish and by Wiebe *et al.* (1990) for large plankters. When the scattering is dominated by animals of one size, the process appears to work reasonably well for some purposes, including biomass estimation, but it should be recognized that its accuracy is subject to the same conditions that constrain any single-frequency technique. Those conditions include an inability to distinguish between changes in size and changes in abundance.

The second method involves measurement of volume backscattering at multiple acoustic frequencies. This technique works best when the scatterers in the acoustic sample volume are not distinguishable as individuals, such as, for small plankton. The Multifrequency Acoustic Profiling System (MAPS) is one possible implementation of this method. A truncated version of the fluid-sphere scattering model is used to relate the acoustic scattering at each frequency to the size-abundance spectrum of the zooplankters in the measurement volume. Measurements of volume-scattering strength at multiple frequencies are combined to estimate the zooplankton size-abundance spectrum. The MAPS is described in considerably more detail in Holliday *et al.* (1989).

Because neither the target-strength measurement method—which has been most successfully applied to fish, fish larvae, and larger zooplankton—nor the multifrequency technique—which works best for small zooplankton—will cover the entire range of animal sizes of interest, we choose to employ both in an overlapping mode for our hypothetical instrumentation suite for this ecosystem study. The dual-beam method and the split-beam methods are basically equivalent. The dual-beam method appears, at present, to require slightly simpler beam forming and signal processing than does the split-beam method. The split-beam method has slightly better performance and the potential for an improved data acquisition rate. For this application, partly to minimize electrical power consumption and partly

to improve reliability by minimizing electronic complexity, we choose to use the dual-beam method. Rather than have two separate instruments, to reduce the duplication of much of the electronics in the dual-beam system, we choose to modify one of the acoustic channels in the multifrequency system to allow the parallel implementation of a dual-beam device.

Selection of instrument characteristics

Acoustic volume-scattering at any particular frequency incorporates information from all of the organisms present in the scattering volume. Thus, in addition to assuring that our multifrequency instrument can detect scattering from the smallest organism of concern, we must take into account both the largest and smallest scatterers present in transforming volume-scattering estimates into size-abundance or biomass estimates. *Nephrops megalops*, the largest organism in the LHPR samples from Station 3C (Figure 1), occurred in five of the eight LHPR samples analyzed from our "typical" station in the central Irish Sea. The largest density (biovolume) for this species at any depth was $0.6 \text{ mm}^3 \cdot \text{m}^{-3}$. The average biovolume was $0.3 \text{ mm}^3 \cdot \text{m}^{-3}$. Based on the observed abundance of *N. megalops* and the measured size distribution (Figure 2), an upper size limit of 4 mm ESR seems a reasonable choice.

To assure that organisms larger than 4 mm ESR do not often contribute to our measurements of volume-scattering strength, thereby masking scattering from the smaller organisms present, we choose a system acoustic beam width and pulse length such that the volume sampled by the multifrequency acoustic instrument is sufficiently small that the probability of observing a large, rare organism is small. A sampling volume on the order of 0.05 m^3 would appear to satisfy this criterion. We further determine to eliminate the contribution of individual measurements of volume-scattering strengths (single occurrences) that exceed the mean plus twice the standard deviation of the last eight pings and the next eight pings. This requires memory and a data-processing lag. The samples eliminated would be logged and considered an index of larger organism abundance.

Current electronics technology and the absorption of sound in sea water sets the lower limit on the size range to which our hypothetical instrument will be sensitive. Building acoustic transducers that operate at frequencies above about 10 MHz is mechanically more challenging than is construction of lower-frequency transducers. At 10 MHz, performance is also usually limited by thermal noise in the preamplifier. Unless one makes an extraordinary effort, this sets a practical lower limit of about 0.050 mm ESR on the plankton sizes to be measured.

To select the frequency band in which we would like to operate, we must know the dependence of scattering strength on frequency for the organisms. The theoretical dependence of target strength on acoustic frequency is displayed in Figure 3 for the sizes at each end of our hypothetical size spectrum. To obtain this curve we used the truncated fluid-sphere scattering model. Although improvements on this model would be welcomed, it predicts—relatively well—the major features of the scattering from small plankton. This has been demonstrated by comparing the multifrequency method with a high-volume pumping system to obtain estimates of plankton size (Pieper and Holliday, 1984; Costello *et al.*, 1989).

There is some evidence that this model is not an adequate description of scattering from larger organisms with more complex shapes (e.g., adult euphausiids). Several models are the subject of current research that might be applied to larger and more complex shapes (Stanton, 1988, 1989a, 1989b, 1990). For

the purposes of transforming multifrequency volume-scattering strengths to size-abundance distributions, we would use the truncated fluid-sphere model for plankters smaller than about 4 mm ESR, with recognition that there is potential for increased errors in abundance at the high end of the range. As additional mathematical models are developed and validated, their inclusion in algorithms used to convert volume-scattering strengths to size-abundance estimates is straightforward (Holliday, 1977).

The highest frequency required is set by a need to define the transition from Rayleigh to geometric scattering for the smallest animal of interest. For an ESR of about 0.050 mm, this means a frequency of at least 10 MHz (Figure 3, lower curve). The lowest frequency used should be on the Rayleigh slope (Rayleigh, 1945) for the largest animal we expect to contribute to the scattering under the conditions set out above, which is, 4 mm ESR. This places the lowest frequency at about 50 kHz (Figure 3, upper curve).

Because today's technology will not support construction of a sufficiently wide-band, single-acoustic source, we must select a number of discrete frequencies in this band (50 kHz

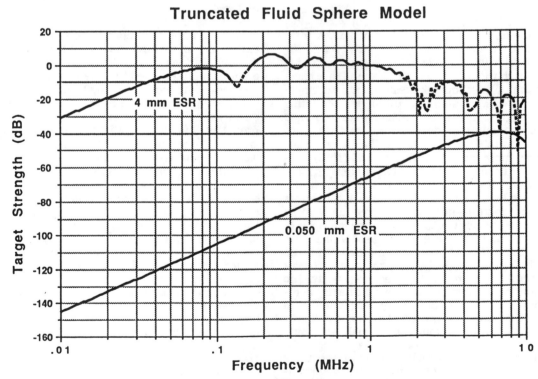

Figure 3. Frequency dependence of target strength in decibels (dB) from two sizes of zooplankters, 0.050 mm and 4 mm equivalent spherical radius (ESR).

to 10 MHz) for use in our hypothetical instrument. The number of frequencies selected will determine the number of independent sizes at which we can make estimates of abundance, that is, the size resolution of the system. There is not a simple relation between the frequencies chosen and the sizes. The scattering at each frequency contributes to the abundance estimate at each size. The number of frequencies chosen will have a major, though nonlinear, impact on system cost. In determining the number of frequencies, one must recognize that there is no basic requirement for even-spacing of the sizes at which abundance is calculated. This allows one to increase the system's size resolution in regions for which more definition is required, at the expense of sizes that contain no, or few, animals. This can be done in post-data-acquisition processing and does not directly influence system hardware design. The accuracy of the size-abundance estimates does, however, depend on allowing all of the sizes actually present to be in the calculation.

If one knows something of the size distribution a priori, one can simply select a set of sizes that will span and define that distribution well. If one has no a priori information, an initial even-spacing will allow a crude estimate of distribution, and a second calculation can be made with a selection of sizes that is appropriate to obtain an improved estimate. In practice, with the MAPS we have used the same size-resolution pattern in almost all cases. This pattern is denser on the small end of the spectrum, where organisms are more abundant, and more sparsely sampled at larger sizes.

If cost, system calibration, and maintenance were not of concern, it is likely that we would choose as many as twenty or even thirty frequencies. However, based on the distribution of sizes in the sample from Station 3C and our stated objective involving tracking the growth of cohorts over time, we determine that 11 frequencies will give us sufficient resolution for our hypothetical experiment. This will allow us to (barely) resolve the principal peaks in the size spectrum (Figure 2).

The frequencies selected must be capable of faithfully representing the dominant size-dependent features of the scattering curve (Figure 3) for all sizes that contribute to the measured scattering. The most important of these is the definition of the transition from Rayleigh scattering (at low frequencies) to geometric and resonance scattering (at high frequencies). There is no a priori requirement that the frequencies be spaced evenly over the frequency band. In fact, we have found it advantageous to select a logarithmic spacing in our work with the MAPS. Following this experience, we select 50 kHz, 85 kHz, 144 kHz, 245 kHz, 416 kHz, 707 kHz, 1.2 MHz, 2 MHz, 3.5 MHz, 5.9 MHz and 10 MHz as the design frequencies for the multifrequency system.

Because the dual-beam system will share components with one channel of the multifrequency system, we must select its operating frequency from those given above. This system is included to make estimates of abundance and size for larger organisms, such as micronekton and fish. Thus, we can take advantage of the larger target strengths of these organisms and the lower absorption losses at the low end of the multifrequency band. There is a minimum in the acoustic ambient noise spectrum in the ocean near 100 kHz. To take advantage of a potential performance gain, we select 85 kHz as the operating frequency for the dual-beam system.

Modes of deployment

In order to achieve a sensitivity to small zooplankton, the multifrequency sensor described above makes use of relatively high acoustic frequencies and small sample volumes. The combination of these characteristics requires that the sensor be placed reasonably close to the animals one wishes to study. The effective operating range for an echo-ranging acoustic system will depend not only on the size distributions of the animals, but on their abundances and spatial distributions, thus no single number can describe all conditions that might be encountered. However, based on our experience with the MAPS, which has similar system characteristics and has been used to examine distributions of plankton similar to those expected in the Irish Sea, one could expect operating ranges of a few meters when acoustically sampling nauplii and small copepods.

The hypothetical study we have described for the Irish Sea suggests a requirement for two different modes of deployment for the acoustic instrument described above. One would involve the deployment of an array of instrumented moorings as shown in Figure 1 (Stations 3A through 3G). The moorings would be designed to achieve a lifetime of at least 3 years, with battery replacement and servicing (transducer cleaning, preventative maintenance) on a 6-month schedule. During the first year, intensive conventional sampling, including the use of routine echo integration for nekton, would be carried out at 3-month intervals near and between the moorings. During these cruises, zooplankton spatial pattern would be defined with a high spatial resolution by a combination multifrequency dual-beam acoustic system deployed on a towed body. The body would be equipped with a capability for operator depth control and would be operated in a "tow-yo" mode (surface to bottom to surface oblique tows). The acoustic sensor would be functionally and electronically identical to the one employed on the moorings.

Data collected along transects around and between specific stations could be used to describe the three-dimensional distribution of plankton in relation to the local physical environment. Directed samples from such cruises would be used to associate discrete peaks in acoustically measured size spectra with species. Data from those cruises would also provide information on variability and trends in the records from the moorings that might be associated with advection of spatially heterogeneous distributions as opposed to temporal changes in more uniform distributions. In subsequent years, the high-resolution surveys could be conducted during the two servicing cruises each year.

Because the examination of small organisms precludes sensing at long distances, a mooring would either have to contain multiple copies of the sensor (at different depths) or means would have to be provided to profile the water column with a single sensor. The latter choice seems to be a better one in terms of instrument acquisition cost and maintenance. We envision a captive adaptation of the Slocum float device recently described by Stommel (1989). Isopycnal float technology was significantly advanced during the 1980s (Rossby *et al.*, 1985; Rossby *et al.* 1986). For the Irish Sea application, the concept would involve an instrumented programmable float whose depth is controlled by varying the float volume about neutral buoyancy.

In our adaptation of this concept, the float would be constrained in the horizontal plane by the anchor wire on a mooring. The float would carry the acoustic sensor package, temperature, conductivity, depth, and light sensors, as well as a battery module sufficient to operate the system for at least 6 months. Communications between the surface buoy and the float would be via an acoustic link at a frequency near 100 kHz; therefore, there would be no need for an electrical connection to the surface float. The float size would be dictated primarily by the weight of the sensor payloads. The energy needed to vary the float position in the water column is small, largely independent of the size of the instrument payload, and would be easily within reason for a moored environment. The battery power required would be determined almost entirely by the energy requirements of the sensor electronics.

In some environments, fouling would cause significant external changes in float buoyancy. It would be necessary to include suffi-

cient buoyancy reserve in the depth control system to allow the on-board computer to compensate for marine fouling. Parking the system at its maximum depth between profiles could be of some use in reducing the rate of fouling.

The surface buoy on the mooring would contain the necessary communications electronics to allow communication with a shore-based computer system. Two-way communication would allow partially processed data to be transmitted to shore each day and would also allow the investigator to reconfigure the data-acquisition rate in order to interact with the observed variations in plankton distributions, increasing the sampling rate when needed to resolve transient phenomena and decreasing the rate to conserve electrical power when unplanned schedule changes delay servicing.

In the Irish Sea, there is adequate shoreline topographic relief to support communications by VHF radio (line-of-sight). A coverage similar to that indicated by the circles in Figure 1 could be achieved with as few as three inexpensive shore stations. Standard telephone computer modems would be used for access to the shore-based communications stations from an arbitrarily located laboratory. Suggestions for communications technologies for use in places not readily adaptable to line-of-sight methods are discussed below.

Ancillary system components

There is an increasing recognition that the micro- and fine-scale distribution of physical properties may be important in influencing the distribution and recruitment success of zooplankton, micronekton, and nekton. The measurement of physical parameters may also be useful in analyzing related zooplankton physiological related rates (e.g., feeding, growth, reproduction).

It is reasonably simple to measure water temperature to the accuracy needed to interpret most biological zooplankton and nekton distributional data. It is more demanding, but still possible, to measure temperature with an accuracy appropriate for use in assessing water-column motions by using the principles of physical oceanography. Neither of these temperature measurements requires an excessive amount of electrical energy, and one or the other should be implemented as an ancillary sensor on the acoustic package. For cost reasons, one might choose to take a middle ground, measuring temperature to an accuracy of about 0.05° C within about a second. This would be more than sufficient for most biological rate calculations and also would suffice for at least crude estimates of water-column density and dynamics. It would not be sufficient to describe small-scale turbulence.

In the coastal zone of the Irish Sea, salinity variations caused by terrestrial runoff can have a major local influence on stratification and hydrography. Thus, even though one must be concerned about fouling, it would be necessary to include a conductivity sensor in the instrument package. Because commercially available sensors have largely ignored power consumption, the sensor should be turned on and off between profiles.

There are not presently any widely accepted low electrical energy methods of measuring fluorescence, an indicator of phytoplankton abundance. There are, however, some promising attempts to use natural fluorescence as an index of phytoplankton abundance and gross photosynthetic rate (Kiefer *et al.*, 1989; Chamberlin *et al.*, 1990). Such a technique has the potential of falling into the category of low-power instrumentation. This is one possible addition to the instrumentation discussed in this chapter and argues strongly for providing a few extra data channels to be left for future expansion.

Measurements of downwelling light and incident surface light could also be useful. Additions of these measurements, as well as wind speed and direction at the surface buoy, could be readily accomplished at little additional expense.

The addition of a small number of care-fully placed moorings that make use of a bottom-mounted Acoustic Doppler Current Profiler (ADCP) would be invaluable in defining advective transport and current patterns. Telemetry for data from these sensors from an attached surface buoy would be essentially the same as described above for the bio-acoustic sensors. These devices are available commercially.

Potential data products

In the summer of 1989, a cooperative research project was carried out in the central Irish Sea by investigators from the Ministry of Agriculture, Fisheries, and Food (MAFF) Laboratory in Lowestoft, Suffolk, U.K.; the University of Southern California in Los Angeles, California; Nova University in Dania, Florida; and Tracor Applied Sciences in San Diego, California. The MAPS was deployed at the stations labeled 3A through 3G as illustrated in Figure 1. Although the mode of deployment was a simple cast of the instrument at each location as opposed to the moored mode suggested above, the data from that cruise comprise a snapshot that is illustrative of the data products one might obtain from an array of moored instruments at those locations.

In order to provide a physical context for the description of the plankton distributions that follow, a cross section of water temperature for the transect is illustrated in Figure 4.

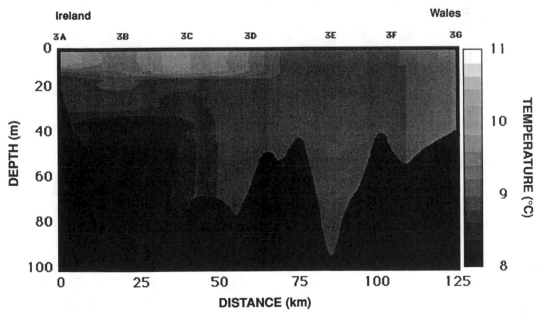

Figure 4. A temperature cross section for a transect between Stations 3A and 3G (Figure 1) in the Irish Sea. The data were collected with the MAPS on May 8 and 9, 1985.

Differences in the intensity of tidal mixing on the Irish and Welsh coasts allow the annual formation of an area of weak stratification on the western end of the transect. In addition to the relative warmth of this shallow feature, the MAPS measurement of salinity confirmed that the water in this surface lens was relatively fresh, a feature thought to result from coastal runoff (White *et al.*, 1988).

Measurements of acoustic volume-scattering strengths were made at twenty-one frequencies spaced approximately logarithmically between 100 kHz and 10 MHz. These data were mapped, by methods described in Holliday (1977), into estimates of zooplankton size-abundance spectra at each station. The result was a series of contoured cross-section plots at 21 independent sizes ranging from 0.025 mm ESR to 4 mm ESR. Three of those are illustrated here to demonstrate the type of data one can obtain from modern zooplankton acoustics. The first cross-section plot illustrates the two-dimensional spatial distribution for organisms with a size of 0.225 mm ESR (Figure 5). An LHPR sample collected near Station 3C contained *Pseudocalanus elongatus* adults and immatures, *Calanus helgolandicus* (C II), and a variety of other copepodites at sizes near 0.225 mm ESR. Again at Station 3C, the collection from the LHPR tow indicated that the second size-class, 0.625 mm ESR, for which a spatial distribution

is illustrated, was dominated by adult *C. helgolandicus* (Figure 6). *Natantia* zoea were present at 0.925 mm ESR (Figure 7).

The spatial distribution for the total biovolume along this transect, obtained by integrating over all of the sizes detected by the MAPS, indicated that although there were thin layers of organisms throughout the water column, much of the biological activity was in the upper 10 m (Figure 8). Coastal concentrations were evident at Stations 3A and 3B on the Irish side and at Stations 3F and 3G on the Welsh side of the sea. A biomass peak was also found in the transition zone between the area of weak stratification on the Irish side and the well-mixed zone to the east. If a moored system such as is described above were in place in this location today, software and reasonably priced computing and display hardware is available that would allow sequential display of a series of such plots as are displayed in Figures 5 through 8 in a "movie" format, allowing one to follow the seasonal and annual evolution of plankton biomass distributions in this environment.

Acoustic data from all depths at Stations 3C were pooled and the composite size spectrum computed (Figure 9). The data from the LHPR oblique tow suggests some species associations with the peaks that are evident in the size spectrum. Male, female, and immature *P.*

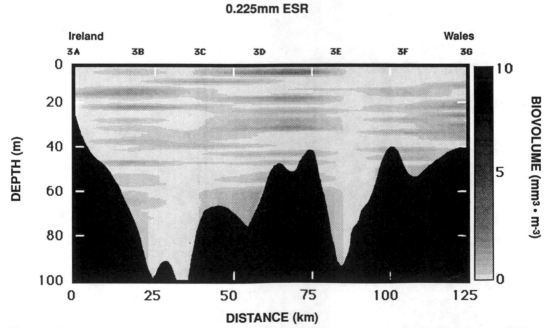

Figure 5. Zooplankton biovolume distribution along the transect from station 3A to 3G for the 0.225 mm ESR size class, estimated acoustically with the MAPS. The data were collected on May 8 and 9, 1985.

0.625mm ESR

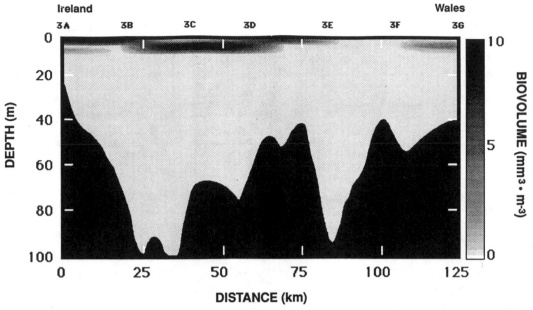

Figure 6. Zooplankton biovolume distribution along the transect from station 3A to 3G for the 0.625 mm ESR size class, estimated acoustically with the MAPS. The data were collected on May 8 and 9, 1985.

elongatus were captured that were of a proper size to contribute to the biomass associated with the peak centered at 0.150 mm ESR (peak A). *C. helgolandicus* (C I through C V) and a variety of unidentified copepodites were also evident in the samples, which could explain the distribution of biomass on the larger side of this peak. Adult *C. helgolandicus* were present at sizes appropriate to explain the peak near 0.650 mm ESR (peak B). The peak C, near 1 mm ESR, could be evidence of the presence of *Natantia* zoea, which were present in the sample at those sizes. *Nephrops* larvae and megalops were collected that could explain the peaks near 1.75 (peak D) and 3 mm ESR (peak E), respectively. The system design offered above in response to the hypothetical study would be (just) capable of resolving these peaks in the size-abundance spectrum and allowing one to attempt to follow cohorts within this assemblage of zooplankton over time.

In the final example (Figure 10), a contour is presented for acoustic estimates of zooplankton biovolume by size along the entire transect. Because we only have an LHPR sample at Station 3C, we refrain from making species associations for the other stations, noting only that there are striking coherences between some of the same size-classes at multiple locations.

One can conceive of numerous additional data displays that could be generated from acoustic and ancillary measurements similar to those illustrated above. For example, time sequences of simple or multiple coherences between size classes or total biovolume and measurements of temperature, salinity, light, and chlorophyll distributions could provide information on the magnitudes and phasing of plankton responses to their environment. Such displays could also form the basis for advances in our understanding of the impact of transient events, such as mixing caused by storms.

Technical Challenges

There will undoubtedly be advances in acoustic sensor technology during the 1990s. As in most disciplines that depend on field observations, the development of new sensors should always be encouraged. During the next decade, however, the most significant advances could come from improved engineering of old concepts to make them less expensive and more operator-tolerant. None of those improvements would be free, and, because little basic science is involved in the engineering effort, there is little chance that the needed advances will come entirely from within the science commu-

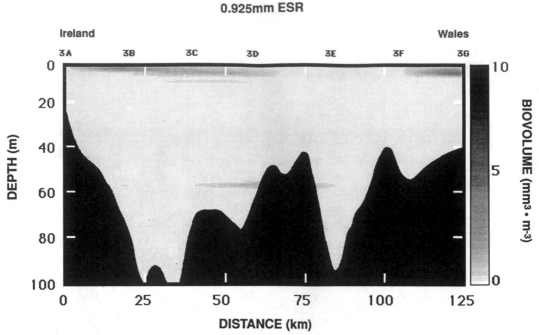

Figure 7. Zooplankton biovolume distribution along the transect from station 3A to 3G for the 0.925 mm ESR size class, estimated acoustically with the MAPS. The data were collected on May 8 and 9, 1985.

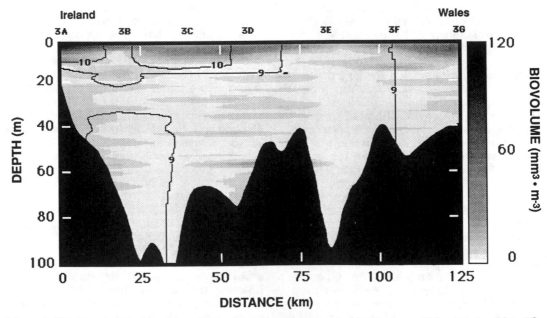

Figure 8. Total zooplankton biovolume distribution and temperature c° along the transect from station 3A to 3G, estimated acoustically with the MAPS. The spatial distribution of the total biovolume was obtained by integrating over all of the sizes detected by the MAPS. The data were collected on May 8 and 9, 1985.

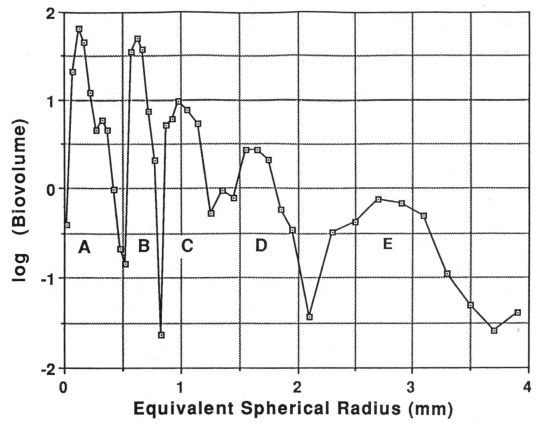

Figure 9. The MAPS (acoustic) estimate of the size spectrum at Station 3C in the Irish Sea. The labels A–E identify peaks in the size-abundance spectrum that were associated with particular species or assemblages of species. (See the text for details).

nity. Concise statements of needs and requirements from the potential user-communities backed by reliable, long-term funding will be needed to stimulate the interest of the engineering community.

There are substantial opportunities for improvements in mathematical modeling of scattering from marine plankton, micronekton, and fish. One can and should argue strongly for continued and even enhanced funding in these areas.

In many cases, the sensor technology that exists today is underutilized. Education of the user community in existing technology and stimulation of a wider employment of that technology should have a major role in plans to cope with the monitoring and management of LMEs in the 1990s.

The major technological challenges facing the scientist interested in improving our ability to sample LMEs appear to be involved with data communication, the display and visualization of multidimensional data, and its interpretation.

With respect to data communication from sensors distributed in an LME, there are three promising developments that appear to be evolving for our use in this decade. One of these is the advent of commercial systems for meteor burst communications (Stix, 1990). These systems, long available for military communications, are also currently in use for transmitting environmental data on such parameters as precipitation and snowpack from remote field sites. A second mode of communications, a cellular telephone network, is currently available for limited coverage in some offshore areas (e.g., the Gulf of Mexico). A third option is a system of 77 low-altitude, nongeosynchronous satellites with worldwide coverage for two-way personal mobile communications. Projected battery-powered handset prices are estimated to be about $3,500 (U.S.) and per minute connect costs are projected to be in the range of $1.50 to $3.00 (U.S.).

If these plans proceed to completion, inexpensive two-way communication with oceanographic sensors will be well within the budgets of most individual principal investigators. The ability to have one's computer dial individual sensors, at will, on a mooring, drifter, or ship, should open numerous opportunities for investigators interested in measuring biological as well as physical parameters in the world's oceans.

Summary and Conclusions

The sampling system described above is simply an illustration of the kind of capability that can be implemented by acousticians in response to needs voiced and supported by fisheries scientists, biological oceanographers, and resource managers during the next decade. Although the system described sounds ambitious in its complexity and scope, much of the technology needed to implement the hypothetical system exists today. It is much less ambitious than many of the successful technological programs of the 1970s and 1980s (e.g., the Voyager project).

Much of the more difficult sensor development technology has been accomplished. The MAPS contains most of the features of the multifrequency system described. The system in use by Greene *et al.* (1991) is one of several possible implementations of the dual-beam instrument. Isopycnal float technology contains

the essential ingredients from which the suggested profiling platform could be developed. Acoustic Doppler Current Profilers (ADCP) for measuring ocean-current profiles are commonly available to physical oceanographers. The line-of-sight telemetry for the system is off-the-shelf, as are meteor burst communications systems for use where longer distances or larger areas must be covered.

In summary, the technologies required to implement the system described are the legacy of the 1980s. The application of that technology to problems in the study and management of LMEs during the 1990s will largely depend on our success in convincing the public of the need to manage those resources from a scientifically sound base of data, made available in a timely manner, in adequate quantities.

Acknowledgments

Although the responsibility for the material presented in this paper is mine, I would like to acknowledge the fact that one's ideas are usually influenced by or are the direct result of interactions with professional colleagues, acquaintances, and friends. Consequently, I would especially like to thank Rick Pieper, David Doan, Charles Greenlaw, Gary Kleppel, John Dawson, Peter Ortner, Paul Smith, and Mike Reeve for numerous conversations over

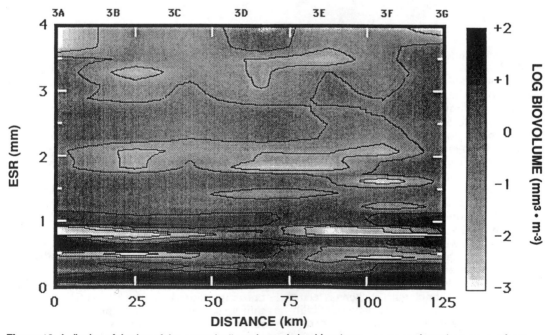

Figure 10. A display of the log of the acoustically estimated size-biovolume spectrum along the transect from station 3A to 3G (Figure 1) in the Irish Sea on May 8–9, 1989.

the last 2 decades about what might or might not be useful and possible in our attempts to introduce new acoustic technology into addressing the problems of sampling for fisheries scientists and biological oceanographers. I would also like to thank Gary Hitchcock for several informative discussions about the technology of free and tethered near-neutral buoyancy floating instrumentation packages. Ron Mitson and Keith Brander of the MAFF Fisheries Laboratory at Lowestoft, U.K., provided invaluable assistance in the collection of the data from the Irish Sea. Bernie Zahuranec (Office of Naval Research, Oceanic Biology) and Phil Taylor (National Science Foundation, Biological Oceanography) have provided funding under Contract N00014-81-C-0562 for this work and have also provided numerous opportunities for contact with other key individuals in the biological oceanography community both in the United States and overseas.

References

Anderson, V. C. 1950. Sound scattering from a fluid sphere. J. Acoust. Soc. Am. 22:426–431.

Arnone, R. A., Nero, R. W., Jech, J. M., and De Palma, I. 1990. Acoustic imaging of biological and physical processes within Gulf Stream meanders. EOS 71:982.

Balls, R. 1948. Herring fishing with the echometer. J. Cons. Int. Explor. Mer 15:193–206.

Baraclough, W. E., Lebrasseur, R. J., and Kennedy, O. D. 1969. Shallow scattering layer in the subarctic Pacific Ocean: Detection by high frequency echosounder. Science 166:611–613.

Bary, B. McK. 1966a. Backscattering at 12 kc/s in relation to biomass and numbers of zooplanktonic organisms in Saanich Inlet, British Columbia. Deep-Sea Res. 13:655–666.

Bary, B. McK. 1966b. Qualitative observations of 12 kc/s sound in Saanich Inlet, British Columbia. Deep-Sea Res. 13:667–677.

Chamberlin, W. S., Booth, C. R., Kiefer, D. A., Morrow, J. H., and Murphy, R. C. 1990. Evidence for a simple relationship between natural fluorescence, photosynthesis and chlorophyll in the sea. Deep-Sea Res. 37(6):951–973.

Clay, C. S., and Medwin, H. 1977. Acoustical oceanography: Principles and applications. Wiley-Interscience, New York.

Costello, J. A., Pieper, R. E., and Holliday, D. V. 1989. Comparison of acoustic and pump sampling techniques for the analysis of zooplankton distributions. J. Plankton Res. 11(4):703–709.

Craig, R. E., and Forbes, S. T. 1969. Design of a sonar for fish counting. Fiskerdir. Skr. Ser. Havunders 15:210–219.

Duvall, G. E., and Christiansen, R. J. 1946. Stratification of sound scatterers in the ocean. J. Acoust. Soc. Am. 20:254.

Ehrenberg, J. E. 1972. A method for extracting the fish target strength distribution for acoustic echoes. Proc. IEEE Int. Conf. Eng. Ocean Environ. 1:61–64.

Ehrenberg, J. E. 1974. Two applications for a dual beam transducer in hydroacoustic fish assessment systems. Proc IEEE Int. Conf. Eng. Ocean Environ. 1:152–155.

Ehrenberg, J. E. 1983. A review of in situ target strength estimation techniques. FAO Fish. Rep. 300:90–95.

Ehrenberg, J. E. 1989. A review of target strength estimation techniques. In: Underwater acoustic data processing. pp. 161–176. Ed. by Y. T. Chan. Kluwer Academic Publishers, Dordrecht, The Netherlands.

Ehrenberg, J. E., Carlson, T. J., Traynor, J. J., and Williamson, N. J. 1981. Indirect measurement of the mean acoustic backscattering strength of fish. J. Acoust. Soc. Am. 69:955–962.

Ehrenberg, J. E., and Lytle, D. W. 1972. Acoustic techniques for estimating fish abundance. IEEE Trans. Geosci. Electron. 10:138–145.

Eyring, C. F., Christiansen, R. J., and Raitt, R. W. 1948. Reverberation in the sea. J. Acoust. Soc. Am. 20:462–475.

Foote, K. G. 1987. Fish target strengths for use in echo integrator surveys. J. Acoust. Soc. Am. 82(3):981–987.

Foote, K. G., Aglen, A., and Nakken, O. 1986. Measurement of fish target strength with a split-beam echo sounder. J. Acoust. Soc. Am. 80(2):612–621.

Forbes, S., and Nakken, O. 1972. Part 2 of: The use of acoustic instruments for fish detection and abundance estimation. Manual of methods for fisheries resource survey and appraisal. FAO Man. Fish. Sci. 5. FAO, Rome.

Greene, C. H., and Wiebe, P. H. 1990. Bioacoustical oceanography: New tools for zooplankton and micronekton research in the 1990s. Oceanography 3:12–17.

Greene, C. H., Wiebe, P. H., and Burczynski, J. 1989. Analyzing distributions of zooplankton and micronekton using high-frequency, dual-beam acoustics. Progr. Fish. Acoust. 11:45–54.

Greene, C. H., Wiebe, P. H., Miyamoto, R. T., and Burczynski, Jr. 1991. Probing the fine-structure of the ocean sound-scattering layers with ROVERSE technology. Limnol. Oceanogr. 36:193–204.

Greenlaw, C. F. 1979. Acoustical estimation of zooplankton distributions. Limnol. Oceanogr. 24:226–242.

Harden-Jones, F. R., and McCartney, B. S. 1962. The use of electronic sector-scanning sonar for following the movements of fish shoals: Sea trials on RRS Discovery II. J. Cons. Int. Explor. Mer 27:141–149.

Hersey, J. R., and Backus, R. H. 1954. Sound scattering by marine organisms. In: The sea. Vol. 1. Ed. by M. N. Hill. Interscience, New York.

Hewitt, R. P., Smith, P. E., and Brown, J. C. 1976. Development and use of sonar mapping for pelagic stock assessment in the California Current area. Fish. Bull. U.S. 74(2):281–300.

Holliday, D. V. 1972. Resonance structure in echoes from schooled pelagic fish. J. Acoust. Soc. Am. 51(4):1322–1334.

Holliday, D. V. 1974. Doppler structure in echoes from schools of pelagic fish. J. Acoust. Soc. Am. 55(6):1313–1322.

Holliday, D. V. 1977. Extracting bio-physical information from the acoustic signatures of marine organisms. In: Oceanic sound scattering prediction. pp. 619–624. Ed. by N. R. Anderson and B. J. Zahuranec. Plenum Press, New York.

Holliday, D. V. 1980. Use of acoustic frequency diversity for marine biological measurements. In: Advanced concepts in ocean measurements for marine biology. pp. 423–460. Ed. by F. P. Diemer, F. J. Vernberg, and D. Z. Mirkes. The Belle W. Baruch Library in Marine Science, No. 10. Univ. of South Carolina Press, Columbia, SC.

Holliday, D. V. 1985. Active acoustic characteristics of nekton. In: Biology and target acoustics of marine life. pp. 115–144. Ed. by J. W. Foerster. U.S. Naval Academy, Annapolis, MD.

Holliday, D. V. and Pieper, R. E. 1980. Volume scattering strengths and zooplankton distributions at acoustic frequencies between 0.5 and 3 MHz. J. Acoust. Soc. Am. 67:135–146.

Holliday, D. V., Pieper, R. E., and Kleppel, G. S. 1989. Determination of zooplankton size and distribution with multifrequency acoustic technology. J. Cons. Int. Explor. Mer 46:52–61.

Johnson, M. W. 1948. Sound as a tool in marine ecology, from data on biological noises and the deep scattering layer. J. Mar. Res. 7:443–458.

Johnson, R. K. 1977a. Acoustic estimation of scattering layer composition. J. Acoust. Soc. Am. 61:1636–1639.

Johnson, R. K. 1977b. Sound scattering from a fluid sphere revisited. J. Acoust. Soc. Am. 61:375–377.

Kalish, J. M., Greenlaw, C. F., Pearcy, W. G., and Holliday, D. V. 1986. The biological and acoustical structure of sound scattering layers off Oregon. Deep-Sea Res. 33(5):631–653.

Kiefer, D. A., Chamberlin, W. S., and Booth, C. R. 1989. Natural fluorescence of chlorophyll a: Relationship to photosynthesis and chlorophyll concentration in the western South Pacific gyre. Limnol. Oceanogr. 34(5):868–881.

Love, R. H. 1969. Maximum side aspect target strength of an individual fish. J. Acoust. Soc. Am. 46:746–752.

MacLennan, D. N., and Forbes, S. T. 1984. Fisheries acoustics: A review of general principles. Rapp. P.-v. Reun. Cons. int. Explor. Mer 184:7–18.

McNaught, D. C. 1968. Acoustical determination of zooplankton distributions. Proc. 11th Conf. Great Lakes Res. 1968:76–84.

Midttun, L., and Nakken, O. 1971. On acoustic identification, sizing and abundance estimation of fish. Fiskerdir. Skr. Ser. Havunders. 16:36–48.

Nakken, O., and Olsen, K. 1977. Target strength measurements of fish. Rapp. P.-v. Reun. Cons. int. Explor. Mer 170:52–69.

Northcote, T. G. 1964. Use of a high-frequency echosounder to record distribution and

migration of Chaoborus larvae. Limnol. Oceanogr. 9:87–91.

Peterson, M. L., Clay, C. S., and Brandt, S. B. 1976. Acoustic estimates of fish density and scattering function. J. Acoust. Soc. Am. 60:618–622.

Pieper, R. E. 1979. Euphausiid distribution and biomass determined acoustically at 102 kHz. Deep-Sea Res. 26:687–702.

Pieper, R. E., and Holliday, D. V. 1984. Acoustic measurements of zooplankton distributions in the sea. J. Cons. Int. Explor. Mer 41:226–238.

Pieper, R. E., Holliday, D. V., and Kleppel, G. S. 1990. Quantitative zooplankton distributions from multifrequency acoustics. J. Plankton Res. 12:433–441.

Rayleigh, Lord. 1945. Theory of sound. 2d ed. Dover Publications, New York.

Richter, K. E. 1985a. Acoustic determination of small scale distributions of individual zooplankters and zooplankton aggregations. Deep-Sea Res. 32:163–182.

Richter, K. E. 1985b. Acoustic scattering at 1.2 MHz from individual zooplankters and copepod populations. Deep-Sea Res. 32:149–161.

Rossby, H. T., Dorson, D., and Fontaine, J. 1986. The RAFOS system. J. Atmos. Oceanic Tech. 3:672–679.

Rossby, H. T., Levine, E. R., and Connors, D. N. 1985. The isopycnal Swallow float—A simple device for tracking water parcels in the ocean. Prog. Oceanogr. 14:511–525.

Sameoto, D. D. 1976. Distribution of sound scattering layers caused by euphausiids and their relationship to chlorophyll a concentrations in the Gulf of St. Lawrence Estuary. J. Fish. Res. Bd. Can. 33:681–687.

Sameoto, D. D. 1980. Quantitative measurements of euphausiids using a 120-kHz sounder and their in situ orientation. Can. J. Fish. Aquat. Sci. 37:693–702.

Simard, Y., de Ladurantaye, R., and Therriault, J.-C. 1986. Aggregation of euphausiids along a coastal shelf in an upwelling environment. Mar. Ecol. Prog. Ser. 32:203–315.

Stanton, T. K. 1985a. Density estimates of biological sound scatterers using sonar echo peak PDFs. J. Acoust. Soc. Am. 78(5):1868–1873.

Stanton, T. K. 1985b. Volume scattering: Echo peak PDF. J. Acoust. Soc. Am. 77(4):1358–1366.

Stanton, T. K. 1988. Sound scattering by cylinders of finite length. I. Fluid cylinders. J. Acoust. Soc. Am. 83:55–63.

Stanton, T. K. 1989a. Simple approximate formulas for backscattering of sound by spherical and elongated objects. J. Acoust. Soc. Am. 86:1499–1510.

Stanton, T. K. 1989b. Sound scattering by cylinders of finite length. III. Deformed cylinders., J. Acoust. Soc. Am. 86:691–705.

Stanton, T. K. 1990. Sound scattering by spherical and elongated shelled bodies. J. Acoust. Soc. Am. 88:1619–1633.

Stanton, T. K., and Clay, C. S. 1986. Sonar echo statistics as a remote-sensing tool: Volume and seafloor. IEEE J. Oceanic Eng. OE-11(1):79–96.

Stix, G. 1990. Meteoric messages. Sci. Am. 263(3):167.

Stommel, H. 1989. The Slocum mission. Oceanography 2(1):22–25.

Sund, O. 1935. Echo sounding in fisheries research. Nature 135:956.

Traynor, J. J., and Ehrenberg, J. E. 1979. Evaluation of the dual beam fish target strength method. J. Fish. Res. Board Can. 36(9):1065–1071.

White, R. G., Hill, A. E., and Jones, D. A. 1988. Distribution of *Nephrops norvegicus* (L.) larvae in the western Irish Sea: An example of advective control on recruitment. J. Plankton Res. 10(4):735–747.

Wiebe, P. H., Greene, C. H., Stanton, T. K., and Burczynski, J. 1990. Sound scattering by live zooplankton and micronekton: Empirical studies with a dual beam acoustical system. J. Acoust. Soc. Am. 88:2346–2360.

Chapter 29

Application of Molecular Techniques to Large Marine Ecosystems

Dennis A. Powers

Introduction

During the past 2 decades there have been revolutionary advances in the biological sciences as the result of spectacular discoveries in molecular biology and the development of sophisticated biological instrumentation. Although these new technologies have been rapidly incorporated into the biomedical research community (e.g., see Powers, 1989; 1990a; 1991), marine biology and biological oceanography have been slow to adopt these powerful new approaches. Notable exceptions are those marine biologists who use cellular and molecular techniques to address biomedically oriented questions, employing marine organisms as model systems. The extraordinary successes of these marine biologists suggest that the application of cellular and molecular approaches to marine ecology and biological oceanography also have the potential for revolutionary advances. In addition to advancing our fundamental understanding of marine organisms, molecular technologies could be invaluable in addressing process-oriented problems in the ocean sciences for which solutions have not been readily achieved, such as biogeochemical processes, recruitment processes, anthropogenic environmental impacts, and ecological and evolutionary changes that are expected to be associated with global warming.

Marine ecologists and biological oceanographers are now beginning to utilize modern molecular techniques in their research, and, as a result, these fields are on the threshold of an exciting new frontier (e.g., see Yentsch *et al.*, 1988; Powers, 1989; 1990a; 1991; Powers *et al.*, 1990; Arnheim *et al.*, 1990; Ward, 1990; Falkowski and LaRoche, 1991). By coupling highly sophisticated instrumentation such as satellite remote-sensing—which permits synoptic monitoring of chemical, physical, and biological parameters over large areas—with the power of modern molecular tools, these marine scientists will be able to address questions that were previously unapproachable. Exploitation of modern molecular tools, coupled with automated molecular analyses, promises to revolutionize our fundamental knowledge of marine organisms and biological oceanography, as well as enormously expand our understanding of the role of the oceans in regulating global processes.

The purpose of this chapter is to alert oceanographers to the potential for applying molecular techniques toward resolving major oceanographic problems that have proven intractable in the past. For example, a major problem that has occupied biological oceanographers for the better part of a century is the inability to understand and predict spatial and temporal biological variability in the ocean. During the past several decades, advances in technologies such as remote sensing, submersible technologies, shipboard sampling methods, towed instrumentation, collecting nets, automated photographic technologies, chemical and physical sensors, and acoustic arrays have tremendously improved the ability of oceanographers to monitor changes in the species composition of marine communities and biogeochemical cycles in relation to a variety of physical and chemical variables. As a result of these technological advances, oceanographers have been able to record the rise and fall of population densities and species compositions of several marine ecosystems, but the mechanisms responsible for this variability remain unknown. Because many marine species, directly or indirectly, comprise the food supply of many countries, play an important role in the global carbon cycle, and are sensitive to the impacts of human activities, the causes and predictability of such variability are of societal concern, as well as of scientific interest. In fact, the potential scientific and economic significance of biological variability in the oceans is so great that it can be concluded that understanding the mechanisms responsible for these processes and developing the capability to predict those changes is the *most important problem in biological oceanography*. The techniques and instrumentation employed by molecular biologists and biochemists can provide powerful tools to help solve these important problems.

Biological Variability: A Long-Standing Central Problem

Because most marine organisms produce orders-of-magnitude more offspring than are needed to replace the parents, the vast majority of them must die before they become reproductive adults. Individuals that do survive are referred to as recruits, and recruitment is the process or number of offspring brought into the adult population each year. Transitions between life-history stages entail differential mortality, the magnitude of which often can be staggering. In fact, the probability of survival between egg and adult is typically less than a fraction of 1%. It is usually assumed that death is random, especially in relation to genetic background, and that survival is a stochastic process driven primarily by physical and chemical forces. Differential survival is responsible for changes in species composition and the genetic architecture of marine ecosystems. The ultimate goal of oceanographers is to delineate the detailed interactions between ocean physics, chemistry, and population dynamics, so as to be able to predict the biological variability in large marine ecosystems.

Predicting biological variability in the ocean requires a thorough understanding of the interrelationships among and between biotic and abiotic components. To obtain this information, it is necessary to understand the mechanisms that regulate primary and secondary production. It is also imperative that we learn how the physical environment affects interactions between and among biotic and abiotic constituents; how organic and inorganic material are transformed into biomass; how biotic and abiotic components of LMEs interact to regulate biodiversity; how environmental and biological variables regulate the expression of genetic potential; how genes and their products regulate physiological acclimation and evolutionary adaptation; and how gene flow between populations is regulated.

These and other important questions can be addressed in LMEs if adequate methods can be developed to rapidly identify marine taxa and assess the physiological status of marine organisms. In some cases, conventional methods can be employed, but, more often, either current technology is inadequate or no effective method exists. In such cases, the application of molecular techniques that use such tools as isozymes, deoxyribonucleic acid (DNA) probes, nucleic acid sequencing techniques, and immunological methodologies can be particularly useful. I shall provide a few examples of this approach below.

Identification of Marine Taxa

Identification and classification of marine organisms is one of the most important, yet time-consuming and often problematic, tasks faced by marine ecologists and biological oceanographers. Accurate species identification is fundamental to the understanding of community dynamics, biogeochemical processes, and ecosystem stability. In some cases, organisms are categorized by functional groups (e.g., microorganisms, phytoplankton, and zooplankton); in other cases, it may be critical to obtain information at the species or population levels. Some marine taxa can be distinguished by conventional methods, whereas others, especially planktonic forms, are either difficult or impossible to identify without the use of molecular methods. In addition, many taxa that are morphologically distinguishable as adults have larval forms that are either difficult or impossible to identify.

This inability to rapidly distinguish marine taxa creates a major barrier to understanding the structure and interrelationships within and between LMEs. Even when identification and/or classification is possible by conventional approaches, the procedures are often so slow that it is impossible to couple the biological data with physical and chemical information within a reasonable time frame. This temporal incompatibility intrinsically limits attempts to develop a unified approach to solving large-scale oceanographic problems. Clearly, the ability to rapidly identify and classify taxa would remove a major stumbling block for biological oceanographers, fisheries biologists, and marine ecologists.

Molecular techniques can be particularly useful in the identification and description of taxonomic units, including species, subspecies,

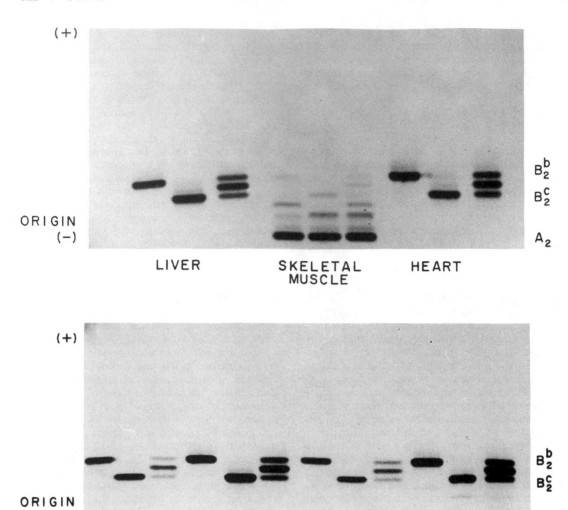

Figure 1. Tissue distribution of glucose phosphate isomerase (GPI) isozymes in the teleost *Fundulus heteroclitus*. The liver and heart express the GPI-B isozyme; the skeletal muscle predominantly expresses the GPI-A isozyme. The GPI-B isozyme has five allelic alternatives and this figure shows the expression of two alleles (*GPI-B^b* and *GPI-B^c*) which are represented by three allelic isozymes (GPI-B$^b_2$, GPI-B$^c_2$, and the heterozygote GPI-B^b/B^c).

populations, and demes. Such genetic data can contribute to an understanding of the evolutionary, ecological, and developmental processes that structure populations and the scale(s) at which these processes work. These techniques can sometimes provide species identification with greater accuracy and speed than can conventional approaches. Moreover, some methods can be applied to partially digested gut contents in predator-prey studies and to the identification of early life-history stages of organisms that cannot be determined by conventional techniques.

There are a number of molecular and biochemical methods that can be applied to the identification of species and/or the detection of genetic variation within species. All of these methods rely on the ability to discriminate between variants of specific genes (DNA) or their gene products (such as ribonucleic acid [RNA] and proteins).

Isozymes and allozymes

During the better part of 3 decades, electrophoresis coupled with histochemical staining for the identification of protein electromorphs has uncovered a wealth of genetic variation at the molecular level, both within species and between species. Variations in electrophoretic mobility of homologous proteins reflect their amino acid sequence differences, which, in turn, indicate differences in nucleotide sequences of the genes that encode these proteins. Such electrophoretic variations are reflected as either multilocus isozymes or allelic isozymes. Multilocus isozymes indicate one or more ancient gene duplications followed by evolutionary divergence of the duplicated locus. These multilocus isozymes are usually expressed within all members of a given species, but the level of expression may vary among tissues of the same individual (e.g., heart, brain, skeletal muscle, eye, and gut). In some cases, the tissue-specific expression of multilocus isozymes has been shown to complement the metabolic needs of the tissue in which the isozymes are expressed.

Figure 1 depicts the distribution of two glucose phosphate isomerase isozymes (GPI-A_2 and GPI-B_2) in liver, skeletal muscle, and heart tissues from three different individual teleost fish (*Fundulus heteroclitus*). Figure 1 shows that the GPI-A_2 isozyme is the major isozyme expressed in skeletal muscle, and the GPI-B_2 isozyme is the only glucose phosphate isomerase expressed in liver and heart. Figure 1 also shows no detectable electrophoretic variation between individuals for the GPI-A_2 isozyme, but significant variation between individuals for the GPI-B_2 isozyme. This latter type of genetic variation (i.e., within a species) at a specific locus is referred to as an allelic isozyme (often referred to as an allozyme). Buth (1990) has recently reviewed the genetic principles and interpretative approach to be used when electrophoretic data is employed by fisheries biologists.

When there is no genetic variation at a particular locus within a species (e.g., GPI-A_2 in Figure 1), but substantial variation between species, multilocus isozymes can be used to rapidly detect species differences between larvae or other life stages that may be ambiguous from a morphological perspective. For example, the white perch, *Morone americana*, is extremely difficult to distinguish from *Morone saxatilis* in early larval stages, and studies of larval recruitment of these species are difficult. Biochemical methods have been developed to assess the relative portion of each species in a given collection of larval fish (Morgan, 1975; Sidell *et al.*, 1978). These methods provide relatively fast and unambiguous species identifications, whereas morphological identification has a significant associated error.

Perhaps even more importantly, isozymes have also revealed cryptic species that were not morphologically distinguishable at any life-history stage. A striking example is represented by the work of Shaklee and Tamaru (1981) with Hawaiian bonefish (*Albula*). They identified two sympatric sibling species that were isozymically distinct at 58 out of 84 loci. Other examples of isozymically detected cryptic species and populations have been described in oysters, commercially exploited squids, clupeids in the southwestern Pacific, lizardfish, mackerels, and other fish species (reviewed by Shaklee, 1983; Shaklee *et al.*, 1990).

Hybridization between species is another example of the power of isozyme methods. Natural hybridization is much more common in fish than in other vertebrates (reviewed in Campton, 1987; 1990). The detection of hybridization by morphological criteria suffers from many shortcomings. The basic problem is that interspecific hybrids of fishes are often not morphologically intermediate to their parental taxa. The analysis of isozymes encoded by multiple diagnostic loci is the most sensitive and reliable method to identify hybridization. Apparently, hybridization is not as common in marine as it is in freshwater species. However, the use of these and other molecular techniques (e.g., mitochondrial DNA [mtDNA]) in marine species is likely to reveal more hybridization than has previously been thought to exist (e.g., She *et al.*, 1987; Campton, 1990).

Fisheries biologists use the "stock" concept as a basis to manage commercially important marine organisms. Shaklee *et al.* (1990, p. 174) defines a stock as "a panmictic population of related individuals within a single species that is genetically distinct from other such populations." Most evolutionary biologists consider the stock as the fundamental genetic unit of evolution. Therefore, the identification of stocks and the impact of harvest-induced mortality (Fetterolf, 1981) are critical elements in any management regime. Although morphological, life-history characteristics and other variables have been useful in such studies, the analyses of isozymes and their allelic variants have also been widely employed in the study of fish population structure and stock identification. Moreover, they have often helped resolve problems that could not be resolved by conventional approaches.

The advantages of using isozymes over morphological and other classical variables are that (i) the biochemical phenotype is essentially unaffected by the environment, (ii) the biochemical phenotype of each individual is stable through time, and (iii) the observed genetic variation is usually caused by a single gene whose alleles are codominantly expressed (Ayala, 1975).

Allozymic description of stock composition of marine fishes has become routine in the last 20 years, since the pioneering review of de Ligny (1969). Ryman and Utter (1987) have extensively reviewed the voluminous literature on fish population genetics and fisheries management. The ingeniously entitled paper, "Which witch is which?," by Fairbairn (1981) exemplifies the value of these studies. Isozyme analysis of samples from three management areas of witch flounder revealed a total of six genetically distinct stocks that differed with respect to age structure, time of spawning, individual growth rate, and temperature and depth of capture. Existing practices had managed these fish as large, homogeneous stock units.

An extension of this approach also allows estimating the geographical origin of fish caught in mixed-stock fisheries. It is essential to be able to identify the origin and proportional contribution of different stocks to a mixed fishery. However, this approach has been very difficult to apply (Larkin, 1981). Statistical techniques have been developed that allow the use of allelic isozyme variants to estimate the contribution of different stocks to mixed-stock fisheries (reviewed in Pella and Milner, 1987). Shaklee *et al.* (1990) recently reviewed how isozyme techniques could be used to establish the existence of multiple stocks of Spanish mackerel in Australia and Papua New Guinea. Shaklee and his colleagues (Shaklee and Salini, 1985; Shaklee *et al.*, 1990) have also shown that there were several stocks of Australian Barramundi (*Lates calcarifer*), and they have used isozyme techniques to define their inter-stock boundaries. Utter *et al.* (1987) have used isozyme techniques to demonstrate the presence of several chinook salmon fisheries in the Pacific Northwest. Shaklee *et al.* (1990) have reviewed the extensive isozyme literature and shown that several distinct chum salmon stocks also contribute to mixed fisheries in the Pacific Northwest.

The potential for genetic differentiation between populations of a species depends upon a number of variables, including migration rate, the number of individuals within a population, and natural selection at different loci. Determination of genetic exchange between populations has been estimated by a number of experimental and theoretical approaches. Release-recapture studies have been used to examine the migration of individuals, but there are a number of practical restrictions, questionable assumptions, and theoretical constraints that have limited the usefulness of this approach.

In addition to describing the genetic architecture of marine species, allelic isozymes can be used to estimate the amount of gene flow between populations. Allendorf and Phelps (1981) used population genetic theory (Wright, 1969) and computer simulation to show that the amount of allelic divergence, as measured by Wright's F_{st}, between subpopulations is a function of the absolute number, rather than the fraction, of migrants exchanged. This is an important finding because it emphasizes the need to know the population size in order to estimate reproductive isolation from allozyme data. They also cautioned that the use of larval data to draw conclusions about divergence of reproducing adults could be greatly misleading.

Slatkin (1985) has developed a powerful method for using the frequency of rare alleles to estimate gene flow between populations by determining the average number of migrants exchanged. This approach is both useful and relatively insensitive to changes in any parameter except the number of migrants exchanged and the number of individuals sampled per population.

Waples (1987) has used Wright's F_{st} and Slatkin's (1985) methods to estimate gene flow in 10 species of marine shore fishes. Estimates of gene flow were highly correlated with the dispersal ability of the species. Waples concluded that genetic differentiation among these fishes was primarily determined by gene flow and genetic drift, rather than by natural selection.

Sometimes the gene frequencies of certain enzyme-synthesizing loci are correlated with directional changes in specific environmental variables such as temperature, salinity, and oceanic circulation. In such cases, extensive studies have been launched in order to elucidate if one or more of these environmental parameters have acted as a selective force to favor one allelic alternative over the other. Two examples of this approach are provided by the works of Koehn and his colleagues (reviewed by Koehn and Hilbish, 1987) on the leucine amino peptidase (*Lap*) locus in the mussel, *Mytilus edulis*, and our work on the heart-type lactate dehydrogenase locus (*Ldh-B*) from the fish, *Fundulus heteroclitus* (reviewed by Powers, 1987; 1990b; Powers *et al.*, 1991).

In both examples, a combination of genetics, physiology, and biochemistry was used to demonstrate that selection was acting on these loci. In the former case, Koehn's group showed that salinity and/or temperature were differentially affecting the allelic isozymes of the *Lap-1* locus (e.g., Koehn *et al.*, 1976; Koehn and Immerman, 1981; Koehn and Siebenaller, 1981). In the latter case, we showed that the *Ldh-B* allelic isozymes—whose gene frequencies changed along the east coast of North America (Place and Powers, 1978)—were structurally and functionally different (Place and Powers, 1979, 1984; Crawford and Powers, 1989, 1990; Crawford *et al.*, 1989) and that these differences allowed us to make predictions concerning differential cell metabolism, developmental rates (DiMichele and Powers, 1982a, 1984, 1991; DiMichele *et al.*, 1991), and swimming performance (DiMichele and Powers, 1982b). Those predictions were substantiated by experimentation and shown to be of selective value in field hatching studies (DiMichele, *et al.*, 1986) and laboratory temperature-selection experiments (DiMichele and Powers, 1991).

In addition to our work on the heart-type *Ldh-B* locus, we have also examined the functional aspects of the allelic isozymes of several other loci in *Fundulus heteroclitus*, including: the heart-type glucose phosphate isomerase (*Gpi-B*) (Van Beneden and Powers, 1989), the hexose-6-phosphate dehydrogenase locus (*H6pdh-A*) (Ropson and Powers, 1989), the cytosolic nicotinamide adenine dinucleotide phosphate (NADP)-dependent isocitrate dehydrogenase locus (*Idh-B*) (Gonzalez-Villasenor and Powers, submitted), the cytosolic malate dehydrogenase (*Mdh-A*) (Cashon and Powers, submitted), and aspartate amino transferase (*Aat-B*) (Ropson *et al.*, in press). In every case, there were significant differences between the allelic isozymes in one or more of the following functional parameters: thermal stability, enzyme concentration, substrate affinity, substrate or product inhibition, inhibition by another metabolite, and catalytic efficiency.

The importance of genetic variation in critical traits such as survival, growth, development, and age at first reproduction has been demonstrated in a variety of marine organisms. The classic work by Ricker (1981) demonstrated alterations in several important life-history characteristics in five species of Pacific salmon (*Oncorhynchus*) that are at least partially caused by changes in the genetic composition in these species because of commercial harvesting. A positive association between the amount of genetic variation and growth rate has been found in many marine species of mollusks (reviewed in Zouros, 1987).

Nevo and his co-workers (1985, 1986) have shown that resistance to pollution in marine organisms is associated with genetic variation. It has also been shown that rate of development and hatching of the fish *Fundulus heteroclitus* is directly related to the genetic background of the following loci: *Ldh-B*, *Mdh-A*, *Gpi-B*, and phosphoglucomutase (*Pgm-A*) (DiMichele and Powers, 1982a, 1984; DiMichele *et al.*, 1986; DiMichele and Powers, 1991). Moreover, these loci have also been shown to

be related to differential survival at high temperature (DiMichele and Powers, 1991).

Another example of this approach is the work of Vrijenhoek and his colleagues (Vrijenhoek *et al.*, in press), who have shown dramatic differences in survival among populations of the fish genus *Poeciliopsis*. Their laboratory and field selection experiments revealed natural selection on allelic isozyme variants marked by four different enzyme-encoding loci. Using acute cold, heat, and anoxia (variables that mimic seasonal environmental stress), they demonstrated that enzyme heterozygosity and survival were intrinsically linked.

Some researchers have demonstrated the utility of isozyme techniques to describe the transport of organisms by physical processes in the ocean environment. For example, Heath and Walker (1987) used allozymes to describe the pattern of drift and to identify the geographical origin of spawning of larval herring (*Clupea harengus*) netted in the North Sea. Bucklin *et al.* (1989) used allelic isozyme markers to describe transport of a calanoid copepod in coastal filaments off the west coast of the United States. Isozymes have also been used to study genetic variation within and between species of other copepods (Frost, 1989; Sevigny *et al.*, 1989). In addition to zooplankton, isozymes have been useful in examining phytoplankton species and communities. For example, Gallagher (1982) demonstrated that isolates of the diatom *Skeletonema costatum* could be distinguished by their allelic isozymes, and Gallagher and Alberte (1985) later showed that these isolates were physiologically distinguishable as well. Gallagher (1980; 1982) was able to use isozymes to ascribe semiannual algal blooms in Narragansett Bay to genetic variants of *Skeletonema costatum*. Similar techniques have also been used to identify and map macroalgal species of economic importance (McMillan, 1982).

Mitochondrial DNA (mtDNA) and chloroplast DNA (cpDNA)

Although multilocus isozymes and allelic isozymes are useful for characterizing some species and populations, the lack of observable genetic variation in other species of economic and scientific importance has restricted the application of this technique to a wider application of recruitment and management problems. In part, this is because isozyme methods grossly underestimate the extent of genetic variability arising from changes in the isopolar amino acids and in the nucleotides that are not reflected in the corresponding protein sequence.

Recent molecular tools employed in population biology studies have led to the discovery of genetic variability that was unexpected on the basis of isozyme data. Not only do these methods permit analysis of a broader scope of genetic diversity, but the sensitivity of these methods permits the study of egg and juvenile stages, tissue biopsies of adults, and even single cells.

Endonuclease restriction digests of mitochondrial DNA (mtDNA) in marine organisms and chloroplast DNA (cpDNA) in marine plants and algae are beginning to be routinely employed. The general approach is as follows: DNA is isolated from a given taxonomic unit (individual or species, for example); each sample is divided into a series of test tubes to which is added one or more sequence-specific DNA "cutting" enzyme(s), referred to as endonuclease restriction enzymes; the fragmentated DNA is subjected to electrophoresis, and the resulting electrophoretic patterns are visualized by one of several methods. These electrophoretic patterns are referred to as restriction fragment length polymorphisms (RFLP).

A typical RFLP of mtDNA from a single fish, treated with 18 different restriction enzymes, is shown in Figure 2. The RFLP patterns can be compared directly, or this type of data can be used to construct restriction maps. Because mtDNA is usually between 16 and 19 kilobases in length, it is relatively easy to map with an array of 4- and 6-base sequence-specific endonuclease restriction enzymes. For example, RFLP data such as that illustrated in Figure 2, and other RFLP data obtained from digesting mtDNA with a series of endonuclease restriction enzymes, were used to construct a restriction map (Figure 3) of the *Fundulus* mtDNA that had been cloned into a lambda phage (Gonzalez-Villasenor *et al.*, 1986).

Because mtDNA is generally maternally inherited, the variation within and between populations can be studied, and matriarchal lineages can be traced. This variation is reflected both as differences in restriction sites and/as differences in the size of the mtDNA. Chloroplast DNA is larger in size (120 to 200 kilobases) than mtDNA, but has proven valuable for phylogenetic and population studies of phytoplankton and macrophytes. Although RFLP patterns of mtDNA and cpDNA reflect varying degrees of natural selection and historically relevant evolutionary incidents, they may prove extremely valuable for identifying the geographical origin of individuals and populations. Such information can assist in stock identification, fisheries management, transport studies of plankton, and evolutionary studies.

There are a number of examples where mtDNA has already provided insight concerning the population structure of marine species that could not be gained from isozyme studies. For example, the Atlantic eel of the genus *Anguilla* is known to spawn in the Sargasso Sea, and the leptocephalus larvae migrate thousands of kilometers to metamorphose in estuarine waters. Given this life history, it would seem a foregone conclusion that genetic uniformity would be expected over vast regions of the Americas and Europe. However, this has been a debatable point for many years.

The evidence from isozyme studies has been equivocal. Williams *et al.* (1973) found little interlocality variation among elvers, but some interlocality differences among adults in North American eels. In more extensive studies (Koehn and Williams, 1978; Williams and Koehn, 1984), it was argued that the small gene-frequency differences among populations were a result of natural selection. European scientists argued to the contrary.

Comparini and Rodino (1980) have concluded on the basis of allozyme frequency differences that North American and European eels should be considered separate species. However, Williams and Koehn (1984) argued that their work with North American eels suggests that the amount of allele frequency divergence between European and North American eels indicates only partial reproductive isolation between eels from the two continents.

Avise and his colleagues (Avise *et al.*, 1986) resolved this conflict with a classic paper on mtDNA differentiation in North Atlantic eels. Their restriction site polymorphism study of mtDNA showed no genetic divergence among the eels (*Anguilla rostrata*) along the coast of North America and suggested they were all members of a single panmictic population. However, their study showed that samples of the European eel, which they referred to as *Anguilla anguilla*, were significantly different from those studied along the North American coast.

Bermingham and Avise (1986) used mtDNA RFLPs to study the zoogeography of four species of freshwater fish: *Amia calva*, *Lepomis punctatus*, *Lepomis gulosus*, and *Lepomis microlophus*. They found that within each species, major mtDNA phylogenetic discontinuities distinguished populations from different geographical areas. From these data, they concluded that dispersal and gene flow were inadequate to override the historically driven geographical changes in sea level.

Allelic isozyme studies of populations of the teleost *Fundulus heteroclitus* uncovered significant directional changes with latitude in gene frequency (i.e., clines) and in degree of genetic diversity (Cashon *et al.*, 1981; Powers and Place, 1978; Powers *et al.*, 1986, 1991; reviewed by Powers, 1990b; Ropson *et al.*, 1990). Directional changes in genetic characters with geography (i.e., clines) have classically been described by two general models: primary and secondary intergradation. In the primary intergradation model, adaptation to local conditions along an environmental gradient and/or genetic drift may lead to genetic differences along the gradient. Gene flow may not eliminate these differences, either because it is too small or because of nonrandom dispersal along the gradient. In the secondary intergradation model, populations are first separated by some barrier that prevents gene flow. Next, either adaptation to local conditions or genetic drift produces genetic differences between these disjunct populations. Finally, when the barrier is removed, the formerly disjunct populations interbreed, producing a cline in gene frequencies between them. The main difference between these two models, therefore, is the need for the previous existence of isolating barriers to gene flow in the latter.

The present day spatial allelic isozyme patterns of *Fundulus heteroclitus* could have arisen by either type of intergradation (Powers *et al.* 1986, 1991). One cannot distinguish between these on the basis of classical zoogeographical data unless it is available within a few hundred generations of an alleged secondary contact (Endler, 1977). However, analysis of mtDNA can provide insight concerning secondary intergradation thousands of years after a secondary contact.

Several populations of *Fundulus heteroclitus* were analyzed by studying mtDNA fragments obtained by digestion with each of 17 restriction endonucleases (Gonzalez-Villasenor and Powers, 1990). Analysis of the mtDNA restriction fragment data indicated an intergrade zone at or near 41° N latitude. This conclusion was based on the fact that the mtDNA restriction patterns of fish at specific localities could be interrelated by a network of single-nucleotide base changes. However, populations on each side of 41° N latitude required many nucleotide changes. Previously, Cashon *et al.* (1981) had suggested that the last glacial event (approximately 20,000 years ago) might have helped shape the present allelic isozyme clines, because several showed sharp gene-frequency changes near the Hudson River, which are associated with the edge of the last major glacial advance. Although the mtDNA data of Gonzalez-Villasenor and Powers (1990) showed a sharp disjunction consistent with that hypothesis, the nucleotide differences between the "northern" and "southern" mtDNA haplotype assemblages suggested a divergence time closer to 1 million

years ago—inconsistent with the last glacial advance. Because the Chesapeake and Delaware bays were only rivers during the last glacial advance, Smith *et al.* (submitted) examined the mtDNA haplotypes of approximately 700 individual *Fundulus* from 20 populations within these bays to determine if remnant "northern" mtDNA haplotypes could be detected. Not only were the "northern" mtDNA haplotypes detected, but they were distributed in a clinal fashion up the bays and tributaries. The "northern" mtDNA haplotypes were common in the head waters of the bays and tributaries, while the "southern" mtDNA haplotypes dominated in the lower bays and rivers. The data of Smith *et al.* (submitted) clearly indicates that the intergrade zone has been temporally unstable for many thousands of years and has previously existed at least as far south as the mouth of the Chesapeake Bay prior to the last glaciation. A Wagner parsimony network and bootstrap analysis identified some mtDNA haplotypes that were transitional between the "northern" and "southern" forms. Although these intermediate forms were found in the Chesapeake Bay, they were predominantly found in rivers and estuaries several hundred miles south of the present-day mouth of the Chesapeake Bay. The presence of these apparent intermediate mtDNA haplotypes suggests that the intergrade zone has existed much further south than the current mouth of the Chesapeake Bay (perhaps as far south as Florida) and that this species may have retreated and advanced during more than one glacial cycle.

Reeb and Avise (1990) used mtDNA RFLP analysis to dissect an interesting genetic discontinuity in a species of commercially important oyster, *Crassostrea virginica.* They showed that the mtDNA clones could be grouped into two distinct genetic arrays that characterized oysters collected north versus south of a region on the Atlantic midcoast of Florida. However, unlike the case of *Fundulus heteroclitus*, the oyster allelic isozyme data showed no such genetic discontinuity. Reeb and Avise (1990) concluded that genetic discontinuities in continuously distributed species with high dispersal capability, as in oysters, are probably the result of various historical events coupled with environmental influences, such as ocean currents affecting gene flow. In addition, Avise and his colleagues have used mtDNA RFLP analyses to study genotypic diversity in a variety of other marine and freshwater organisms, including eels, green sea turtles, menhaden fish, bowfin fish, sunfish, horseshoe crabs, oyster toadfish, and catfish (Bermingham and Avise, 1986; Saunders *et al.*, 1986; Avise and Shapiro, 1986; Avise *et al.*, 1987, 1990; Avise, 1986, 1987, 1989,

1990; Bowen *et al.*, 1989; Bowen and Avise, 1990; Meylan *et al.*, 1990).

There have been many attempts to identify discrete populations of striped bass, *Morone saxatilis*. Some studies have focused on morphological features (e.g., Setzler *et al.*, 1980), and others have used allelic isozyme variants (Morgan *et al.*, 1973; Grove *et al.*, 1976; Sidell *et al.*, 1980; Rogier *et al.*, 1985). Although these studies have been able to delineate populations from the Hudson River, Chesapeake Bay, and Albemarle Sound, attempts to discriminate between spawning grounds of the Chesapeake Bay have been less compelling and have tended to generate opposing conclusions. A major problem with these studies was that striped bass are among the most homozygous vertebrates known and thus limit one's ability to discriminate between stocks. Chapman (1987, 1989) and Chapman and Brown (1990) have used mtDNA RFLP studies to resolve this problem. Although they found little variation between individuals or populations in the restriction sites, there were significant differences in the RFLPs as a result of variations in the size of the mtDNA. The mtDNA size-variation studies provided evidence for discrete stocks of striped bass along the East coast, within the Chesapeake Bay, and along the Gulf coast. It is noteworthy that these mtDNA studies within the Chesapeake Bay and between the bay and the Dan River in North Carolina are consistent with previous morphological and some isozyme studies.

Saunders *et al.* (1986) used mtDNA analysis to show that the horseshoe crab, *Limulus polyphemus*, could be separated into at least two different groups and that the area of divergence was associated with a zoogeographic boundary between warm-temperate and tropical marine faunas. Thomas *et al.* (1986) used mtDNA to study intra- and interspecific variation for rainbow trout and five species of salmon. Although the small sample size limited the intraspecies

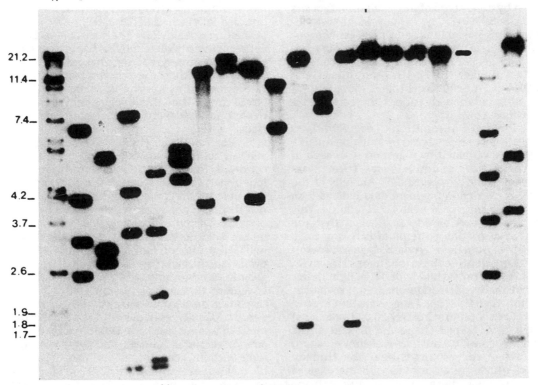

Figure 2. Autoradiograph of a single fish's mtDNA electromorph pattern generated by digestion with each of 18 restriction endonucleases. The visualization of the mtDNA fragments was attained by Southern analysis followed by hybridization with the complementary DNA of the radiolabeled probe. Bacteriophage lambda fragments produced by digestion with *Hin*dIII (A) and *Eco*RI + *Cla*I (B) were used as markers. Restriction endonucleases: (1) *Eco*RI; (2) *Hin*dIII; (3) *Bcl*I; (4) *Ava*I; (5) *Hin*cII; (6) *Bst*EII; (7) *Kpn*I; (8) *Sma*I; (9) *Xho*I; (10) *Sst*I; (11) *Bgl*I; (12) *Pvu*II; (13) *Pst*I; (14) *Bam*HI; (15) *Bgl*II; (16) *Cla*I; (17) *Sal*I; (18) *Xba*I. (This figure taken from Gonzalez-Villasenor *et al.*, [1986] with permission.)

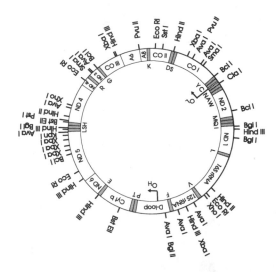

Figure 3. Restriction map of mtDNA of *Fundulus heteroclitus*. The map positions of the restriction sites were determined by a time-course digestion of ^{32}P-end-labeled mtDNA followed by electrophoresis, autoradiography, and determination of the fragment sizes. The various restriction sites identified on the map are common to the vast majority of individuals in populations from as far north as the coastal waters off Maine and as far south as the coastal waters of Florida. There are also a number of sites that are unique to each locality (see Gonzalez-Villasenor and Powers, 1990). The lines of the outer and inner circle represent the light (L) and heavy (H) strands of the mtDNA, respectively. The origins of replication for the light and heavy strands are identified as O_L and O_H. The various coding regions of the mtDNA are identified between the two strands by a series of symbols. For example, cytochrome b and the cytochrome oxidase subunits I, II, and III are identified as Cyt b, CO I, CO II, and CO III, respectively. The small and large subunits of the ribosomal RNAs are denoted as 12S rRNA and 16S rRNA. The transfer RNAs are shown as stippled areas with the standard single-letter amino acid symbols identifying their functional specificity. The symbols A6 and A8 identify the coding regions for the subunits 6 and 8 of the mitochondrial ATPase complex. The regions with ND before a number represent unidentified open reading frames for proteins that have not been identified.

studies, interesting interspecific divergences were observed. Similar intra- and interspecific mtDNA studies have been performed on three trout species. Wilson *et al.* (1985) demonstrated significant divergence between all populations and the expected evolutionary divergence between trout species.

Kornfield and Bogdanowicz (1987) studied the mtDNA of Atlantic herring, *Clupea*

harengus, from the Gulf of Maine and the Gulf of St. Lawrence. They concluded that their data did not support the notion that these were separate genetic stocks.

The Atlantic salmon (*Salmo salar*) fishery is a classic example of a mixed fishery problem. These fish, which spawn in both Europe and North America, comprise a major fishery off the coast of Greenland during the summer. In order to properly manage this fishery, it is important to assess the relative European and North American contributions to the fishery. Although isozyme and meristic techniques provided insight concerning this issue (e.g., Reddin, 1986; Stahl *et al.*, 1987), the use of mtDNA analysis has proven extremely valuable and is unequaled in its ability to distinguish between European and North American stocks (reviewed by Bermingham, 1990).

Mitochondrial DNA analysis has been used to examine the genetic structure and interrelationships among anadromous and landlocked populations of the rainbow smelt, *Osmerus mordax* (Baby *et al.,* 1991). Significant geographical variability was found between anadromous and landlocked populations. Of the two major anadromous mtDNA haplotypes, only one was represented in the landlocked populations. In addition, St. Lawrence smelt were found to be genetically distinct from adjacent populations, which supported the hypothesis that the genetic architecture of populations reflect the number of larval retention zones rather than the spawning sites.

Because many species of fish are cultured and heavily stocked in rivers, the ability to discriminate between hatchery-raised and wild populations is important in assessing the impact of introduced fry on wild stocks. Danzman *et al.* (1991) have used mtDNA to discriminate between wild and hatchery-reared populations of brook char, *Salvelinus fontinalis,* in Ontario using mtDNA analysis. Their results open the possibility of developing a better management tool for introduced, hatchery-raised stocks of fish. Similar approaches are being attempted in other areas of Canada, the United States, and other countries. Ferguson *et al.* (1991) have used the mtDNA approach developed by Danzman and his colleagues (Danzman *et al.,* 1991) combined with allelic isozyme techniques to differentiate natural populations of brook char from Cape Race and Newfoundland. Partti-Pellinen *et al.* (1991) have used a similar approach to study mtDNA variation between *Salvelinus* species in Finland, including Arctic char (*Salvelinus alpinus*), brook char (*Salvelinus fontinalis*), and lake char (*Salvelinus namaycush*). Mitochondrial DNA has also been used to study the reproductive success of transplanted fish stocks. For example, Pastene *et al.* (1991) used a combination

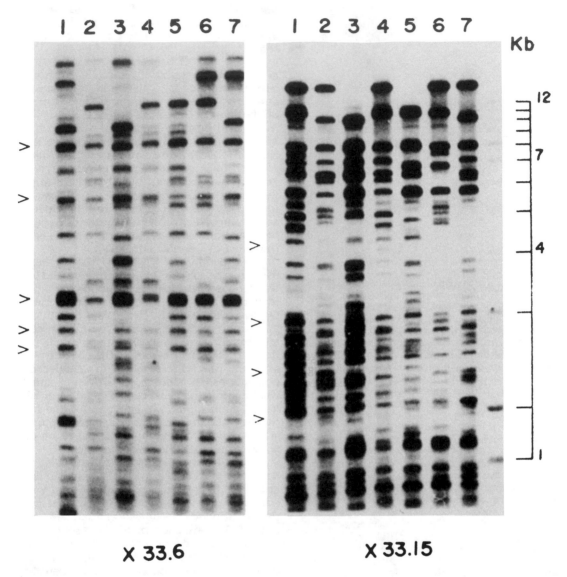

Figure 4. DNA fingerprints of fragment patterns of sibling largemouth bass that were Southern-blotted and hybridized to human minisatellite probes 33.6 and 33.15. The arrows in the figure indicate fragments common to all sibs. (This figure taken from Whitmore et al., 1990, with permission.)

of mtDNA and isozyme markers to examine the reproductive success of transplanted stocks of an amphidromous fish, *Plecoglossus altivelis*, in a heavily stocked Japanese river. They were able to discriminate between wild and introduced populations and assess the contribution of each to the fishery.

In addition to the analysis of RFLPs of mtDNA, the use of DNA sequence analysis, including mtDNA, is being increasingly em-

ployed to study the phylogenetic relationships among marine taxa. Although this approach has been slow and rather tedious in the past, the recent introduction of the polymerase chain reaction method (see below and review by Marx, 1988) to increase the amount of a specific piece of DNA promises to dramatically increase the applicability of this approach in the near future.

Restriction fragment length polymorphism analyses have also been used to study marine algae and macrophytes. Chesnick *et al.* (1991) used mtDNA RFLP analysis to assess cryptomonad phylogenetics. They used mtDNA from six freshwater cryptophytes to determine the evolutionary associations between host cells without interference from their photosynthetic endosymbionts. Because *Cryptophytes* have mtDNA genomes that are relatively small (35 to 70 kilobases) compared to terrestrial plants, their mtDNA genomes are ideal for both species and population studies.

The chloroplast genome organization of a chromophytic alga, *Ochromonas danica,* has been studied by Li and Cattolico (1992). They found that the chloroplast DNA (cpDNA) is about 121 kilobases and has an inverted 15 kilobase (Kb) duplication that separates the small (20 Kb) and large (71 Kb) single-copy regions. Eleven chloroplast-specific genes have been mapped onto the circular cpDNA genome.

Fain *et al.* (1988) used cpDNA analysis to detail the genetics and evolution of kelps. Goff and Coleman (1988) examined similar questions in the red algae. In addition to analysis of RFLPs of mtDNA and cpDNA, DNA sequence analysis is being employed increasingly to study phylogenetic relationships among marine taxa. Although the methods of DNA sequencing have been very slow and rather tedious, automation and the polymerase chain reaction (see below) offer the possibility of studying very small marine organisms and tissue biopsies from living organisms of large size, as well as museum specimens captured and processed years earlier for display.

Genomic DNA and mtDNA "fingerprinting"

Individual-specific "fingerprints" of human DNA have been successfully employed for parenthood verification. Tandem-repetitive regions of DNA, called minisatellites, are dispersed throughout the genome of a number of organisms. Jeffreys *et al.* (1985) showed that a subset of human minisatellites shared a common 10–15 base-pair core that had hypervariable regions. Later, they demonstrated that a hybridization probe could detect highly polymorphic minisatellites that could be used as DNA fingerprints specific to an individual. DNA fingerprinting is now commonplace in biomedical research and is routinely employed in a variety of legal situations. This methodology has been extensively applied to the analysis of genomic and mitochondrial DNA of a variety of marine organisms, including corals, sea urchins, sea turtles, whales, and other marine mammals (e.g., Britten *et al.*, 1978; Bowen *et al.*, 1989; Baker *et al.*, 1990; McMillan and Miller, 1990; Palumbi and Wilson, 1990).

In addition, the same general approach has proved useful when applied to marine and freshwater fishes (e.g., Avise *et al.*, 1989; Fields *et al.*, 1989; Whitmore *et al.*, 1990). For example, Whitmore *et al.* (1990) used DNA fingerprinting (Figure 4) to study sibling largemouth bass (*M. salmoides*). They showed that Southern blots of bass DNA hybridized to human minisatellite DNA probes yielded patterns that were different for each individual (Figure 4) but that sibs were more similar than fish from wild stocks.

Wirgin and Maceda (1991) developed striped bass–specific DNA probes, 10 to 20 base pairs in length, which they used in combination with general DNA fingerprinting and mtDNA probes to study the genetic structure of striped bass populations. One of the probes they developed allowed them to distinguish between Gulf of Mexico and Atlantic coast striped bass and among striped bass representative of several of the Atlantic coast breeding stocks. Turner and his colleagues (Turner *et al.*, 1991) have begun using repetitive DNA sequences to study the population structure of several fresh- and saltwater fish species. Taggart *et al.* (1991) isolated hypervariable single-locus probes from an Atlantic salmon (*Salmo salar*), genomic library. These probes, and some of those developed for human DNA fingerprinting, are being used to study populations of salmon from Irish rivers. Taggart *et al.* (1991) presented this work at a Fisheries Society of British Isles (FSBI) symposium on fish genetics and taxonomy held in Belfast, Ireland (July 22–26, 1991). At that symposium, several investigators presented papers that used DNA fingerprinting and sequencing techniques to address population and conservation problems involving both cultured and wild stocks of marine and freshwater fish. For example, Goodier and Davidson (1991) presented a paper that identified a 450-base pair tandemly repeated component in the genome of Atlantic salmon, which they cloned and sequenced. This element contained several recognition sites for restriction enzymes that will make them useful for the analysis of population structure. Bentzen and Wright (1991) reported that they had cloned and sequenced a variety of highly polymorphic salmon minisatellites of 9 to 65 base pairs in length and microsatellites that were 4 bases or less. Many other investigators reported the development of clones for other fish species, and, still others, the use of mammalian DNA probes in DNA fingerprinting of a variety of marine and freshwater species. Clearly, the use of the fish clones reported at the FSBI symposium and others currently being developed in laboratories throughout the world should provide powerful tools for analyzing the genetic architecture of fish populations and provide critical information for the effective management and conservation of this valuable natural resource.

DNA amplification employing the polymerase chain reaction (PCR)

Although analysis of mtDNA, cpDNA, genomic DNA, and other such methods offer very powerful approaches for addressing fundamental problems in the marine sciences, they all require an adequate amount of DNA for analysis. When very small organisms (microscopic) or only a few cells are available, obtaining enough DNA can be a limitation. Under such conditions, the desired DNA is usually cloned and analyzed directly or used as a molecular probe.

Usually, when a significant amount of a specific DNA sequence length is needed, it is cloned into an active replicon, identified by some specific method, and produced in large quantities in bacteria. However, the initial cloning and identification methods can be very time consuming, especially when the target sequence is not abundant and the starting sample is a complex mixture of DNAs.

A revolutionary approach to the detection and characterization of specific DNA sequences, the polymerase chain reaction (PCR), was developed commercially by the Cetus Corporation as a simple alternative to the cloning of specific genes. An important modification of the method has expanded its usefulness (Saiki et al., 1988; Stoflet et al., 1988). The current procedure employs a thermostable bacterial DNA polymerase and specific oligonucleotide primers to replicate a target DNA sequence in vitro (Figure 5). From as little as a single molecule of the target sequence, enough material for standard analytical procedures such as restriction mapping, hybridization, or DNA sequencing is produced in about 3 hours. A commercially available microprocessor-controlled device automatically takes the reaction through multiple cycles of DNA denaturation, primer annealing, and DNA synthesis. The target sequence may be amplified from many samples in parallel through this automated process.

Problems that can be addressed by PCR include the following: (i) rapid detection and identification of microorganisms of very low frequency of occurrence, such as one or a few cells in a liter of water, or minute, individual fish eggs, or fish and invertebrate larvae; (ii) rapid analysis of individual genomes for population studies; (iii) detection and analysis of "rare events," such as gene rearrangements that occur in a small fraction of cells in a tissue sample or field collection; (iv) analysis of a rare DNA sequence in a complex sample mixture, such as in total DNA extracted from a mixed phytoplankton or zooplankton community; (v) examination of symbiotic and parasitic relationships; and (vi) estimation of water quality for public health purposes by detection of pathogenic viruses, bacteria, and parasites. The application of PCR to terrestrial and aquatic organisms and has been reviewed by Arnheim et al. (1990). They provide many examples of the use of PCR technology in biosystematics, population biology, conservation biology, ecology, developmental biology, and experimental genetics.

Recently, Rowan and Powers (1991a; 1991b) used PCR amplification and DNA sequencing to clarify a long-standing taxonomic problem by establishing the systematic relationships between a variety of zooxanthellae endosymbionts and free-living dinoflagellates. The procedure they developed used symbiont-specific primers, which, in turn, allowed the amplification of tiny amounts of the symbiont DNA in the presence of the animal host DNA without the amplification of the latter. This type of specificity is one of the great strengths of the PCR procedure.

Highly conserved regions of mtDNA—such as 12S ribosomal RNA (rRNA), 16S rRNA, cytochrome oxidase, cytochrome b, and other loci—are routinely being amplified via PCR to generate fragments of mtDNA that are studied for population and species variation by RFLP and/or analysis of the DNA sequences.

For example, Palumbi and Benzie (1991) used PCR to amplify the cytochrome oxidase (CO I) gene encoded in the mtDNA genome of several Penaeid shrimp species and compared their evolutionary relatedness. They found greater genetic divergence than had been expected. Differences at silent sites between subgenera of shrimp were larger than between families of mammals. Because the morphology of the shrimp showed little difference between taxa, but at the molecular level genetic variation was extensive, the authors concluded that conventional morphological measurements underestimated the true extent of genetic divergence and suggested that either evolution at the molecular level was significantly accelerated or stabilizing selection over extended time periods has constrained evolutionary rates of change in morphology.

Baker et al. (1990) used mtDNA analysis to study the migration of humpback whales (Megaptera novaeangliae) in the Atlantic and Pacific oceans. These whales migrate over 10,000 km each year between summer feeding grounds in temperate or near-polar seas and winter breeding grounds in shallow tropical

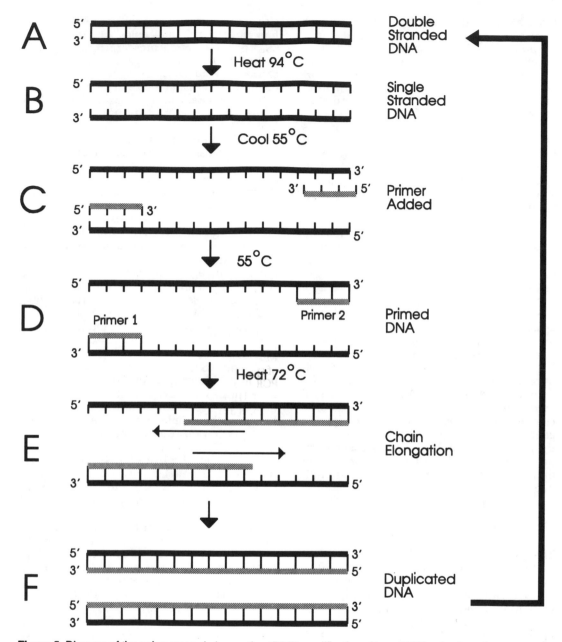

Figure 5. Diagram of the polymerase chain reaction (PCR) amplification of target DNA using homologous oligonucleotide primers. Letters A through F depict the various stages in generating a single replication of a DNA target. As each cycle is repeated, the number of duplicated DNA strands will double.

waters. The investigators used mtDNA analysis to ascertain whether the migration patterns were reflected in the genetic structure of the populations. They found significant differences between populations inhabiting the Atlantic and Pacific oceans, as well as differences between subpopulations within each ocean basin system. The migration patterns were ap-

parently a consequence of the maternally directed fidelity to migratory patterns and ultimate destinations.

Similarly, coding and noncoding regions of the genomic DNA are amplified and studied for variation within and between species. The amplification of genomic ribosomal genes are the most common target for these types of studies (see below). In fact, "universal" primers for these and other genes are used for PCR amplification of DNA from single cells of marine vertebrates and invertebrates in summer courses taught at the Marine Biological Laboratory in Woods Hole, Massachusetts, and at the Hopkins Marine Station of Stanford University in Pacific Grove, California.

Because much of the PCR procedure is automated (specifically, amplification of the target sequence), it may be possible to automate the entire procedure, including extraction of the DNA, generation of primers, and analysis of the PCR product. In fact, there is a potential to couple the PCR machine with an automated method for analysis of the PCR product.

Because PCR techniques appear to work on a variety of preserved samples and even on partially degraded samples obtained from gut contents of predators, it is possible not only to analyze fresh specimens, but also to analyze samples fixed in the field or even museum specimens collected a hundred years or more ago. After amplification, the DNA can be studied by restriction mapping, or it can be sequenced (e.g., Gyllenstein and Erlich, 1988; Higuchi and Ochman, 1989). Paabo *et al.* (1988) were able to extract and sequence mtDNA for brain tissue that was 7,000 years old. The PCR approach is virtually taking over more classical cloning approaches by enriching for specific genes via PCR amplification and then cloning the amplified product or directly characterizing it via endonuclease restriction analysis or direct sequencing of the PCR generated product.

Ribosomal RNA (rRNA) genes

The analysis of specific rRNA genes is a very useful tool for the identification and classification of both eukaryotic and prokaryotic species. It is particularly useful for microorganisms, planktonic organisms, and others for which conventional identification is either difficult or impossible. In addition, quantitative analysis of homologous sequences within and between species can provide information on the composition of microbial and planktonic communities, as well as insight concerning evolutionary relationships (e.g., Olsen *et al.*, 1986; Pace *et al.*, 1986).

Approximately 60% of the total mass of the ribosomes within living cells consists of rRNA. The rRNA can be obtained from prokaryote ribosomes as linear, single-stranded molecules. Ranging from largest to smallest, with reference to their sedimentation coefficients, three characteristic sizes of rRNA are found in prokaryotes: 23S, 16S, and 5S. Eukaryotes have larger ribosomes than prokaryotes and have four types of rRNA molecules: 28S, 18S, 7S, and 5S. The rRNA molecules are multiple-copy genes and have some regions that are highly conserved across taxa. Because they are found in all living cells and their function is presumably also conserved, it is reasoned that their sequences should provide useful taxonomic information.

With some exceptions, the 5S rRNA from most microbial species contains about 115–120 nucleotides, which can be rapidly sequenced by standard methods (Donis-Keller *et al.*, 1977; Peattie, 1979). Several hundred 5S rRNA sequences have been determined, and analysis of the data indicate that prokaryote 5S rRNA has a highly conserved structure. When variation does occur, it is not randomly distributed, that is, some positions change more freely than others. Although this method has been used successfully to investigate the phylogenetic relationships of several types of microbial communities (Stahl *et al.*, 1984; 1987), the paucity of independently varying nucleotide positions of 5S rRNA has limited its usefulness in broader phylogenetic studies.

The 16S rRNA is a more appropriate size for broader phylogenetic analysis; however, because of its larger size, the complete sequence of 16S rRNA cannot be determined easily by the simple, direct RNA-sequencing methods previously used for 5S rRNA sequencing.

Through the use of DNA cloning and nucleotide sequencing techniques, the nucleotide sequences of a large number of 16S rRNAs from a significant variety of prokaryote and eukaryote organisms have been determined. Examination of 16S rRNA sequences from prokaryotes and 18S rRNA from eukaryotes has revealed regions of constant or nearly constant nucleotide sequences across a wide variety of taxa. The oligonucleotides of these conserved regions can be synthesized and used as primers to provide a fast and direct approach for the determination of these rRNA sequences. Application of the PCR method to rRNA genes has created an explosion of sequences from a variety of marine organisms. A new sequence can be generated in a few days instead of in the weeks or months previously required.

This approach has been used to address systematic, evolutionary, and oceanographic problems involving a large number of prokaryotic and eukaryotic taxa, including bacteria, phytoplankton, dinoflagellates, macroalgae, intertidal invertebrates, teleosts, sharks, and marine mammals (e.g., Sogin *et al.*, 1986; Sogin and Gunderson, 1987; Hillis, 1987; Field *et al.*, 1988; DeLong *et al.*, 1989; Giovannoni *et al.*, 1990; Arnheim *et al.*, 1990; Baker *et al.*, 1990;

Sogin, 1990; Ward *et al.*, 1990; Palumbi and Benzie, 1991; Rowan and Powers, 1991a, 1991b, 1992; Bernardi *et al.*, 1992).

Nucleic acid hybridization and DNA sequencing

It is possible to identify single organisms, even single cells, using DNA or RNA oligonucleotide probes that target unique nucleotide sequences. This technique is particularly useful for discriminating among organisms that lack distinguishing morphological features, such as microorganisms and many larval metazoans.

A novel approach for counting and identifying species in a community containing a variety of species is the application of in situ hybridization. Radiolabeled nucleic acid probes with complementary sequences can hybridize with DNA inside fixed cells or in fixed thin sections of tissues. These probes can then be visualized by autoradiography, enzyme-linked probes or another type of reporter group. This approach is particularly attractive because it is easily applicable to field situations.

Use of fluorescently tagged 16S rRNA-targeted oligonucleotide probes for identification of species in microbial populations in seawater has been demonstrated in whole cells (DeLong *et al.*, 1989). Examination of the sample with epifluorescence microscopy allows the observer to identify and enumerate cells containing the target molecules. Similar approaches have been used to identify organisms with a specific functional capacity using immunological probes (Yentsch *et al.*, 1988; Currin *et al.*, 1990).

Genomic rRNA provides a convenient target molecule because many copies occur in all cells and because different sequence regions within the molecule have been found to vary in relation to phylogenetic specificity. Thus, analyses can be designed to discriminate among major groups of organisms (e.g., eukaryotes, Archaebacteria, and eubacteria) or to detect specific species within a genus or within a closely related group of organisms. Because cellular rRNA content is related to growth rate in prokaryotes, it is hypothesized that ultimately it should be possible to measure the growth rate of a single cell based on its relative fluorescence (DeLong *et al.*, 1989). Similar approaches are also being used to discriminate between eukaryote taxa and to identify early life-history stages.

Ideally, one would like to extract DNA from a planktonic community and selectively hybridize the community DNA with species-specific DNA probes in order to quantitatively determine the fractional composition of the various species within the community. Lee and Fuhrman (1990) have developed a relatively simple method of total community DNA hybridization that allows one to determine within reason the composition of some microbial communities. With this approach, they determined a 20% to 50% similarity between samples from Long Island Sound, the Sargasso Sea, and the Caribbean Sea, whereas microbes from a coral reef community were only 10% similar to their nonreef counterparts. Although this technique appears useful for crude determinations, it falls significantly short of the ideal, wherein one would be able to accurately determine species composition via species-specific DNA probes coupled with hybridization with DNA-isolated from a mixed microbial community. On the other hand, further refinements of the procedure and the identification of highly specific DNA probes might significantly improve the usefulness of this technique in the future, but there are a number of practical limitations on this approach that will probably cause it to fall far short of the ideal situation alluded to above.

Finally, coupling these and other probes and reporter groups with automated technology (e.g., flow cytometry, see below) will provide quantitative information in the appropriate space and time scales to allow integration with the chemical and physical measurements of other oceanographers.

Immunochemical methods

Immunochemical detection of species- or population-specific antigens (such as proteins or carbohydrates on the surface of or within organisms is very useful as a molecular probe to identify marine taxa. Polyclonal and monoclonal antibodies can often be used in food-chain studies in the identification of minute larvae, and in the analysis of population structure.

Polyclonal antibodies, which are most useful for interspecies studies, are prepared by immunizing an animal with a protein that is specific for a particular species. On the other hand, a general protein or even fragments of an organism of a particular species can be injected. Afterward, the polyclonal serum can be made more specific by absorbing out the antibodies that cross-react with other species, thereby yielding a restricted serum that would react only with the species for which it was developed.

Figure 6 illustrates the general approach for generating polyclonal antibodies against ground nauplii of one species followed by absorption of the antiserum by tissue extracts of

Polyclonal Antibody Production

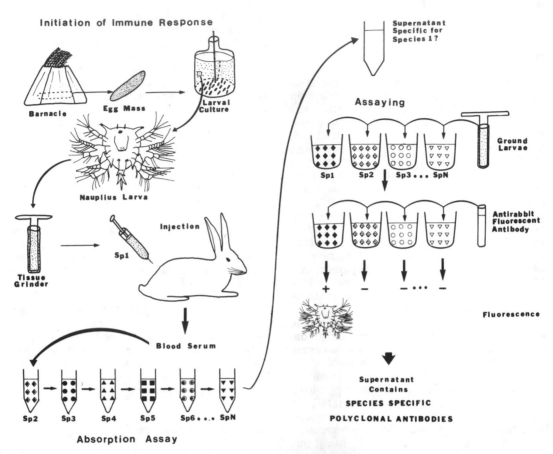

Figure 6. Schematic for the generation of polyclonal antibodies against ground nauplius larvae, with subsequent absorption to reduce cross-reactivity and increase specificity. The antisera is made more specific to species 1 (Sp1) by absorption of the antisera with the antigens of other potentially cross-reacting species (Sp2.....SpN). (Courtesy of Ms. Kristi Miller.)

several other species, thereby yielding an antiserum restricted to the original antigen. The absorbed serum of the immunized animal can be used to detect small amounts of the species in question and can be visualized by coupling the antiserum with a fluorescent tag or another identifiable label. In addition, proteins isolated from large adults can be used as antigens to generate antibodies that are useful in detecting and quantitating species-specific larvae when the antigenic determinants are shared between adults and larvae. This approach can allow the identification of minute larvae that are undetectable by less-sensitive methods. Polyclonal antibodies have been particularly useful in food-chain studies.

Immunological assays of gut contents of predators are only one of several approaches used to establish predator-prey interactions. The main advantage of dietary immunoassays is their ability to identify visually indistinguishable gut contents. For example, the incidence of unrecognized prey in the gut contents of deep-sea predators caused Feller *et al.* (1985) to use immunological techniques to determine the gut contents of North Atlantic grenadiers (*Coryphaenoides armatus*) that were taken from 2,500 meters. They used 32 polyclonal antisera, generated by injection of whole-organism extracts of species representing 10 common deep-sea phyla, to clearly demonstrate the superiority of the immunological approach over classical observational methods. Although this qualitative approach provided important insight concerning the deep-sea food web of grenadiers, the antisera were not species-specific. Feller (1986) later used a similar approach to study the natural predators of the hard clam, *Mercenaria mercenaria*, by generating whole-organism antisera and assaying the gut contents of its invertebrate predators. Feller (1986) used immunoelectrophoretic methods that revealed unique antigens for several life-history stages of the clam, including veligers, newly settled spat, juveniles, and adults. Grisley and Boyle (1988) used immunochemical assays to study the gut contents of *Octopus vulgaris* and the hepatopancreas of shrimp. They showed that the amount of time after ingestion affected the antigenicity of the prey. Hunter and Feller (1987) showed that whole-organism extracts often cross-reacted with antigens from related taxa and required some method to reduce cross-reactivity. Gallagher *et al.* (1988) developed monospecific antisera to soft-bottom benthic taxa so as to increase the quality of information obtained by this approach. Whereas the initial applications of predator-prey immunoassays were qualitative and not highly specific in nature, recent applications have employed more quantitative approaches (Feller *et al.*, 1990; Hentschel and Feller, 1990). Combining the absorption method alluded to above, coupled with the assessment of predator-prey relationships, is now the mainstream approach in such studies.

Ward and Perry (1980) were the first to use immunological methods to identify aquatic nitrifying bacteria, and Campbell *et al.* (1983) used a similar approach to identify serotypes of cyanobacteria. In both cases, the method was also used to map the distribution of these organisms in water masses over broad geographical regions. Since these early studies, numerous other studies have taken this approach on a host of marine bacteria, phytoplankton, and zooplankton (Brock, 1985; Theilacker *et al.*, 1986; Currin *et al.*, 1990; also see collected papers in Yentsch *et al.*, 1988). In addition, coupling of immunochemical methods with automated instrumentation has yielded methods for the rapid detection of the spatial and temporal variability of planktonic species (see Automation of Molecular Methods, below). Moreover, the development of antibodies against key enzymes, such as ribulose-1,5-bisphosphate carboxylase and peridinin/chlorophyll a, has opened the possibility of measuring primary production and other metabolically important processes in exciting new ways (see review by Ward, 1990).

Because polyclonal antibodies are usually of inadequate specificity for population studies and require preabsorption with all possible cross-reacting antigens, monoclonal antibodies are often preferred. Preparation of a series of specific monoclonal antibodies generally involves the following steps: (i) several mice are immunized with desired antigens (purified or partially purified), and spleen cells are prepared from these mice several weeks after immunization; (ii) the spleen cells are fused to mouse myeloma cells, and the hybrid cells are selected and propagated on hypoxanthine=aminopterin=thymidine (HAT) selective medium; (iii) monoclonal antibody-producing hybridoma clones are screened by an immunobinding assay, using purified antigen as a probe. This general procedure is illustrated in Figure 7. Each monoclonal antibody is then characterized extensively in order to determine its specificity for the respective antigen.

Whereas the antibody provides the taxonomic specificity, detection is usually provided by either a direct or indirect coupling to one of the numerous reporter systems. Sometimes a radiolabeled antibody (e.g., ^{125}I-IgG) is used as the reporter. However, some investigators prefer to use a fluorescent label, an enzyme-linked assay (e.g., with alkaline phosphatase), or another method to detect the specificity of the reaction. Whatever the method, this approach is potentially one of the most powerful and practical for application to field studies and particularly for the identification of phytoplankton, zooplankton, eggs, larvae, microbes, and parasites.

Recently, Miller *et al.* (1991) produced species-specific monoclonal antibodies that distinguish planktonic larvae of intertidal barnacles. One antibody distinguished two morphologically identical larval species of the genus *Chthamalus* (*Chthamalus dalli* and *Chthamalus fissus*), but it cross-reacted with another barnacle species (*Pollicipes polymerus*). However, a separate species-specific monoclonal antibody reacted only with *Pollicipes polymerus*, so it was possible to distinguish between the three species by using a combination of the two antibodies, in series. With these antibodies, a series of fluorochrome-tagged antibody assays can be used to distinguish between these three barnacle species. This system could be done manually by fluorescence microscopy, or it could be coupled to an automated system via flow cytometry (see later section on flow cytometry). The paper by Miller *et al.* (1991) provides a road map to the development of similar detection systems for any number of zooplankton problems. If automated (e.g., via flow cytometry), these methods open the possibility for study of the movement of larvae in relation to oceanic currents, which would provide important insight into the mechanisms that regulate population density and composition of zooplankton communities and the genetic architecture of their populations.

Monoclonal Antibody Production

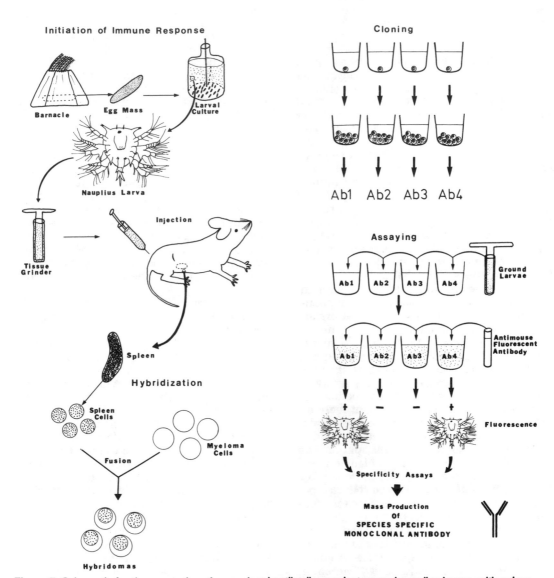

Figure 7. Schematic for the generation of monoclonal antibodies against ground nauplius larvae, with subsequent screening procedures. The antibodies produced by the cloning of single cells (e.g. cells 1,2,3 and 4) produce antibodies 1,2,3 and 4 respectively which are symbolized as Ab1, Ab2, Ab3, and Ab4. (Courtesy of Ms. Kristi Miller.)

Assessment of Physiological Status

Variation in physiological condition (e.g., nutritional, immunological, developmental, reproductive) in relation to environmental parameters (e.g., food, temperature, oxygen, dissolved nutrients, and light) is believed to account for much of the variability in population dynamics and community composition in marine food webs. Although current molecular methods can be readily adapted to identify genetic variation within and between biological assemblages, assessment of physiological status and functional diversity of individual organisms within these assemblages is not easily accomplished. Development of appropriate diagnostic tests requires considerable fundamental knowledge about the basic biology of the target organisms. Many physiological features of marine organisms are a product of their environment. As such, an understanding of how environmental parameters influence the basic biochemistry, physiology, and molecular biology of these organisms is necessary for the development and exploitation of molecular technologies for diagnostic markers of physiological state.

The responsiveness of gene expression to environmental factors, which is well documented in some terrestrial and model organisms, is poorly understood in most marine organisms, and particularly in those of major ecological and economic importance. Understanding the molecular details of these regulatory processes could allow the development of oligonucleotide, immunological, or other molecular probes to assess responses quantitatively and to achieve an understanding of the temporal scales of change. Clearly, obtaining extensive, basic biological and biochemical information on key indicator organisms, and on organisms that are critical links in oceanic and coastal food webs, is important for dissecting the physiological mechanisms governing biological variability in the oceans. Were this information available, it would also provide fundamental insight into the mechanisms of speciation and evolution in response to changing environmental conditions and, as such, provide a framework for understanding the impact of global climate change on the marine biota. In addition, such data will be essential for the construction of numerical models of predictive and management value. In fact, the general dearth of such information for the vast majority of marine organisms is the major impediment for developing appropriate techniques to assess physiological status and relating physiology to biogeochemical cycles and other dynamic ocean processes.

Techniques that would assess growth and reproductive and nutritional status, and signal a response to environmental stress would be particularly useful. At the present, very few attempts have been made to develop these molecular techniques for marine organisms. We shall address a very few of these possibilities below (most of which have yet to be proven useful for field studies). This area of research is in its infancy and there is a tremendous need for fundamental research on this topic. The potential applicability of molecular techniques to detect the physiological status of marine organisms is almost without limits.

Nutritional status

There are a number of variables that are used to determine nutritional status. Length to weight, length to width, hepato-somatic index, color, relative parasite infestation, gut contents, and other measurements are routinely used to assess relative nutritional status. In addition, the types and concentration of digestive enzymes, in conjunction with gut contents, have been used as crude indicators of food preference, diet, and general nutritional status (e.g., Patton *et al.*, 1975; Sargent *et al.*, 1979; Seiderer *et al.*, 1987).

In attempts to understand phytoplankton-zooplankton food webs, a number of useful laboratory studies have been done, but adequate field studies have been rare. Although field studies on phytoplankton allow one to rapidly measure carbon fixation and other quantitative parameters to assess physiology and growth, such studies on zooplankton have been so variable that they are of questionable use. The quantitation of digestive enzymes (Cox *et al.*, 1983) has been used to determine long-term changes in copepod feeding. These methods, coupled with gut fluorescence (Mackas and Bohrer, 1976; Dagg and Wyman, 1983) and conventional methods, have sometimes proven useful. However, these and related methods are very time consuming, and the results are often affected by confinement. Hakanson (1984) developed a rapid and accurate method that relies on lipid analysis in copepods (*Calanus pacificus*) as an index of feeding conditions. He found that wax ester content was an excellent index of feeding over a several-day period, and triglyceride content was a good index of recent feeding activity. A recent paper (Hakanson, 1987) showed that, using this approach, it is possible to relate lipid content (presumably feeding conditions) of copepods to field phytoplankton pigment.

Salmonids subjected to long periods of starvation initially catabolize nonessential digestive enzymes (Moon, 1983; Loughna and Goldspink, 1984), after which lipid reserves are mobilized (Love, 1980; Mommsen *et al.*, 1980). During starvation, increased concentrations of lipolytic enzymes are commonly found in a variety of fish species (Zammit and Newsholme, 1979; Black and Skinner, 1986). Although growth hormone activity increases lipolytic enzyme activity, it was still surprising when Wagner and McKeown (1986) found that growth hormone levels of starved fish were greater than those of fed controls.

Therefore, decreases in digestive enzymes reflect early phases of starvation; and increases in lipolytic enzymes, and perhaps growth hormones, reflect longer-term starvation. It is interesting that Barrett and McKeown (1988) have shown that swimming amplifies this latter effect in starved fish, such that growth hormone levels are almost four times greater than in the exercised fed controls. It follows that measurements of these enzymes and the determination of plasma growth hormone levels might be useful indicators of a fish's nutritional status as long as proper standards and controls were available. However, it is important to point out that growth hormone levels alone would be very misleading because enhanced levels of this and other hormones are also associated with differential development and enhanced growth of well-fed organisms. Moreover, growth hormone and prolactin levels are also affected by life stage and exposure of some species to changes in environmental salinity. In addition, starvation in fish is accompanied by a dramatic reduction in protein synthesis, with the white muscle tissue being the most sensitive, while liver and gills are essentially unaffected. Although a suite of indicators might help define various nutritional states, there needs to be much more research to define the interrelationships between the various parameters before they become standards for fieldwork. The coupling of conventional methods for assessing nutritional status with the biochemical indicators described below (digestive enzymes, lipolytic enzymes, RNA/DNA ratios, RNA/protein ratios, and perhaps hormones and other indicators) might be useful, but the details and applicability of such an approach is yet to be defined and adequately documented.

Growth status

Growth rate studies are normally done in the laboratory by regulating food intake and measuring size and/or weight on a periodic basis. However, these procedures are time consuming and are not practical for field studies. In recent years, a number of attempts have been made to develop simple biochemical techniques to assess growth that could be applied to field samples. RNA/DNA ratios, RNA content per individual, tissue-specific RNA concentration, and RNA/protein ratios have all been used as indicators of growth, metabolism, and physiological condition. Although these techniques have limitations, they are extremely useful if proper controls and standards are employed. For example, the most popular technique, RNA/DNA ratios, is useful when it is restricted by species, size, life-history stage, environmental temperature, and physical activity of the individuals being tested (reviewed by Bulow, 1987). Thus, although these techniques have limited potential for interspecies application, they can be very useful if limited to intraspecies studies of similar life-history stages.

When laboratory growth rate studies are done on fish with different amounts of growth hormone, growth rates are easily differentiated (Sekine *et al.*, 1985; Agellon *et al.*, 1987). Recently, it has been shown that increased growth rate stimulated by increased growth hormone is directly related to increased RNA/DNA ratio, testosterone, and other variables (Danzmann, *et al.*, 1990; R.G. Danzman, per. com.). However, those studies were done under a defined feeding regime and at constant environmental conditions (e.g., temperature). Houlihan and his colleagues (1988, 1989) have shown that an increase in protein ration results in a linear increase in protein synthesis and protein degradation in cod, whereas in starving fish there is a constant level of protein synthesis and a 60% recycling of proteins. The obvious questions are: Can one estimate the relative growth rate or growth potential of an organism sampled from the field by quantitating protein synthesis versus degradation, RNA/DNA ratio, RNA/protein ratio, hormone levels, and/or other parameters? If so, which tissues should be used, and which life-history stages are the best indicators? Can those data be related to nutritional status? To date, these questions remain unanswered. Clearly, research to explore the potential of these and other molecular techniques to assess growth and nutrition status might make recruitment studies much more definitive, but, at present, the basic biology is simply not available.

Reproductive status

Classically, gametogenic capacity is estimated by counting changes in gamete numbers or gonad weight (i.e., gonadosomatic index). Although these are useful for estimating reproductive capacity, gamete counts are excessively labor intensive and gonad weight indices are often misleading. Moreover, such approaches are limited to a few taxa. A possible alternate method for detecting reproductive status might be the assessment of specific reproductive genes, such as the gene for vitellogenin in vertebrate females.

During oogenesis in fish, the egg-yolk precursor protein (vitellogenin) is synthesized in the liver, secreted into the vascular system, and then deposited in the developing oocytes as lipovitellin and phosvitin (Chen, 1983). Therefore, the reproductive status of female fish in a population might be determined by measuring the levels of vitellogenin in the serum, the rate of vitellogenin synthesis in the liver, or the accumulation of vitellogenin mRNA in the liver. Levels of vitellogenin in serum samples can be easily determined quantitatively by rocket immunoelectrophoresis. This method can detect vitellogenin levels as low as 0.05 mg/ml in serum and can give reliable quantitation of vitellogenin in both hormone-induced and reproductively active female fish. The rates of vitellogenin synthesis in livers of reproductively active females can be determined by either a radioimmunoprecipitation method (Chen, 1983) or by RNA-DNA hybridization. It should be emphasized that this approach is only a possibility that needs exploration and appropriate documentation for key model species.

Stress assessment

When faced with environmental stress, organisms must adapt in order to maintain homeostasis. As a result, animals and plants have devised a host of molecular, physiological, and behavioral strategies in order to accomplish this task. Because aquatic poikilotherms usually respond directly to environmental parameters such as temperature, pH, metals, salinity, and pollutants, they have been excellent models for the study of adaptation to environmental stress.

In all organisms studied, acute exposure to high-temperature stress results in the rapid activation of a specific family of genes and the efficient and preferential synthesis of a novel set of gene products referred to as "heat-shock" proteins (hsps) (Atkinson and Walden, 1985). A role for hsps in thermal resistance adaptation stems from the observation that induction of hsps will allow cells to transiently tolerate subsequent exposure to temperatures that would be lethal in the absence of hsp synthesis and accumulation (Atkinson and Walden, 1985). There are positive correlations between the temperatures at which hsps are maximally induced, the upper lethal temperature of an organism, and the range of temperatures normally encountered in the habitat of the

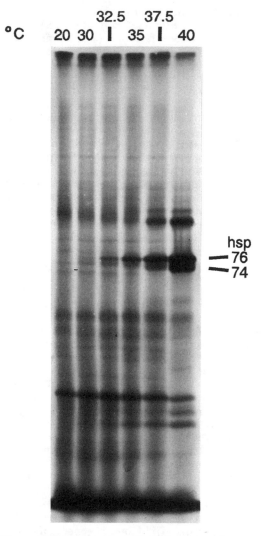

Figure 8. Erythrocytes from the blood of the teleost, *Fundulus heteroclitus,* were incubated at each temperature for 1 hour, followed by a 2-hour labeling period. Labeled proteins were resolved by SDS-polyacrylamide gel electrophoresis and visualized by fluorography. (From Koban *et al.,* 1991.)

organism. For example, salmonid fishes are stenothermal and are adapted to relatively cold temperatures. Cell lines derived from these fish maximally synthesize hsps at temperatures below 30° C (Kothary and Candido, 1982; Gedamu *et al.,* 1983). Hepatocytes of a eurythermal and warm-temperature adapted species, such as channel catfish, will maximally synthesize hsps at temperatures approaching 40° C (e.g., Koban *et al.,* 1987). A variety of tissues of the eurythermal teleost *Fundulus heteroclitus* demonstrates hsp synthesis consistent with the upper temperature of its thermal environment, but with regulatory responses that varied between tissues (e.g., Koban *et al.*, 1991). For example, *Fundulus* erythrocytes actively synthesize hsp 74 and 76 at 40° C (Figure 8), however, liver tissue maximizes hsp 76 synthesis at 37.5° C but turns off synthesis at 40° C (Figure 9).

The molecular basis by which hsps provide thermal stress resistance is not known, but the universal expression of hsps and their conserved structures across various taxa from bacteria to humans suggest that these proteins must be fundamental in the maintenance of cellular homeostasis. In addition to heat stress, a variety of other perturbations of homeostasis—such as heavy metal exposure, viral infections, synthesis of defective proteins, and oxygen radical formation—will activate hsp genes (Schlesinger *et al.*, 1982; Atkinson and Walden, 1985).

Because increases in hsp synthesis are a good indication of physiological stress, it seems reasonable to assume that research focused on quantitative estimates of hsp in target species would be a useful indicator for environmental stress in recruitment studies. Several molecular techniques are available that can be applied to that end, including hsp immunoblotting, immunoprecipitation, rocket immunoelectrophoresis, RNA blotting, and Northern analysis (i.e., electrophoresis followed by RNA-DNA hybridization).

It is important to point out that the strong evolutionary conservation of the hsp 70 gene complex has allowed the use of antibodies and complementary DNA (cDNA) probes from diverse taxa (e.g., fruit flies, fish, and man) to be used to probe proteins and RNA across taxonomic groups. Therefore, a general method developed for one species should have broad applicability to the detection of these stress proteins in other species. However, appropriate standards and protocols must be developed for each taxonomic group.

Isozyme, mtDNA, cpDNA, immunological, and other methods have already proven their usefulness in detecting genetic variation within and between species. The challenge for the future is to implement these methods and expand their usefulness. An even greater challenge, however, is to develop techniques to quantitatively assess the physiological (e.g., nutritional, growth, reproductive, stress) status of marine organisms under field conditions.

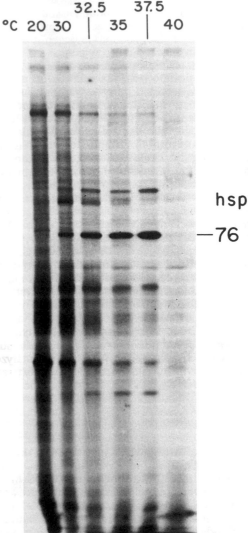

Figure 9. Temperature profile of hsp synthesis by the liver of the teleost *Fundulus heteroclitus*. Liver slices were incubated with L(^{35}S)= methionine at 20° C (control) and heat stress/shock temperatures of 30°, 32.5°, 35°, 37.5°, and 40° C. Labeled proteins were resolved by SDS-polyacrylamide gel electrophoresis and visualized by fluorography. Hsp 76 is identified at the right of the figure. (From Koban *et al.*, 1991.)

Automation of Molecular Methods

Although precise and highly accurate, the use of molecular techniques to identify marine taxa and physiological status can be as tedious or more time consuming than conventional approaches. Therefore, molecular methods need to be automated to facilitate the rapid identification of marine taxa and the quantitation of the physiological, nutritional, reproductive, and immunological status of marine organisms. This combination of automated technology with molecular and cellular methods has begun to be applied to marine microorganisms. In principle, however, an automated approach could be coupled to any molecular probe and to a large variety of methodologies.

Clearly, automated diagnostic methods developed to identify microorganisms in humans can be adapted for the identification of microorganisms or larger organisms in samples of ocean water. Similarly, methods developed to detect low concentrations of hormones or other biomolecules in human blood samples (for example, in order to assess the physiological status of patients) can be adapted to assess the physiological status of finfish or shellfish.

A number of automated techniques and procedures that are routinely employed in the biomedical field should be readily available for transfer into the marine sciences. For example, there are automated two-dimensional gel methods with computerized image analysis that could be used to assess the expression of stress proteins (Koban *et al.*, 1991). Although DNA mapping and sequencing has been used to study genes from a few marine organisms, the automatic procedures employed in the human genome project have never been applied to critical marine issues. Similarly, computational tools routinely used by biomedical researchers to analyze DNA and protein sequences are only rarely used by marine scientists. The use of automatic confocal microscopes and other image-processing tools is rare among the marine sciences community, and automated molecular recognition elements are essentially never used. The only automated cellular and molecular instrument that has been embraced by a select group of biological oceanographers is the use of the flow cytometer, which has been used to identify and quantitate microorganisms. Rather than enumerate the virtues of the many automated procedures that could be used by the marine sciences community, I will elaborate on the flow cytometer as an example of automation that is already being used by sea-going oceanographers.

Although flow cytometry is still being developed for sea-going use, it promises to open a new dimension for researchers interested in planktonic communities. Coupled with satellite remote sensing and acoustical methods, it could eventually revolutionize our understanding of primary and secondary productivity in the ocean.

Flow cytometry has been used for cell analyses in biomedical research for the better part of 2 decades (see Shapiro, 1988). This technology allows one to measure fluorescence (either autofluorescence from cellular pigments or induced fluorescence from fluorescent stains) and light-scatter properties of individual cells in a suspension. In a typical system, three or more signals can be measured simultaneously on each cell at a rate of about 1,000 cells sec^{-1}. Cells are pumped through a nozzle in the flow cytometer such that each cell is contained in a droplet, and as it passes a laser detector complex, the droplet is charged positive, negative, or not at all. As the droplets pass a pair of charged deflection plates, they are separated and quantitated. The entire process is controlled by a computer that contains the sorting logic, the signal processing, the quantitative display of the cell distribution, and related tasks. This general procedure is illustrated in Figure 10.

Automatic cell analysis and sorting technology have already been applied to several groups of marine microorganisms, including selected phytoplankton and bacterial species (Olson *et al.*, 1983, 1986; Yentsch *et al.*, 1983; Chisholm *et al.*, 1986, 1988a). Moreover, the flow cytometer has been used successfully aboard ship to carry out a number of important systematic, physiological, and ecological studies of marine organisms in LMEs (e.g., Olson *et al.*, 1985, 1986, 1990a, 1990b, 1990c; Chisholm *et al.*, 1988a, 1988b; Neveux *et al.*, 1989).

Over the past several years, the sensitivity of flow cytometers has been extended to the submicron particle range, revealing the presence of some very abundant picoplankton populations in the North Atlantic (Chisholm *et al.*, 1988b) that were essentially unknown until discovered by flow cytometry.

A new generation of multichannel flow cytometers is being developed with the capacity to identify automatically and to quantitate an array of planktonic species as part of a shipboard multi-instrument water sampler. Flow cytometers can be used at sea to quantitatively separate selected species of plankton by either their characteristic fluorescence spectra, their size, or by specific reporter groups. New types of multidetector flow cytometers that can handle a larger variety of planktonic species have the potential to quantitatively identify numerous planktonic species simultaneously.

Applications of this analytical approach include the following: the determination of the species compositions of planktonic communities within and between LMEs; the analysis of food webs; the determination in the field of growth characteristics of individual species within complex planktonic communities; water-quality monitoring and control; and the enumeration of different microorganisms, many of which play critical roles in the biogeochemical cycles controlling greenhouse gases.

Summary

Recent advances in molecular biology have revolutionized biomedical research. Similarly,

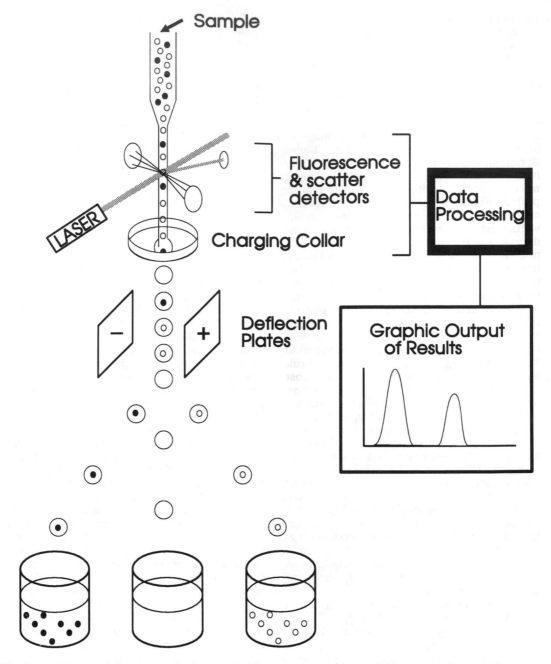

Figure 10. Schematic for an automatically controlled flow cytometer, with sample data output illustrated in square in lower right corner.

the application of molecular techniques to questions in the ocean sciences could revolutionize marine biology and biological oceanography. Present barriers to the rapid implementation of these molecular methodologies into the marine sciences include inadequate molecular training of biological oceanographers, deficiencies in molecular instrumentation at many oceanographic facilities, and the general dearth of fundamental knowledge about the vast majority of marine organisms that is required to develop appropriate molecular methodologies. As these barriers are eliminated, it will be possible to use molecular probes to trace water masses, to examine energy and carbon flow in ocean and coastal food webs, to unravel the complexities of trophic dynamics, and to delineate the intricacies of marine community structures. Eventually, implementation of molecular methods will allow researchers to resolve oceanographic questions that have proven intractable in the past, including the causes of biological variability in the ocean and the implications of this variability to long-term global climate change. The development of automated molecular capabilities alluded to in this paper promises scientific advances equivalent to or exceeding those achieved by satellite remote sensing. Clearly, we are on the verge of an exciting new frontier in the marine sciences.

References

Agellon, L. B., Emery, C. J., Jones, J. M., Davies, S. L., Dingle, A. D., and Chen, T. T. 1987. Promotion of rapid fish growth by a recombinant fish growth hormone. Can. J. Fish. Aquat. Sci. 45:146–151.

Allendorf, F. W., and Phelps, S. R. 1981. Use of allelic frequencies to describe population structure. Can. J. Fish. Aquat. Sci. 38:1507–1514.

Arnheim, N., White, T., and Rainey, W. E. 1990. Application of PCR: Organismal and population biology. BioScience 40(3):174–182.

Atkinson, B. G., and Walden, D. B. 1985. Changes in eukaryotic gene expression in response to environmental stress. Academic Press, Orlando, FL.

Avise, J. C. 1986. Mitochondrial DNA and the evolutionary genetics of higher animals. Phil. Trans. Royal Soc. Lond. B312:325–342.

Avise, J. C. 1987. Identification and interpretation of mitochondrial DNA stocks in marine species. In: Proc. Stock Identification Workshop. pp. 105–136. Ed. by H. Kumpf and H. L. Nakamura. U.S. Dept. of Commerce, National Oceanographic and Atmospheric Administration. Washington, D.C.

Avise, J. C., Arnold, J., Ball, R. M., Bermingham, E., Lamb, T., Neigel, J. E., Reeb, C. A., and Saunders, N. C. 1987. Intraspecific phylogeography: The mtDNA bridge between population genetics and systematics. Annu. Rev. Ecol. Syst. 18:489–522.

Avise, J.C. 1990. Flocks of African Fishes. Nature. 347:512–513.

Avise, J. C., Bowen, B. W., and Lamb, T. 1989. DNA fingerprints for hypervariable mitochondrial genotypes. Mol. Biol. Evol. 6(3):258–269.

Avise, J. C., Helfman, G. S., Saunders, N. C., and Hales, S. 1986. Mitochondrial DNA differentiation in North Atlantic eels: Population genetic consequences of an unusual life history pattern. Proc. Nat. Acad. Sci. 83:4350–4354.

Avise, J. C., and Shapiro, D. Y. 1986. Evaluating kinship of newly-settled juveniles within social groups of coral reef fish, *Anthias squamipinnis.* Evolution 40:1051–1059.

Ayala, F. J. 1975. Molecular evolution. Sinauer Associates, Inc., Sunderland, MA.

Baby, M. C., Bernatchez, L., and Dodson, J. J. 1991. Genetic structure and relationships among anadromous and landlocked populations of rainbow smelt, *Osmerus mordax* (Mitchell), as revealed by mtDNA restriction analysis. J. Fish Biol. 39(A):61–68.

Baker, C. S., Palumbi, S. R., Lambertsen, R. H., Weinrich, M. T., Calambkidis, J., and O'Brien, S. J. 1990. The influence of seasonal migration on geographic distribution of mtDNA haplotypes in humpback whales. Nature 344:238–240.

Barrett, B. A., and McKeown, B. A. 1988. Sustained exercise augments long-term starvation increases in plasma growth hormone in the steelhead trout, *Salmon gairdneri.* Can. J. Zool. 66:853–855.

Bentzen, P., and Wright, J.M., 1991. Cloning and characterization of variable number tandem repeat loci in salmonid fishes. Symposium on Biochemical Genetics and Taxonomy of Fish, Queen's University of Belfast, Northern Ireland July 22–26, 1991.

Bermingham, E. 1990. Mitochondrial DNA and the analysis of fish population structure. In: Electrophoretic and isoelectric focusing techniques in fisheries management. pp. 197–222. Ed. by D. H. Whitmore. CRC Press, Boca Raton, FL.

Bermingham, E., and Avise, J. C. 1986. Molecular zoogeography of freshwater fishes in the southeastern United States. Genetics 113:939–965.

Bernardi, G., Sordino, P., and Powers, D. A. 1992. Nucleotide sequence of the 18S rRNA gene from the killifish, *Fundulus heteroclitus,* the spiny dogfish, *Squalus acanthias,* and the prickly shark, *Echinorhinus cookei,* and their molecular phylogeny. Mol. Mar. Biol. Biotechnol. 1(3):187–194.

Black, D., and Skinner, E. R. 1986. Features of the lipid transport system of fish as demonstrated by studies on starvation in the rainbow trout. J. Comp. Physiol. B 156:497–502.

Bowen, B. W., and Avise, J. C. 1990. The genetic structure of Atlantic and Gulf of Mexico populations of sea bass, menhaden, and sturgeon: The influence of zoogeographic factors and life history patterns. Mar. Biol. 107:371–381.

Bowen, R. W., Meylan, A. B., and Avise, J. C. 1989. An odyssey of the green sea turtle: Ascension Island revisited. Proc. Nat. Acad. Sci. 86:573–576.

Britten, R. J., Cetta, A., and Davidson, E. H. 1978. The single-copy DNA sequence polymorphism of the sea urchin *Strongylocentrotus purpuratus*. Cell 15:1175–1186.

Brock, V. 1985. Immuno-electrophoretic studies of genetic relations between populations of *Mytilus edulis* and *M. galloprovincialis* for the Mediterranean, Baltic, East and West Atlantic, and East Pacific. In: Proceedings of the nineteenth European marine biology symposium. pp. 515–523. Ed. by P. E. Gibbs. Cambridge University Press, England.

Bucklin, A., Rienecker, M. M., and Mooers, C. N. K. 1989. Genetic markers of zooplankton transport in coastal filaments off the U.S. west coast. J. Geophys. Res. 94:8277–8288.

Bulow, F. J. 1987. RNA-DNA indicators of growth in fish: A review. In: The age and growth of fish. pp. 45–64. Ed. by R. C. Summerfelt and G. E. Hall. Iowa State Press, Ames.

Buth, D. G. 1990. Genetic principles and the interpretation of electrophoretic data. In: Electrophoretic and isoelectric focusing techniques in fisheries management. pp. 1–21. Ed. by D. H. Whitmore. CRC Press, Boca Raton, FL.

Campbell, L., Carpenter, E. J., and Iacono, V. J. 1983. Identification and enumeration of marine *Chroococcoid* cyanobacteria by immunofluorescence. Appl. Environ. Microbiol. 46:553–559.

Campton, D. E. 1987. Natural hybridization and introgression in fishes: Methods of detection and genetic interpretations. In: Population genetics and fisheries management. pp. 161–192. Ed. by N. Ryman and F. M. Utter. University of Washington Press, Seattle.

Campton, D. E. 1990. Application of biochemical and molecular genetic markers in analysis of hybridization. In: Electrophoretic and isoelectric focusing techniques in fisheries management. pp. 107–130. Ed. by D. H. Whitmore. CRC Press, Boca Raton, FL.

Cashon, R.E., Van Beneden, R.J., and Powers, D.A. 1981. Biochemical genetics of *Fundulus heteroclitus*. IV. Spatial variation in gene frequencies of Idh-A, Idh-B, 6-Pgdh-A, and Est-S, Biochem. Genetics, 19:715–728.

Cashon, R., and Powers, D. A. Submitted. Functional characterization of the cytoplasmic malate dehydrogenase allelic isozymes from the teleost fish *Fundulus heteroclitus*.

Chapman, R. W. 1987. Changes in the population structure of male striped bass, *Morone saxatilis*, spawning in three areas of the Chesapeake Bay from 1984–1986. Fish. Bull. U.S. 85:167–170.

Chapman, R. W. 1989. Spatial and temporal variation in mitochondrial DNA haplotype frequencies in the striped bass 1982 year class. Copeia 1989:344–354.

Chapman, R. W., and Brown, B. L. 1990. Mitochondrial DNA isolation methods. In: Electrophoretic and isoelectric focusing techniques in fisheries management. pp. 107–130. Ed. by D. H. Whitmore. CRC Press, Boca Raton, Florida.

Chen, T. T. 1983. Identification and characterization of estrogen-responsive gene products in the liver of the rainbow trout. Can. J. Biochem. Cell Biol. 61:802–810.

Chesnick, J. M., Kugrens, P., and Cattolico, R. A. 1991. The utility of mtDNA restriction fragment length polymorphisms in *Cryptomonad* phylogenetic assessment. Mol. Mar. Biol. Biotechnol. 1(1):18–26.

Chisholm, S. W., Armbrust, E. V., and Olson, R. J. 1986. The individual cell in phytoplankton ecology: Cell cycles and applications of flow cytometry. In: Photosynthetic picoplankton. pp. 343–369. Ed. by T. Platt and W. K. W. Li. Dept. Fish and Oceans. Can. Bull. Fish. Aquat. Sci. 214. Ottawa, Canada.

Chisholm, S. W., Olson, R. J., and Yentsch, C. M. 1988a. Flow cytometry in oceanography: Status and prospects. EOS Trans. Am. Geophys. Union. vol. 69 No. 18.

Chisholm, S. W., Olson, R. J., Zetter, E. R., and Armbrust, E. V. 1988b. A novel free-living prochlorophyte abundant in the oceanic euphotic zone. Nature 334:340–343.

Comparini, A., and Rodino, E. 1980. Electrophoretic evidence for two species of *Anguilla leptocephali* in the Sargasso Sea. Nature 287:435–437.

Cox, J. L., Williamson, S., and Harding, L. 1983. Consequences of distributional heterogeneity of *Calanus pacificus* grazing. Bull. Mar. Sci. 33:213–226.

Crawford, D. L., Costantino, H. R., and Powers, D. A. 1989. Lactate dehydrogenase-B cDNA from the teleost *Fundulus heteroclitus*: Evolutionary implications. Mol. Biol. Evol. 6(4):369–383.

Crawford, D. L., and Powers, D. A. 1989. Molecular basis of evolutionary adaptation in two latitudinally extreme populations of *Fundulus heteroclitus*. Proc. Nat. Acad Sci. 86:9365–9369.

Crawford, D. L., and Powers, D. A. 1990. Molecular adaptation to different thermal environments: Genetic and physiological mechanisms. In: Molecular evolution. pp. 213–222. Ed. by M. T. Clegg and S. J. O'Brian. Wiley-Liss Inc., New York.

Currin, C., Paerl, H. W., Suba, G., and Alberte, R. S. 1990. Immunofluorescence detection and characterization of N_2-fixing microorganisms in aquatic environments. Limnol. Oceanogr. 35(1):59–71.

Dagg, M. J., and Wyman, K. D. 1983. Natural ingestion rates of the copepods *Neocalanus plumchrus* and *N. cristatus* calculated from gut contents. Mar. Ecol. Progr. Ser. 13:37–46.

Danzmann, R.G., Van der Kraak, G.J., Chen, T.T., and Powers, D.A., 1990. Metabolic Effects of Bovine Growth Hormone and Biosynthetic Rainbow Trout Growth Hormone in Rainbow Trout Reared at a High Temperature, Canadian Journal of Fisheries and Aquatic Sciences, 47:1292–1301.

Danzman, R. G., Ihssen, P. E., and Hebert, P. D. N. 1991. Genetic discrimination of wild and hatchery populations of brook char, *Salvelinus fontanilis* (Mitchell), in Ontario using mitochondrial DNA analysis. J. Fish Biol. 39(A):69–77.

de Ligny, W. 1969. Serological and biochemical studies on fish populations. Oceanogr. Mar. Biol. 7:411–513.

DeLong, E. F., Wickham, G. S., and Pace, N. R. 1989. Phylogenetic stains: Ribosomal RNA-based probes for the identification of single cells. Science 243:1360–1364.

DiMichele, L., Paynter, K., and Powers, D. A. 1991. Lactate dehydrogenase-B allozymes directly effect development of *Fundulus heteroclitus*. Science 253:898–900.

DiMichele, L., and Powers, D. A. 1982a. Ldh-B genotype-specific hatching times of *Fundulus heteroclitus* embryos. Nature 296:563–564.

DiMichele, L., and Powers, D. A. 1982b. Physiological basis for swimming endurance difference between Ldh-B genotypes of *Fundulus heteroclitus*. Science 216:1014–1016.

DiMichele, L., and Powers, D. A. 1984. Developmental and oxygen consumption rate differences between Ldh-B genotypes of *Fundulus heteroclitus* and their effect on hatching time. Physiol. Zool. 57:52–56.

DiMichele, L., and Powers, D. A. 1991. Allozyme variation, developmental rate, and differential mortality in the teleost, *Fundulus heteroclitus*. Physiol. Zool. 64(6):1426–1443.

DiMichele, L., Powers, D. A., and DiMichele, J. 1986. Developmental and physiological consequences of genetic variation at enzyme synthesizing loci in *Fundulus heteroclitus*. Am. Zool. 26:201–208.

Donis-Keller, H., Maxam, A., and Gilbert, W. 1977. Mapping adenines, guanines and pyrimidines in R or A. Nucleic Acid Res. 4:2527–2538.

Endler, J.A. 1977. Geographic Variation, Speciation and Clines. Princeton University Press, Princeton, NJ.

Fain, S. R., Druehl, L. D., and Baillie, D. C. 1988. Repeat and single copy sequences are differentially conserved in the evaluation of kelp chloroplast DNA. J. Phycol. 24:292–302.

Fairbairn, D. J. 1981. Which witch is which? A study of the stock structure of witch flounder (*Glyptocephalus cynoglossus*) in the Newfoundland region. Can. J. Fish. Aquat. Sci. 38:782–794.

Falkowski, P. G., and LaRoche, J. 1991. Molecular biology in studies of ocean processes. Int. Rev. Cytol. 128:261–303.

Feller, R. J. 1986. Immunological detection of *Mercenaria mercenaria* in a predator and preparation of size-class specific antibodies. The Veliger 28(4):341–347.

Feller, R. J., Hentschel, B. T., and Ferguson, R. B. 1990. Immunoelectrophoretic assay of mixed species meals: An example using penaeid shrimp. In: Trophic relationships in the marine environment. pp. 588–596. Ed. by M. Barnes and R. N. Gibson. Proc. 24th Europ. Mar. Biol. Symp. Aberdeen Univ. Press, Aberdeen, Scotland.

Feller, R. J., Zagursky, G., and Day, E. A. 1985. Deep-sea food web analysis using cross-reacting sera. Deep-Sea Res. 32(4):485–497.

Ferguson, M. M., Danzman, R. G., and Hutchings. J. A. 1991. Incongruent estimates of population differentiation among brook char, *Salvelinus fontinalis*, from Cape Race, Newfoundland, Canada, based upon allozyme and mitochrondrial DNA variation. J. Fish Biol. 39(A):79–85.

Fetterolf, C. M., Jr. 1981. Foreword to the Stock Concept Symposium. Can. J. Fish. Aquat. Sci. 38:iv–v.

Field, K. G., Olsen, G. J., Lane, D. J., Giovannoni, S. J., Ghiselin, M. T., Raff, E. C., Pace, N. R., and Raff, R. A. 1988. Molecular phylogeny of the animal kingdom based on 18S ribosomal RNA sequences. Science 239:784–753.

Fields, R. D., Johnson, K. R., and Throgaard, G. H. 1989. DNA fingerprints in rainbow trout detected by hybridization with DNA of bacteriophage M13. Trans. Am. Fish. Soc. 118:78–81.

Frost, B. W. 1989. A taxonomy of the marine calanoid copepod genus *Pseudocalanus*. Can. J. Zool. 67(3):233–250.

Gallagher, J. C. 1980. Populations of *Skeletonema costatum* (Bacillariophyceae) in Narragansett Bay. J. Phycol. 16:464–474.

Gallagher, J. C. 1982. Physiological variation and electrophoretic banding patterns of genetically different seasonal populations of *Skeletonema costatum* (Bacillariophyceae). J. Phycol. 18:148–162.

Gallagher, J. C., and Alberte, R. S. 1985. Photosynthetic and cellular photoadaptive characteristics of three ecotypes of the marine diatom, *Skeletonema costatus* (Grev.). Cleve. J. Exp. Mar. Biol. Ecol. 94(1–2):233–250.

Gallagher, J. C., Jumars, P. A., and Taghon, G. L. 1988. The production of monospecific antisera to soft-bottom benthic taxa. In: Immunological approaches to coastal, estuarine and oceanographic questions. pp. 120–132. Ed. by C. M. Yentsch, F. C. Mague, and P. K. Horan. Springer-Verlag, New York.

Gedamu, L., Culham, B., and Heikkila, J. J. 1983. Analysis of the temperature-dependent temporal pattern of heat-shock-protein synthesis in fish cells. Biosci. Rep. 3:647–658.

Giovannoni, S. J., Britschgi, T. B., Moyer, C. L., and Field, K. G. 1990. Genetic diversity in Sargasso Sea bacterioplankton. Nature 345:60–63.

Goff, L. J., and Coleman, A. W. 1988. The use of plasmid DNA restriction endonuclease patterns in delineating red algal species and populations. J. Phycol. 24:357–368.

Gonzalez-Villasenor, L. I., Burkhoff, M., Corces, V., and Powers, D. A. 1986. Characterization of cloned mitochondrial DNA from the teleost *Fundulus heteroclitus* and its usefulness as an interspecies hybridization probe. Can. J. Fish. Aquat. Sci. 43:1866–1872.

Gonzalez-Villasenor, L. I., and Powers, D. A. 1990. Mitochondrial DNA restriction site polymorphisms in the teleost *Fundulus heteroclitus* supports secondary intergradation. Evolution 44(1):27–37.

Gonzalez-Villasenor, L. I., and Powers, D. A. Submitted. Functional characterization of the cytoplasmic NADP-dependent isocitrate dehydrogenase allelic isozymes from the teleost fish *Fundulus heteroclitus*.

Goodier, J.L., and Davidson, W.S. 1991. A repetitive element in the Atlantic salmon genome. Symposium on Biochemical Genetics and Taxonomy of Fish, Queen's University of Belfast, Northern Ireland July 22–26, 1991.

Grisley, M. S., and Boyle, P. R. 1988. Recognition of food in octopus digestive tract. J. Exp. Mar. Biol. Ecol. 118:7–32.

Grove, T. L., Berggren, T. S., and Powers, D. A. 1976. The use of innate genetic tags to segregate spawning stocks of striped bass (*Morone saxatilis*). Estuarine Process. 1:166–176.

Gyllenstein, U., and Erlich, H. 1988. Generation of single-stranded DNA by the polymerase chain reaction and its application to direct sequencing of the HLA-DQa locus. Proc. Nat. Acad. Sci. 85:7652–7660.

Hakanson, J. L. 1984. The long and short term feeding condition in field-caught *Calanus pacificus*, as determined from the lipid content. Limnol. Oceanogr. 29:794–804.

Hakanson, J. L. 1987. The feeding condition of *Calanus pacificus* and other zooplankton in relation to phytoplankton pigments in the California Current. Limnol. Oceanogr. 32(4):881–884.

Heath, M. R., and Walker, J. 1987. A preliminary study of the drift of larval herring (*Clupea harengus* L.) using gene-frequency data. J. Cons. Int. Explor. Mer 43:139–145.

Hentschel, B. T., and Feller, R. J. 1990. Quantitative immunoassay of the proventricular contents of white shrimp *Penaeus setiferus* Linnaeus: A laboratory study. J. Exp. Mar. Biol. Ecol. 139:85–99.

Higuchi, R., and Ochman, H. 1989. Production of single-stranded DNA templates by exonuclease digestion following PCR. Nucleic Acid Res. 17:5865–5872.

Hillis, D. M. 1987. Molecular vs. morphological approaches to systematics. Annu. Rev. Ecol. Syst. 18:23–42.

Houlihan, D. F., Hall, S. J., and Gray, C. 1989. Effects of ration on protein turnover in cod. Aquaculture 79:103–110.

Houlihan, D. F., Hall, S. J., Gray, C., and Nobel, B. S. 1988. Growth rates and protein turnover in Atlantic cod, *Gadus morhua*. Can. J. Fish. Aquat. Sci. 45:951–964.

Hunter, J., and Feller, R. J. 1987. Immunological dietary analysis of two penaeid shrimp species from a South Carolina tidal creek. J. Exp. Mar. Biol. Ecol. 107:61–70.

Jeffreys, A. J., Wilson, V., and Thein, S. L. 1985. Individual-specific "fingerprints" of human DNA. Nature 316:76–79.

Koban, M., Graham, G., and Prosser, C. L. 1987. Induction of heat-shock protein synthesis in teleost hepatocytes: Effects of acclimation temperature. Physiol. Zool. 60:290–296.

Koban, M., Yup, A. A., Agellon, L. B., and Powers, D. A. 1991. Molecular adaptation to the thermal environment: Heat-shock response of the eurythermal teleost *Fundulus heteroclitus*. Mol. Mar. Biol. Biotechnol. 1(1):1–17.

Koehn, R. K., and Hilbish, T. J. 1987. The adaptive importance of genetic variation. Am. Sci. 75:134–141.

Koehn, R. K., and Immerman, F. W. 1981. Biochemical studies of aminopeptidase polymorphism in *Mytilus edulis*. I. Dependence of enzyme activity on season, tissue and genotype. Biochem. Genet. 19:1115–1142.

Koehn, R. K., Milkman, R., and Mitton, J. B. 1976. Population genetics of marine pelecypods. IV. Selection, migration and genetic differentiation in the blue mussel *Mytilus edulis*. Evolution 30:2–32.

Koehn, R. K., and Siebenaller, J. F. 1981. Biochemical studies of aminopeptidase polymorphism in *Mytilus edulis*. II. Dependence of reaction rate on physical factors and enzyme concentration. Biochem. Genet. 19:1143–1162.

Koehn, R. K., and Williams, G. C. 1978. Genetic differentiation without isolation in the American eel, *Anguilla rostrata*. II. Temporal stability of geographic patterns. Evolution 32:624–637.

Kornfield, I., and Bogdanowicz, S. M. 1987. Differentiation of mitochondrial DNA in Atlantic herring, *Clupea harengus*. Fish. Bull. U.S. 85(3):561–568.

Kothary, R. K., and Candido, E. P. M. 1982. Induction of a novel set of polypeptides by heat shock or sodium arsenite in cultured cells of rainbow trout, *Salmo gairdneri*. Can. J. Biochem. 60:347–355.

Larkin, P. A. 1981. A perspective on population genetics and salmon management. Can. J. Fish. Aquat. Sci. 38:1469–1475.

Lee, S., and Fuhrman, J.A., 1990. DNA hybridization ot compare species compositions of natural bacterioplankton assemblages. Appl. Environ. Microbiol. 56 (3):739–746.

Li, N., and Cattolico, R. A. 1992. *Ochromonas danica* (Chrysophyceae) chloroplast genome organization. Mol. Mar. Biol. Biotechnol. 1(2):154–174.

Loughna, P. T., and Goldspink, G. 1984. The effects of starvation upon protein turnover in red and white myotomal muscle of rainbow trout, *Salmo gairdneri*. J. Fish Biol. 25:223–230.

Love, R. M. 1980. The chemical biology of fishes. Vol. 2. Academic Press, New York.

Mackas, D. L., and Bohrer, R. 1976. Fluorescence analysis of zooplankton gut contents and an investigation of diet feeding patterns. J. Exp. Mar. Biol. Ecol. 25:77–85.

Marx, J. L. 1988. Multiplying genes by leaps and bounds. Science 240:1408–1410.

McMillan, C. 1982. Isozymes in seagrasses. Aquat. Bot. 14:231–243.

McMillan, J., and Miller, D. J. 1990. Highly repeated DNA sequences in the scleractinian coral genus *Acropora*: Evaluation of cloned repeats as taxonomic probes. Mar. Biol. 104:483–487.

Meylan, A. B., Bowen, B. W., and Avise, J. C. 1990. A genetic test of "natal homing" versus "social facilitation" in green turtle migration. Science 248:724–727.

Miller, K. M., Jones, P., and Roughgarden, J. 1991. Monoclonal antibodies as species-specific probes in oceanographic research: Examples with intertidal barnacle larvae. Mol. Mar. Biol. Biotechnol. 1(1):35–47.

Mommsen, T. P., French, C. J., and Hochachka, P. W. 1980. Sites and patterns of protein and amino acid utilization during the spawning migration of salmon. Can. J. Zool. 58:1785–1799.

Moon, T. W. 1983. Metabolic reserves and enzyme activities with food deprivation in immature American eels, *Anquilla rostrata* (LeSueur). Can. J. Zool. 61:802–811.

Morgan, R. P., II. 1975. Distinguishing larval white perch and striped bass by electrophoresis. Chesapeake Sci. 16:68–70.

Morgan, R. P., II, Koo, T. S. Y., and Krantz, G. E. 1973. Electrophoretic determination of populations of striped bass, *Morone saxatilis*, in the Chesapeake Bay. Trans. Am. Fish. Soc. 102:21–32.

Neveux, J., Vaulot, D., Courties, C., and Fukai, G. E. 1989. Photosynthetic bacteria associated with the deep chlorophyll maximum of the Sargasso Sea. C. R. Acad. Sci. (Paris) 308. Series III. pp. 9–14.

Nevo, E., Noy, R., Lavie, B., Beiles, A., and Muchtar, S. 1986. Genetic diversity and resistance to marine pollution. Biol. J. Linn. Soc. 29:139–144.

Nevo, E., Noy, R., Lavie, B., and Muchtar, S. 1985. Levels of genetic diversity and resistance to pollution in marine organisms. FAO Fisheries Report No. 352 Supplement. pp. 175–182, New York, NY.

Olsen, G. J., Lane, D. J., Giovannoni, S. J., and Pace, N. R. 1986. Microbial ecology and evolution: A ribosomal RNA approach. Am. Rev. Microbiol. 40:337–65.

Olson, R. J., Chisholm, S. W., Altabet, M., Zettler, E. R., and Dusenberry, J. 1990a. Spatial and temporal patterns of prochlorophyte picoplankton in the North Atlantic Ocean. Deep-Sea Res. 37(6A):1033–1051.

Olson, R. J., Chisholm, S. W., Zettler, E. R., and Armbrust, E. V. 1990b. Analysis of *Synechococcus* pigment types in the sea using single and dual beam flow cytometry. Deep-Sea Res. 35(3):425–440.

Olson, R. J., Chisholm, S. W., Zettler, E. R., and Armbrust, E. V. 1990c. Pigments, size and distribution of *Synechococcus* in the North Atlantic and Pacific Oceans. Limnol. Oceanogr. 35:45–58.

Olson, R. J., Frankel, S. L., Chisholm, S. W., and Shapiro, H. M. 1983. An inexpensive flow cytometer for the analysis of fluorescence signals in phytoplankton: Chlorophyll and DNA distributions. J. Exp. Mar. Biol. Ecol. 68:129–144.

Olson, R. J., Vaulot, D., and Chisholm, S. W. 1985. Marine phytoplankton distributions measured using shipboard flow cytometry. Deep-Sea Res. 10:1273–1280.

Olson, R. J., Vaulot, D., and Chisholm, S. W. 1986. Effects of environmental stresses on the cell cycle

of two marine phytoplankton species. Plant Physiol. 80:981–925.

Paabo, S., Gifford, J. A., and Wilson, A. C. 1988. Mitochondrial DNA sequences from a 7000-year-old brain. Nucleic Acid Res. 16:9775–9786.

Pace, N. R., Stahl, D. A., Lane, D. J., and Olsen, G. T. 1986. The analysis of natural microbial populations by ribosomal RNA sequences. Adv. Microbiol. Ecol. 9:1–55.

Palumbi, S. R., and Benzie, J. 1991. Large mitochondrial DNA between morphologically similar Penaeid shrimp. Mol. Mar. Biol. Biotechnol. 1(1):27–34.

Palumbi, S. R., and Wilson, A. C. 1990. Mitochondrial DNA diversity in the sea urchins *Strongylocentrotus purpuratus* and *S. drobachiensis*. Evolution 44:403–415.

Partti-Pellinen, K. A., Elo, K., Palva, T. K., Tuunainen, P., and Kakumaki, M. O. K. 1991. Mitochondrial DNA variation of *Salvelinus* species found in Finland. J. Fish Biol. 39(A):87–92.

Pastene, L.A., Numachi, K., and Tsukamoto, K. 1991. Examination of reproductive success of transplanted stocks in an amphidromous fish. *Plecoglossus altivelis*, (Temmink et Schlegel) using mitochondrial DNA and isozyme markers. Fish Biol. 39(A):93–100.

Patton, J. S., Nevenzel, J. C., and Benson, A. A. 1975. Specificity of digestive lipases in hydrolysis of wax esters and triglycerides studied in anchovy and other selected fish. Lipids 10(10):575–583.

Peattie, D. A. 1979. Direct chemical method for sequencing RNA. Proc. Nat. Acad. Sci. 76:1760–1765.

Pella, J. J., and Milner, G. B. 1987. Use of genetic marks in stock composition analysis. In: Population genetics and fisheries management. pp. 247–276. Ed. by N. Ryman and F. M. Utter. University of Washington Press, Seattle.

Place, A. R., and Powers, D. A. 1978. Genetic bases for protein polymorphism in *Fundulus heteroclitus*. I. Lactate dehydrogenase (Ldh-B), malate dehydrogenase (Mdh-A), phosphoglucoisomerase (Pgi-B), and phosphoglucomutase (Pgm-A). Biochem. Genet. 16:577–591.

Place, A. R., and Powers, D. A. 1979. Genetic variation and relative catalytic efficiencies: Lactate dehydrogenase B allozymes of *Fundulus heteroclitus*. Proc. Nat. Acad. Sci. 76(5):2354–2358.

Place, A. R., and Powers, D. A. 1984. Kinetic characterization of the lactate dehydrogenase (Ldh-B4) allozymes of *Fundulus heteroclitus*. J. Biol. Chem. 259(20):1309–1318.

Powers, D. A. 1987. A multidisciplinary approach to the study of genetic variation within species. In: New directions in ecological physiology. pp. 102–130. Ed. by M. E. Feder, A. F. Bennett, W. W. Burggren, and R. B. Huey. Cambridge University Press, England.

Powers, D. A. 1989. Fish as model systems. Science 246:352–358.

Powers, D. A. 1990a. Marine and freshwater biotechnology: A new frontier. In: Biotechnology:

Science, education and commercialization. pp. 41–70. Ed. by I. Vasil. Elsevier Press, New York.

Powers, D. A. 1990b. The adaptive significance of allelic isozyme variation in natural populations. In: Electrophoretic and isoelectric focusing techniques in fisheries management. pp. 323–339. Ed. by D. H. Whitmore. CRC Press, Boca Raton, Florida.

Powers, D. A. 1991. Evolutional genetics of fish. Adv. Genet. 29:119–228.

Powers, D. A., Allendorf, F., and Chen, T. T. 1990. Application of molecular techniques to the study of marine recruitment problems. In: Large marine ecosystems: Patterns, processes, and yields. pp. 104–121. Ed. by K. Sherman, L. M. Alexander, and B. D. Gold. AAAS Symposium. AAAS, Washington, DC.

Powers, D. A., Lauerman, T., Crawford, D., and DiMichele, L. 1991. Genetic mechanisms for adapting to a changing environment. Ann. Rev. Genet. 25:629–659.

Powers, D. A., and Place, A. R. 1978. Biochemical genetics of Fundulus heteroclitus (L.). I. Temporal and spatial variation in gene frequencies of Ldh-B, Mdh-A, Gpi-B, and Pgm-A. Biochem. Genet. 16(5,6):593–607.

Powers, D. A., Ropson, I., Brown, W. C., Van Beneden, R., Cashon, R., Gonzalez-Villasenor, L. I., and DiMichele, J. 1986. Genetic variation in Fundulus heteroclitus: Geographic distribution. Am. Zool. 26:131–144.

Reddin, D. G. 1986. Discrimination between Atlantic salmon (Salmo salar) of North American and European origin. J. Cons. Int. Explor. Mer 43:50–64.

Reeb, C. A., and Avise, J. C. 1990. A genetic discontinuity in a continuously distributed species: Mitochondrial DNA in the American oyster, Crassostrea virginica. Genetics 124:397–406.

Ricker, W. E. 1981. Changes in the average size and average age of Pacific salmon. Can. J. Fish. Aquat. Sci. 38:1636–1656.

Rogier, C. G., Ney, J. J., and Turner, B. J. 1985. Electrophoretic analysis of genetic variability in a landlocked striped bass population. Trans. Am. Fish. Soc. 114:244–249.

Ropson, I., and Powers, D., A. 1989. Allelic isozymes of hexose-6-phosphate dehydrogenase from the teleost Fundulus heteroclitus: Physical characteristics and kinetic properties. Mol. Biol. Evol. 6(2):171–185.

Ropson, I., Brown, D., and Powers, D.A. 1990. Biochemical genetics of Fundulus heteroclitus (L.) VI. Geographical variation in the gene frequencies of 15 loci. Evolution 44(1):16–26.

Ropson, I., Van Elken, H., and Powers, D. A. In press. Purification and functional characterization of the allelic isozymes of aspartate amino transferase from the teleost Fundulus heteroclitus. Mol. Mar. Biol. Biotechnol.

Rowan, R., and Powers, D. A. 1991a. A molecular genetic classification of zooxanthellae and the evolution of animal algal symbioses. Science 251:1348–1350.

Rowan, R., and Powers, D. A. 1991b. The molecular genetic identification of symbiotic dinoflagellates (zooxanthellae). Mar. Ecol. Prog. Ser. 71:65–73.

Rowan, R., and Powers, D. A. 1992. Ribosomal RNA sequences and the diversity of symbiotic dinoflagellates. Proc. Nat. Acad. Sci. 89:3639–3643.

Ryman, N., and Utter, F. (Editors). 1987. Population genetics and fisheries management. University of Washington Press, Seattle.

Saiki, R. K., Gelfand, D. H., Stoffe, S., Scharf, S. J., Higuchi, R., Horn, G. T., Mullis, K. B., and Erlich, H. A. 1988. Primer-directed enzymatic amplification of DNA with a thermostable DNA polymerase. Science 239:487–491.

Sargent, J. R., McIntosh, R., Bauermeister, A., and Blaxter, J. H. S. 1979. Assimilation of the wax esters of marine zooplankton by herring (Clupea harengus) and rainbow trout (Salmo gairdnerii). Mar. Biol. 51:203–207.

Saunders, N. C., Kessler, L. G., and Avise, J. C. 1986. Genetic variation and geographic differentiation in mitochondrial DNA of the horseshoe crab, Limulus polyphemus. Genetics 112:613–627.

Schlesinger, M. J., Ashburner, M., and Tissieres, A. (eds.) 1982. Heat shock, from bacteria to man. Cold Spring Harbor Laboratory Press, Cold Spring Harbor, NY.

Seiderer, L. J., Davis, C. L., Robb, F. T., and Newell, R. C. 1987. Digestive enzymes of the anchovy Engraulis capensis in relation to diet. Mar. Ecol. (Prog. Ser.) 35:15–23.

Sekine, S. T., Mizukami, T., Nishi, T., Kuwana, Y., Saito, A., Sato, M., Itch, S., and Kawauchi, H. 1985. Proc. Nat. Acad. Sci. 82:4306–4310.

Setzler, E. M., et al.. 1980. Synopsis of biological data on striped bass, Morone saxatilis. FAO Synopsis 121.

Sevigny, J. M., McLaren, I. A., and Frost, B. W. 1989. Discrimination among and variation within species of Pseudocalanus based on the GPI locus. Mar. Biol. 102(3):321–327.

Shaklee, J. B. 1983. The utilization of isozymes as gene markers in fisheries management and conservation. In: Isozymes: Current topics in biological and medical research. Vol. 11. pp. 213–247. C. Markert, J. Scandolis, G. Whitt (eds.) Alan R. Liss Publ. Co., New York.

Shaklee, J. B., Phelps, S. R., and Salini, J. 1990. Analysis of fish stock structure and mixed-stock fisheries by the electrophoretic characterization of allelic isozymes. In: Electrophoretic and isoelectric focusing techniques in fisheries management. pp. 173–196. Ed. by D. H. Whitmore. CRC Press, Boca Raton, FL.

Shaklee, J. B., and Salini, J. 1985. Genetic variation and population subdivision in Australian barramundi, Lates calcarifer (Bloch). Aust. J. Mar. Freshw. Res. 36:203–208.

Shaklee, J. B., and Tamaru, T. 1981. Biochemical and morphological evolution of Hawaiian bone fishes. Syst. Zool. 30:135–146.

Shapiro, H. M. 1988. Practical flow cytometry. 2d ed. Alan R. Liss Publ. Co., New York.

She, J. X., Autem, M., Kotulas, G., Pasteur, N., and Bonhomme, F. 1987. Multivariate analysis of genetic exchanges between *Sole aegyptiaca* and *Solea senegensis* (Teleosts, Soleodae). Biol. J. Linn. Soc. 32:357–371.

Sidell, B. D., Otto, R. G., and Powers, D. A. 1978. A biomedical method for distinction of striped bass and white perch larvae. Copeia 2:340–343.

Sidell, B. D., Otto, R. G., Powers, D. A., Karweit, M., and Smith, J. 1980. A reevaluation of the occurrence of subpopulations of striped bass (*Morone saxatilis*, Walbaum) in the upper Chesapeake Bay. Trans. Am. Fish. Soc. 109:99–107.

Slatkin, M. 1985. Rare alleles as indicators of gene flow. Evolution 39:53–65.

Smith, M. W., Chapman, R. W., and Powers, D. A. Submitted. Glacial and sea level changes shaped the geographic distributions of *Fundulus heteroclitus* subspecies mitochondrial genomes. Mol. Biol. and Evolution.

Sogin, M. L. 1990. Amplification of ribosomal RNA genes for molecular evolution studies. In: PCR Protocols. A guide to methods and applications. pp. 307–314. Ed. by M. A. Innis, D. H. Gelfand, J. J. Sninsky, and T. J. White. Academic Press, New York.

Sogin, M. L., Elwood, H. J., and Gunderson, J. H. 1986. Evolutionary diversity of eukaryotic small-subunit rRNA genes. Proc. Nat. Acad. Sci. 83:1383–1387.

Sogin, M. L., and Gunderson, J. H. 1987. Structural diversity of eukaryotic small subunit ribosomal RNAs. Ann. New York Acad. Sci. 503:125–139.

Stahl, D. A., Lane, D. J., Olsen, G. J., Heller, D. J., Schmidt, T. M., and Pace, N. R. 1987. Phylogenetic analysis of certain sulfide-oxidizing and related morphologically conspicuous bacteria by SS ribosomal ribonucleic acid sequences. Int. J. Syst. Bacteriol. 37:116–133.

Stahl, D. A., Lane, D. J., Olsen, G. J., and Pace, N. R. 1984. Analyses of hydrothermal vent-associated symbionts by ribosomal RNA sequences. Science 224:409–411.

Stoflet, E. S., Koeberi, D. D., Sarkar, G., and Sommer, S. S. 1988. Genomic amplification with transcript sequencing. Science 239:491–494.

Taggart, J., Prodhol, P., and Ferguson, A. 1991. Application of single locus minisatellite DNA probes to the study of population genetics of salmon. Symposium on Biochemical Genetics and Taxonomy of Fish, Queen's University of Belfast, Northern Ireland, July 22–26, 1991.

Theilacker, G. H., Kimball, A. S., and Trimmer, J. S. 1986. Use of an ELISPOT immunoassay to detect euphausiid predation on larval anchovy. Mar. Ecol. Prog. Ser. 30:127–131.

Thomas, W. K., Withler, R. E., and Beckenbach, A. T. 1986. Mitochondrial DNA analysis of Pacific salmonid evolution. Can. J. Zool. 64:1058–1064.

Turner, B. J., Elder, J. F., and Laughlin, T. F. 1991. Repetitive DNA sequences and the divergence to fish populations: Some hopeful beginnings. J. Fish Biol. 39(A):131–142.

Utter, F., Teel, D., Milner, G., and McIsaac, D. 1987. Genetic estimates of stock compositions of 1983 chinook salmon harvests off the Washington coast and the Columbia River. Fish. Bull. U.S. 85:13–21.

Van Beneden, R.J., Cashon, R.E., and Powers, D.A. 1981. Biochemical genetics of *Fundulus heteroclitus* III, inheritance of isocitrate dehydrogenase (Idh-A and Idh-B), 6-phosphogluconate dehydrogenase (6-Pgdh-A), and serum esterase (Est-S) polymorphisms. Biochem. Genetics. 19:701–714.

Van Beneden, R. J., and Powers, D. A. 1989. Glucose phosphate isomerase allozymes from the teleost *Fundulus heteroclitus*. Mol. Biol. Evol. 6(2):155–170.

Vrijenhoek, R.C., Pfeiler, and Wetherington, J.D., Selection in a desert stream-dwelling fish, *Poeciliopsis monacha*. Evolution 46:in press.

Wagner, G. F., and McKeown, B. A. 1986. Development of a salmon growth hormone radioimmunoassay. Gen. Comp. Endocrinol. 62:452–458.

Waples, R. S. 1987. A multispecies approach to the analysis of gene flow in marine shore fishes. Evolution 41:385–400.

Ward, B. B. 1990. Immunology in biological oceanography and marine ecology. Oceanography 3:30–35.

Ward, B. B., and Perry, M. J. 1980. Immunofluorescent assay for the marine ammonium-oxidizing bacterium *Nitrosococcus oceanus*. Appl. Environ. Microbiol. 39:913–918.

Ward, D. M., Weller, R., and Bateson, M. M. 1990. 16S rRNA sequences reveal numerous uncultured microorganisms in a natural community. Nature 345:63–65.

Whitmore, D. H., Cotton, R., and Sheridan, K. 1990. DNA fingerprinting. In: Electrophoretic and isoelectric focusing techniques in fisheries management. pp. 81–106. Ed. by D. H. Whitmore. CRC Press, Boca Raton, FL.

Williams, G. C., and Koehn, R. K. 1984. Population genetics of North Atlantic catadromous eels (*Anquilla*). In: Evolutionary genetics of fishes. pp. 529–560. Ed. by B. J. Turner. Plenum Press, New York.

Williams, G. C., Koehn, R. K., and Mitton, J. B. 1973. Genetic differentiation without isolation in the American eel, *Anquilla rostrata*. Evolution 27:192–204.

Wilson, G. M., Thomas, W. K., and Beckenbach, A. T. 1985. Intra- and inter-specific mitochondrial DNA sequence divergence in *Salmo*: Rainbow, steelhead, and cutthroat trouts. Can. J. Zool. 63:2088–2094.

Wirgin, I. I., and Maceda, L. 1991. Development and use of striped bass-specific RFLP probes. J. Fish Biol. 39(A):159–167.

Wright, S. 1969. Evolution and the genetics of populations. Vol. 2. The theory of gene frequencies. Univ. of Chicago Press, Chicago, IL.

Yentsch, C. M., Horan, P. K., Muirhead, K., Dortch, Q., Haugen, E. M., Legendre, L., Murphy, L. S., and Zahurenec, M. 1983. Flow cytometry: A powerful technique with potential applications in aquatic sciences. Limnol. Oceanogr. 28:1275–1287.

Yentsch, C. M., Mague, F. C., and Horan, P. K. 1988. Immunochemical approaches to coastal, estuarine and oceanographic questions. Springer-Verlag, New York.

Zammit, V. A., and Newsholme, E. A. 1979. Activities of enzymes of fat and ketone-body metabolism and effects of starvation on blood concentrations of glucose and fat fuels in teleost and elasmobranch fish. Biochem. J. 184:313–322.

Zouros, E. 1987. On the relation between heterozygosity and heterosis: An evaluation of the evidence from marine mollusks. In: Isozymes: Current topics in biological and medical research. Vol. 15. pp. 255–270. Ed. by N. C. Rattazi, J. G. Scandalios, and G. S. Whitt. Alan R. Liss Publ. Co., New York.

Chapter 30

Application of Satellite Remote Sensing and Optical Buoys/Moorings To LME Studies

James A. Yoder and Graciela Garcia-Moliner

Introduction

The Coastal Zone Color Scanner (CZCS) (along with six other scientific instruments) was launched on the *Nimbus-7* satellite in October 1978. The purpose of the CZCS was to determine if satellite sensors could generate accurate images of near-surface chlorophyll concentrations. The CZCS mission was planned for a 1-year proof-of-concept mission only, but operated until June 1986. The *Nimbus-7* is still in orbit, but the CZCS sensor no longer works (Yoder *et al.*, 1988).

CZCS measured sunlight radiance backscattered out of the ocean at five spectral bands centered at 443, 520, 550, 670, and 750 nm and infrared emissions at a sixth band centered at 11,500 nm, across a swath 2,200 km wide, with a maximum spatial resolution (pixel size at nadir) of 0.8 km^2. The *Nimbus-7* orbit returned the sensor to the same location over the earth every 1–2 days; but temporal coverage at this resolution was rarely realized because of cloud cover, and because the CZCS sensor was only operated approximately 10% of the time to conserve power for other instruments on the satellite.

Many steps are required to process raw CZCS data into chlorophyll images. A particularly difficult problem was to extract the radiance signal backscattered from the ocean from sensor measurements dominated by radiances resulting from atmospheric scattering. Atmospheric scattering by molecules and aerosols (dust particles) accounted for 80% to 90% of CZCS measurements. Thus, the accuracy with which the atmospheric signal is removed ultimately determines the accuracy of chlorophyll estimates (Gordon, 1987). The final version of the atmospheric correction algorithm used by National Aeronautics and Space Administration (NASA) investigators today was implemented in 1988 (Gordon *et al.*, 1988), approximately 10 years after launch!

CZCS measurements yield estimates of mean chlorophyll *a* plus phaeopigment concentrations in the upper attenuation depth, defined as the inverse of the attenuation coefficient for visible irradiance (i.e., 1/K) (Gordon *et al.*, 1983). The abbreviation, "CZCS-Chl," is generally used to refer to this estimate. The CZCS-Chl algorithm presently used by NASA investigators is based on power curve relations between chlorophyll and the ratio of 550 nm:443 nm radiance reflectance (ratio of upwelling to downwelling radiance) for chlorophyll concentrations less than 1.5 mg/m^3 and the 550 nm:520 nm ratio for chlorophyll concentrations greater than 1.5 mg/m^3 (Gordon *et al.*, 1983). CZCS chlorophyll algorithms rapidly lose accuracy and sensitivity at concentrations greater than 4.0 mg/m^3.

Data Availability

Insufficient computer speed, prohibitively high mass- storage requirements, and the lack of accurate and well-tested algorithms inhibited the smooth functioning of an image distribution system for the CZCS mission in the late 1970s and early 1980s. Technology finally caught up with user requirements in the late 1980s. Satellite oceanography groups at NASA/ Goddard Space Flight Center and the University of Miami collaborated on the design and implementation of a sophisticated processing and archiving system for the CZCS global data set consisting of 66,000 individual scenes (images) encompassing more than 800 gigabytes (1 gigabyte = 10^9 bytes) of raw data (Feldman et al., 1989).

Users of the new CZCS archive initially locate available imagery using a videodisc browse system, which provides a list of image file names corresponding to individual CZCS scenes meeting the user-defined search criteria. Digital data for these scenes are retrieved by the University of Miami satellite image-processing software package, which normally runs on VAX or microVAX minicomputers. The system retrieves the 4-km resolution digital CZCS-Chl images, or digital images of the individual CZCS bands corrected for Rayleigh scattering, from optical discs. Each digital optical disc holds approximately 4,000 CZCS scenes, so that the entire global data set can be stored easily on-line in an optical disc jukebox, which holds up to 50 discs.

Another product featured in the new archive is CZCS-derived chlorophyll values (and radiances corrected for Rayleigh scattering for the individual CZCS bands) at 20-km spatial resolution and composited at daily, weekly, and monthly intervals (binned to a fixed, linear lati-tude-longitude array of dimension 1024 x 2048). This data product is particularly useful for studying large-scale phenomena such as the spring bloom in the North Atlantic, equatorial up-welling, or the effects of monsoon winds over the Indian Ocean (Platt and Sathyendranath, 1988; McClain et al., 1990). The central archive is located at the Goddard Space Flight Center, which is the only location where 1-km spatial resolution, level-1 (calibrated radiances, but not atmospherically corrected) data can be acquired. Access to the videodisc browse system, as well as the 4-km and 20-km resolution global digital images, is also available from other institutions in the United States, including University of Rhode Island; Bigelow Laboratory for Ocean Sciences; University of Miami; University of South Florida; University of California at Santa Barbara; Jet Propulsion Laboratory, Pasadena, California; Oregon State University; and University of Washington (Feldman et al., 1989).

Another important new development is the availability of software packages for relatively inexpensive computer workstations. An example is the PC-Seapak package, which can process CZCS and Advanced Very High Resolution Radiometer (AVHRR) imagery from raw

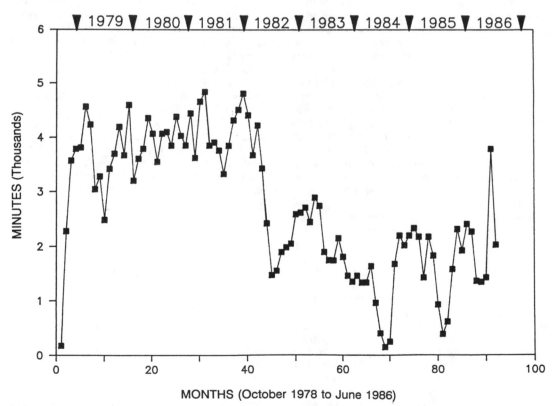

Figure 1. Minutes of CZCS operation per month for the duration of the mission. (From McClain et al., 1990.)

data to finished products and is implemented with an IBM-AT compatible, 386-class microcomputer workstation (Firestone *et al.*, 1989, 1990). Although lacking the full power of minicomputer image-processing systems (e.g., the Miami VAX-based program), microcomputer-based image analysis software packages such as PC-Seapak have most of the important capabilities of more sophisticated software and are attractive because of their relatively low implementation cost (approximately $20,000 for a complete hardware-software system).

Problems with CZCS Imagery

Figure 1 shows minutes of CZCS data collected per month for the duration of the mission. This figure illustrates two of the important limitations of CZCS imagery. First, if all possible data had been collected over the global ocean, then data collection should have exceeded 16,000 minutes per month—or approximately four times the maximum monthly collection rate during the mission. Maximum data collection was not possible because the CZCS had to share power with the other five instruments on the satellite. Secondly, data collection dropped off sharply beginning in early 1982. Thus, many regions of the ocean were poorly covered during the mission, particularly after that time. The net effect of the results depicted in Fig. 1 is that image time-series of many ocean areas are poorly resolved and thus poorly suited for conventional time-series statistical analyses.

As discussed above, correction for atmospheric aerosol concentrations is an important step in CZCS processing. The effects of Saharan dust over the tropical Atlantic and Asian dust over the North Pacific can be particularly difficult to remove accurately from the raw satellite data.

Another important limitation of CZCS imagery is that reflectances for only three spectral bands (443, 520, and 550 nm) are available for pigment and other algorithms. With respect to determining ocean chlorophyll concentrations, most of the information is in the 443 nm band (Gordon, 1987). The accuracy of the chlorophyll algorithm discussed above is greatly affected by the amount of nonpigment substances in the water that absorb irradiance at 443 nm. Yellow substances (humic and fulvid compounds), which can be present in high abundance in coastal waters (Carder *et al.*, 1989), are strong absorbers at 443 nm. Regions with high concentrations of these substances will appear in CZCS imagery as areas rich in chlorophyll. High concentrations of detached coccoliths and suspended sediment also confound the CZCS chlorophyll algorithm because

of their strong backscattering characteristics, particularly at 520 and 550 nm. Users of CZCS imagery should be aware of these limitations and should be careful when interpreting CZCS imagery from areas where the confounding effects discussed above may be significant.

CZCS Imagery and LMEs

The West Coast Time Series (WCTS) is an excellent example of how CZCS imagery can be used to study LMEs—in this case, the California Current system off the U.S. west coast. CZCS imagery for the WCTS was collected at the Scripps Satellite Oceanography Facility, La Jolla, California, and processed at the Jet Propulsion Laboratory using software developed by the University of Miami. CZCS-chlorophyll imagery at 1-km spatial resolution was mapped onto standard grids covering the region from Baja California, Mexico, to British Columbia, Canada. Image time-series were widely distributed to ocean scientists interested in both basic and applied research.

Coastal and oceanic waters off the west coast of North America are strongly influenced by the California Current flowing from north to south, and by wind-driven upwelling. One of the first important findings from the WCTS was the discovery of chlorophyll-rich plumes (filaments) extending hundreds of kilometers offshore. CZCS imagery showed that these plumes had characteristic time-space scales related to the onset of upwelling-favorable winds and to topographic irregularities along the coast (Abbott and Zion, 1987). Other applications of the WCTS imagery included the study of mesoscale eddies (Haury *et al.*, 1986; Simpson *et al.*, 1986), relation of larval recruitment of benthic organisms to the presence of the aforementioned filaments (Roughgarden *et al.*, 1988), estimation of seasonal and spatial patterns in primary production (Eppley *et al.*, 1985; Perry *et al.*, 1989), and effects of El Niño on chlorophyll concentrations (Fiedler, 1984). Several investigators used WCTS imagery to quantify mesoscale phytoplankton spatial variability within the California LME (Pelaez and McGowan, 1986; Denman and Abbott, 1988; Michaelsen *et al.*, 1988; Smith *et al.*, 1988; Thomas and Strub, 1989). Among the significant findings reported in these publications are the following: (i) CZCS-derived chlorophyll estimates patterns having 50-km length scales de-correlate after 8–10 days; (ii) mesoscale eddies and geostrophic currents are dominant sources of variability in CZCS-chlorophyll at length scales greater than 10 km; (iii) large-scale changes of CZCS-derived chlorophyll patterns are associated with the timing of the spring transition to upwelling-favorable winds, which is characterized by high interannual variability; and (iv) there is danger in forming conclusions about large-scale spatial patterns based on just a few years of satellite (or other) data.

Bio-Optical Moorings

One of the difficulties in working with CZCS imagery is that image time-series are not well resolved. For most ocean areas, daily coverage for 1-week periods or longer is very rare. Imagery from future ocean color missions will be better resolved in time than CZCS because, unlike CZCS, future sensors will be operated continuously in the sunlit portions of their orbits. Nevertheless, cloud cover will interfere with coverage, and, as a result, many parts of the global ocean will be poorly covered. In addition, vertical pigment distributions cannot be directly resolved by satellite color scanners. However, spectroradiometers, transmissometers, fluorometers, and sophisticated data-logging systems suitable for long-term moorings are becoming generally available to oceanographers (e.g., Booth et al., 1987; Booth and Smith, 1988). Bio-optical moorings provide a means for improving temporal and vertical coverage of pigment and particle concentrations at specified locations at the same time that future ocean color scanners generate chlorophyll images with excellent horizontal spatial resolution. Combining data sets from satellite imagery and bio-optical moorings to study ocean dynamics is an exciting possibility for the future.

Bio-optical moorings have been deployed in only a few instances. One of the first major deployments was by the Biowatt (Bio-optical Variability on Seasonal Time Scales in the Sargasso Sea) Program, an interdisciplinary effort sponsored by the U.S. Office of Naval Research. A mooring was deployed in the North Atlantic Ocean at 34°N, 70°W for three consecutive periods from February 28 through November 23, 1987 (Dickey, 1991; Dickey et al., 1991). Incident irradiance, chlorophyll fluorescence, beam transmission, and other parameters were measured at up to eight depths with 4-minute sampling resolution. The results showed rapid (2-day) development of spring stratification concurrent with increases in fluorescence and beam attenuation coefficient, significant diurnal changes in bio-optical properties, and the effects of mesoscale eddies as they advect past the mooring (Dickey et al., 1991).

Similar bio-optical moorings were deployed in the North Atlantic just south of Iceland in 1989 during a follow-on study to Biowatt and in cooperation with the first field study of the Joint Global Ocean Flux Study (JGOFS) (Dickey, 1990). Investigators from the United States are also planning to deploy such moorings as part of the U.S. JGOFS time-series in Hawaii and Bermuda. NASA investigators are planning bio-optical moorings that include spectroradiometers to validate and calibrate the U.S. follow-on ocean color scanner to CZCS, now tentatively scheduled for launch in 1993. This mission will provide an opportunity to develop sampling strategies to take advantage of the combined capabilities of bio-optical moorings and satellite ocean color measurements.

Table 1. Comparison of CZCS Sea-WiFS and OCTS (improvements noted).

Category	CZCS	SeaWiFS & OCTS	Improvement
1. Spectral Bands (nm)	–	–	–
a. Atmospheric correction	670	670,765 & 865	Better coastal data
b. Ocean algorithms	443, 520 & 560	412, 443, 490, 510 & 555	Improved algorithms for chlorophyll a
2. Data collection (hours/days)	<=2	>=10	Continuous global coverage
3. Monitor sensor calibration & drift?	no	yes	Better quality data
4. Time to distribute data products to scientists	10 yrs.	10–14 days	–
5. Maximum spatial resolution (pixel size)	1 x 1 km	1 x 1 km	–

Sea Wide-Field Sensor and Ocean Color and Temperature Sensor

To replace the CZCS, NASA is planning a new satellite ocean color instrument tentatively scheduled for launch in 1993. The new sensor, Sea Wide-Field Sensor (Sea-WiFS), will have the same five visible and near-infrared bands as CZCS, as well as three additional bands in the visible and near-infrared. The additional spectral bands will allow improved chlorophyll estimates for oceanic and coastal waters.

The Japanese are planning the Ocean Color and Temperature Sensor (OCTS) for their Advanced Earth Observing Satellite (ADEOS) to be launched in 1995. The specifications for OCTS are similar to those for Sea-WiFS, and additional thermal infrared bands are included to provide concurrent estimates of sea-surface temperature (Table 1).

Conclusion

Satellite ocean color research during the 1980s focused on technique development. Algorithms, processing capability, and the CZCS data are now available to many more research users than was possible just a few years ago. These data are presently being used to study LMEs around the world, including the California Current System, Middle Atlantic Bight, and South Atlantic Bight—off the west and east coasts of North America. These studies constitute the groundwork for the mid-1990s, when data sets derived from sophisticated new in situ and satellite sensors become available to the international oceanographic community.

References

Abbott, M. R., and Zion, P. M. 1987. Spatial and temporal variability of phytoplankton pigment off northern California during coastal ocean dynamics experiment 1. J. Geophys. Res. 92:1745–1755.

Booth, C. R., Mitchell, B. G., and Holm-Hansen, O. 1987. Development of moored oceanographic spectroradiometer. Biospherical Technical Reference 87–1. Biospherical Instruments Inc., San Diego.

Booth, C. R., and Smith, R. C. 1988. Moorable spectroradiometers in the Biowatt experiment. SPIE, 925 (Ocean Optics IV): 176–188.

Carder, K. L., Steward, R. G., Harvey, G. R., and Ortner, P. B. 1989. Marine humic and fulvic acids: Their effects on remote sensing of ocean chlorophyll. Limnol. Oceanogr. 34:68–81.

Denman, K. L., and Abbott, M. R. 1988. Time evolution of surface chlorophyll patterns from cross-spectrum analysis of satellite color. J. Geophys. Res. 93:6789–6998.

Dickey, T. D. 1990. Moored instrument observations offer time-series insights for studies of ocean fluxes. U.S. JGOFS Newsletter 1:3. Woods Hole Oceanographic Institution, Woods Hole, MA.

Dickey, T. D. 1991. The emergence of concurrent high resolution physical and bio-optical measurements in the upper ocean and their applications. Reviews of Geophysics 96:8643–8663.

Dickey, T. D., Marra, J. Granata, T., Langdon, C., Hamilton, M., Wiggert, J., Siegel, D., and Bratkovich, A. 1991. Concurrent high resolution bio-optical and physical time-series observations in the Sargasso Sea during the spring of 1987. J. Geophys. Res. 96:8643–8663.

Eppley, R. W., Stewart, E., Abbott, M. R., and Heyman, U. 1985. Estimating ocean primary production from satellite chlorophyll: Introduction to regional differences and statistics for the Southern California Bight. J. Plankton Res. 7:57–70.

Feldman, G. C., et al. 1989. Ocean color. Availability of the global data set. EOS—The Oceanography Report. Vol. 70 (June 6, No. 23). pp. 634–635 and 640–641. American Geophysical Union. Washington, DC.

Fiedler, P. C. 1984. Satellite observations of the 1982–1983 El Niño along the U.S. Pacific coast. Science 224:1251–54.

Firestone, J. K., Fu, G., Chen, J., Darzi, M., and McClain, C. R. 1989. PC-Seapak: A state-of-the-art image display and analysis system for NASA's oceanographic research program. Preprints Fifth Intl. Conf. Interactive Information and Processing Sys. for Meteor., Oceano. and Hydrology, Anaheim, CA. Jan. 29–Feb. 3, 1989. Amer. Meteor. Soc., Boston.

Firestone, J.K., Fu, G., Darzi, M., and McClain, C.R. 1990. NASA's SEAPAK software for oceanographic data analysis. Preprint Volume, 6th Intl. Conf. on Interactive and Information Processing Systems for Meteorology, Oceanography, and Hydrology. Amer. Meteorol. Soc., Boston. pp. 260–267.

Gordon, H. R. 1987. Calibration requirements and methodology for remote sensors viewing the ocean in the visible. Remote Sens. Environ. 22:103–126.

Gordon, H. R., Brown, O. B., Evans, R. H., Brown, J. W., Smith, R. C., Baker, K. S., and Clark, D. K. 1988. A semianalytic radiance model of ocean color. J. Geophys. Res. 93:10909–10924.

Gordon, H. R., Clark, D. K., Brown, J. W., Brown, O. B., Evans, R. H., and Broenkow, W. O. 1983. Phytoplankton pigment concentrations in the Middle Atlantic Bight: Comparison of ship determination and CZCS estimate. Appl. Opt. 22(1):20–36.

Haury, L. R., Simpson, J. J., Pelaez, J., Koblinsky, C. J., and Wiesenhahn, D. 1986. Biological consequences of a recurrent eddy off Point Conception, California. J. Geophys. Res. 91:12937–12956.

McClain, C. R., Esaias, W. E., Fedlman, G. C., Elrod, J., Endres, D. Firestone, J., Darzi, M., Evans, R.,

1990. Physical and biological processes in the North Atlantic during the first GARP experiment. J. Geophys. Res. 95:18027–18048.

Michaelsen, J., Zhang, X., and Smith, R. C. 1988. Variability of pigment biomass in the California current system as determined by satellite imagery. 2. Temporal variability. J. Geophys. Res. 93:10883–10896.

Platt, T. and Sathyendranath, S. 1988. Oceanic primary production: Estimation by remote sensing at local and regional scales. Science 241:1613–1620.

Pelaez, J., and McGowan, J. A. 1986. Phytoplankton pigment patterns in the California Current as determined by satellite. Limnol. Oceanogr. 31:927–50.

Perry, M. J., Bolger, J. P., and English, D. C. 1989. Primary production in Washington coastal waters. In: Coastal oceanography of Washington and Oregon. pp. 117–138. Ed. by M. R. Landry and B. M. Hickey. Elsevier, Amsterdam.

Roughgarden, J., Gaines, S., and Possingham, H. 1988. Recruitment dynamics in complex life cycles. Science 241:1460–1466.

Simpson, J. J., Koblinsky, C. J., Pelaez, J., Haury, L. R., and Wiesenhahn, D. 1986. Temperature-plant pigment-optical relations in a recurrent offshore mesoscale eddy near Pt. Conception, California. J. Geophys. Res. 91:12919–12936.

Smith, R. C., Zhang, X., and Michaelsen, J. 1988. Variability of pigment biomass in the California Current system as determined by satellite imagery. 1. Spatial variability. J. Geophys. Res. 93:10863–10882.

Thomas, A. C., and Strub, P. T. 1989. Interannual variability in phytoplankton pigment distributions during the spring transition along the west coast of North America. J. Geophys. Res. 94:18095–18117.

Yoder, J. A., Esaias, W. E., Feldman, G. C., and McClain, C. R. 1988. Satellite ocean color-status report. Oceanogr. Mag. 1:18–20.

Index